To our parents

# Introductory Microbiology

Trevor Gross
Principal Lecturer
Department of Biological Sciences
The Manchester Metropolitan University
Manchester
UK

Jane Faull
Department of Biology
Birkbeck College
University of London
London
UK

Steve Ketteridge
Staff Development Officer
Queen Mary and Westfield College
University of London
London
UK

and

Derek Springham
Queen Mary and Westfield College
University of London
London
UK

**CHAPMAN & HALL**
University and Professional Division
London · Glasgow · Weinheim · New York · Tokyo · Melbourne · Madras

**Published by Chapman & Hall, 2–6 Boundary Row, London SE1 8HN, UK**

Chapman & Hall, 2–6 Boundary Row, London SE1 8HN, UK

Blackie Academic & Professional, Wester Cleddens Road, Bishopbriggs, Glasgow G64 2NZ, UK

Chapman & Hall GmbH, Pappelallee 3, 69469 Weinheim, Germany

Chapman & Hall USA., One Penn Plaza, 41st Floor, New York NY 10119, USA

Chapman & Hall Japan, ITP- Japan, Kyowa Building, 3F, 2–2–1 Hirakawa-cho, Chiyoda-ku, Tokyo 102, Japan

Chapman & Hall Australia, Thomas Nelson Australia, 102 Dodds Street, South Melbourne, Victoria 3205, Australia

Chapman & Hall India, R. Seshadri, 32 Second Main Road, CIT East, Madras 600 035, India

First edition 1995



Typeset in 11.5/14 pt Garamond by ROM Data Corp. Ltd., Falmouth, Cornwall

**Printed in Great Britain by the Alden Press, Osney Mead, Oxford**

ISBN 0 412 45300 2



A catalogue record for this book is available from the British Library

Library of Congress Catalog Card Number: 94-70974

# Contents

**Preface** xi
**Acknowledgements** xiii

**1 Introduction to the microbial world** 1
Microorganisms and size 3
Microorganisms as experimental subjects 5
Microorganisms in industry 6
Microorganisms in the environment 7
Microorganisms and disease 9
Further reading 9
Questions 10

**2 Microbial structure** 12
Structure and function of eukaryotes and prokaryotes 14
Structure and function of eukaryotic and prokaryotic organelles 19
Cell walls 21
Motility 22
Motility in eukaryotes 24
The bacterial cell surface 26
Bacterial endospores 27
Chapter summary 28
Further reading 28
Questions 29

**3 Microbial growth** 31
Major and minor chemical elements 31
Microbial growth factors 34
Growing microorganisms in the laboratory 37
Batch culture 43
Continuous culture 46
Synchronous growth 49
The effect of environmental conditions on microbial growth 51
Control of microbial growth 53
Chapter summary 58
Further reading 59
Questions 60

**4 Energy production in microorganisms 62**
Energy and thermodynamics 62
Energy from chemical reactions: oxidation–reduction 66
Fermentations 72
Alternative pathways for the fermentation of glucose 76
Respiration 80
Anaerobic respiration 89
Photosynthesis 90
The Calvin cycle 101
Archaebacterial photosynthesis 104
Chemolithotrophy: energy from the oxidation of inorganic electron donors 105
Chapter summary 106
Further reading 107
Questions 107

**5 Biosynthesis 110**
Anaplerosis 111
Biosynthesis of low-molecular-weight compounds 113
Synthesis of macromolecules 117
Polypeptide synthesis 118
Polynucleotide synthesis 126
The eubacterial cell wall 135
Chapter summary 146
Further reading 148
Questions 148

**6 Coordination of metabolism 150**
Control of enzyme activity 150
Control of enzyme synthesis 155
Induction of the *lac* operon 159
Alternative mechanisms of controlling protein synthesis 170
Gene regulation in *Saccharomyces cerevisiae*, a eukaryotic microorganism 171
Chapter summary 172
Further reading 172
Questions 173

**7 The microbial genome: organization, mutation and repair 175**
Advantages of using microorganisms for genetic studies 176
DNA is the genetic material 177

Organization of the hereditary material 181
Mutation 184
Mutagenesis: making mutations happen 191
Selection of mutants 195
Reversal of mutation 197
Recombination 201
DNA repair systems 203
Chapter summary 207
Further reading 207
Questions 208

**8 Bacterial gene transfer and genetic engineering 211**
Bacterial plasmids 211
Conjugation and the F factor 214
Transduction: bacteriophage-mediated gene transfer in bacteria 222
Gene cloning and genetic engineering 229
Eukaryotic microorganisms and genetic engineering 236
Chapter summary 237
Further reading 238
Questions 238

**9 Eukaryotic microorganisms 241**
Structure and growth of protistan microorganisms 244
Structure of the fungi 244
Structure of the protistan algae 250
The protozoa 254
Reproduction in the fungi and the Protista 260
Growth and reproduction in the algae 275
Growth and reproduction in the protozoa 280
Chapter summary 284
Further reading 285
Questions 286

**10 Viruses 287**
Distinctive properties of viruses 287
Major groups of viruses 294
Types of viral infections 295
Structure of virus particles 301
The viral genome 310
Assaying viruses 319
Virus multiplication 328

Chapter summary 338
Further reading 339
Questions 340

**11 Microbial biotechnology 342**
Alcoholic beverages 343
Other yeast products 352
Fermented foods 356
Antibiotic production 359
Production of human insulin by genetic manipulation 373
Production of hepatitis B vaccine by gene manipulation 377
Enzyme production 379
Steroid transformations 380
Organic acids 382
Amino acids 385
Purification of organic wastes 385
Chapter summary 389
Further reading 390
Questions 391

*Answers to selected questions* 393

*Index* 403

# Preface

In view of the wide range of microbiology texts currently on offer from bookshops, one might reasonably question the need for another. In light of this, the authors of an entirely new microbiology book have an obligation to explain why they were inspired to make their own contribution.

In writing the present text, we have tried to produce a genuine introduction to the biology of microorganisms, presenting fundamental information and ideas about microorganisms in an understandable, straightforward fashion, without resorting to over-simplification. At the same time, we hope that we have given sufficient, detailed information to inspire the readers to progress to more advanced and specialized texts. These objectives are relatively easy to reach with a book on microbiology, because most undergraduates in biological sciences have a natural interest in the subject, particularly in its applied aspects.

In writing the book we had in mind two particular kinds of readers. We hope that it will constitute an attractive introduction to students in the early years of a specialist degree or diploma course in microbiology. But we have also written the book for students who are taking courses in a wide range of biological disciplines. Although these readers will not necessarily become microbiologists and do not require the advanced, detailed treatment of an honours microbiology undergraduate, they do require a sound and concise grounding in the basics of the subject with clear links to microbiological applications. In contrast to many of our competitors, we have covered all the major groups of microorganisms and resisted the temptation to concentrate only on the bacteria. The emphasis is on those aspects of microorganisms which separate them from plants and animals.

Although an ultimately rewarding experience, preparation of the text has proved difficult and time-consuming. We are indebted to our editor, Dominic Recaldin, whose advice and enthusiasm have been major sources of encouragement throughout the enterprise. The comments of a number of reviewers were extremely valuable, particularly when we were taking our first tentative steps preparing drafts of the first chapters. I would like to acknowledge Dr Peter Gowland's contribution during the early stages of production. Especially warm thanks go to Jane Faull, Derek Springham and

Steve Ketteridge who contributed chapters on eukaryotic microorganisms, microbial biotechnology and viruses respectively. I am further indebted to Steve Ketteridge for reading the drafts of many of the chapters and for his helpful advice. Finally, I thank my wife, Gerry, for typing a large proportion of the manuscript, and for her constant encouragement and cups of tea!

Trevor Gross

# Acknowledgements

Chapman and Hall gratefully acknowledge the following sources for redrawn figures.

From: The Microbes (1987)
Paul J. Van Demark and Barry L. Batzing, Benjamin–Cummings Publishing Co. Inc.
Figures: 1.1, 2.9, 2.12, 3.2, 3.3, 3.4, 3.5, 3.6, Box 3.4, 7.1, 7.2, 8.4, Box 8.1, 8.3, 8.8

From: Microbial Ecology (1987)
R.M. Atlas and R. Bartha, Benjamin–Cummings Publishing Co. Inc.
Figure: 1.2

From: Cell & Molecular Biology (1986)
P. Scheeler and D. Bianchi, J. Wiley & Sons Ltd
Figure: 2.1

From: General Microbiology (1988)
Robert F. Boyd, Times-Mirror/Mosby College Publishing
Figures: 2.2, 2.4, 2.8, 2.11, 4.11

From: Microbiology (1986)
R.J. Canon and J.S. Colomé, West Publishing Co.
Figures: 2.3, 2.5, 2.7

From: Schaum's outline of theory and problems of biochemistry (1988)
P.W. Kuchel *et al.*, McGraw-Hill Book Co.
Figures: 4.1, 4.13

From: Photosynthesis (1972)
D.O. Hall and K.K. Rao, Studies in Biology No. 37., Edward Arnold Ltd
Figure: 4.18

From: Biochemistry of Bacterial Growth (1982)
J. Mandelstam, K. McQuillen and I. Dawes, Blackwell Scientific Publications
Figures: Box 5.3, 5.28

From: Molecular Biology (1986)
D. Freifelder, Jones & Bartlett, Boston
Figures: 5.6, 5.7, 5.8, 5.9, 5.10, 5.11, 5.12, 5.13, 5.15, 5.16, 5.17, 5.18, 5.19, 7.4, 7.6, 7.24, 7.25, 7.26

From: Physiology of the bacterial cell (1990)
F.C. Neidhardt, J.L. Ingraham and M. Schaechter Sinauer Associates Inc.
Figure: 6.11

From: Genetics
Ursula Goodenough, Saunders College Publishing
Figure: 7.3

From: Genetics: a molecular approach (1989)
T.A. Brown, Van Nostrand Reinholt (International)
Figure: Box 7.4

# Introduction to the microbial world

# 1

All living things below a size visible to the unaided eye are grouped artificially into a vast and motley collection called microorganisms. The existence of such organisms could only be guessed at before the development of the first primitive microscopes in the 17th century. Microorganisms are among the most successful living things. Their 'club' claims among its members the bacteria, the yeasts, many algae and fungi, and all the protozoa.

Microorganisms are everywhere: in the air, in water, in the soil, in animals and plants, and even in and on other microorganisms. Their size belies their economic importance. Many cause debilitating or devastating disease in man, in cultivated crops, and in domesticated animals. Others are immensely beneficial and we put some of them to work to make a range of foods and drinks which support enormous industries. Soil microorganisms are indispensable cogs in the natural cycles which break down organic materials into nutrients usable by other organisms.

What are the size limits of this artificial 'kingdom'? One of the biggest organisms is a protozoan called *Amoeba*, which is about 1 millimetre across. You could probably just detect it, in a good light, moving through a drop of water. At the other end of the scale, the foot-and-mouth disease virus is about 0.01 micrometres across: that is, about 100 000 times smaller than the giant *Amoeba* (Figure 1.1).

Being small has its advantages. One definite benefit claimed by single-celled organisms is that their surface is large compared with their volume. In fact, the smaller the cell, the greater its surface area compared with its volume. The bigger the surface is, the more effective it will be in exchanging materials, taking in nutrients, and expelling wastes. So a single-celled organism can take in the nutrients it needs and expel its wastes with greater efficiency (Table 1.1).

It is not their size alone, however, but their simplicity which holds the key to microbial success in the living world. The vast majority of microorganisms consist of a single cell. They have no need for bloodstreams or translocation pathways to move nutrients from one place to another. Within the minuscule volume of a single cell are all the ingredients the organism needs. Growth and reproduction, once underway, can give rise to astronomical numbers of offspring at what, in terms of human generation times,

A word here about dimensions might be helpful. As you know, a metre is divided into 1000 millimetres (mm). Similarly, a millimetre is divided into 1000 micrometres (μm). You will have deduced, therefore, that there are one million micrometres in a metre. The dimensions of sub-cellular particles, such as viruses, are usually expressed in nanometres (nm). There are 1000 nanometres in one micrometre.
Only bacteria and protozoans are all microscopic. Algae and fungi, although they have some microscopic members, also have species which can grow very large indeed. Some algae, the great kelps of the Pacific Ocean, are among the largest organisms in the world, growing many metres in length.

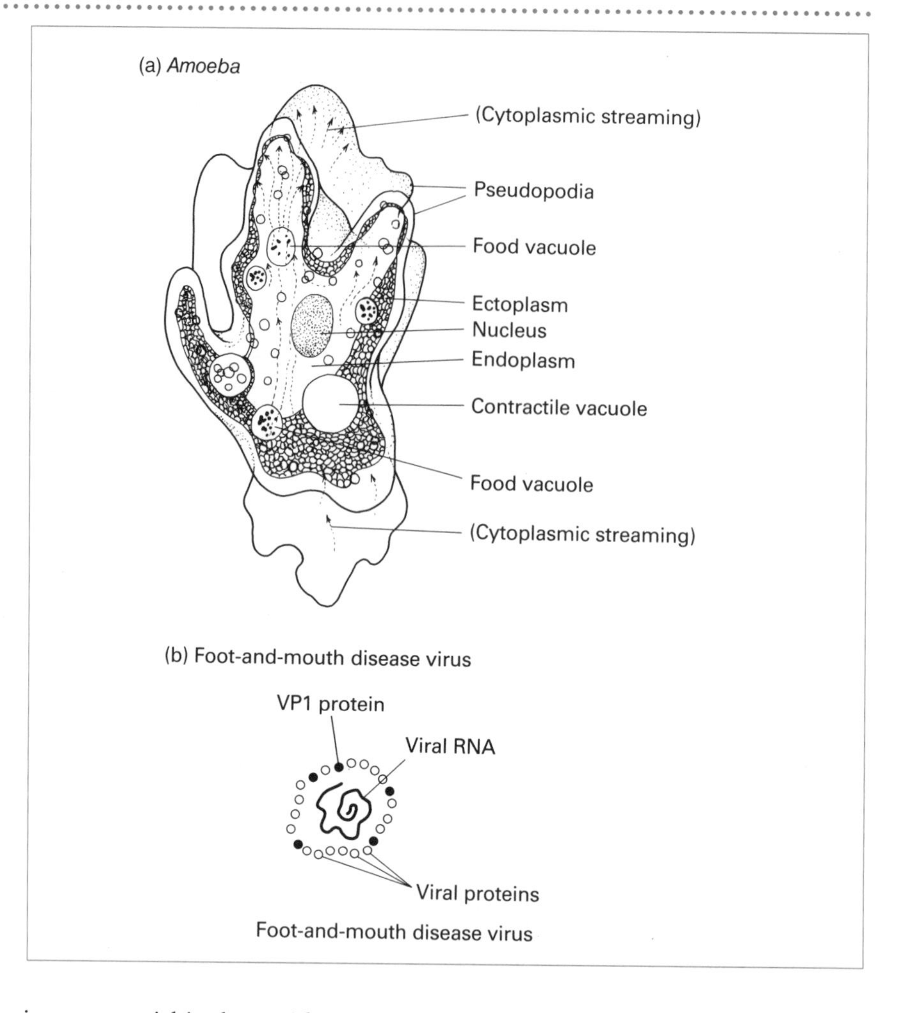


**Figure 1.1** The giant *Amoeba* and the foot-and-mouth disease virus. Question: these figures are not to scale. If they were, how long would the *Amoeba* drawing be? Remember the giant *Amoeba* is 1 mm long and the foot-and-mouth virus, 10 nm. Answer: a staggering 1.5 km!

is an astonishingly rapid rate. Unlike sophisticated parents far up the evolutionary tree, which mature and eventually die even when food is plentiful, microorganism 'parents' live along with their offspring as long as food lasts. As a result, populations double at regular intervals and soon become astronomical in size.

One consequence of rapidly increasing populations is to enhance the potential of useful mutations. Mutations are rare (and beneficial ones rarer still) but rapid cell division will soon produce mutations in significant numbers, ready to exploit conditions which might later become unsuitable to unmutated ancestors.

While all single-celled species are microorganisms, not all microorganisms are single-celled. Species such as *Volvox*, for example, comprise groups of individuals linked in colonies.

Microorganisms use a variety of modes of nutrition which contribute to their success as a group. Many species of bacteria, as well as algae, are **autotrophic** and are able to use carbon dioxide as their only source of carbon. Others live as **saprophytes** on dead or decaying organic matter, digesting complex molecules into simpler ones which are absorbed into

**Table 1.1** As a cell increases in size, its surface/volume ratio decreases. This can be shown by looking at three hypothetical, spherical organisms with radii of 1 μm, 10 μm, and 100 μm.

| *Radius* (μm) | *Surface area (mm2)* | *Volume* (μm$^3$) | *Surface area / Volume* |
|---|---|---|---|
| 1 | 12.6 | 4.2 | 3 |
| 10 | 1260 | 4200 | 0.3 |
| 100 | 126 000 | 4 200 000 | 0.03 |

A hypothetical, spherical organism with a radius of 1 μm has a surface/volume ratio of 3. One with a radius of 1 mm (1000 μm) has a surface/volume ratio of 0.003. The smaller organism has, therefore, a thousand times more surface area per unit volume than the larger one. The large surface area/volume ratio allows for very rapid intake of nutrients and expulsion of wastes. This leads to rapid growth and reproduction. It accounts for much of the evolutionary success of microorganisms.

their cells as food. Yet others are **parasites** and obtain the foods they need directly from their living hosts.

Some microorganisms are remarkably versatile. *Euglena*, a single-celled alga, can live as an autotroph in the light and as a saprophyte in the dark. You might deduce from this that *Euglena* sometimes behaves like a plant and at other times like an animal, and you would be right. Many microorganisms display a mixture of animal and plant characteristics (Chapter 9). Certainly, many species of bacteria, protozoa and microscopic algae can move about by means of various hair-like appendages called **cilia** or **flagella**. Yet many types of bacteria and microscopic algae are autotrophic. The former is very much an animal characteristic and the latter a plant one.

## Microorganisms and size

The property which unites all microorganisms is their minute size. The smallest sub-cellular particles are only a few nanometres across and the largest protozoans may attain diameters approaching 1 millimetre; in other words, there is a size range of something like 100 000-fold between the smallest virus and the largest protozoans. Table 1.2 lists some representative microorganisms and their sizes.

Although small, microorganisms are very diverse and are, therefore, considered to be a successful group. If we accept that microorganisms are small and successful, is it then possible to point to ways in which small size can lead to success of a particular microorganism? Is there any advantage in being small? Are there any disadvantages?

Some definitions:
**Autotrophic** organisms are those that can synthesize all their cell material from carbon dioxide.
**Photosynthetic autotrophs** derive their energy for this synthesis from sunlight; green plants, algae and some bacteria fall into this category.
**Chemosynthetic autotrophs** are represented only by a small number of bacteria. They derive their energy for carbon dioxide fixation from the oxidation of inorganic molecules.
**Heterotrophic** organisms include animals, fungi, protozoans and most bacteria. These require an organic carbon source.
**Photosynthetic heterotrophs** derive energy from sunlight and are represented by only a small number of bacteria.
**Chemosynthetic heterotrophs** require organic molecules as both carbon and energy sources. Higher animals, including you and me, protozoans, fungi and most bacteria fall into this group.

**Table 1.2** *Comparative sizes of representative microorganisms*
The largest viruses (pox viruses) are approximately 0.250 μm (250 nm) in diameter. The smallest (foot-and-mouth disease virus) are approximately 0.01 μm (10 nm) in diameter.

| *Microbial shape* | *Genus and/or species or type* | *Size (length)* | *Microbial group* |
|---|---|---|---|
|  | Saprosphira | 500 μm | Bacterium |
| 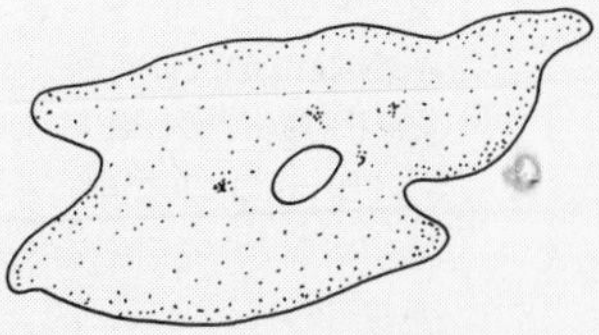 | Giant *Amoeba* | 1 mm | Protozoa |
| 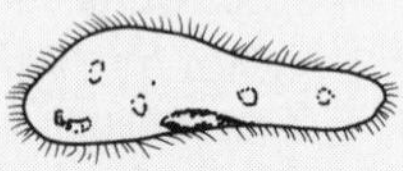 | *Paramecium* | 300 μm | Protozoa |
| 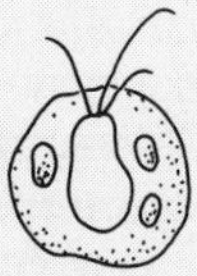 | *Chlamydomonas* | 25 μm | Alga |
| 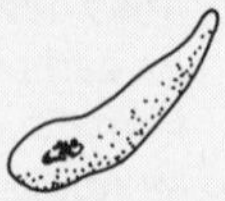 | Malaria parasite | 15 μm | Protozoa |
| 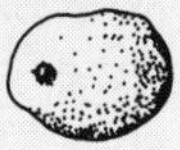 | Yeast (*Saccharomyces cerevisiae*) | 10 μm | Fungus |
|  | *Treponema pallidum* | 10 μm | Bacterium |
| | *Escherichia coli* | 3 μm | Bacterium |
| | Mycoplasma | 0.2 μm | Bacterium |

It can be argued that the success of microorganisms is primarily a reflection of the capacity of a microbial individual for rapid growth and multiplication. This, in turn, produces astronomical numbers of individuals. The occurrence of such large populations of individuals means that although mutation produces variant forms at a very low frequency, the total number of variants produced will be significant.

Let us consider a population of $10^9$ bacteria. Given that the frequency with which mutations arise is about 1 in $10^8$, such a population would produce approximately 10 mutant individuals. These variants might be better suited than the non-mutant individuals for growth and multiplication under different environmental conditions. In contrast to microorganisms, larger organisms do not produce such large populations and variants will arise much less rapidly than in a population of a particular microorganism.

Once a variant has arisen, it is then capable, provided that the appropriate environmental conditions exist, of rapid growth and multiplication, producing a large population of variant individuals able to dominate a particular ecological niche.

In 1993 the scientific community was rocked by the discovery of a giant bacterium (80 × 600 μm) found on the surgeon fish of Australia. This bacterium, *Epulopiscium fishelsoni*, far exceeds the maximum size believed possible for a prokaryotic organism.

## Microorganisms as experimental subjects

The study of microorganisms also contributes indirectly to our well-being because microorganisms have a number of characteristics which make them easier to investigate experimentally than more complex organisms. For this reason, they are often used as model systems to investigate genetic and biochemical processes common to themselves and more complex animals and plants. In contrast, microorganisms are of very little value for the investigation of biological processes which are only associated with complex animals and plants, such as the transmission of nerve impulses or the formation of wood.

Some of the advantages of using microbes as experimental subjects are as follows:

1. Many are relatively inexpensive to grow and maintain in laboratory cultures.
2. So far, no one has protested against the use of microorganisms as experimental organisms and there appear to be no ethical objections to experimenting with such organisms.
3. Microbes exist as essentially homogeneous populations of cells. That is, we do not encounter the situation that we find with multicellular

With the possible exception of man, the common gut bacterium *Escherichia coli* is the most extensively studied living organism.

organisms which consist of a large variety of different cell types. Biochemical investigations of plant and animal cells require preliminary dissection and separation of different cell types. This is often not the case with unicellular microbes because all the cells are the same.

4. Microbial populations rapidly increase in size because of the very short times between successive generations. Some bacteria can reproduce as often as every 20 minutes. This means that genetic experiments can be set up on a particular day and the offspring will be produced either the next day or within a few days. This contrasts with higher animals where it takes much longer (perhaps years) to carry out a comparable study.

## Microorganisms in industry

Although microorganisms have been essential for the production of dairy products and alcoholic drinks for thousands of years, it is extremely unlikely that ancient peoples knew of their involvement in such processes. Microbes already present in raw materials, such as those in milk, produced yoghourt and cheese, and in grape sugar produced wine. Further batches of raw material were then fermented using a portion of a previous batch to start the process again.

In contrast to these traditional uses of microorganisms in a somewhat uncontrolled fashion, modern industrial microbiology or microbial biotechnology has grown into a huge industry, in which the biochemical activities of microbes have been manipulated and harnessed to the formation of an unimaginably wide range of commercially valuable products. This has been made possible by the parallel development of appropriate fermentation techniques and computer technology.

The use of *Clostridium acetobutylicum* for acetone production from cereal grain was developed by Chaim Weizmann in Manchester, England during World War I. A grateful British government agreed to support the establishment of the State of Israel after World War II as an expression of its gratitude to Weizmann, who became Israel's first President.

The modern era of industrial microbiology is said to have begun with the production of a number of organic solvents, such as acetone, isopropanol and butanol, by the bacterium *Clostridium acetobutylicum*. Acetone provided the raw material for the production of the high explosive trinitrotoluene (TNT) and aircraft varnish when the usual source of acetone from the German chemical industry was no longer available.

A further major boost to the use of microbes as tireless slaves of biotechnology came with the development of the techniques of genetic engineering. By means of these techniques, genes from any organism, including humans, can be incorporated into bacteria or other microbes. Once the genes are incorporated, engineered bacteria can then be used to synthesize a variety of therapeutic polypeptides including hormones and antibodies. The uses of microorganisms in industrial and biotechnological processes are covered in detail in Chapter 11.

## Microorganisms in the environment

Much of our knowledge of microbes is derived from studies of pure cultures of microorganisms in the laboratory. It is important, therefore, to emphasize that all microorganisms also exist outside the laboratory environment, alongside more obvious animals and plants where they constitute members of the biosphere's **biota**.

The number of microbial individuals which occupy the earth's surface is almost impossible to imagine, but they outnumber animals and plants many times over. Let us look at some statistics to give some idea of the kinds of numbers concerned. A single gram of garden soil would contain between 10 and 20 million bacteria, half a million fungi, 50 000 algae and 30 000 protozoans. Although we chose the soil as an example of the number of microbes found in the environment, they are also found in the air, in water and in association with larger animals and plants.

One of the most important functions that microbes perform in nature is the recycling of a variety of chemical elements; they are indispensable components of these **geochemical cycles**. Microorganisms are responsible for the cycling of carbon, nitrogen, phosphorus and sulphur, all of which are essential components of living organisms. As an example, the role of microbes in the nitrogen cycle is summarized in Figure 1.2.

The ultimate source of the nitrogen required for the synthesis of amino acids, nucleic acids and other nitrogen-containing compounds is the earth's atmosphere. Nitrogen gas ($N_2$) accounts for nearly 80% of the atmosphere. Four major processes are required to make nitrogen available to living organisms and then to return it to the atmosphere. These are nitrogen fixation, ammonification, nitrification and denitrification.

**Nitrogen fixation** can be accomplished only by a number of prokaryotic organisms, some of which are free-living (including a number of blue–green bacteria) and some which form symbiotic relationships with higher plants. The best known example of a symbiotic relationship is that between bacteria of the genus *Rhizobium* and the root nodules of leguminous plants. As a result of nitrogen fixation, nitrogen from the atmosphere is incorporated into the proteins and nucleic acids of nitrogen-fixing bacteria.

To be made available to plants and animals, these microbes must die.

**Biosphere**: the surface of the earth and its atmosphere which contains living organisms.
**Biota**: the earth's complement of living organisms.

The rumen of a cow, which enables it to digest the cellulose of grass and other plants, contains about half a kilogram each of bacteria and protozoans. Each gram of rumen contents contains $10^{10}$–$10^{11}$ bacteria and $10^5$–$10^6$ protozoans. Human faeces from healthy individuals contain about $10^{11}$ bacteria per gram. Methane-producing bacteria (methanogens) pump $180 \times 10^{12}$ grams of methane into the atmosphere each year.

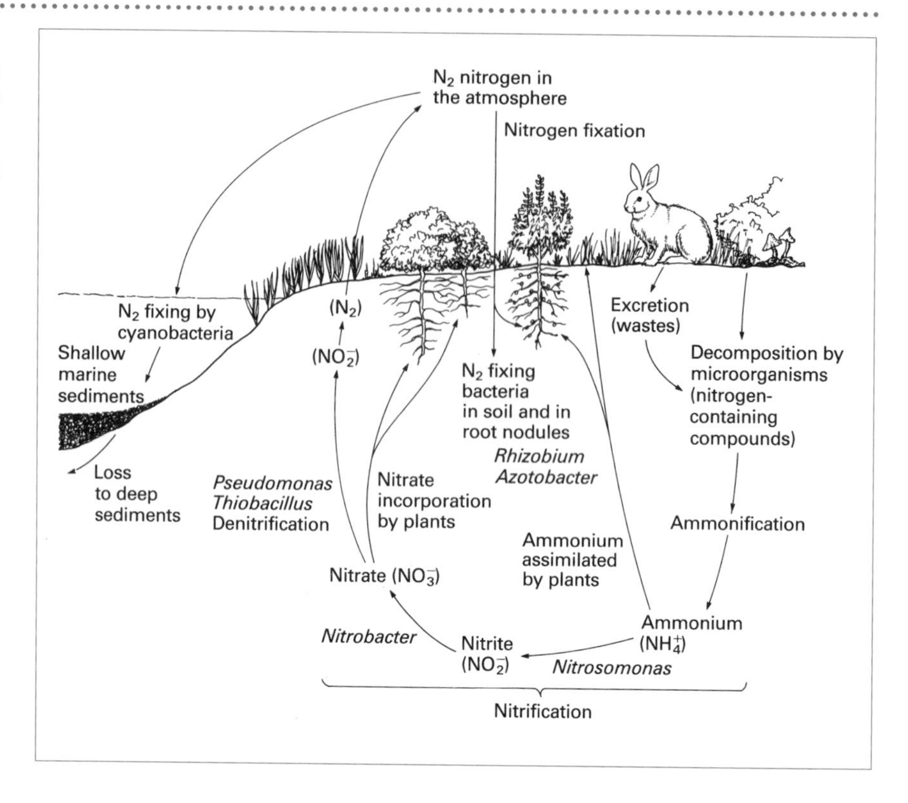


**Figure 1.2** The nitrogen cycle. The nitrogen cycle is an example of a geochemical cycle. In nature it ensures an adequate supply of nitrogen compounds to living organisms. Reactions are carried out by microorganisms and include organic and inorganic forms of nitrogen.

When this happens, protein amino acids are enzymatically deaminated (Chapter 6) to release fixed nitrogen as the ammonium ions ($NH_4^+$). This is **ammonification**.

Although some plants use inorganic ammonium compounds as their nitrogen source, it is the nitrate ion ($NO_3^-$) which is the major form of nitrogen used by plants. To produce nitrate from ammonium, two oxidation steps are carried out by **nitrifying bacteria**. The ammonium ion is initially oxidized to the nitrite ion ($NO_2^-$) by bacteria called *Nitrosomonas*, *Nitrospira*, *Nitrosococcus* and *Nitrosolobus*. In the second stage, nitrite is oxidized to nitrate by *Nitrobacter*. In this way the nitrate required as the major source of nitrogen for plants is produced.

To complete the nitrogen cycle, nitrate must be reduced to nitrogen gas, a process called **denitrification**. This is achieved by several kinds of bacteria which are able to replace oxygen by nitrate and thereby continue to respire under anaerobic conditions, using nitrate in place of oxygen as terminal electron acceptor (Chapter 4). Denitrifying bacteria are found in the heterotrophic genera *Pseudomonas*, *Bacillus* and *Thiobacillus*. A number of chemolithotrophic bacteria such as *Paracoccus denitrificans* are also denitrifiers. We shall look at these in Chapter 4.

Although we have used the nitrogen cycle to illustrate the importance of microorganisms in the cycling of elements, their role is equally essential in the other geochemical cycles.

## Microorganisms and disease

A major characteristic of some microorganisms is that they cause diseases of humans, animals and plants. Since we have considerable interest in sustaining long, healthy lives, there is an enormous impetus to the study of pathogenic microorganisms and the mechanisms by which they cause disease. Infectious disease is one of the main reasons why the development of microbiological science has taken place. Importantly, this aspect of microbiology also encourages financial backing for microbiological research.

Diseases of animals and plants are also important from the human perspective because we are dependent upon these organisms to satisfy our food requirements. If a staple component of our diet becomes scarce as a result of infectious disease, both shortages and starvation may follow.

Why are some microbes disease-causing or **pathogenic** and others not? Pathogenic microorganisms are often much less robust than their free-living relatives. They may require a supply of a particular complex organic molecule for survival which can only be provided by the host plant or animal. In addition, the body of a human or other warm-blooded animal can provide the conditions which a pathogen may require to survive and which may not exist in the outside world.

Medical microbiology is a major branch of microbiology which is highly relevant to the allied science of immunology. Although it is beyond the scope of this text, immunology is the study of the immune system, which combats disease through the medium of host defence mechanisms, including antibody production.

One of the first demonstrations that a microbe caused a disease came in 1836 when the English clergyman, Miles Berkeley, showed that a fungal mould caused the disease potato blight.

The bacterium *Neisseria gonorrhoeae* causes a venereal disease called **gonorrhoea**. It is such a fragile organism that it is completely unequipped for life in the outside world. It is transmitted from one individual to another by sexual intercourse. In this way it is able to increase in number and spread without leaving the environment of the human body. In the laboratory the 'gonococcus' may be cultured on 'chocolate agar' under appropriate conditions of temperature and humidity and in the presence of 5–10% carbon dioxide. The appearance of 'chocolate agar' is caused by the presence of blood which is heated until it turns the brown colour of milk chocolate. This has the effect of bursting the red blood corpuscles so that essential growth factors are released into the culture medium.

## Further reading

R.M. Atlas and R. Bartha (1987) *Microbial Ecology*. 2nd Edition. The Benjamin-Cummings Publishing Company, Inc., Menlo Park, California, USA. – a very useful introduction to microorganisms in nature.

R.F. Boyd and B.H. Hald (1986) *Basic Medical Microbiology*. 3rd Edition. Little, Brown and Company, Boston, Mass., USA. – an introduction to microbes and infectious diseases.

Industrial Microbiology, *Scientific American* (1981) **245**: entire issue. A useful, well illustrated review of the subject.
H.A. Lechavalier and M. Solotorovsky (1974) *Three centuries of Microbiology*. Dover Publications, Inc., New York, USA – microbiology in a historical context.
R. Reid (1975) *Microbes and Men*. British Broadcasting Corporation. Based on the TV series – an inspiration!

## Questions

**1.** Short-answer questions:
(i) Why has microbiology evolved as a subject for study?
(ii) What characteristic do all microorganisms share?
(iii) List four foods or drinks which require microbes for their production.
(iv) Explain the advantages and disadvantages of being very small.
(v) Name a colonial alga.
(vi) Name a nitrifying bacterium.
(vii) What is meant by the expression 'nitrogen fixation'?

**2.** Multiple-choice questions:
(i) Methanogenesis is (a) growth on methane as carbon source, (b) growth on methane as energy source, (c) methane production, (d) methanol production.
(ii) The cow's rumen enables it to (a) break down cellulose, (b) excrete cellulose, (c) synthesize cellulose, (d) photosynthesize.
(iii) *Nitrobacter* is (a) a nitrogen fixer, (b) a nitrifier, (c) a denitrifier, (d) all of these, (e) none of these.
(iv) *Rhizobium* is (a) a free-living nitrogen fixer, (b) a symbiotic nitrogen fixer, (c) a free-living denitrifier, (d) a symbiotic denitrifier.
(v) The causative organism of gonorrhoea is (a) a virus, (b) a bacterium, (c) a protozoan, (d) a fungus.
(vi) The causative organism of AIDS is (a) an alga, (b) a parasitic worm, (c) a bacterium, (d) a virus.
(vii) Which of the following diseases is not caused by a bacterium? (a) tuberculosis, (b) influenza, (c) anthrax, (d) syphilis.

**3.** Fill in the gaps:
Microbes are often used as ___________ tools for the investigation of biological processes. The reasons for this are that they grow __________ and reproduce after a short period of time. This means that ___________ of genetic ___________ may be obtained within a very short time period. Microorganisms all occur in very ___________ populations so that very rare ___________ individuals may be found relatively easily. Most microbes are also ___________ and large amounts of ___________ cell populations can readily be produced. In contrast plants and animals consist of ___________ populations of cells and biochemical investigation may require time-consuming ___________ and separation of different cell types.

Choose from: mutant, unicellular, homogeneous, dissection, heterogeneous, large, crosses, experimental, results, rapidly.

# 2 Microbial structure

In this chapter we start by showing that all living organisms consist of cells which are structurally organized along one of two basic plans described as **eukaryotic** and **prokaryotic**. From here, we go on to describe the structural characteristics of both types of cell which enable them to carry out the activities necessary for day-to-day survival, growth and reproduction. In doing this we shall see that some microorganisms (fungi, algae and protozoa) make extensive use of membranes to localize different metabolic activities. In contrast, we shall see that bacteria are non-compartmentalized but are still capable of growth and reproduction in a very wide range of environments.

Here we compare and contrast prokaryotic and eukaryotic fine structure; eukaryotic morphology, growth and reproduction are covered in detail in Chapter 9.

There are two fundamentally different types of cellular organization in microbes. Bacteria constitute one group and are described as prokaryotic. Bacteria lack a nuclear envelope and their genetic material lies free in the cytoplasm. They also lack mitochondria and chloroplasts, even though some of them are capable of respiration and photosynthesis.

All other cellular microorganisms possess a true nucleus in which the chromosomes are surrounded by a nuclear envelope. They are called eukaryotes (*eu* = true, *karyon* = nucleus). Eukaryotes possess mitochondria and, if they photosynthesize, chloroplasts as well. The differences between prokaryotic and eukaryotic microbes are summarized in Table 2.1.

It is believed that prokaryotes are more primitive than eukaryotes and first appeared on the earth some 3.5 billion years ago. Eukaryotes are thought to have evolved from ancestors of modern bacteria which formed an association with ancestors of the present-day eukaryotic cytoplasm. According to this **endosymbiotic theory**, free-living prokaryotes then became resident components of the cytoplasm of these primitive eukaryotes. As a consequence, some of the eukaryotic cells produced were able to photosynthesize and respire.

There are a number of pieces of experimental evidence which support this endosymbiotic theory and these are summarized below.

**Table 2.1** Comparison of prokaryotic and eukaryotic cells

| | *Prokaryotes* | *Eukaryotes* |
|---|---|---|
| *Microbial groups* | Eubacteria (including cyanobacteria and actinomycetes) and Archaebacteria | Algae, fungi, protozoa |
| *Structure and organization of the cytoplasm* | | |
| Cytoplasmic (plasma) membrane | Lacks sterols | Sterols present |
| Internal membranes | Relatively simple and restricted to certain bacteria, e.g. photosynthetic membranes | Complex and including endoplasmic reticulum and Golgi apparatus |
| Ribosomes | 70S | 80S (except for those of mitochondria and chloroplasts which are 70S) |
| Membranous organelles | Absent | Several including chloroplasts and mitochondria |
| Photosynthetic pigments | No chloroplasts but may be located within internal membranes, e.g. chlorosomes of green sulphur bacteria | In chloroplasts |
| Cell walls | Peptidoglycan characteristic of eubacteria. Archaebacteria contain typical polysaccharides, protein or glycoprotein. Very few bacteria lack a cell wall | Found in algae and fungi but not protozoa. Typically polysaccharide |
| Endospores | Heat resistant – produced by some eubacteria (e.g. *Bacillus* and *Clostridium* | Absent |
| Gas vesicles | Present in some | Absent |
| *Nuclear structure and function* | | |
| Nuclear envelope (membrane) | Absent | Present |
| Nucleolus | Absent | Present |
| DNA | One major 'chromosome' consisting of a single circular molecule of naked (no histones) DNA. Accessory chromosomes or plasmids may also be present | Present in several chromosomes complexed with histones |
| Cell division | Binary fission – no mitosis | Mitosis occurs |
| Sexual reproduction | No meiosis. Conjugation in some bacteria. Recombination usually involves only part of the genome | Meiosis occurs and recombination involves whole genome |
| Introns and exons | Very rare | Common |
| *Motility* | | |
| Flagella and cilia | No cilia. Flagella of submicroscopic dimensions. Flagella rigid and rotate | Flagella or cilia present. Visible using light microscope. 9 + 2 arrangement of microtubules. Flexible, and exhibit whiplash movement |
| Non-flagellar movement | Gliding motility; gas vesicle mediated | Cytoplasmic streaming and amoeboid movement; gliding motility |
| *Microtubules* | Absent | Widespread in flagella, cilia basal bodies, cytoskeleton |
| *Size* | Usually < 2 μm in diameter | Larger, usually 2–100 μm in diameter |

**Box 2.1.**

**The endosymbiotic theory**
Symbiosis is a mutually beneficial association and presumably the prokaryotic component was protected from adverse environmental conditions and surrounded by a hospitable environment. The eukaryotic cytoplasm benefits from energy released by what are now mitochondria or chloroplasts. The use of the prefix 'endo' emphasizes that the prokaryotic component of the association is located **inside** the eukaryotic cytoplasm. A number of pieces of experimental evidence support this endosymbiotic theory:

1. Both mitochondria and chloroplasts contain DNA which codes for some of their respective protein components. The DNA of these organelles resembles that of bacteria in being devoid of histones and having the ends of the molecule joined to each other to give a looped configuration.
2. Ribosomes are found in both types of organelle and these are of the bacterial type (70S ribosomes). Ribosomes of the eukaryotic cytoplasm are larger (80S).
3. Antibiotics which inhibit protein synthesis in eubacteria often inhibit synthesis in chloroplasts and mitochondria but do not affect protein synthesis in the eukaryotic cytoplasm.
4. Determinations of the base sequence of ribosomal RNA from a variety of eukaryotic and prokaryotic organisms have added further support to the idea that chloroplasts and mitochondria have evolved from eubacteria.

It seems likely that both organelles have evolved by the gradual loss of more and more genetic information. They have become very specialized in producing ATP and dependent on their eukaryotic host cell. The result of this process has been the production of the chloroplasts and mitochondria of present-day eukaryotes.

## Structure and function of eukaryotes and prokaryotes

### Membrane systems

Both prokaroytic and eukaryotic organisms contain membranes which are similar in overall structure. Membranes of eukaryotic microorganisms serve to compartmentalize the cell contents into organelles such as mitochondria, thus allowing concentration of specific metabolites at certain locations. Prokaryotic organisms contain only a single membranous structure, the **cytoplasmic membrane** or **plasma membrane**, which eukaryotic cells also possess. Although only 4–5 nm thick, the cytoplasmic membrane separates the cell contents from the external environment. If it is broken, the cell cytoplasm leaks into the environment and the cell dies. The cytoplasmic membrane also constitutes the permeability barrier of the cell and allows it to concentrate desired nutrients and excrete waste products.

Membranes also allow the various components of a complex biochemical process, such as respiration, to be aligned in a fashion such that maximum efficiency is achieved.

Although they perform a variety of functions, all biological membranes have the same general structure based on the formation of a **lipid bilayer**

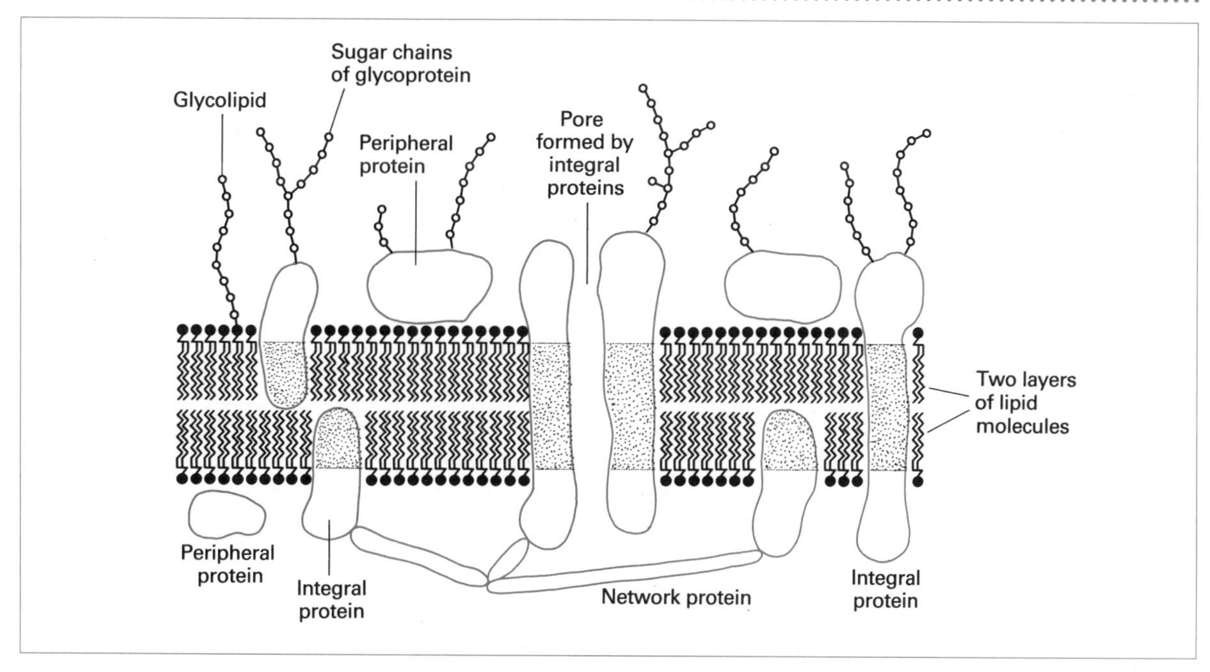


**Figure 2.1** Membrane structure. The diagram shows two layers of lipid molecules (lipid bilayer). The hydrophilic 'heads' containing phosphate groups point towards the watery media of the cell cytoplasm and the surrounding environment. The hydrophobic fatty acid 'tails' are directed away from these aqueous regions. A number of proteins are embedded in the lipid layer and the overall arrangement is described as a **fluid mosaic.**

(Figure 2.1). Everyday experience will tell you that lipids such as vegetable oils, fats and waxes do not readily mix with water and this provides a clue to the potential suitability of lipids in serving a permeability function. Phospholipid molecules consist of two parts, a fatty acid portion which is hydrophobic (water-hating) and which, therefore, does not readily associate with water molecules, and a glycerol phosphate part which is hydrophilic (water-loving), and which, therefore, readily associates with water molecules.

Because they consist of hydrophobic and hydrophilic portions, phospholipid molecules tend to form lipid bilayers spontaneously. In such a bilayer, the hydrophobic fatty acid portions point towards each other and produce a hydrophobic environment. The hydrophilic parts are exposed to the aqueous external environment. Under the electron microscope, preparations of cytoplasmic membrane in which the hydrophilic component of the membrane has been stained with an electron-dense material appear as two, dark, thin lines separated by a lighter area. This confirms the overall configuration of lipid molecules of the cytoplasmic membrane, which also contains a number of major proteins (Figure 2.1).

The proteins of membranes such as the cytoplasmic membrane or mitochondrial membrane may be associated with only one side of the membrane, or may be completely embedded in the phospholipid matrix. Because of this, the inner and outer sides of the cytoplasmic membrane have different properties; in other words, membranes exhibit 'sidedness' and this property is of considerable importance in membrane function.

The overall structure of a membrane is maintained by hydrogen bonds and hydrophobic interactions. Positively charged ions such as $Mg^{2+}$ and $Ca^{2+}$ (cations) help to stabilize the structure by forming ionic bonds with negatively charged phospholipids.

Eukaryotic membranes are distinguished from those of almost all prokaryotes because they contain **sterols**. Sterols are flat, rigid molecules, whereas lipids are flexible. Because eukaryotic cells are larger than prokaryotic cells, their cytoplasmic membrane has to withstand greater physical stress and a more rigid cytoplasmic membrane is required to keep the cell stable and functional.

The only group of prokaryotes that contain sterols in their cytoplasmic membranes are **mycoplasmas**. These bacteria lack cell walls and the greater stresses imposed upon the cell membrane necessitate a sterol-containing structure.

### The cytoskeleton

The cytoplasm of eukaryotic cells contains an extensive network of microfilaments and microtubules. The filaments contain a number of proteins, including actin and myosin. The main protein components of the microtubules are called tubulins and are found in association with other proteins whose identity is, as yet, unknown. The cytoskeleton has a number of functions, including the maintenance of cell shape, positioning cell organelles and cell motility. The cytoskeleton of *Trypanosoma* (a flagellate protozoan causing sleeping sickness in man) forms an internal helical 'corset', giving the organism its characteristic shape.

The creeping movement shown by *Amoeba* is thought to require depolymerization and repolymerization of the cytoskeleton, emphasizing its dynamic structure (Chapter 9).

Possession of a cytoskeleton appears to be mainly a eukaryotic characteristic, although spiral-shaped bacteria, called **spirochaetes**, possess microtubule-like proteins which run from one end of the cell to the other and appear to be responsible for the characteristic bending and flexing motion seen in these bacteria.

The difference in the sterol content of the cytoplasmic membrane of eukaryotes and most prokaryotes means that a number of agents (filipin, nystatin and candicidin), which react with sterols, disrupt eukaryotic and mycoplasmal membrane structure, but do not interfere with those of most bacteria.

### Chloroplasts

All photosynthetic eukaryotic organisms, including microscopic algae, (Chapter 9), contain chlorophyll-containing organelles called **chloroplasts**. These structures, in contrast to mitochondria, vary greatly in size and are typically much larger than bacteria. The structure of a typical chloroplast is shown in Figure 2.2.

The thylakoid membrane is, for the most part, impermeable to ions and

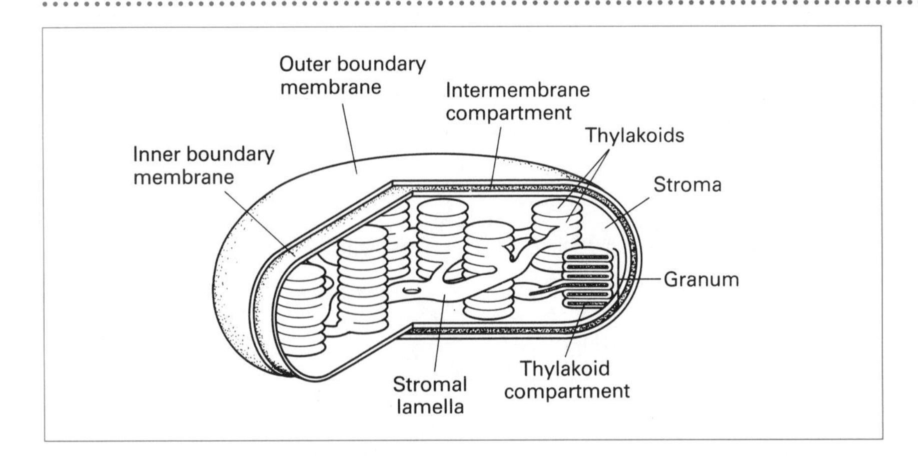


**Figure 2.2** Structure of chloroplasts. These structures are found in photosynthetic eukaryotes. They have two outer membranes and many stacked membranes (forming grana) in the stroma.

other metabolites and is therefore well-suited for its role in proton translocation and the generation of a proton motive force which can be used to drive the synthesis of ATP (Chapter 4). The green algae and green plants resemble each other in that thylakoids are stacked to produce discrete structural units called **grana**.

The stroma of chloroplasts contain enzymes of the Calvin cycle and are responsible for reduction of $CO_2$ to organic material (Chapter 4). These include ribulose bisphosphate carboxylase, one of the key enzymes of the Calvin cycle and one of the most abundant enzyme proteins found in nature. The high permeability of the outer chloroplast membrane allows glucose and ATP synthesized during photosynthesis to diffuse into the cytoplasm, where they may be used to produce cell material. Although some bacteria are photosynthetic, they do not possess chloroplasts. Instead, components of the photosynthetic machinery are located on a variety of membranous structures formed by modification of the cytoplasmic membrane (Figure 2.3).

## Mitochondria

The processes of respiration and oxidative phosphorylation (the mechanism of ATP synthesis during respiration) are localized in membrane-bound organelles in eukaryotic organisms. Individual mitochondria are approximately the same size as rod-shaped bacteria; that is to say, approximately 2–3 μm long and 1 μm in diameter. The number of mitochondria per cell varies and may be as many as 1000 in a liver cell and as few as two in a yeast cell.

The mitochondrial membranes lack sterols and tend to be less rigid than the cytoplasmic membrane. Furthermore, the outer mitochondrial membrane is more permeable than the cytoplasmic membrane and allows the

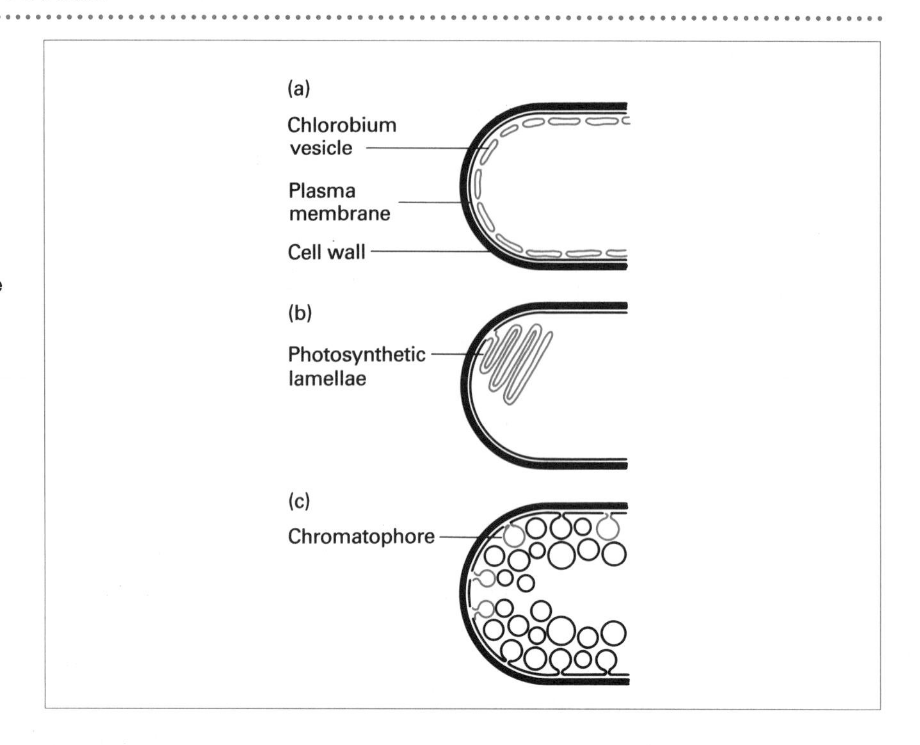


**Figure 2.3** Photosynthetic membranes of bacteria. (a) Chlorobium vesicles or chlorosomes are found in the green sulphur bacteria. These vesicles lie just under the cytoplasmic membrane but are not continuous with it. (b) The stacks of folded sheet-like membranes formed in the purple sulphur bacteria are continuous with the cytoplasmic membrane. (c) The chromatophores of the purple non-sulphur bacteria are continuous with the cytoplasmic membrane.

passage of molecules with molecular weights less than about 10 000. As a result, ATP synthesized within the mitochondrion can move into the cytoplasm, where it is consumed in energy-requiring reactions.

The inner mitochondrial membrane is highly folded to produce **cristae**, which project into the mitochondrial lumen. Located within the inner membrane are the sites of respiratory enzyme action, the enzymes involved in ATP production, and specific transport proteins which regulate the passage of metabolites into and out of the matrix of the mitochondrion (Figure 2.4). Using a high-resolution electron microscope it is possible to see that cristae contain small round particles attached to the membrane by

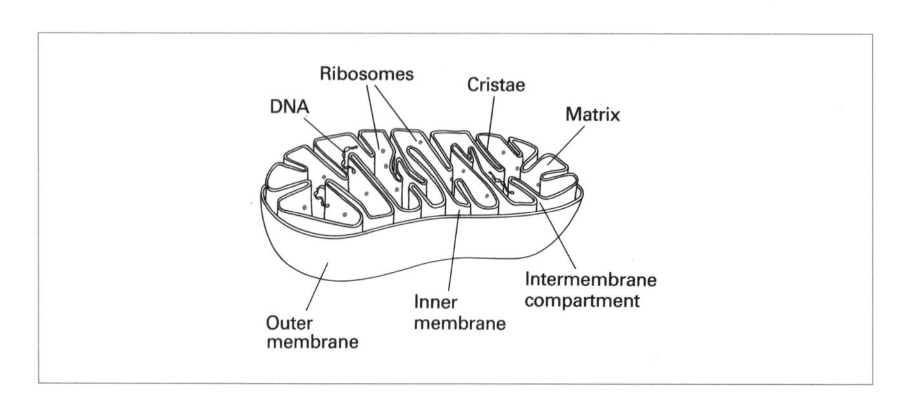


**Figure 2.4** Structure of mitochondria. These double-membraned organelles are found in eukaryotic cells only. They are responsible for respiration. The inner membrane is folded (invaginated) to form **cristae**. Most of the respiratory enzymes are located in the cristae and in the matrix. Mitochondria are thought to have evolved from free-living prokaryotes and contain circular DNA and 70S ribosomes.

short stalks. These structures are complexes of the enzyme ATP synthetase which catalyse ATP synthesis in the mitochondrion during oxidative phosphorylation.

The mitochondrial matrix contains enzymes responsible for the oxidation of pyruvate and fatty acids together with certain of the TCA cycle enzymes (Chapter 4). The space between the two membranes contains a number of proteins required for conversion of ATP into other high-energy phosphate intermediates used in the matrix or in the cell cytoplasm.

## Structure and function of eukaryotic and prokaryotic organelles

### Nuclei and nucleoids

As you now know, eukaryotic organisms possess a nuclear envelope which surrounds the genetic material. The envelope consists of two membranes which fuse at a number of places to produce pores. Through these pores, contact is established between the nuclear contents and the cytoplasm. The outer membrane of the nuclear envelope carries ribosomes and forms a continuous structure with the **endoplasmic reticulum** and **cytoplasmic membrane**.

The nuclear envelope encloses a number of **chromosomes** in association with the chromosomal proteins, such as histones. Histones are positively charged proteins which neutralize the negatively charged phosphate groups of the DNA. This is important because it means that during nuclear division the genetic material can be compacted to form cylindrical chromosomes which take part in the division process. This can occur only if the negative charges of DNA are neutralized by the positive charges of histone proteins.

In eukaryotes, the DNA exists as a single linear molecule to which histones and other proteins are attached (Figure 2.5). This contrasts with bacterial DNA, which occurs in a circular form. Although the size of a nucleus, the number of chromosomes and number of genes are characteristic of a particular species, different microorganisms vary greatly in all these aspects. The number of chromosomes can vary from just a few to many hundreds. Furthermore, even if two cells contain the same quantity of DNA, the number of genes may vary considerably because some genes may be present in a number of copies in one organism but as only a single copy in another.

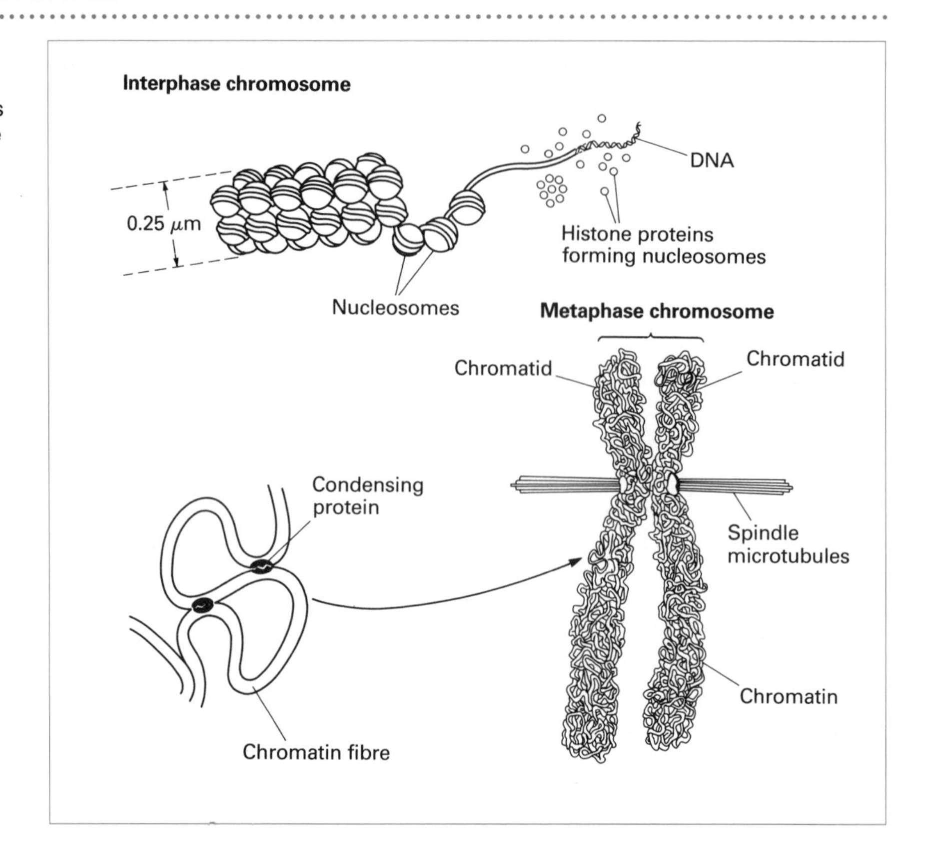


**Figure 2.5** Structure of the eukaryotic genome. During interphase (when the nucleus is not dividing), chromosomes are only 0.025 μm thick. At the metaphase stage of nuclear division the chromosomes condense. Their diameters increase to 0.5 μm and they become visible with a light-microscope.

Prokaryotic microorganisms differ from eukaryotic ones in their lack of a nuclear envelope. Bacterial DNA is not associated with proteins and, for this reason, it is sometimes described as 'naked' DNA. For the same reason, some biologists recommend that the term chromosome be reserved for the eukaryotic structure. The major 'chromosome' of *Escherichia coli* is between 1.1 mm and 1.3 mm in length and is approximately six hundred times longer than the cell itself. For this reason, the DNA is highly coiled (sometimes described as **supercoiled**), so that it can fit into the confines of the cell. It may also be concentrated in the centre of the cell as a 'nucleoid', a term used to emphasize the absence of a nuclear envelope.

In addition to the main 'chromosome', bacteria often contain one or more accessory 'chromosomes' called **plasmids**. Plasmids are molecules of circular DNA just like the main 'chromosome' but contain genes which carry out functions required only under certain circumstances. The bacterium *Shigella dysenteriae*, which causes bacillary dysentery, contains a plasmid which confers resistance to a number of antibacterial agents including a number of antibiotics. Resistance to antibiotics would be a useful characteristic for bacteria competing in an environment containing organisms which produce antibiotics. Plasmids are discussed in detail in Chapter 8, where their use in genetic engineering is reviewed.

## Cell walls

The possession of a cell wall is often said to be a plant-like characteristic and its absence, an animal-like one. Although the concept of animals and plants is not entirely applicable to microorganisms, it can be useful. Protozoans lack a cell wall, whereas bacteria, algae and fungi have one. The structure and synthesis of the bacterial cell wall is described in Chapter 5. We shall restrict ourselves to eukaryotic cell walls here.

First, we look at the algal cell wall. This is made of fibres of **cellulose** which form a strong wall surrounding the whole cell, although the wall lacks any division into layers. Cellulose is a straight-chain polymer of glucose, in which carbon atoms 1 and 4 of adjacent glucose residues are linked by a type of chemical bond called a β 1–4 linkage (Figure 2.6). In addition to cellulose, **hemicellulose**, which is a polysaccharide of glucose and other sugars, forms fibres which are hydrogen-bonded to the cellulose. Pectins containing galacturonic acid and rhamnose give a jelly-like consistency to the cell wall because of their high water content.

In some algae, the cell wall may also contain a number of additional types of polysaccharides. These are called xylans, mannans and alginic acids. Some algal cell walls contain minerals, for example calcium carbonate. The algae known as diatoms have a cell wall of silica, protein and polysaccharide.

Algal cell walls are permeable to low-molecular-weight compounds such as water, ions and gases but are impermeable to larger molecules.

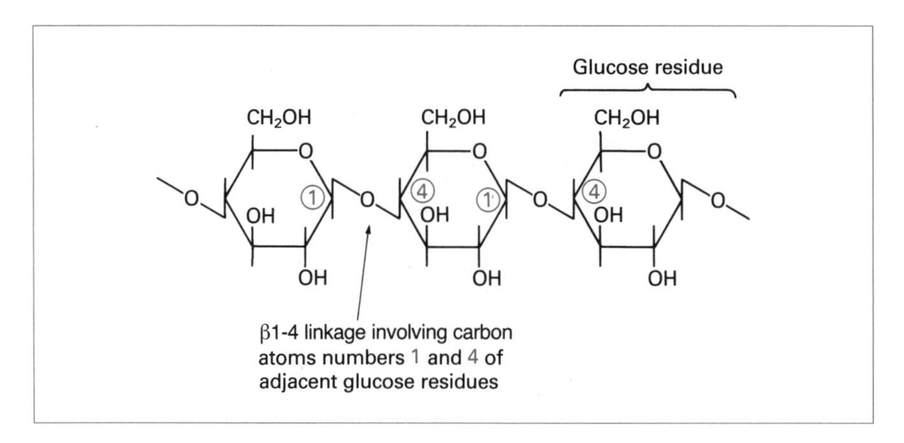


**Figure 2.6** Structure of cellulose.

## Motility

### Motility in bacteria

Many bacteria move by means of hair-like structures called flagella. As a result, they can move towards favourable environments (nutrients, light for photosynthetic bacteria and so on) and away from less favourable ones such as toxic substances.

Here we first describe the structure of the bacterial flagellum and then look at the way in which bacteria move. Bacterial flagella are thin (20 nm diameter) and rigid and rotate very much like a ship's propeller. They cannot be seen under the light microscope unless special stains are used to increase their diameter.

Bacterial flagella are rigid and coiled in a spiral fashion. They are made up of a protein called **flagellin**, which is organized into flagellar subunits. At the base of the flagellum there is a basal body, which rotates the flagellum to cause movement of the cell. The structure of the bacterial flagellum is shown in Figure 2.7.

Rotation of the flagellum is driven by the basal body, which acts as a motor. The energy required for rotation of the flagellum comes from the proton motive force generated during either respiratory or photosynthetic electron transport (Chapter 4). Through the action of the basal body, each flagellum is caused to rotate in an anticlockwise direction at about 12 000

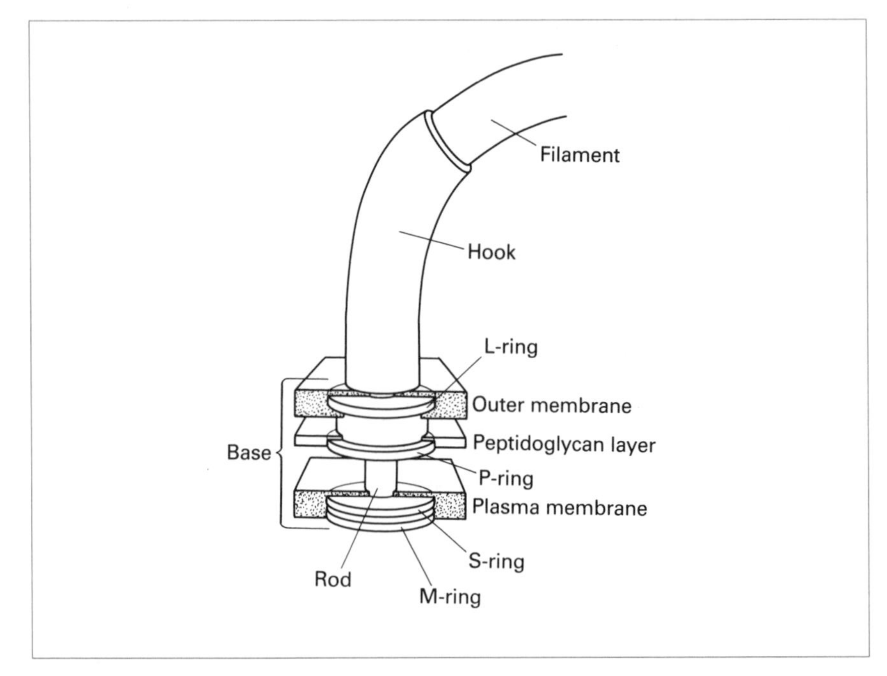


**Figure 2.7** Structure of bacterial flagellum. This diagram of the base of a bacterial flagellum shows the three main components of the structure found in the Gram-negative eubacteria such as *Escherichia coli* and *Salmonella typhimurium*. These are filament, hook and basal body. The basal body anchors the flagellum to the cytoplasmic membrane and cell wall. Rotation of the hook and filament causes the cell to move.

rpm. This rotation of the flagellum causes the bacterium to move at a speed of 20 μm to 80 μm per second, which is about 10 lengths per second and compares favourably with speeds of much larger organisms.

### Chemotaxis in bacteria

Bacteria respond to chemicals in the environment. They swim towards attractants and away from repellants. **Chemotaxis** is the word used to describe this primitive behavioural response which suggests that bacteria can sense changes in concentration of a variety of compounds in their environment. They can do this because they possess a number of protein **chemoreceptors** located near the cell surface. These combine with attractants or repellants which are present in the environment. A stimulated chemoreceptor is then able to pass a message to the flagellum basal body to determine the direction of flagellum rotation and its duration.

Before we describe how bacteria respond to a stimulus, let us look at their movement. Swimming is made up of two types of action, called **runs** and **twiddles**. Bacteria such as *Escherichia coli* and *Salmonella typhimurium* have flagella distributed over the whole of the cell surface (they are said to show **peritrichous** flagellation). When the flagella rotate in an anticlockwise direction, they come together to form a **flagellar bundle**, which itself rotates anticlockwise (Figure 2.8). This rotation moves the cell forward, and a run is produced. After about a second of this smooth swimming, the run stops, the flagella start to rotate in the opposite direction and the bundle falls apart. The cell goes into free-fall for about a tenth of a second and tumbles through the surrounding medium in an uncontrolled manner called a twiddle. After a twiddle, the direction in which the cell points is random. The direction of the following run is also random. This kind of movement is shown in Figure 2.9 and, in the absence of a stimulus, the cell does not move very far from its original position.

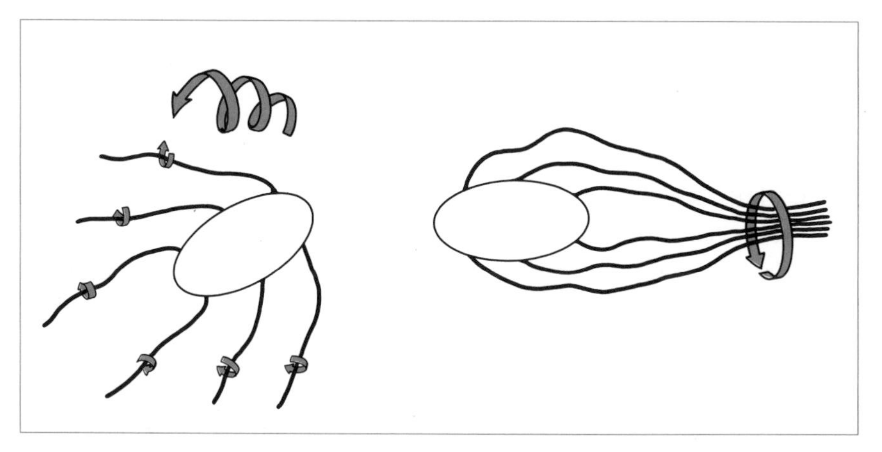

**Figure 2.8** Flagellar rotation and bacterial movement. During anticlockwise rotation, peritrichous flagella form a bundle behind the cell. Rotation of the flagellar bundle causes the cell to move forwards (a **run**). When the direction of rotation changes (becomes clockwise), the bundle flies apart and the cell tumbles through the surrounding medium, producing a **twiddle**.

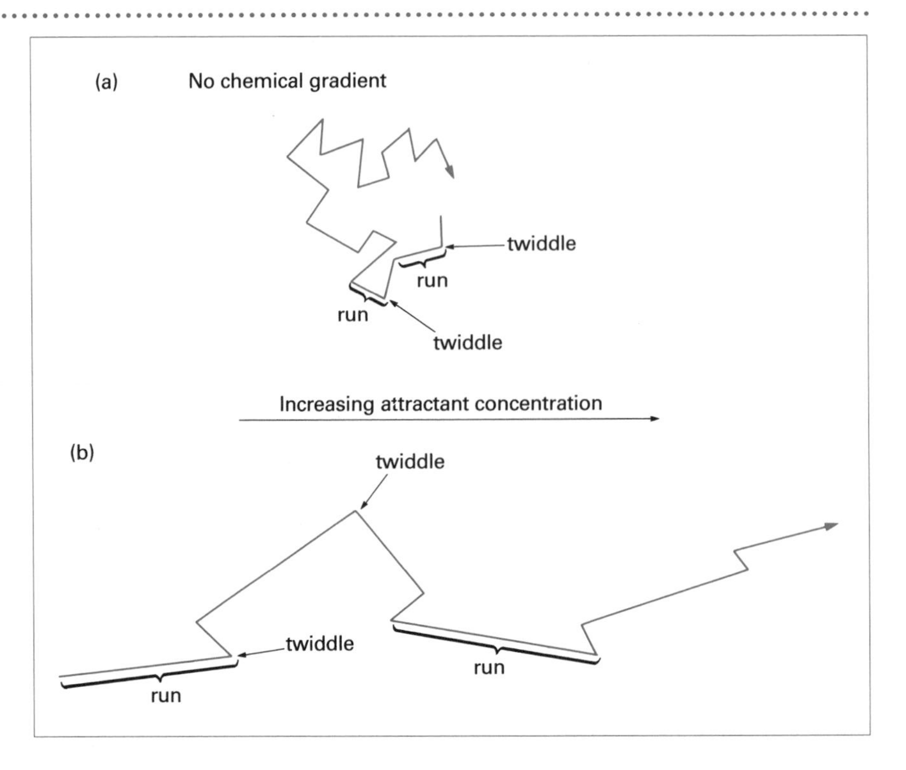


**Figure 2.9** Chemotaxis in *E. coli.* (a) The diagram shows the random movements of a cell in a liquid where there is no chemical gradient. Runs and twiddles (tumbles) occur frequently and the cell makes no directional progress. (b) The cell moves towards a chemical attractant: that is to say, movement is up a chemical gradient. Although runs occur initially in a random direction, those towards higher concentrations of attractant tend to become longer, and twiddles occur less frequently.

A chemical attractant, such as a nutrient, if added to the medium, will diffuse through it and produce a concentration gradient. Under these conditions, movement becomes less random and the bacterium tends to move towards higher concentrations of attractant. This is because the runs towards the attractant become longer and the twiddles less frequent. This type of response occurs because the bacterium detects changes in concentration over a period of time. If it senses an increase in the concentration after a period of swimming in a particular direction, the duration of runs towards the attractant increases and the number of twiddles falls.

## Motility in eukaryotes

Many kinds of eukaryotic microorganisms also move about by means of flagella and cilia. Both structures are constructed on the same basic plan, but cilia are much shorter than flagella and occur in much larger numbers. Cilia are restricted to one group of protozoans, the ciliates, of which *Paramecium* is an example (see also Chapter 9).

Eukaryotic flagella are long, flexible structures which move in a whiplash fashion (Figure 2.10). They are much thicker than bacterial flagella and

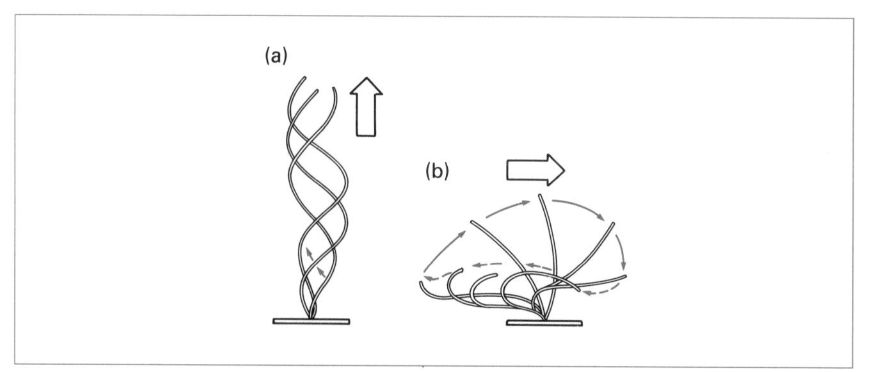


Figure 2.10 Movement of eukaryotic flagella and cilia. During flagellar movement (a) successive waves of motion pass towards the tip of the flagellum (whiplash movement) and the cell is driven in the opposite direction. A cilium (b) moves like an oar in a rowing boat and remains more or less rigid. During the recovery stroke, the cilium is more curved to produce least resistance. Cilia occur in large numbers on ciliates such as *Paramecium* where they beat in a coordinated fashion called **metachronal rhythm**.

easily observed using a light microscope. They have a complex structure consisting of a number of **microtubules** (Figure 2.11). Each microtubule is composed of protein, **tubulin**, with subunits arranged in a helical fashion along the tubule axis. The flagellum is caused to move by the coordinated movement of microtubules towards or away from the base of the flagellum. Eukaryotic microorganisms move at a rate of between 30 and 250 μm per second.

Whereas an individual microorganism possesses only a relatively small number of flagella, cilia are found in much larger numbers. A single *Paramecium* may have 10 000 or more cilia. In contrast to flagella, cilia are more or less rigid and beat in a coordinated fashion called **metachronal rhythm**. Each cilium beats about 20 times per second and produces a top speed of 2.5 mm per second.

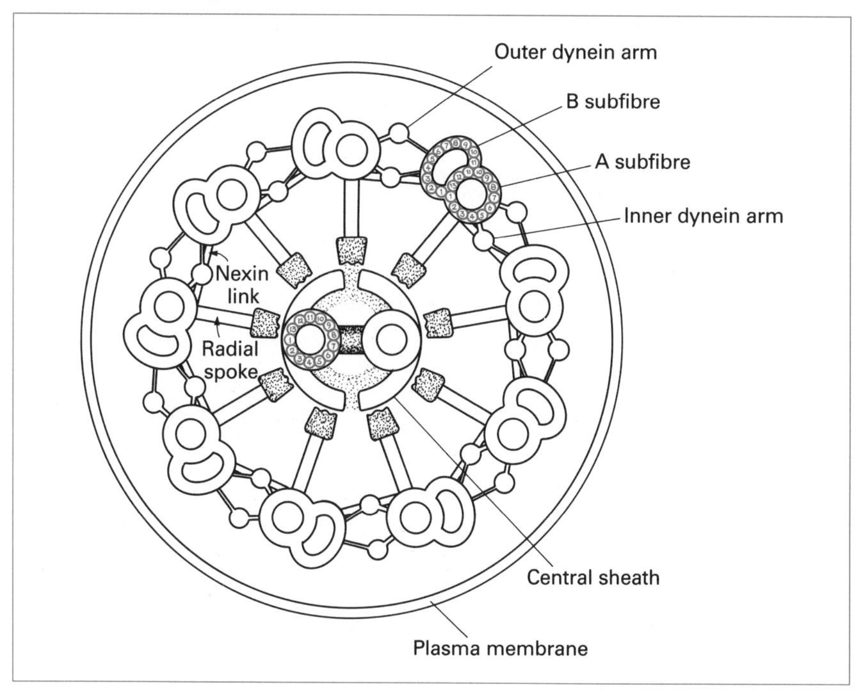


Figure 2.11 Structure of a eukaryotic flagellum seen in cross-section. The 9 + 2 arrangement of nine peripheral pairs of microtubules plus one central pair is shown. This arrangement is characteristic of all eukaryotic flagella and cilia. The radial spokes occur at 24-, 32- and 40-nm intervals along the shaft of the flagellum. Nexin links occur every 86 nm and link together pairs of peripheral microtubules. Outer dynein arms are located every 24 nm and inner ones at the same intervals as radial spokes. Both nexin and dynein are proteins. Dynein arms are thought to bend the flagellum and nexin arms help to maintain location of subfibres in the flagellum.

### Amoeboid movement

Amoebae and slime moulds possess neither flagella nor cilia, but show a characteristic type of movement produced by their cytoplasm, called **cytoplasmic streaming**. Such cytoplasmic flow produces projections, called **pseudopodia**, of the flexible cell surface. The cytoplasmic flow and production of pseudopodia causes the cell to move (Figure 9.11).

Amoeboid movement is driven by ATP which causes bundles of actin fibres, located immediately below the cytoplasmic membrane, to move against each other. Movement of these bundles causes cytoplasmic streaming but the exact mechanism remains a mystery.

## The bacterial cell surface

Some bacteria possess additional hair-like structures on their cell surface, called **fimbriae**. These are shorter than flagella but are more numerous. Like flagella, they consist of a protein. There is some doubt about the function of fimbriae but they may enable an organism to stick to a surface.

**Pili** are similar to fimbriae but tend to be longer and fewer in number. The function of the **sex pilus** of *Escherichia coli* has been widely studied. It brings together two cells during the conjugation process (a form of bacterial mating) before transfer of genetic material from the donor to the recipient cell. Some bacteriologists believe that the sex pilus also acts as a **conjugation tube**, through which the DNA is transferred from cell to cell (Chapter 8). Pili of a different type are concerned with the attachment of pathogenic bacteria to human tissues.

### The glycocalyx

The glycocalyx consists of a number of polysaccharides, usually in association with glycoprotein, secreted on to the outer surface of the bacterial cell wall. The nature of the glycocalyx varies from one kind of bacterium to another and it may be thick or thin, rigid or flexible. A variety of functions have been assigned to the glycocalyx, depending on the bacterium concerned. In some, they aid invasion of a host organism by binding a pathogenic bacterium to a specific tissue. In others, the glycocalyx is seen as a capsule and may hinder the engulfing (**phagocytosis**) of the bacterium by a host organism's phagocytic immune defence cells. Such an encapsulated strain of a pathogenic bacterium (for example, the S strain of *Streptococcus pneumoniae* examined in Chapter 7) is likely to be more virulent than a non-encapsulated strain. Finally, because the glycocalyx contains water, it may prevent desiccation.

## Bacterial endospores

Many living organisms produce structures called **spores**. In many filamentous fungi, such as *Aspergillus* and *Penicillium*, spores are produced in very large numbers and their function is a reproductive one. The asexual reproductive spores of fungi are produced externally and this aids the dispersal and spread of the fungus (Chapter 9).

The bacterial endospore is not a reproductive structure. The prefix 'endo' emphasizes the location of these structures inside the cell. The endospore is resistant to harsh environmental conditions, including high temperatures and the presence of toxic chemicals. Furthermore, a bacterial cell produces only one endospore. Bacteria which produce endospores belong to the genera *Bacillus* and *Clostridium*, common in the soil.

The structure of the endospore (Figure 2.12) is more complex than that of the vegetative cell which produces it. Dipicolinic acid (DPA), together with high concentrations of calcium ions, have been found in all endospores examined but not in vegetative cells. It has been suggested, therefore, that a complex of calcium and DPA is responsible for their heat resistance.

The production of endospores by some bacteria is the reason why an autoclave is used to produce sterile culture media. Temperatures in the region of 121°C are produced in an autoclave because water is boiled under pressure and this is high enough to kill the most resistant endospores. Strips containing endospores of *Bacillus stearothermophilus* are commonly used to check that an autoclave is working efficiently.

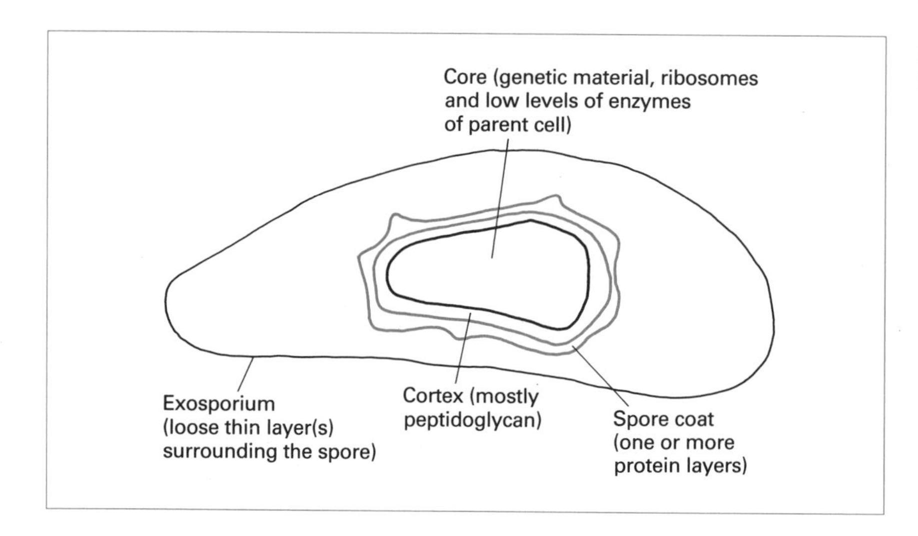


**Figure 2.12** The layers of a bacterial endospore.

## Chapter summary

We started this chapter by describing the way in which the chloroplasts and mitochondria of eukaryotic organisms are thought to have evolved from prokaryotic ancestors. We went on to examine the experimental evidence which supports this endosymbiotic theory.

Once we had established that living cells are organized according to one of two basic plans, known as prokaryotic and eukaryotic, the features distinguishing the two types of cell were described, together with the importance of the presence or absence of membrane-bound organelles (chloroplasts, mitochondria and nucleus).

We examined the structure of a number of eukaryotic organelles and their prokaryotic equivalents. The location of the genetic material in nuclei of eukaryotes and nucleoids of bacteria was examined and we saw that only the former was surrounded by a double-membraned envelope.

A number of the structures located at the surface of microbial cells, including cell walls and the glycocalyx, were described. A comparison was made between the structure and function of prokaryotic and eukaryotic flagella. The primitive pattern of behaviour in bacteria, chemotaxis, was described.

The structure and function of other surface filamentous structures, including bacterial fimbriae and pili, were described and the role of the endospore produced by some bacteria was examined.

## Further reading

B. Alberts, D. Bray, J. Lewis, M. Raff, K. Roberts and J.D. Watson (1989) *The Molecular Biology of the Cell.* 2nd Edition. Garland Publishing, New York – a comprehensive account covering microorganisms and multicellular organisms.

C. De Duve (1984) *A Guided Tour of the Living Cell.* Scientific American Books, New York – a readable, well-illustrated account in the *Scientific American* mould.

P. Sheeler and D. Bianchi (1986) *Cell and Molecular Biology.* 3rd Edition. John Wiley and Sons Inc., New York – a concise, well-illustrated account.

C. Smith and E.J. Wood (1991) *Molecular and Cell Biology. Cell Biology.* Chapman & Hall, London – an up-to-date treatment of this important area.

## Questions

1. Multiple-choice questions:
   (i) Which of the following do not possess a nuclear envelope? (a) algae, (b) bacteria, (c) fungi, (d) protozoa, (e) eukaryotes.
   (ii) Which of the following are prokaryotic? (a) all fungi, (b) eubacteria and archaebacteria, (c) Gram-negative bacteria only, (d) protozoa and algae, (e) viruses.
   (iii) Amoeboid movement is accomplished by which of the following? (a) flagella, (b) cilia, (c) pseudopodia, (d) axial filament, (e) cilia and flagella.
   (iv) Which of the following structures are found in bacteria? (a) mitochondria, (b) chloroplasts, (c) endoplasmic reticulum, (d) ribosomes, (e) Golgi apparatus.
   (v) Metachronal rhythm is used to describe: (a) movement of bacterial flagella, (b) movement of protozoan flagella, (c) flexing and bending movement of spirochaetes, (d) amoeboid movement, (e) beating of protozoan cilia.
2. Short-answer questions:
   (i) How does the function of the bacterial endospore differ from that of fungal conidia (asexual spores)?
   (ii) Describe how the endosymbiotic theory explains the evolutionary origin of chloroplasts and mitochondria.
   (iii) Describe the differences in structure between prokaryotic and eukaryotic flagella.
   (iv) What is a nucleoid?
   (v) Describe **two** functions of bacterial pili.
3. Fill in the gaps:
   ____________ is the term used to describe the movement of bacteria in response to chemicals in the environment. Nutrients are often ____________ and the bacterium moves ____________ such substances. ____________ substances are repellants and the bacterium moves ____________. Bacteria which exhibit such behaviour possess one or more hair-like __________ and which __________ very much like a ship's propeller. During a chemotactic response, the cell moves by a series of ____________ and ________. The former is accomplished by movement of flagella in an ____________ direction and the latter occurs when the direction changes. As it moves through the surrounding medium,

the presence of attractants or repellants is sensed by __________ on the cell surface.

The eukaryotic flagellum is much ____________ than the bacterial one and can therefore be seen using the ____________ microscope. In cross-section a complex arrangement of fibrils in a __________ + ____________ arrangement can be seen. These cause the flagellum to produce a ____________ type of movement, a process which requires ____________ in the form of ATP.

Choose from: twiddles, anticlockwise, thicker, 9, whiplash, energy, 2, light, receptors, runs, rotate, away, towards, chemotaxis, required, toxic, flagella.

# Microbial growth 3

Microorganisms, like all living organisms, grow and reproduce. In this way they ensure continuation of their kind and provide the raw material on which the evolutionary process operates. To grow, any organism must have a supply of nutrients to produce new cell material, enzymes and cofactors. In addition, an energy source is required to fuel cellular activities. Some organisms can harvest the sun's rays for this purpose, but others require a chemical energy source.

In this chapter, we consider the range of nutrients which microorganisms require to support growth and reproduction. We examine the techniques available for the laboratory culture of microorganisms. Finally, we look at the ways in which the growth of microorganisms can be controlled through the use of sterilization techniques, antibiotics and disinfection.

## Major and minor chemical elements

Although the chemical and structural diversity of microorganisms is awesome, only a small number of elements are required in relatively high concentrations by all forms of organism (Table 3.1). Carbon, hydrogen, oxygen and nitrogen are the major components of the compounds found in microorganisms. Sulphur is present in proteins as a constituent of two amino acids, cysteine and methionine. Additionally, sulphur is a constituent of a number of cofactors required for enzyme function. Phosphorus is found in nucleic acids, phospholipids and coenzymes such as flavin adenine dinucleotide (FAD) and nicotinamide adenine dinucleotide (NAD), both of which take part in energy metabolism (Chapter 4).

A number of metal ions, including potassium ($K^+$), magnesium ($Mg^{2+}$), calcium ($Ca^{2+}$) and iron ($Fe^{2+}$ and $Fe^{3+}$), are cofactors for enzymes or components of important biological molecules. The functions of these metal ions are summarized in Table 3.1.

In addition to these major elements, microorganisms require other elements in small amounts. In some instances, the amounts needed are very

**Table 3.1** Nutrients required in relatively large amounts by microorganisms

| *Element* | *Source* | *Function in metabolism* |
|---|---|---|
| Carbon (C) | $CO_2$, organic compounds | major components of cell material |
| Hydrogen (H) | $H_2$, $H_2O$, organic compounds | |
| Oxygen (O) | $O_2$, $H_2O$, CO, organic compounds | |
| Nitrogen (N) | $NH_4^+$, $NO_3^-$, $N_2$, organic compounds | |
| Phosphorus (P) | Phosphates | Constituent of nucleotides, nucleic acids and phospholipids |
| Sulphur (S) | Sulphates, sulphides, thiosulphates, elemental sulphur, organic compounds | Constituents of cysteine, methionine and number of coenzymes and cofactors |
| Potassium (K) | Potassium salts | Cofactor of some enzymes (for example, pyruvate kinase) |
| Magnesium (Mg) | Magnesium salts | Cofactor of many enzymes; ribosome stability |
| Calcium (Ca) | Calcium salts | Cofactor for exoenzymes (for example, amylases, proteases); resistance of bacterial endospores |
| Iron (Fe) | Ferrous and ferric salts | Component of cytochromes, iron sulphur proteins; cofactor of some enzymes |
| Sodium (Na) | Sodium salts | Required for survival of halophilic archaebacteria |

small indeed. In fact, these so-called **trace elements** may often be present in water as impurities and a requirement is sometimes difficult to demonstrate in the laboratory. For this reason, new trace elements are added to the list from time to time. Traces of manganese and zinc are essential to all microorganisms, but other trace elements are not universally required (Table 3.2).

**Table 3.2** Nutrients required in minute amounts (trace elements) by microorganisms

| *Element* | *Function in metabolism* |
|---|---|
| Zinc (Zn) | Stability of a number of enzymes (for example, alcohol dehydrogenase, DNA and RNA polymerases) |
| Manganese (Mn) | Activator of some enzymes; component of superoxide dismutases (required for oxygenic photosynthesis) |
| Molybdenum (Mo) | Component of molybdoflavoproteins required for assimilatory nitrate reduction; required for nitrogen fixation |
| Selenium (S) | Required for formate-utilizing bacteria as component of formate dehydrogenase |
| Cobalt (Co) | A component of vitamin $B_{12}$, used for (a) methylation reactions, including methionine biosynthesis, and (b) reduction of ribonucleotides to deoxyribonucleotides (c) in transmolecular rearrangements |
| Copper (Cu) | Required for activity of respiratory enzymes such as cytochrome oxidase |
| Nickel (Ni) | Component of hydrogenase enzymes |
| Tungsten (W) | Required by formate-utilizing bacteria |

Most of the elements required occur in nature as salts ($MgSO_4$, $CaSO_4$, and so on) and are taken in solution into the cell as either positively or negatively charged components of the appropriate salt. For example, magnesium sulphate can act as a source of magnesium as the magnesium ion ($Mg^{2+}$), and of sulphur as the sulphate ion ($SO_4^{2-}$).

Carbon, hydrogen and oxygen are obtained from either organic or inorganic compounds in the environment. Microorganisms are particularly versatile in their use of organic molecules and they are sometimes said to show **catabolic infallibility**. This emphasizes that all naturally occurring organic compounds can be used by one kind of microbe or another. Although the same organic molecule can often supply all three of these elements – glucose is an example – they are sometimes provided by two different molecules. For example, many photosynthetic organisms obtain carbon and oxygen from carbon dioxide, and hydrogen from water.

Nitrogen amounts to about 10% of the dry weight of a microorganism and is required in large amounts. Nitrogen is found naturally in a number of inorganic molecules as the ammonium ($NH_4^+$), nitrate ($NO_3^-$) and nitrite ($NO_2^-$) ions, as molecular nitrogen ($N_2$) and in organic molecules such as proteins, nucleic acids and their breakdown products. The ammonium ion can be used by nearly all microorganisms, and nitrate ions by many. Nitrogen gas can be used by nitrogen-fixing bacteria. In all these examples, nitrogen is reduced to the ammonium ion ($NH_4^+$) before it is incorporated into cell material.

Sulphur is usually absorbed by microorganisms as sulphate ions ($SO_4^{2-}$) and reduced to sulphide ($S^{2-}$) before it is incorporated into organic molecules. A number of other sulphur-containing compounds can act as energy sources in certain bacteria (Chapter 4). These include elemental sulphur ($S^0$) and the thiosulphate ($S_2O_3^{2-}$) ion.

## Microbial growth factors

The large molecules (**macromolecules**) which make up the cell material of microorganisms are assembled from the kinds of building blocks used by all living things. Polypeptides are built from the same twenty amino acids and nucleic acids from the same nucleotides. It is in the *origin* of the building blocks that we see variation between individuals or groups of individuals. For example, algae are able to synthesize all their cell constituents from carbon dioxide and do not require an organic substrate as a carbon source at all. By contrast, lactic acid bacteria and protozoans need a complex mixture of a number of organic molecules to satisfy the same requirement.

Among the bacteria, some, such as *Escherichia coli*, can use a single organic molecule like glucose to provide a carbon skeleton from which all cell constituents may be produced. *E. coli* possesses all the genes required to specify the necessary biosynthetic enzymes for amino acid and nucleotide formation. However, many bacteria do not produce the same range of enzymes seen in *E. coli*. These bacteria need to find the molecules they cannot synthesize ready-made in the environment. Substances which must be supplied ready-made in the environment are called **growth factors** and fall into three groups: vitamins, amino acids, and purines and pyrimidines.

Required growth factors differ from one strain of bacterium to another. Lactic acid bacteria, for example, have a rather lower biosynthetic capacity than *E. coli* and require a large number of growth factors. The vitamins required as growth factors for a number of kinds of bacteria are shown in Table 3.3.

**Table 3.3** Microbial growth factors: vitamins

| *Vitamin* | *Growth factor for* | *Function* |
|---|---|---|
| *p*-Aminobenzoic acid | *Chlamydomonas moewusii*<br>*Clostridium acetobutylicum* | Constituent of folic acid and derivatives |
| Folic acid and derivatives | *Leuconostoc citrovorum*<br>*Strep. faecalis*<br>*Lactobacillus casei* | Transfer of 1-C units |
| Biotin and derivatives | Yeast<br>*Lactcbacillus plantarum* | Carboxylation reactions |
| Nicotinic acid and derivatives | *Lactobacillus plantarum*<br>*Leuconostoc mesenteroides* | Constituent of coenzymes of oxido-reduction reactions ($NAD^+$, $NADP^+$) |
| Pantothenic acid and derivatives | *Acetobacter suboxydans*<br>Yeast | Constituent of CoA (acyl group carrier) |
| Iron porphyrins and derivatives | *Mycobacterium tuberculosis* | Component of catalase, cytochr omes, etc. |
| Riboflavin and derivatives (Vitamin $B_2$) | *Lactobacillus helveticus* | Constituent of coenzymes of oxido-reduction reactions (FMN, FAD) |

**Table 3.3** *continued*

| *Vitamin* | *Growth factor for* | *Function* |
|---|---|---|
| Thiamine and derivatives (Vitamin $B_1$) 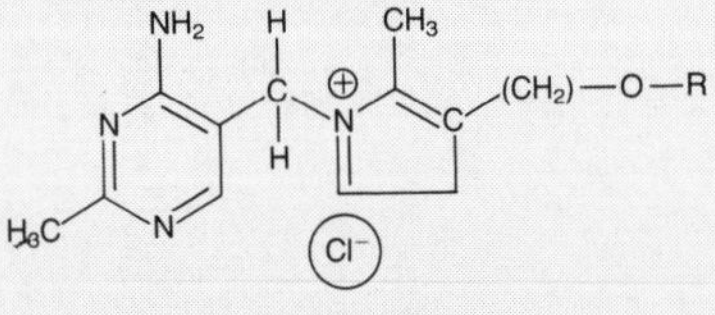<br>R = H or $PP_i$ | *Staph. aureus*<br>*Lactobacillus fermenti* | Carboxylation reactions |
| Pyridoxine and derivatives (Vitamin $B_6$) 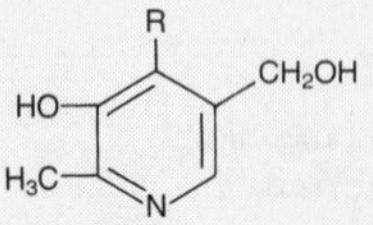<br>R = -CHO pyridoxal<br>-$CH_2OH$ pyridoxine<br>-$CH_2NH_2$ pyridoxamine | *Clostridium welchii* | Reactions of amino acids |
| Lipoic (thioctic) acid 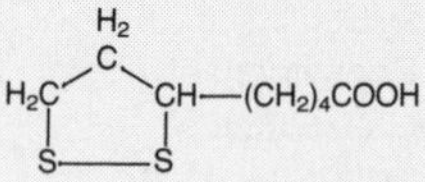 | *Strep. faecalis* | Transfer of acyl groups from α-ketoacid decarboxylation |
| Quinones (vitamin K group) 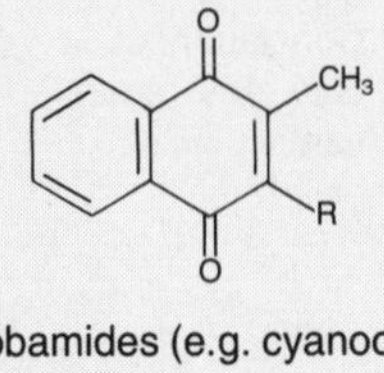 | *Fusiformis nigrescens* | Electron transport systems |
| Cobamides (e.g. cyanocobalamin: vitamin $B_{12}$) 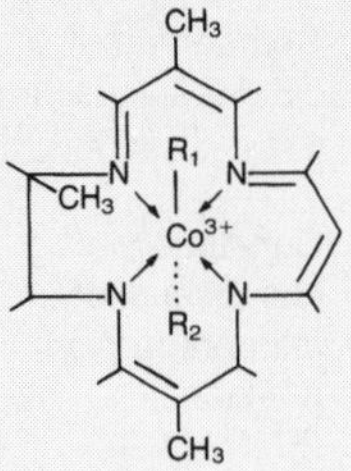<br>This shows the functional group only, not the whole molecule! | *Lactobacillus leishmanii*<br>*Euglena gracilis* | Rearrangements and synthesis of $-CH_3$ groups |

## Growing microorganisms in the laboratory

If a microorganism grows in a particular location in a natural environment, it is safe to conclude that nutrients and conditions are appropriate for that particular species. However, for rigorous investigations in microbial biology, individual types of microorganism need to be isolated from their natural environment and cultured in the laboratory.

Microorganisms are grown in the laboratory using suitable **culture media** to provide nutrients essential for growth. With the exception of viruses, two categories of culture media are available. **Complex media** contain one or more sources of nutrients, and their exact composition is not critical. Defined or **synthetic media** are those in which the components are known precisely and are present in relatively pure form.

Complex media have the advantages of being relatively inexpensive and easy to prepare. They are also appropriate for the culture of microorganisms whose exact nutritional requirements are unknown. Defined media are more appropriate for nutritional, genetic and physiological studies on unknown species where experimental conditions need to be standardized.

Defined media for different kinds of microorganisms vary tremendously in their complexity. Media for autotrophic organisms may consist of solutions of just a few mineral salts, whereas those for lactic acid bacteria and some animal pathogens are much more complicated, containing amino acids, vitamins, purines, pyrimidines, coenzymes and fatty acids. Nutrients required in only minute concentrations are normally not deliberately added to defined media because they are usually present as contaminants of the other media components. Examples of defined media are shown in Table 3.4.

Nutrient broth is an example of a complex medium. It contains 0.3% beef extract (a water extract of beef containing sugars, amino acids and minerals) and 0.5% peptone. Peptone is prepared by hydrolysing milk or beef protein and contains amino acids and peptides as sources of nitrogen, carbon and energy. If agar is added to nutrient broth to give a concentration of 1.5% to 2% of the final volume of medium, it solidifies into a jelly when cold. Thin layers of agar jelly are conveniently prepared in containers called Petri dishes, which are ideal for growing microorganisms under closely controlled laboratory conditions.

### Preparing pure cultures

In nature, populations of microorganisms are usually mixed; that is, they consist of a number of species. For laboratory investigations, however, it is usual to study a **pure culture** (a population of a single species). How do we isolate a pure culture from a mixture of species growing in a natural environment?

One way to do this is to spread a mixed population onto a solid culture medium so that individual cells become separated. Each cell multiplies to produce a population of cells visible to the naked eye as a **colony**. Unless two or more cells from different organisms were located very closely together, a colony will contain cells derived from one original cell. If a sample is now transferred to fresh medium and the procedure repeated two or three times, pure cultures are produced even when a mixed colony is produced from two or more cells (Figure 3.1).

**Table 3.4** Examples of chemically defined culture media

| | | | |
|---|---|---|---|
| 1. For *Escherichia coli* | | | |
| $K_2HPO_4$ | 7.0 g | | |
| $KH_2PO_4$ | 2.0 g | | |
| $(NH_4)_2SO_4$ | 1.0 g | | |
| $MgSO_4$ | 0.1 g | | |
| $CaCl_2$ | 0.02 g | | |
| Glucose | 10 g | | |
| Distilled water to 1 litre | | | |
| 2. For *Lactobacillus* | | | |
| Glucose | 10 g | Adenine, guanine and uracil | 10 mg of each |
| Sodium acetate | 6 g | | |
| $NH_4Cl$ | 2.5 g | Pyridoxamine-HCl | 0.4 mg |
| $(NH_4)_2SO_4$ | 2.5 g | Riboflavin, thiamine, niacin and pantothenic acid | 0.2 mg of each |
| $KH_2PO_4$ | 0.5 g | | |
| $K_2HPO_4$ | 0.5 g | *p*-Aminobenzoic acid | 40 µg |
| NaCl | 5 mg | Folic acid | 20 µg |
| $FeSO_4$ | 5 mg | Biotin | 0.2 µg |
| $MnSO_4$ | 5 mg | Distilled water to | 1 litre |
| Anginine, cystine, glycine, histidine, proline, hydroxyproline, tryptophan, tyrosine | 0.2 g of each | | |
| Alanine, aspartic acid, glutamic acid, isoleucine, leucine, lysine, methionine, norleucine, phenylalanine, serine, threonine and valine | 0.1 g of each | | |
| Glutamine | 25 mg | | |

This technique is useful when you want to grow the maximum number of species from a particular habitat or if you want to isolate the dominant species. To isolate species present in smaller numbers a different approach is required. **Enrichment techniques** employ conditions designed to favour the growth of the desired species (Table 3.5). Enrichment is usually undertaken in liquid culture media from which pure isolates of the enriched species may be obtained by subsequent streaking on to a solid medium. An elaboration of this approach is the addition of ingredients to the culture medium which inhibit unwanted types, but not the sought-after species. For example, McConkey's agar contains bile salts which, although inhibitory to many bacteria, permit the growth of enteric bacteria such as *Escherichia coli.*

## Determination of cell number

A **viable count** is an estimate of the number of living cells present in a microbial suspension. If necessary the suspension is diluted by a known factor, and samples are either inoculated into a liquified (heated) medium containing agar which is allowed to set, or spread on to the surface of a solidified culture medium. The effect of these procedures is to isolate

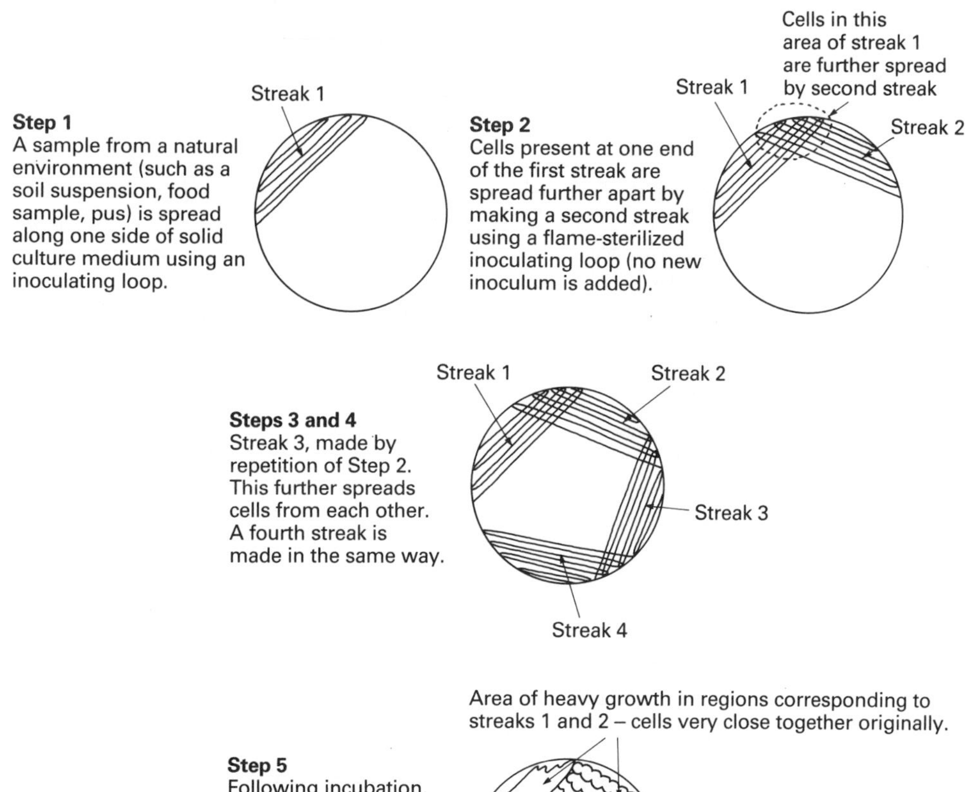


**Figure 3.1** Spread plate techniques for purification of a mixed population of bacteria.

individual cells so that when they divide they produce a visible and distinct colony. Following incubation for an appropriate period of time (this may vary from a few hours to a number of days, depending upon the organism), the number of colonies is counted and, after correction for dilution, can be taken to be equivalent to the number of cells present in the original suspension.

**Table 3.5** Examples of enrichment culture. All of the following media contain appropriate mineral salts and trace elements

| *Organism to be enriched* | *Anaerobic or aerobic conditions* | *N source* | *C source* | *Illumination* |
|---|---|---|---|---|
| Cyanobacteria | Either | $N_2$ | $CO_2$ | Light |
| *Azotobacter* (a nitrogen-fixing bacterium) | Aerobic | $N_2$ | Glucose | Dark |
| Denitrifying bacteria (for example, *Paracoccus denitrificans*) | Anaerobic | $NaNO_3$ | Alcohol, fatty acids, etc. | Dark |
| Fermentative bacteria | Anaerobic | $NaNO_3$ | Glucose | Dark |
| *Nitrosomonas* (a nitrifying bacterium) | Aerobic | $NH_4Cl$ | $CO_2$ | Dark |

This technique needs considerable skill in carrying out dilutions and in the inoculation of the culture medium. In addition, many kinds of bacteria typically form aggregates. *Streptococcus* bacteria form chains of cells. Such groups of cells are not easily disrupted to produce individuals and a visible colony may be the consequence of repeated divisions of more than one cell. For this reason, counts are often expressed as **colony-forming units** (**cfu**)

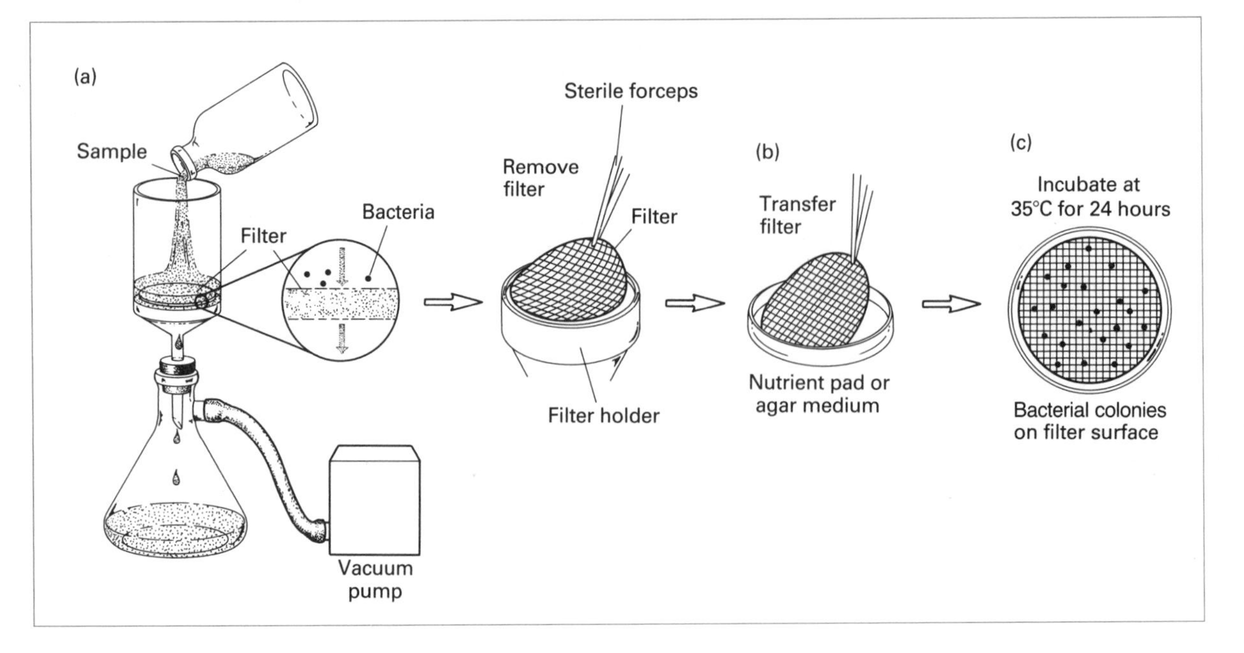


**Figure 3.2** The filtration method for bacterial cell counts.

rather than numbers. These plate-count techniques are most suited to samples which contain large numbers of individuals. However, practical situations arise where microbial numbers are very small; an example is the bacteriological examination of samples of drinking water. In this situation, the **membrane filter technique** is used (Figure 3.2). A known volume of a suspension containing the organism is filtered through a thin membrane, which traps the organism on its surface. The membrane is then transferred to a Petri dish containing solid culture medium. During incubation, nutrients diffuse from the medium to the microbes, which divide and produce individual colonies on the membrane surface. The number of cfu can be determined by counting the number of colonies formed.

A number of techniques are available for estimation of **total counts** which, in contrast to viable counts, do not distinguish living cells from dead cells. In the **direct microscopic count** method, a calibrated microscope slide with a small central indentation, or **well**, is filled with a known volume of a microbial suspension (Figure 3.3). The number of cells is then counted under a light-microscope. If the dilution of the sample and the volume of the suspension in the well are known, a simple calculation allows determination of the total count.

A more rapid technique for determining a total count requires the use of an instrument called a **Coulter counter** (Figure 3.4). This piece of equipment measures the conductivity of an electrolyte when it passes through a small channel. Any non-conducting particles or cells lower the conductivity

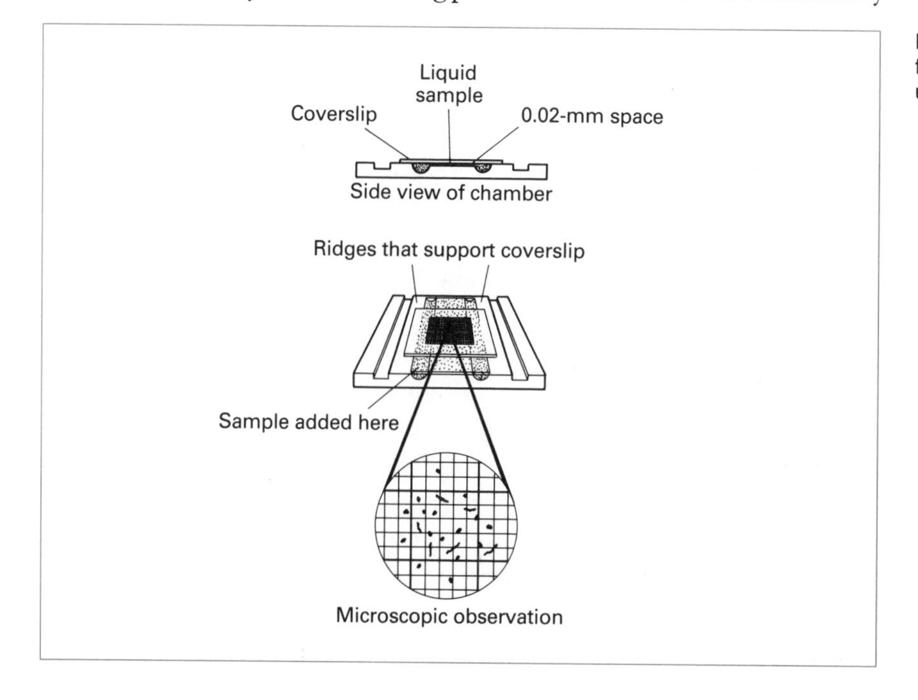


**Figure 3.3** A counting chamber for determining counts of unicellular organisms.

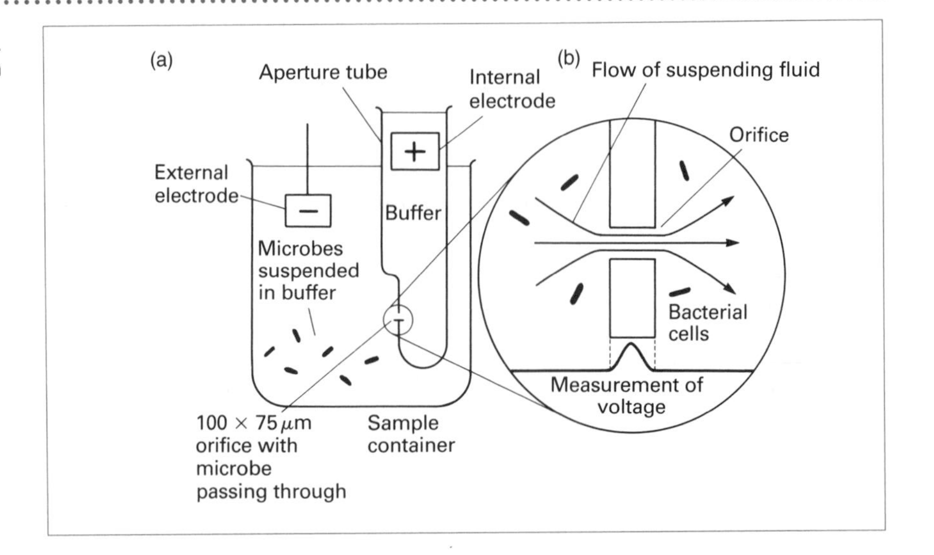


e Coulter counter. ıatic representation e container. (b) E... ment of the orifice.

as they pass through the aperture. From these changes in conductivity, the counter can determine both the number and the size of the organisms in suspension. Although it is less tedious than the direct microscopic count, the main drawback of this technique is the need to keep the suspending medium free of dust and other inanimate particles which might be counted mistakenly as cells.

In addition to methods for direct determination of cell numbers, a variety of techniques are available for estimating cell mass. Dry weight, turbidity and nitrogen content can all be determined experimentally and converted to cell number, if required, using a calibration curve.

Turbidity measurements are widely used in microbiological work because of the ease and speed with which they can be carried out. Particles, including microbial cells, can absorb and scatter light in proportion to their density. For this reason, a suspension of cells appears turbid or cloudy. The degree of turbidity can be measured using a colorimeter or spectrophotometer. When light is passed through a tube containing a microbial suspension, the light reaching the photoelectric recording device is reduced in proportion to the light absorbed or scattered by the interposed cells. Over a range of values, the percentage of light transmitted is inversely proportional to the cell mass. For the sake of convenience, turbidity is usually expressed as **absorbance** ($A$) because it is directly proportional to cell mass. Red light in the region of 660 nm is generally used because the yellowish colour of many culture media does not interfere greatly at these wavelengths.

Although rapid and easy to carry out, turbidimetric methods also have their drawbacks. Cells which form aggregates or clumps cannot be estimated in this way. Furthermore, accuracy is limited to a range of cell counts.

At fewer than $10^6$ cells/ml, turbidities are too low for accurate measurement. At the other extreme, turbidities at cell densities of $10^9$–$10^{10}$/ml are not proportional to cell number and samples must be diluted before accurate turbidimetric estimates can be attempted.

The methods we have so far examined are relatively slow and take one to several days to carry out. For this reason, rapid techniques for determining cell numbers are now available. One such technique relies on measurements of **electrical impedance**. This is based on the observation that microbial growth influences the rate at which the suspending medium conducts an electric current. The time taken for a measurable change in impedance to occur is inversely proportional to the original microbial count.

A variety of alternative techniques for rapid determinations of cell number are available. These are commonly automated for ease of use, particularly in the areas of food, environmental and clinical microbiology.

## Batch culture

Unicellular organisms grow by increase in mass, usually followed by cell division to produce more cells. Many one-celled organisms, including most bacteria, some yeasts and some algae, increase in number by growing to approximately double their original size and then dividing into two equal cells (called **daughter cells**). In a population of unicellular organisms, all the individuals grow and divide in this way as long as the environmental conditions are suitable.

This type of cell growth and multiplication is called **binary fission**, and one cell becomes successively 2, 4, 8, 16 and so on, with no theoretical limit. Under defined culture conditions, time intervals which separate successive divisions are constant and characteristic of a particular species of microorganism. If the logarithm of cell number is plotted against time, a straight line is produced and such a population is said to show **logarithmic** or **exponential growth**. There are two reasons for expressing growth of cells as logarithms. First, a straight-line graph is more convenient to describe in mathematical terms, and second, the size of graph paper needed could become unmanageable or the scale unacceptably small if actual numbers of cells were plotted.

Furthermore, a convenient method for the graphical expression of the growth of bacterial populations is to plot logarithms to the base 2 to express cell numbers. If this is done, each unit increase in $\log_2$ indicates a cell division and doubling of the population.

A mathematical shorthand for expressing the numbers 10, 100, 1000, 10 000, 100 000 is $10^1$, $10^2$, $10^3$, $10^4$ and $10^5$. Alternatively the logarithms ('logs') of the same numbers are, 1, 2, 3, 4 and 5 and can be written: $\log_{10} 10 = 1$, $\log_{10} 100 = 2$, $\log_{10} 1000 = 3$, and so on. Numbers which fall between 0 and 10, 10 and 100 can be obtained from printed tables of logarithms or from an appropriately programmed pocket calculator. In this example we have shown how a number can be expressed as a logarithm to the base of 10 ($\log_{10}$).

[Logarithms to the base 2 ($\log_2$) or any other base can be used in the same way. For example, the numbers 2, 4, 8, 16, 32 can be written as $2^1$, $2^2$, $2^3$, $2^4$, $2^5$ or as $\log_2 2 = 1$, $\log_2 8 = 3$, $\log_2 16 = 4$, and so on.]

| No. of divisions | 0 | 1 | 2 | 3 | 4 |
|---|---|---|---|---|---|
| No. of cells | 1 | 2 | 4 | 8 | 16 |
| $Log_2$ N cells | 0 | 1 | 2 | 3 | 4 |

$$\log_2 n = \log_{10} n \times 3.332$$

The rate of increase in number of cells can also be expressed as **growth rate**, which is simply the reciprocal of the population doubling time. For example a bacterium with a population doubling time of 30 minutes has a growth rate of 2 doublings/hour. The doubling times of a number of bacteria are shown in Table 3.6.

The **specific growth rate** ($\mu$) is also used to describe the growth of a population of cells. In contrast to the population doubling time and growth rate, $\mu$ represents the rate of increase of cell *mass*, not cell number. Specific growth rate is calculated as follows:

$$\mu = \frac{0.693}{t_d}$$

In what circumstances would you expect there to be a short or non-existent lag phase?

**Table 3.6** Minimum doubling times for a number of bacteria in complex media

| *Bacterium* | *Temperature (°C)* | *Doubling time (min)* |
|---|---|---|
| *Vibrio natriegens* | 37 | 9.6 |
| *Bacillus stearothermophilus* | 60 | 8.4 |
| *Escherichia coli* | 40 | 22.8 |
| *Rhodobacter spheroides* | 30 | 132.0 |
| *Mycobacterium tuberculosis* | 37 | 360.0 |

The rate at which a population of cells increases is often characterized by calculation of the **population doubling time** ($t_d$) – also called the **mean generation time** (***g***). This is the time taken for the number of cells in a population to double.

In this way, the increase in number of a population of cells in a batch culture can be characterized in mathematical terms: duration of lag phase, population doubling time and stationary population. This is important because such information has *predictive* value: an industrial microbiologist can predict, for example, the time it will take for a population to reach its stationary phase or the size (number of cells) of an inoculum which will be needed to produce a stationary population within a specified period of time.

When bacteria are cultured in a closed vessel, growth passes through a sequence of phases. A graph showing the growth of a culture of bacteria in a closed vessel is shown in Figure 3.5.

The **lag phase** is a period following inoculation of bacteria into the medium during which there is little or no increase in cell numbers. It is a period of adjustment required when cells are transferred from one set of conditions to another. A lag period can occur, for example, if the bacteria come from an old culture where the cells were in a sickly condition and starved of nutrients. The cells require time to build up nutrients before growth can re-start. There may also be a significantly long lag period if cells are transferred from a rich medium to a poor one. In this case, a delay may occur to allow for synthesis of biosynthetic enzymes which were not required in the first medium.

The **exponential** or **log phase** follows the period of adjustment. During this period the population increases with a characteristic and constant doubling time.

The **retardation phase** is signalled by an increase in the population doubling time. This may be a consequence of the depletion of an essential nutrient or an accumulation of a toxic waste product.

The **stationary** and **decline phases** follow the retardation phase. During

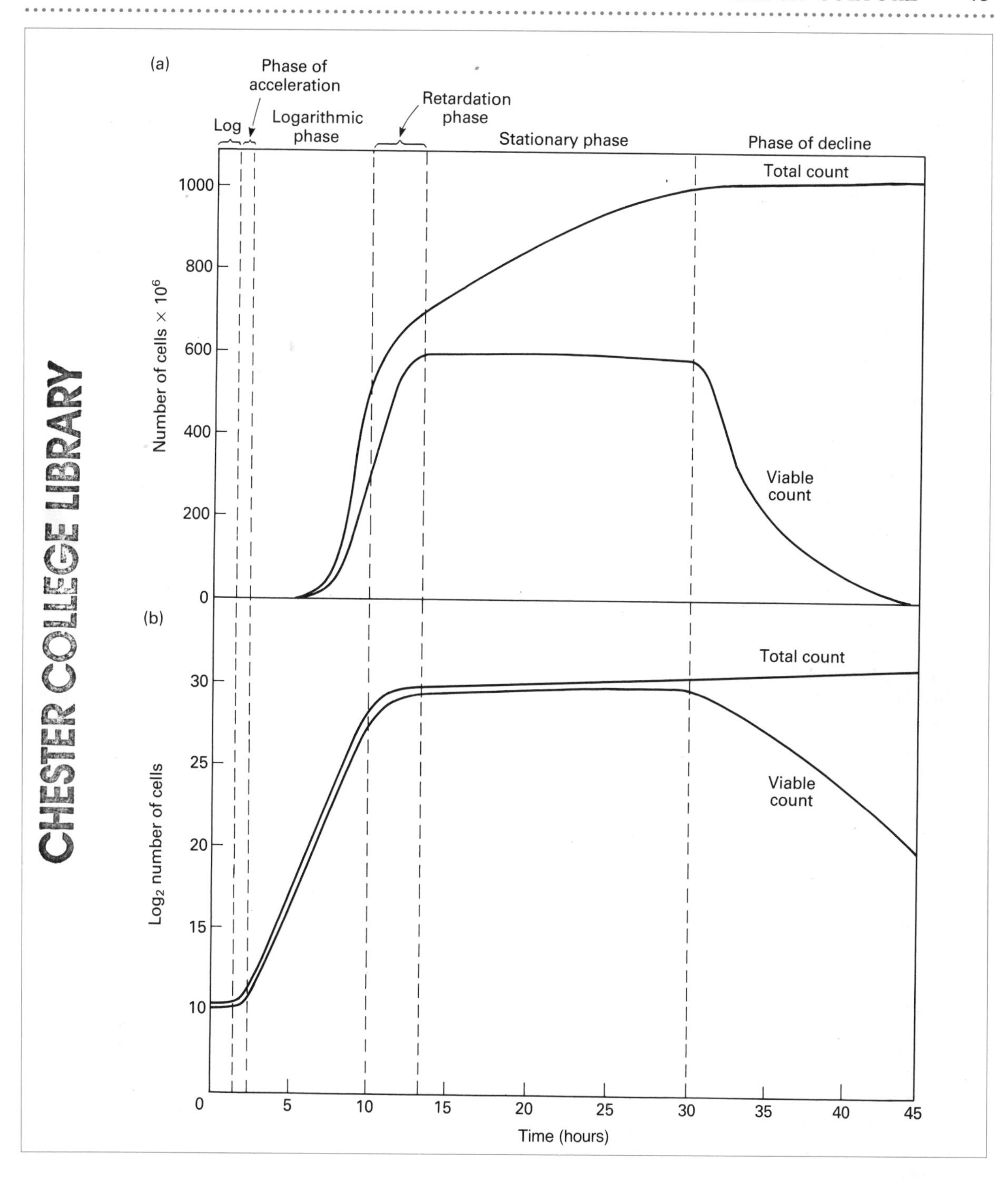




**Figure 3.5** Typical growth curve of *E. coli* in culture at 30°C.

**Figure 3.5** *continued*

Calculations (using Figure 3.5b)

(a) Population doubling time ($t_\alpha$)

An increase of one unit in the value of $\log_2 N$ is equivalent to a doubling of the number of cells present. The time required for this to occur is the population doubling time. From the graph it can be seen that the time taken for $\log_2 N$ to increase from 15 to 25 is 4 hours. The population doubling time is approximately $\frac{8-4}{25-15}$ hours or $\frac{4}{10} \times 60$ minutes

$$t_\alpha = 24 \text{ minutes}$$

(b) Growth rate

$= \frac{60}{24}$ doublings/hour $= 2.5$ doublings/hour

(c) Specific growth rate

$$\mu = \frac{0.693}{t_\alpha} = 0.693 \times \frac{60}{24} = 1.733\text{/hours}$$

Some yeasts multiply by budding-off new cells. The mother cell is composed of older cell material than the bud. But bacteria divide by binary fission. Each bacterial daughter cell is identical and of the same age. Such newly formed cells are spoken of as 'young' and those which are just about to divide as 'old'. So bacterial cells are 'immortal' when environmental conditions are conducive to growth and multiplication.

Some microorganisms grow as thread-like filaments. The filaments of some green algae and blue-green bacteria are chains of individual organisms. Because any one of the cells in the filament is capable of binary fission, the increase in length of the chain is the sum of the growth of individual component cells.

In contrast, fungal hyphae, actinomycete filaments and those of many algae increase in length only at a region just behind the tip of the filament. Further details of eukaryotic growth are described in Chapter 9.

the stationary phase, the number of new cells being produced balances the number dying. During the decline phase, the number of cells dying exceeds the number of new cells generated.

## Continuous culture

Continuous culture approximates more closely to conditions in natural environments than does batch culture. The mouth, urinogenital tract, and a flowing stream are all such examples.

In a continuous culture, a steady supply of fresh medium is added, coupled with simultaneous removal of spent medium. Continuous culture requires a culture vessel fed from a medium reservoir and a second container where spent medium and cells can be collected. Inevitably any spent medium which is removed will also contain a mixture of living and dead cells. The rate at which cells multiply can be controlled according to the rate at which fresh medium is added and old medium removed. Further information on industrial applications of continuous culture are given in Chapter 11.

### The microbiological assay

For a strain of microorganism that requires a growth factor, the final number of individuals produced during batch culture (called the **stationary population**) is proportional to the concentration of growth factor in the culture medium. Within limits, the relationship between concentration of growth factor and stationary population is a straight line. Measurement of the stationary population can be used, therefore, to establish the concentration of a growth factor; this is the basis of a microbiological assay.

For example, suppose we need to determine the level of the vitamin riboflavin in a clinical specimen. We would carry out the assay by adding different volumes of the sample to a range of test tubes containing riboflavin-deficient culture medium. We would then inoculate each tube with the test organism and incubate them. We would establish the degree of growth, most conveniently measured as turbidity, and the riboflavin concentration by referring to a standard curve.

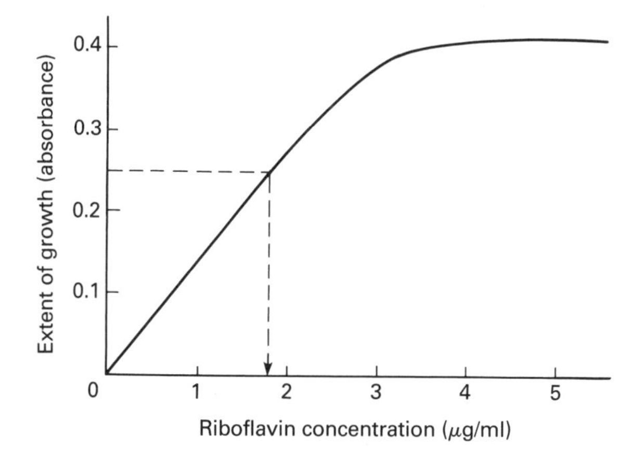


Microbiological assays are widely used in the food and pharmaceutical industries. They are highly sensitive and can detect quantities as low as 1 ng ($10^{-9}$ g).

An unknown sample giving an absorbance reading of 0.25 would contain about 1.8 μg/ml riboflavin.

**BOX 3.1** A standard curve for the microbiological assay of riboflavin using *Lactobacillus casei*.

In continuous cultures, new medium is not added until cell multiplication has started. The rate at which medium flows through the culture vessel is called the **dilution rate**. If the rate of addition is slow, the rate of cell division increases, and exceeds the dilution rate, and the population increases. If the rate of addition of new medium is too fast, the dilution rate will be greater than the **maximum growth rate** possible in the culture. Cell numbers will then steadily decline, a condition known as **washout**.

At dilution rates less than those that induce washout, the growth rate increases until it is equivalent to the dilution rate. In these circumstances, a **steady state** is reached where growth is continuously logarithmic and proportional to the dilution rate. Continuous culture is used for growth studies under steady-state conditions and for commercial exploitation. Laboratory continuous culture is often used in studies of population dynamics.

**BOX 3.2** Continuous culture.
(a) the chemostat;
(b) the turbidostat.

**Types of continuous culture**

There are two standard ways of controlling the flow of medium in a culture vessel during continuous culture operations: by **chemostat** and by **turbidostat**. In a chemostat, the growth rate is controlled by limiting the supply of an essential nutrient. For example, if glucose is the sole carbon and energy source, and it is continuously added to the culture, we can maintain it at a constant limiting concentration in the culture vessel. Provided all other nutrients are present in excess, the growth rate will be determined by the glucose concentration and the rate at which it is added to the fermenter.

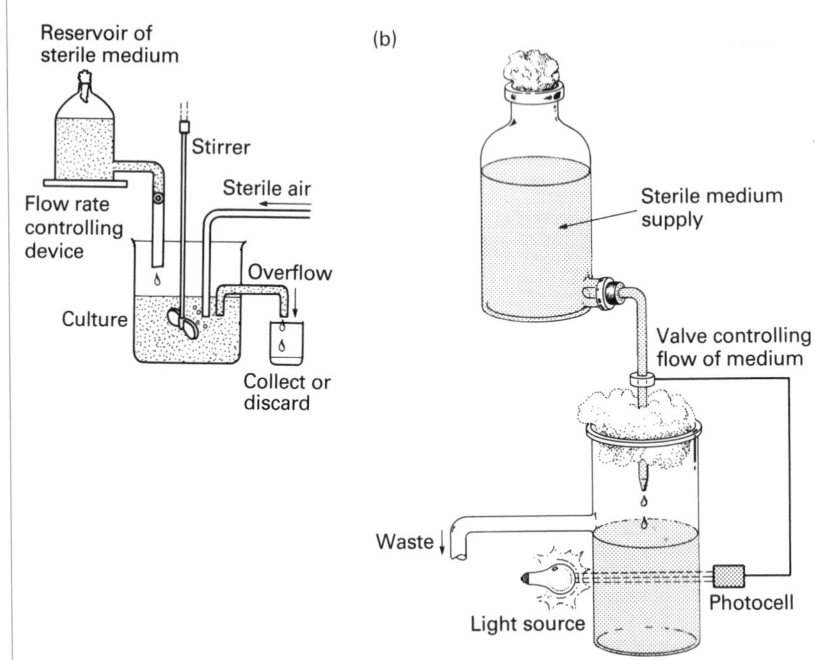


With a turbidostat, a photoelectric recording device determines the number of cells present in terms of their turbidity. If the turbidity increases from a preset value, additional medium is added to the fermenter and culture is removed automatically by the operation of a suitable valve system. In this way a constant cell density is maintained.

The major difference between the two systems is that in a turbidostat, constant growth is maintained by the automatic addition of a nutrient when turbidity falls below a preset value, whereas a chemostat depends upon biological control through the limitation of an essential nutrient.

Mixtures of organisms can be grown in the same vessel and the ways in which they interact followed.

Predatory microorganisms, such as certain protozoans which eat bacteria, can be observed to glean information about the predator–prey relationship. Evolutionary experiments can also be carried out using continuous culture, in which resistance to antibiotics in bacteria containing plasmids can be compared with the resistance of those which do not.

BOX 3.3

**Production of single-cell protein by continuous culture**

Imperial Chemical Industries (ICI) was the first company to develop a continuous process for the commercial production of single-cell protein (SCP). In this process, methanol was used to support the growth of the bacterium *Pseudomonas methylotrophus* in a 3-m-high pilot fermenter with a volume of 37 $m^3$. The pilot plant had a capacity of 1000 tons/year, operated at pH 6.5–6.9 and 34–37°C. On the basis of this pilot study, ICI invested £40 million in 1979 to install a full-scale (1000 $m^3$ capacity) continuous culture system with a capacity of 50 000–70 000 tons/year.

Because of difficulties in separating cells from culture medium, this large scale process for SCP has not yet been placed in full production because of economic reasons. For example, in 1984, the price of soy meal was $125–190 per ton compared with $600 per ton for ICI SCP (Trade name Pruteen).

Continuous culture is also potentially a powerful technique in industrial microbiology because it allows maximum yield of product in minimum time. Techniques have been developed for the production of single-cell protein, antibiotics and organic solvents. (Specific examples are given in Chapter 11.)

## Synchronous growth

Although a great deal of information can be obtained from studies of populations of microbial cells, situations do arise where it is necessary to investigate the biological processes taking place in an individual cell. However, a single bacterial cell is very small and provides very little material for experimental investigation! Even so, there are techniques for producing cultures of microorganisms in which every cell is at about the same point of growth. Such a culture is called a **synchronous culture**. The advantage of the technique is that it makes available large amounts of material from cells which are all at more or less the same stage in their growth cycle. We can then make the assumption that phenomena which occur in whole populations of cells reflect those which occur in each individual cell.

One synchronous culture technique is to manipulate either the composition of the culture medium or the environmental conditions. For example, if we were to keep bacteria that require thymine for growth in a medium which lacks thymine, cells would survive but, because thymine is a constituent of DNA, they would be unable to synthesize new genetic material. Chromosome replication would not take place and the bacteria would not divide. If we then added thymine to the culture, DNA synthesis would occur and all the cells of the population would divide virtually simultaneously (Figure 3.6a).

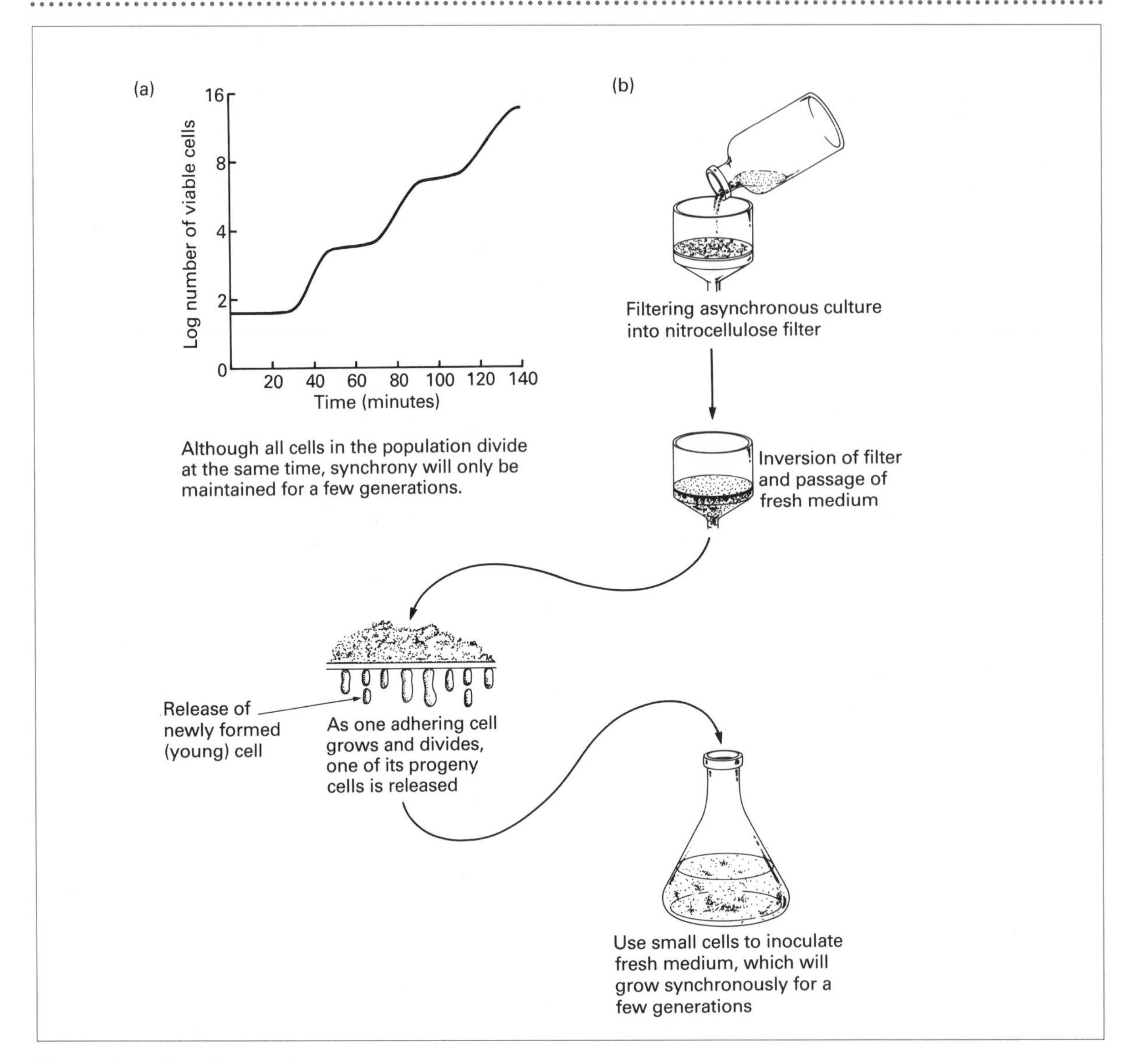


**Figure 3.6** (a) Growth curve of a synchronized culture. (b) The use of a nitrocellulose filter to produce a synchronous culture.

Another technique is to separate cells of the same size from a randomly dividing population. The idea behind this is that cells of the same size are likely to be of the same age and at the same stage in their life-cycle. The most common procedure is to remove the smallest, and therefore the youngest, cells from a culture by filtration. If a culture of unsynchronized cells is filtered through a cellulose nitrate filter, some of the cells adhere to it and continue to grow (Figure 3.6b). If the filter membrane is turned upside down and washed with fresh medium, newly formed cells are washed into the medium. Because these cells have just been formed and are all at the same stage in their life-cycle, they will grow synchronously for a number of generations. One advantage of this technique is that no manipulation of the growth conditions is required and so the cells are physiologically normal.

Once achieved, synchronous growth endures for only a few generations. In spite of this limitation, synchronous cultures are a valuable tool for investigating the bacterial cell cycle and pattern of DNA replication.

## The effect of environmental conditions on microbial growth

So far in our description of microbial growth we have largely ignored the influences of environmental conditions. Now that we have examined the general features of laboratory culture, we can move on to look at the ways in which physical and chemical environmental factors influence the growth of microorganisms. We can also start to make sense of the distribution of microorganisms in the natural world and devise sensible techniques to control the activities of harmful microorganisms.

### Temperature

From your own experience, you will know that temperature can affect living organisms in two ways. If food is left in a warm room, it will go bad more quickly than if it is refrigerated. Enzyme reactions go faster at higher temperatures and the bacterial growth rate increases as a result. Above a certain limit, though, we know that high temperatures kill microorganisms. The reason for this is that the structure of essential cellular components such as proteins (particularly enzymes) and nucleic acids is disrupted at high

**Table 3.7** Temperature ranges supporting growth of a variety of microorganisms

| *Example* | *Temperature range (°C)* | | |
|---|---|---|---|
| | *Minimum* | *Optimum* | *Maximum* |
| Thermophiles: | | | |
| *Bacillus stearothermophilus* | 30 | 55 | 75 |
| Mesophiles: | | | |
| *Escherichia coli* | 10 | 37 | 45 |
| *Streptococcus faecalis* | 0 | 37 | 44 |
| *Saccharomyces cerevisiae* | 1 | 29 | 40 |
| *Fusarium oxysporum* | 4 | 27 | 40 |
| Psychrophiles: | | | |
| *Pseudomonas fluorescens* | –8 | 20 | 37 |
| *Chlamydomonas nivalis* | –36 | 0 | 4 |

The acidity or alkalinity of a solution is expressed as a pH value. Acidity is indicated by values less than pH 7 and alkalinity at values greater than pH 7. Neutrality is indicated by a pH value of precisely 7. pH is the negative logarithm of the hydrogen ion ($H^+$) concentration of a solution, so a pH difference of one unit represents a ten-fold variation in the hydrogen ion concentration of a solution.

pH is an important factor in the preparation of a culture medium. The pH of a laboratory medium is initially adjusted to the optimum pH of the organism to be grown. Furthermore, the medium is usually **buffered** so that the pH of the medium remains constant even if an acidic or alkaline waste product is formed during growth of the organism. A buffer is able to delay changes in pH by either taking $H^+$ ions out of solution if the pH falls below neutrality, or by donating $H^+$ ions to the medium if the pH rises. In complex media, amino acids, peptides and proteins are natural buffers. In synthetic media, mixtures of two phosphate bases together neutralize acids and bases at pH near 7. For example, $K_2HPO_4$ binds excess $H^+$ and $KH_2PO_4$ donates $H^+$.

temperatures. As a result, the organism is unable to carry out the activities which are essential to life.

In nature, microorganisms can be found over a very wide range of temperatures. Bacterial species that inhabit Polar seas can survive temperatures below freezing point. Others, at temperatures above 118°C, isolated from ocean depths in the vicinity of volcanic vents, can endure temperatures above boiling point! The classification of microorganisms according to their optimal growth temperature is given in Table 3.7.

### Acidity and alkalinity (pH)

Nearly all natural environments have pH values between 5 and 9 and most microorganisms grow optimally within this range. Only a very small number of organisms can grow at pH values of less than 2 or greater than 10. Organisms which grow under acid conditions are called **acidophiles** and those requiring alkaline conditions, **alkalophiles**. Although a few acidophilic bacteria are known – those used in vinegar production are an example – most grow optimally at, or near to, neutrality. Fungi grow under more acid conditions, most around pH 5, but some as low as pH 2.

Although the external pH can vary over a wide range of values, the internal pH values of all microorganisms are close to neutrality. In an acid environment, for example, the cell membrane blocks the entry of $H^+$ ions into the cell and excess $H^+$ ions are expelled.

### Water

Water is essential for all living organisms for two reasons. First, all biological reactions take place in an aqueous environment, and second, water is a reactant in a number of enzyme reactions. An important consideration is the amount of water available to a microorganism. Even under very moist conditions, the availability of water may be low because of the semi-permeable nature of the cytoplasmic membrane which surrounds all cells.

Water is able to flow into or out of a cell through this type of membrane. If water flows into a cell, the cell membrane will exert a pressure upon it called the **osmotic pressure**. The magnitude of this is equal to the external pressure that would be required to prevent any inflow of water. Whether water flows into or out of a cell depends upon whether the solute concentration is greater inside the cell or outside. Water flows towards the solution of higher solute concentration because the solution of lower concentration contains *more solvent* (water) molecules and there will be a net movement of these across the membrane into the more concentrated solution which contains fewer water molecules.

This process of **osmosis** can be seen if a bacterial **protoplast** (the cell without its cell wall) is placed in a very dilute medium (of low osmotic pressure). Water will flow into the protoplast causing it to swell and eventually to burst (**lyse**). Conversely, in a concentrated solution (of high osmotic pressure), water will leave the protoplast causing it to dehydrate and contract in size. The usual way of expressing available water is as **water activity** ($a_W$).

Most microorganisms require a minimum water activity of 0.9 for growth, but some fungi can grow at lower values. This is why foods with low $a_W$ values, such as bread, can become spoiled by the growth of moulds, but not by bacteria. However, a bacterium such as *Pseudomonas cepacia* will grow in distilled water containing only traces of contaminants, which it uses as nutrients. At the opposite extreme some bacteria survive only at high salt concentrations found in the Salt Lakes of America or the Dead Sea of Israel. Extreme **halophiles** (salt lovers) such as *Halobacterium* require a sodium chloride concentration of not less than 15%.

A semi-permeable membrane is one through which molecules of a solvent, such as water, can pass, but which cannot be crossed by solute molecules. If two solutions are separated by this type of membrane, solvent molecules will still cross the membrane and dilute the concentration of solute molecules.

Water activity is calculated by dividing the vapour pressure of a substance by the vapour pressure of pure water. Pure distilled water has a water activity of 1.0, and molar solutions of sucrose and sodium chloride both have water activities of 0.98.

### Oxygen

In the microbial world, there is considerable variation in the effect of oxygen on growth. This is determined by the type of energy metabolism the organism pursues. As we shall describe in Chapter 4, many kinds of microorganisms grow only if oxygen is present; these are called **obligate aerobes**. Other species, called **obligate anaerobes**, can grow only in the *absence* of oxygen. Falling between these two extremes is a third group, the so-called **facultative anaerobes**, which can grow whether oxygen is present or not. Some kinds of microorganisms require only low concentrations of oxygen and are called **microaerophilic** organisms, while **aerotolerant** microorganisms are anaerobes which are not killed if oxygen is present. Some lactic acid bacteria fall into this last group.

Algae, most protozoa and fungi and many bacteria are obligate aerobes (Chapter 9). Facultative anaerobes include brewers' yeast (*Saccharomyces cerevisiae*), some protozoans, and many bacteria, including *Escherichia coli*. Some bacteria (for example, *Clostridium tetani* which causes tetanus and *Clostridium botulinum* which causes a sometimes fatal form of food poisoning called botulism) are obligate anaerobes.

During respiratory metabolism, a number of toxic substances are produced. These include hydrogen peroxide ($H_2O_2$), and superoxide ($O_2^-$) and hydroxyl free radicals ($OH^·$). Aerobic organisms possess enzymes to decompose these products. Obligate anaerobes are unable to detoxify these products because they do not have the necessary enzymes and therefore are *poisoned* by oxygen.

## Control of microbial growth

So far we have concentrated on the positive aspects of microbial growth. However, the presence of microorganisms is not always considered desirable, and prevention of their growth is necessary in many practical

Pasteurization of milk is usually carried out speedily and inexpensively by **flash pasteurization** using a heat exchanger: this means that the hot, treated milk is used to heat cold, untreated milk. In this process, the milk's temperature is raised quickly to 71°C, held there for 15 seconds, and then rapidly cooled.

situations. This can be achieved either by using procedures which inhibit growth or by actually killing any organisms present. Killing or removing all living organisms is called **sterilization** and can be accomplished using heat, radiation, filtration and most importantly, chemicals.

**(a) Heat sterilization.** This is an efficient and very widely used procedure. The death of a microbial population occurs more quickly as the temperature is raised. The practical implication of this is that sterilization takes longer at lower temperatures than at higher temperatures.

Bacterial **endospores** (Chapter 2) are much more resistant than vegetative cells, and sterilization methods are designed to eliminate the most resistant of bacterial endospores. Temperatures above 100°C, accomplished by heating water to produce steam under pressure, are required. A popular sterilization regime uses steam at 1.1 $kg/cm^3$ (15 lb/sq in), which gives a temperature of 121°C. Pressurized steam is usually produced in a piece of equipment called an **autoclave**, which works in the same way as a pressure-cooker. At 121°C, autoclaving is normally effective after 10 to 15 minutes.

**Pasteurization** is a technique which uses less extreme heat treatments. Pasteurization was originally developed by Louis Pasteur (a French chemist who lived from 1822 to 1895) to prevent spoilage of wine. Today, we associate pasteurization with milk treatment. It was introduced originally to control the pathogenic bacteria which cause tuberculosis, brucellosis and typhoid. Its main use now is to improve milk's keeping properties.

**(b) Radiation.** This is of great practical value in controlling microorganisms and is of two kinds: **ionizing** and **non-ionizing radiation**. The latter includes ultraviolet light, which is lethal to cells. Ionizing radiation includes X-rays.

Ultraviolet (UV) light with a wavelength of 260 nm is most efficient as a lethal agent, primarily because it causes abnormalities in DNA (Chapter 7). The disadvantage of UV light is that it penetrates materials only to a very slight extent. For this reason it is useful for sterilizing the surfaces of laboratory benches and operating tables in hospitals.

Ionizing radiations appear to have a direct effect on microorganisms. For example, radioactive cobalt emits gamma rays which are powerful enough to break some of the chemical bonds in the sugar–phosphate backbone of the DNA molecule. If enough breaks occur, the DNA cannot be replicated, nor can it function correctly. As a result, the cell will die. Ionizing radiations come into their own for the sterilization of heat-sensitive materials such as foods. They are able to penetrate materials deeply and can be used on foods which have already been packaged. The radioisotope cobalt-60 is most conveniently used for sterilization on a commercial scale.

**(c) Filter sterilization.** This is commonly used for the sterilization of heat-sensitive liquids. The most common type of filter used in microbiology is the **membrane filter**. This is a disc of either cellulose acetate or cellulose nitrate, pierced by very large numbers of microscopic holes. The holes are of such a size as to prevent the passage of microorganisms, but at the same time allow passage of a suspending liquid. This type of filter has a sieve-like action and traps many particles on its surface.

## Chemical control

So far we have looked at a number of physical agents which inhibit or kill microorganisms. In this section, we meet a number of antimicrobial chemicals. Chemicals which are lethal to groups of microorganisms often end in '-cidal' or '-cide' (for example, microbiocidal agent or microbiocide, bactericidal agent or bactericide). Those which inhibit growth but that are not lethal are described as **static agents** (for example, microbiostatic agent, bacteriostatic agent, and so on).

**Selective toxicity** is an important concept in the area of microbial control. Some agents kill many types of cell in a non-selective manner, showing little selective toxicity. Other chemicals are much more poisonous to microbial cells than to mammalian cells, showing a high degree of selective toxicity. This second group of chemicals includes the so-called **chemotherapeutic agents** which are used in the treatment of infectious disease. Here the idea is to kill the microbial cells, but not those of the host.

**(a) Disinfectants and antiseptics.** An important distinction in the use of antimicrobial agents is implied in these two terms. An antiseptic is a substance which can be used on living tissue without damaging it. A disinfectant is an antimicrobial agent which can be used only on non-living material. Sometimes such agents can be used as disinfectants at high concentrations and antiseptics when diluted. A summary of popular antiseptics and disinfectants is given in Table 3.8.

**(b) Chemotherapeutic agents.** In order to combat infectious disease, it is essential that antimicrobial agents are available which can be used internally. Furthermore, such agents must be selectively toxic, threatening the survival of the microbial pathogen but leaving the host relatively unscathed.

**Sulphonamides** are chemotherapeutic agents which exhibit selective toxicity. They are examples of **metabolic analogues**; that is, they bear a close structural resemblance to a naturally occurring metabolite. Sulphonamides are structurally similar to *p*-aminobenzoic acid (PAB), which is itself part of the cofactor called **folic acid**. The sulphonamides bind to the site

Studies on the mode of action of sulphonamide drugs originally gave rise to the concept of **competitive inhibition**, the inhibition of an enzyme by a structurally similar inhibitor competing with the natural substrate for the active site.

**Table 3.8** Disinfectants and antiseptics

| *Agent* | *Uses* |
|---|---|
| Disinfectants | |
| Mercury salts | Disinfection of tables, benches and floors |
| Copper sulphate | Algicide in reservoirs, horticultural antifungal agent |
| Iodine solution | Medical instruments |
| Chlorine | Swimming pools |
| Phenols | In carbolic soap, laboratory disinfectant |
| Cationic detergents | Medical instruments, food and dairy equipment |
| Ethylene oxide gas | Buildings, temperature-sensitive laboratory equipment, including plastics |
| Ozone | Purifying drinking water |
| Antiseptics | |
| Organic mercurials | Skin |
| Silver nitrate | Eyes of newborn to prevent gonorrhoeae |
| Iodine solution | Minor cuts and abrasions |
| Alcohol (70%) | Skin before injection |
| Hexachlorophene | Soaps, lotions, deodorants |
| Cationic detergents | Soaps, lotions |
| Hydrogen peroxide (3%) | Skin |

of the enzyme which synthesizes folic acid from PAB, and so prevent folic acid production. Because they are unable to produce the folic acid required, growth of the pathogenic bacteria is inhibited. The sulphonamides are selectively toxic to bacteria only because bacteria can synthesize their own folic acid whereas mammals cannot, and must have a supply of preformed folic acid in their diet.

**Antibiotics** are antimicrobial agents produced by some microorganisms themselves. Although an enormous number of these agents are known, only about one in a hundred are of practical use in the treatment of infectious disease. Once a successful antibiotic has been described, it can sometimes be modified using chemical methods to produce a **semisynthetic antibiotic**. The original penicillin molecule has been altered in a number of ways to produce derivatives with additional useful properties. For example, methicillin and oxacillin are both acid-stable (and can be taken by mouth rather than by injection), and resistant to the action of the enzyme β-lactamase. (The structure and industrial manufacture of penicillins is described in Chapter 11. The effect of penicillin on bacterial cell wall synthesis is described in Chapter 5.) This enzyme is produced by some strains of bacteria. It allows them to degrade penicillin-type antibiotics and to survive in their presence. Resistance to this enzyme is an extremely valuable property for an antibiotic to possess.

The sensitivity of microorganisms to different antibiotics varies greatly

**Experimental determination of antimicrobial activity**

Although there are several techniques which are used to determine antimicrobial activity, they all depend upon observable inhibition of growth.

In the **tube dilution technique**, a series of test tubes, each containing culture medium and a range of concentrations of the test antimicrobial agent, are inoculated with a similar number of cells of the test organism. Following incubation, the extent of microbial growth can be estimated using a colorimeter or spectrophotometer. The lowest concentration of the antimicrobial agent which inhibits growth is called the **minimum inhibitory concentration (MIC)**. If the conditions under which the test is carried out are standardized, the antimicrobial efficiency of a number of agents can be compared following calculation of their MICs.

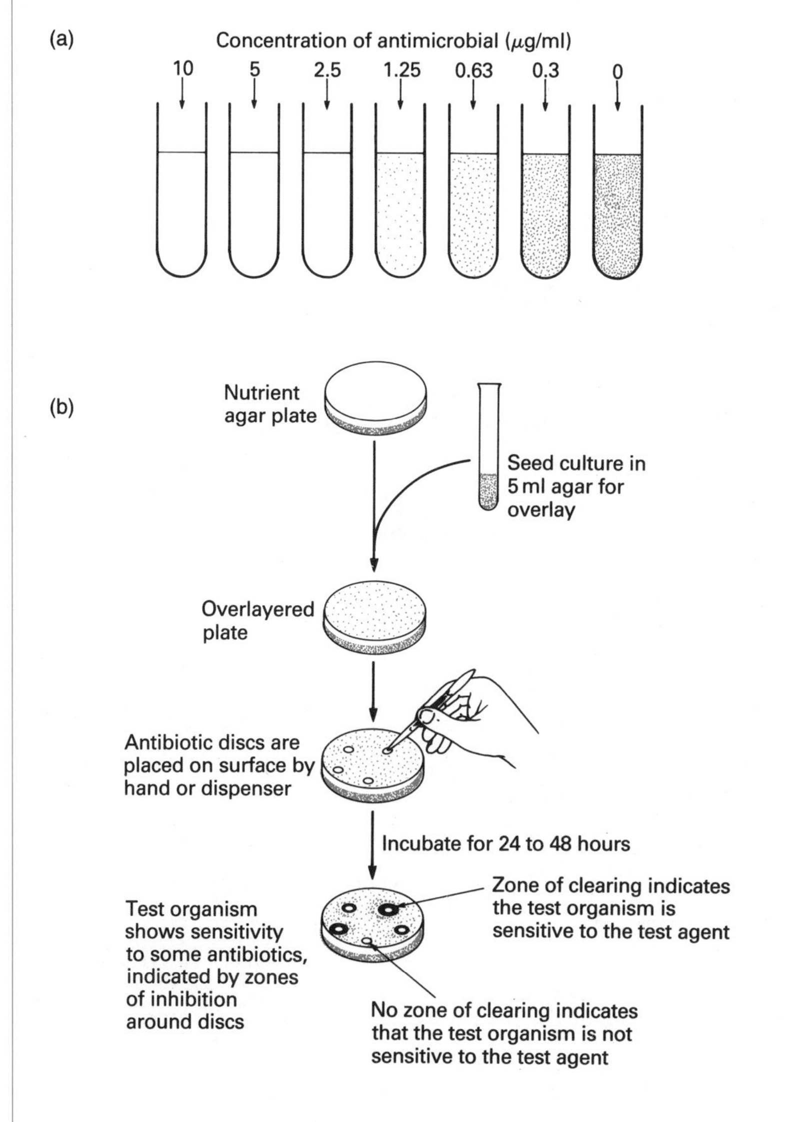


**BOX 3.4** (a) Tube dilution technique. The figure shows the extent of growth after incubation. The lowest concentration of antimicrobial capable of complete inhibition of growth is the MIC, 2.5 μg/ml in this example. (b) Agar diffusion technique.

**BOX 3.4** *continued*

In the **agar diffusion technique**, solidified medium is inoculated with a heavy inoculum of cells so that following incubation a continuous 'lawn' or 'carpet' of microbial growth is produced. Known concentrations of antimicrobial agents are added either to wells cut into the surface of the agar or to filter-paper discs, which are then placed on the surface of the medium. During incubation, the agent diffuses from the well or the filter-paper into the surrounding medium, and the concentration of the agent falls as the distance from the filter-paper or well becomes greater. When the concentration falls below the MIC, growth occurs, but closer to the well, it does not. A zone of inhibition is produced and its diameter can be used as an indicator of the antimicrobial activity of the test agent.

**Table 3.9** Mode of action of some antibiotics

| *Antibiotic* | *Mode of action* |
|---|---|
| Penicillin<br>Cephalosporin<br>Vancomycin | Inhibit eubacterial cell wall synthesis |
| Chloramphenicol<br>Erythromycin<br>Tetracycline<br>Streptomycin | Inhibit bacterial protein synthesis |
| Rifampin<br>Nalidixic acid<br>Trimethoprim<br>Novobiocin<br>Actinomycin D | Inhibit nucleic acid synthesis |
| Polymixin B<br>Nystatin<br>Amphotericin B<br>Gramicidin | Cause membrane damage |
| Rotenone<br>Antimycin A | Inhibits electron transport systems |

but a **broad-spectrum antibiotic** which is active against a wide range of bacteria is of more use than a narrow-spectrum antibiotic which acts only on a small number of species. Examples of antibiotics and their modes of action are shown in Table 3.9.

## Chapter summary

We began by looking at the nutritional requirements of microorganisms. We learned why they were required and the form in which they were obtained from the environment. We showed how a knowledge of a microorganism's nutritional requirements enables us to formulate a culture

BOX 3.5

**Mode of action of streptomycin**

Streptomycin, tetracycline, chloramphenicol and erythromycin are highly potent antibacterial agents that inhibit different stages in prokaryotic protein synthesis. Streptomycin has a twofold action. First, it interferes with the binding of formylmethionyl-tRNA to the ribosome and prevents correct initiation of protein synthesis. Streptomycin also causes misreading of mRNA. For example, if polyU is used as an artificial messenger, isoleucine (AUU) is incorporated in addition to phenylalanine (UUU). As a consequence, faulty proteins, including enzymes, are synthesized. These may lack essential activities, loss of which will lead to death of the bacterial cells concerned. Extensive investigation shows that a single protein of the 30S ribosomal subunit – protein S12 – is the receptor for streptomycin action.

medium for its successful growth in the laboratory. We saw that microorganisms could be grown in either batch or continuous culture and described the phases of growth of a population of unicellular organisms in batch culture.

Having learned of the general features of microbial growth, we went on to examine some of the ways in which growth could be influenced by environmental factors such as temperature and pH.

Finally, we looked at methods used to inhibit or kill unwanted microorganisms. We described a number of physical agents such as heat and radiations and finished the chapter with a brief review of chemotherapeutic agents, including antibiotics.

## Further reading

S.S. Block (ed) (1983) *Disinfection, Sterilization and Preservation.* 3rd edition. Lea and Febiger, Philadelphia, USA – a comprehensive account of this area.

M. Dworkin (1985) *Developmental Biology of the Bacteria.* Benjamin Cummings, Menlo Park, California, USA – a novel approach to the study of bacteria.

G. Gottschalk (1986) *Bacterial Metabolism.* 2nd edition. Springer-Verlag, New York, USA – one of the few available texts in this area.

J. Mandelstam, K. McQuillan and I.W. Dawes (1982) *Biochemistry of Bacterial Growth.* Blackwell Scientific Publications, London – a classic of its kind; although out of print, still available from libraries.

F.C. Neidhardt, J.L. Ingraham and M. Schaechter (1990) *Physiology of the Bacterial Cell.* Sinauer Associates, Sunderland, Mass., USA – a comprehensive, well written text which describes bacterial biochemistry in a physiological context.

## Questions

1. Multiple-choice questions:
   (i) For which of the following is a defined (synthetic) culture medium required? (a) microbiological assay, (b) batch culture, (c) antibiotic assay, (d) viruses.
   (ii) A bacterium which can grow only in the total absence of oxygen is described as (a) a facultative anaerobe, (b) an anaerobe, (c) an obligate aerobe, (d) an obligate anaerobe?
   (iii) The use of atmospheric nitrogen gas ($N_2$) as a nitrogen source is called (a) nitrification, (b) denitrification, (c) nitrogen fixation, (d) nitrate respiration?
   (iv) Microorganisms which grow best around 0 °C are termed (a) mesophiles, (b) thermophiles, (c) xerophiles, (d) psychrophiles?
   (v) During batch culture, a bacterial population doubles most rapidly during (a) the lag phase, (b) the logarithmic (exponential) phase, (c) the stationary phase, (d) the decline phase?
   (vi) A chemostat is used for (a) batch culture, (b) continuous culture, (c) storage of biochemicals, (d) storage of microorganisms.
2. Short-answer questions:
   (i) In what form are microelements (trace elements) usually added to a bacteriological culture medium?
   (ii) Distinguish between the following pairs:
      (a) antibiotic and antiseptic
      (b) pasteurization and sterilization
      (c) microbiostatic agent and microbiocidal agent
      (d) ionizing radiation and non-ionizing radiation
   (iii) Why is there often a lag period following inoculation of a culture medium with bacteria?
   (iv) Why do exponentially growing microorganisms in batch culture eventually stop dividing?
   (v) Explain how you would isolate a pure culture of a soil bacterium.
   (vi) What is enrichment culture?
3. Fill in the gaps:
   (i) In order to treat a clinical infection, an agent which exhibits __________ toxicity is required. This will kill __________ cells but not those of the __________. The first of these __________

agents to be successfully developed were the __________ drugs. These are an example of metabolic __________ which are compounds which have __________ closely resembling __________ compounds.

Sulphanilamide has a structure which closely resembles __________ which is a constituent of __________ acid. This latter compound is a __________ for certain enzyme-catalysed reactions. Sulphanilamide interferes with the formation of this essential compound by __________ inhibiting the enzyme required. Sulphanilamide is effective because the cofactor is synthesized by bacterial enzymes but not by __________.

(ii) Temperature has a complex effect on the growth of a bacterial population. At low temperatures, growth rate __________ as temperature __________. This is because cells carry out a wide variety of __________ reactions and __________-catalysed reactions. The rate at which these occur __________ with temperature. However, biological __________ are composed of complex __________ and their three- __________ structure is determined by many __________ chemical bonds. At higher temperatures these __________ will be __________, the catalyst loses its __________ and __________. As a consequence, the rate of __________ action drops and growth is __________. The combination of these two factors means that graphs of enzyme __________ versus temperature show a characteristic __________.

Choose from: Natural, increases (three times), humans, activity (twice), selective, folic, biochemical, optimum, cofactor, inhibited, benzoic, enzyme, microbial, catalysts, PAB, catalytic, antagonists, derivatives, microbial, proteins, competitively, sulphonamide, dimensional, structure, chemotherapeutic, broken, structures, patient, weak, linkages.

# 4 Energy production in microorganisms

In this chapter we shall examine the ways by which microorganisms satisfy their requirements for energy. We shall review the types of energy sources available in the natural environment and use this as a starting point to describe the metabolic pathways which lead to the production of high-energy compounds, such as adenosine triphosphate (ATP).

Before we look at the details of the processes involved, we must lay the foundations by considering the underlying thermodynamic principles which govern chemical reactions. By doing this, we shall come to understand the rules which govern the energy changes which take place during chemical reactions and the methods microbes use to exploit such changes.

The energy required for cell survival is obtained in three ways. First, photosynthetic microorganisms are able to convert the radiant energy of sunlight into chemical energy. Second, many kinds of microorganisms are able to derive energy from the oxidation of organic molecules. Finally, a relatively small number of bacterial types obtain their energy from the oxidation of inorganic molecules.

From a practical point of view, knowledge of the energy metabolism of microorganisms brings a number of important benefits. It enables us to develop suitable media to culture them in the laboratory, while preventing the growth of unwanted organisms. On an industrial scale, some of the end-products of energy-producing pathways, such as ethanol and lactic acid, are very valuable and support the existence of vast industries (Chapter 11). Other end-products form the bases of diagnostic tests used as aids to identify unknown bacteria in clinical specimens, for example.

> Although energy can exist in a number of interconvertible forms, a single unit may be used to express its magnitude. Although a variety of alternative units exist, in biology the most commonly used units are kilocalorie (kcal) and the kilojoule (kJ). A kilocalorie is defined as the amount of heat energy required to raise the temperature of one kilogram of water by one degree centigrade (Celsius). One kilocalorie is equivalent to 4.184 kJ. In this book, we use the kilojoule (kJ) as the energy unit.

## Energy and thermodynamics

The first law of thermodynamics tells us that energy can be converted from one form to another, but can be neither created nor destroyed.

The second law of thermodynamics relates to the finding that chemical reactions are accompanied by *changes* in energy. During a chemical reaction, energy may be either lost or gained. But only part of the energy released is usable energy; in other words, only part of it can be used to do *work*.

The total energy obtainable from a chemical reaction is called its **enthalpy**, $H$. The *usable* energy obtained from a chemical reaction is its **free energy**, $G$. The amount of non-usable energy is called **entropy** and is normally represented by the letter $S$. The changes in energy content of such a system may be expressed by the following formula where the Greek letter delta ($\Delta$) means 'a change in':

$$\Delta G = \Delta H - T\Delta S$$

where $\Delta G$ represents the change in free energy, $\Delta H$ the change in enthalpy, $T$ is the absolute temperature (add 273 to the temperature in degrees Celsius) and $\Delta S$ is the change in entropy.

In order to standardize the expression of energy changes which occur during chemical reactions, the term $\Delta G^{0\prime}$ or **standard free energy** is often used instead of $\Delta G$. The superscripts $^{0}$ and $'$ indicate that the standard conditions used to calculate free energy are pH 7, 25°C, and all reactants and products are initially at one molar concentrations.

A reaction in which, for example, two compounds A and B are converted to products C and D can be described in thermodynamic terms because there is a change in their energy contents. If, for the reaction $A + B \rightleftharpoons C + D$, $G^{0\prime}$ is negative, free energy will be *released* and the reaction will go forward spontaneously. Such a reaction is called **exergonic**. If $G^{0\prime}$ is positive, the forward reaction will not occur spontaneously but the *reverse* reaction, from right to left, will. This type of reaction is called **endergonic**. Examples of exergonic reactions include oxidation of carbohydrates, fats and proteins, and hydrolytic reactions involving energy-rich molecules such as ATP. Examples of endergonic reactions include those that bring about flagella movement, nutrient transport, and reactions involved in the biosynthesis of macromolecules such as proteins, carbohydrates and nucleic acids (Table 4.1).

## Equilibrium and free energy

A chemical reaction will eventually reach equilibrium and no further overall change in the reaction occurs. At this point, the rate of conversion of reactants A and B to products C and D will be balanced by the rate of conversion of C and D to A and B. This can be expressed conveniently as the **equilibrium constant**, $K_{eq}$, where:

$$K_{eq} = \frac{\text{product of concentrations of products [C] [D]}}{\text{product of concentrations of reactants [A] [B]}}$$

**Table 4.1** Standard free energy changes ($G^{0\prime}$) of some common biochemical reactions (at pH 7.0 and 25°C). Reactions with a positive $G^{0\prime}$ are endergonic and those with a negative value are exergonic and proceed spontaneously from left to right

| *Reaction* | $G^{0\prime}$ *(kJ/mole)* |
|---|---|
| Acetic anhydride + $H_2O$ → 2 acetate | – 91.2 |
| Pyrophosphate + $H_2O$ → 2 phosphate | – 33.4 |
| Glucose-6-phosphate + $H_2O$ → glucose + phosphate | – 13.8 |
| Glutamine + $H_2O$ → glutamate + $NH_3$ | – 14.2 |
| Sucrose + $H_2O$ → glucose + fructose | – 29.3 |
| Glucose + phosphate → glucose-6-phosphate + $H_2O$ | + 13.8 |
| Malate → fumarate + $H_2O$ | + 3.14 |
| Glucose + $6O_2$ → $6CO_2$ + $6H_2O$ | –2870 |

$K_{eq}$ is a reflection of the free-energy change of the components as the following equation shows:

$$\Delta G^{0\prime} = -RT \ln K_{eq}$$

where $R$ is the universal gas constant and ln $K_{eq}$ is the natural logarithm of the equilibrium constant.

In summary, it is important for you to appreciate that all the chemical reactions of a living cell are accompanied by changes in energy. Furthermore, the free energy of a reaction is given a numerical value expressed as calories or joules of energy used up or given off. An exergonic chemical reaction is one in which energy is released and an endergonic one requires energy.

## Enzymes and chemical reactions

The free energy of a chemical reaction is a thermodynamic idea and tells us only about the energy distribution between the reactants when a reaction is at equilibrium. It does not tell us about the *rate* at which the reaction occurs and how long it will take for equilibrium to be reached.

Let us consider a simple chemical reaction in which water is formed from oxygen and hydrogen gases. Although the energetics of the reaction are favourable, with an energy of formation of 238 kJ/mole, it is very probable that, if we mixed the gases $O_2$ and $H_2$, no measurable reaction would occur even over a number of years. The reason is that the formation of water requires the interaction of hydrogen and oxygen *atoms*, whereas gaseous $O_2$ and $H_2$ are *molecules*. So the chemical bonds of the two molecules must be broken to produce oxygen and hydrogen atoms. Such breakages of chemical bonds require energy, referred to as **activation energy**. Activation energy is the amount of energy required to bring all the participants in a chemical

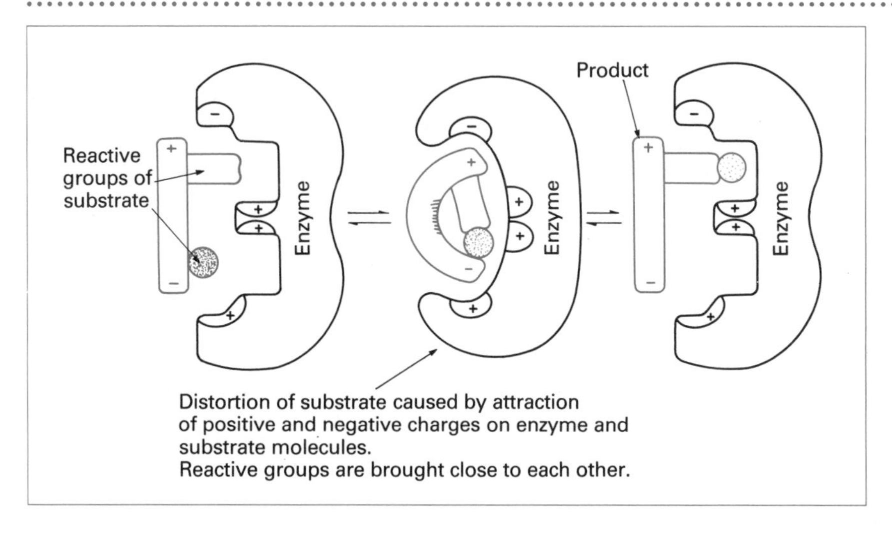


**Figure 4.1** A mechanism of enzyme action. An enzyme speeds up the rate of a biochemical reaction because it can form a combined structure called an enzyme–substrate complex. In this structure, the substrate is distorted by, for example, attraction of charged groups on the enzyme and substrate molecules. Because of the distortion, less energy is required to bring reactive groups together and the reaction is speeded up.

reaction to a reactive state. **Enzymes** are uniquely valuable to living organisms because they reduce the activation energy required to bring reactants into a reactive state (Figure 4.1).

Enzymes, like catalysts, increase the rate of a reaction without being changed at the end of the reaction. Typically, enzymes are able to increase the rate of chemical reactions by up to $10^{12}$ times that which would occur in their absence. And, of course, they accomplish this at ordinary temperatures and pressures, unlike industrial catalysts which often operate at temperatures and pressures which would destroy life.

Almost all known enzymes are proteins. They catalyse only one type of chemical reaction (or one class of closely related chemical reactions). In other words, enzymes are highly specific. Their specificity is a reflection of the precise relationships between the shape of the enzyme and its reactant, or **substrate**.

Although an enzyme is a large molecule, the substrate interacts with only one part of it, called the **active site**. Put simply, the active site has a precise architecture which will accommodate only a molecule or chemical group with a complementary structure. (Think of a lock and its key.) The substrate, S, combines temporarily with the enzyme, E, to form an enzyme–substrate complex, E–S. After the reaction, the product, P, is released and the enzyme returns to its original state:

$$E + S \rightleftharpoons E\text{–}S \rightleftharpoons E + P$$

The formation of such an enzyme–substrate complex aligns reactive groups in such a way as to place strain on specific bonds in the substrate(s). The result is that the amount of activation energy required for a particular chemical reaction to proceed is much reduced (Figure 4.1).

So far we have concentrated on enzyme catalysis of exergonic reactions (those reactions where the free energy of formation of the substrates is greater than that of the products). If an enzyme catalyses an endergonic reaction in which an energy-rich product is formed from an energy-poor substrate, additional free energy must be put into the system to raise the energy level of the substrate to that of the product.

In theory, all enzyme-catalysed reactions are reversible but, in practice, those which are either highly endergonic or those which are highly exergonic proceed in one direction only. If a highly exergonic reaction needs to be reversed during cellular metabolism, a different enzyme altogether is usually required.

To sum up, we have emphasized in this section that before a chemical reaction can occur, the reactants must first be activated by an input of activation energy. The amount of activation necessary is significantly reduced by the use of enzymes. Enzymes are proteins and their specificity arises from the complementary three-dimensional relationship between the active site of the enzyme molecule and its substrate(s).

## Energy from chemical reactions: oxidation–reduction

Both inorganic and organic molecules contain chemical bonds which are sources of energy. Whether this or that particular source of energy is used will depend upon the type of microorganism (Table 4.2).

**Table 4.2** Classification of microorganisms according to carbon and energy source

| | *Type* | *$e^-$-donor* | *Energy source* | *C-source* | *Examples* |
|---|---|---|---|---|---|
| Lithotrophs (autotrophs) | Photolithotroph (photosynthetic autotroph) | Inorganic compounds (bacteria: $H_2S$, $S_2O_3^{2-}$ etc.; algae: $H_2O$) | Sunlight | $CO_2$ | (1) Green S-bacteria (Chlorobiaceae) (2) Purple S-bacteria (Chromatiaceae) (3) Algae |
| | Chemolithotroph (chemosynthetic autotroph) | Inorganic compounds | Oxidation–reduction reactions of inorganic compounds | $CO_2$ | *Thiobacillus thiooxidans* *Nitrosomonas* *Nitrobacter* *Hydrogenomonas* *Beggiatoa* |
| Organotrophs (heterotrophs) | Photoorganotroph (photosynthetic heterotroph) | Organic compounds | Sunlight | $CO_2$ and organic compounds | Non S-bacteria (*Rhodospirillaceae*) |
| | Chemoorganotroph (chemosynthetic heterotroph) | Organic compounds | Oxidation–reduction reactions of organic compounds | Organic compounds | Most microorganisms (i.e. all fungi, all protozoans, most bacteria. e.g. *E. coli*, *Strep. faecalis*, etc. but *not* algae) |

The utilization of chemical energy in living organisms is founded on **oxidation–reduction (redox) reactions**. Although the term oxidation was originally used to describe inorganic reactions involving oxygen, it is now used to describe all chemical reactions involving *removal* of an electron or electrons from a substance. In either case, electrons alone or whole hydrogen atoms may be transferred. In this context remember that a hydrogen atom is made up of one electron and one proton so a hydrogen atom may be oxidized as follows:

$$\underset{\text{hydrogen atom}}{H} \rightarrow \underset{\text{electron}}{e^-} + \underset{\text{proton (hydrogen ion)}}{H^+}$$

However, electrons cannot exist alone and must form part of an atom or molecule. The equation does not represent a complete reaction and is consequently known as a **half-reaction**, which implies a requirement for a second half-reaction. So for an oxidation to occur, a *reduction* reaction must also take place. For the oxidation of hydrogen to occur, the reduction of, for example, oxygen must occur in a second half-reaction:

$$\tfrac{1}{2}O_2 + 2e^- + 2H^+ \rightarrow H_2O$$

When the oxidation and reduction reactions are linked (**coupled**), the overall reaction can be represented as:

$$H_2 + \tfrac{1}{2}O_2 \rightarrow H_2O$$

Biological oxidations of this type always involve one substance which is oxidized, the **electron donor**, and one which is reduced, the **electron acceptor**.

## Reduction potentials

Different substances vary in their tendency to give up electrons and become oxidized or to accept electrons and become reduced. In order to quantify these tendencies, the concept of **reduction potential** ($E^0$) is used. Such potentials are determined electrically by reference to hydrogen gas as a standard.

Conventionally, all reduction potentials are expressed for half-reactions written as reductions, with the oxidant on the left. (An oxidant is a substance which combines with an electron or a hydrogen atom.) Following these conventions, at neutrality the following reduction potentials ($E^{0\prime}$) of our two half-reactions are as follows:

Because reduction potential may be influenced by hydrogen ion concentration (pH), values are quoted for neutrality (pH 7.0) in biology, the reason being that the cytoplasm of cells is neutral or close to neutral.

$$\tfrac{1}{2}O_2 + 2H^+ + 2e^- \rightarrow H_2O \qquad E^{0\prime} = +0.816 \text{ volts}$$

$$2H^+ + 2e^- \rightarrow H_2 \qquad E^{0'} = -0.421 \text{ volts}$$

## Coupled oxidation–reduction reactions

In the two half-reactions we have just looked at, two oxidation–reduction (O–R) pairs can be identified and written as $\frac{1}{2}O_2/H_2O$ and $2H^+/H_2$. Conventionally, the oxidized form (the oxidant) is always written on the left. In coupled redox reactions, it is the reduced substance of an O–R pair (whose reduction potential is more negative) which donates electrons to the oxidized substance of an O–R pair with a more positive potential.

Which is the electron donor and which the electron acceptor in the O–R pairs in our example? Yes, in our example the $2H^+/H_2$ pair has the more negative potential (–0.421 volts) and therefore the reduced substance, $H_2$, donates electrons to the oxidized substance, $O_2$, of the more positive pair, $\frac{1}{2}O_2/H_2O$ (+0.816 volts). In this example then, hydrogen gas will serve as an electron donor and will be oxidized, while oxygen will be the electron acceptor and becomes reduced. Notice finally that although individual half-reactions are always written as reductions, in an oxidation–reduction reaction, one of the two half-reactions is written as an oxidation (Figure 4.2).

## Reduction potentials of some O–R pairs

In addition to the two O–R pairs we have already met, many others are found in biological systems. Different O–R pairs may be placed at appropriate points along a scale according to their reduction potentials. Such an arrangement is sometimes referred to as an **electron tower** (Table 4.3); O–R pairs with the most negative reduction potential are placed at the top and the most positive at the bottom. Furthermore, the reduced substance of the pair with the most negative reduction potential has most potential

**Figure 4.2** Example of a coupled oxidation–reduction reaction.

$H_2 \rightarrow 2e^- + 2H^+$ ...... reaction (i)

electron-donating half-reaction
(written as an oxidation)

$\frac{1}{2}O_2 + 2e^- \rightarrow O^{2-}$ ...... reaction (ii)

electron-accepting half-reaction

$2H^+ + O^{2-} \rightarrow H_2O$

formation of $H_2O$

Overall:

$H_2 + \frac{1}{2}O_2 \rightarrow H_2O$

$H_2$ is the reductant ($e^-$-donor) and becomes oxidized
$O_2$ is the oxidant ($e^-$-acceptor) and becomes reduced

**Table 4.3** The oxidation–reduction potential ($E^{0\prime}$) of some redox pairs in microbial systems

Redox pairs are arranged with the strongest reductants (most negative reduction potentials) at the top and the strongest oxidants (most positive reduction potentials) at the bottom. Electrons removed from redox pairs at the top of the list (or tower) can be passed to acceptors at different levels. The greater the difference in reduction potentials between electron donor and electron acceptor, the greater the energy released.

The free energy released during the transfer of a pair of electrons between two redox pairs may be calculated if their $E^0$ values are known. The equation representing this change is $\Delta G^{0\prime}=-nF\Delta E^{0\prime}$, where $G^{0\prime}$ is the standard free energy at pH = 7.0; $n$ is the number of electrons transferred; F is a Faraday and equals 23 000 cal/volt and $\Delta E^{0\prime}$ is the difference in volts between the two redox pairs. For example, the free energy potential between $NAD^+/NADH$ and $O_2/H_2O$ during the transfer of electrons from NAD to oxygen is given by:

$$\Delta G^{0\prime} = -2 \times 23\,000 \times (+0.82) - (-0.32)$$
$$= -46\,000 \times 1.14$$
$$= -52\,440 \text{ cal/mole}$$

This energy can be stored in the chemical bonds of a variety of high-energy compounds (ATP is the most common of these) until it is required in cellular processes.

| *Redox pairs* | *$E^{0\prime}$ or oxidation–reduction potential (volts)* |
|---|---|
| $CO_2$/formate | −0.432 |
| $H^+/H_2$ | −0.420 |
| $NAD^+/NADH_2$ | −0.320 |
| $S^0/HS^-$ | −0.280 |
| FAD/FADH | −0.22 |
| FMN/FMNH (flavoprotein) | −0.19 |
| $SO_3^-/S$ | −0.11 |
| Fumarate/succinate | +0.03 |
| Ubiquinone ox/ubiquinone red | +0.04 |
| Cytochrome b ox/cytochrome b red | +0.07 |
| Cytochrome $c_1$ ox/cytochrome $c_1$ red | +0.23 |
| Cytochrome o ox/cytochrome o red | +0.280 |
| Cytochrome $a_3$ ox/cytochrome $a_3$ red | +0.385 |
| $NO_3^-/NO_2^-$ | +0.43 |
| $Fe^{3+}/Fe^{2+}$ | +0.770 |
| $O_2/H_2O$ | +0.82 |

energy, and the reduced substance in the pair with the most positive reduction potential has the least potential energy. Conversely, the oxidized substance in the O–R pair with the most negative reduction potential least readily accepts electrons and that of the pair with the most positive potential, most readily accepts electrons.

Electrons may be transferred from more negative electron donors to more positive acceptors at different levels and the difference in electrical potential between them is $E^{0\prime}$. The greater the value of $E^{0\prime}$, the greater the amount of energy released. In other words, $E^{0\prime}$ is proportional to free energy, $G^{0\prime}$. Oxygen, the oxidized member of the $\frac{1}{2}O_2/H_2O$ pair, is the most powerful oxidizing agent and is often the final electron acceptor in biological systems. Oxidation–reduction pairs with intermediate reduction potentials can act as either electron donors or acceptors. In the oxidized ubiquinone/reduced ubiquinone pair (+0.11 volts) the oxidized form may be reduced by the oxidation of methanol in the $CO_2$/methanol pair (−0.38 volts). Alternatively, reduced ubiquinone can reduce ferric ($Fe^{3+}$) iron to ferrous ($Fe^{2+}$) iron in the $Fe^{3+}/Fe^{2+}$ pair (+0.76 volts).

Energy-yielding reactions of living cells often require the oxidation of either an organic or inorganic compound – the electron donor. The electron acceptor receives electrons from the electron donor, and the former consequently becomes reduced and the latter oxidized. The readiness with which different compounds accept or donate electrons is expressed as reduction potentials ($E^{0\prime}$) and the difference in reduction potential between an electron donor and an electron acceptor is proportional to the free energy released $G^{0\prime}$.

**Electron carriers**

In biological systems, electrons do not pass directly from donor to acceptor; instead, electron flow is indirect and involves the participation of a number of intermediates usually called **carriers**. The initial electron donor is called the **primary electron donor** and the final acceptor, the **terminal electron acceptor**. When carrier molecules participate, the energy released still depends on the reduction potentials of the primary donor and the terminal acceptor. Transfer of electrons requires a series of oxidations and reductions of the carrier molecules.

Electron carriers fall into two groups: those which are free to diffuse through the cell cytoplasm, and those which are fixed within a membrane. Examples of freely diffusible carriers include the coenzymes nicotinamide-adenine dinucleotide ($NAD^+$) and its phosphorylated derivative, $NADP^+$. Both of these compounds carry hydrogen atoms rather than electrons alone. For simplicity's sake, both coenzymes are said to transfer two hydrogen atoms to the next carrier in the chain. Strictly speaking, however, both $NAD^+$ and $NADP^+$ each transfer two electrons and one proton, the second proton coming from solution. The reduction of NAD is correctly written as:

$$NAD^+ + 2H^+ + 2e^- \rightleftharpoons NADH + H^+$$

For simplicity, we shall use the notation $NADH_2$ to represent $NADH + H^+$.

Both pairs of oxidized and reduced coenzymes have negative reduction potentials (−0.32 volts) and can, therefore, act as a good electron donors.

**Table 4.4** Some phosphorylated compounds and their functions

| *Compound* | | *Function* |
|---|---|---|
| Phosphoenolpyruvate | O=C–OH, C–O~Ⓟ, =$CH_2$ | ATP synthesis and sugar transport |
| Carbamyl ~ phosphate | | ATP synthesis, biosynthesis of pyrimidines |
| 1.3-Diphosphoglycerate | O=C–O~Ⓟ, H–C–OH, $CH_2$–O–Ⓟ | ATP synthesis |
| Acetyl ~ phosphate | $H_3C$–C(=O)~Ⓟ | ATP synthesis |
| ATP | | Biosynthesis and other cellular functions |
| Glucose-6-phosphate | Ⓟ–O–$CH_2$ (ring structure) | Glycolytic intermediate |
| Glycerol phosphate | $CH_2$–O–Ⓟ, H–C–OH, $CH_2$–OH | Intermediate in lipid metabolism |

The occurrence of two different hydrogen carriers, both with the same reduction potential and in the same cells, is explained by their different metabolic roles. $NAD^+$ is used in energy-generating (catabolic) reactions and $NADP^+$ in biosynthetic (anabolic) reactions.

## Adenosine triphosphate and other high-energy compounds

Having established that energy can be released through a variety of oxidation–reduction reactions, the next problem is how such energy may be stored for use in different cell activities. Most frequently, the energy released is stored in a variety of organic phosphate molecules. These compounds act as intermediates in the conversion of energy into useful work. A common property of these phosphorylated compounds is the presence of phosphate groups which are linked to oxygen atoms (Table 4.4).

The most widespread high-energy compound of living organisms is adenosine triphosphate (ATP). It is synthesized in exergonic reactions and is used as an energy source in endergonic ones such as biosynthetic reactions.

### The oxidation of organic molecules and ATP synthesis

The majority of microbes derive their energy from oxidation–reduction reactions involving organic molecules. That is to say, all protozoans, all fungi and most bacteria oxidize organic molecules to synthesize ATP.

The oxidation of organic molecules proceeds by way of either **fermentation** or **respiration** pathways. Fermentation requires the use of an internal, organic terminal electron acceptor whereas respiration requires an inorganic terminal electron acceptor; this is usually oxygen but, under anaerobic conditions, a nitrate or a sulphate may be used instead.

## Fermentations

Fermentation reactions lead to only a limited oxidation of an organic substrate. They do not need aerobic conditions because the required oxidant is an organic molecule which is produced by the fermentation process itself. Because the substrate is only partially oxidized, much of its potential energy remains locked up in the fermentation products. The amount of ATP synthesized during fermentation reactions is small compared with that produced by complete oxidation in the presence of oxygen (respiration).

Fermentation is a typically bacterial process, although brewer's yeast (a yeast is a unicellular fungus) is a well-known exception to this rule. Although bacteria are able to ferment a variety of organic substrates, most can utilize glucose. Different fermentations produce a wide range of end-products and some of these are commercially valuable. For example, dairy products such as cheese, butter and yoghourt are produced by the activity of bacteria (for example, *Lactobacillus* and *Streptococcus*, which produce lactic acid as a fermentation product (Chapter 11)).

**BOX 4.1** Alternative forms of the formic fermentation, the bases of the methyl red and Voges–Proskauer tests (MRVP tests).

**Products of fermentation in identification**

One of the key characteristics separating the different genera of enteric bacteria is the type and proportion of products formed by the fermentation of glucose. Enteric bacteria exhibit two alternative forms of the **formic fermentation**, either a **mixed-acid fermentation** or a **2,3-butanediol fermentation**. The mixed-acid fermentation is found in the genera *Escherichia*, *Salmonella* and *Shigella*, and the butanediol fermentation in *Klebsiella*, *Erwinia*, *Enterobacter* and *Serratia*.

(a) Mixed acid

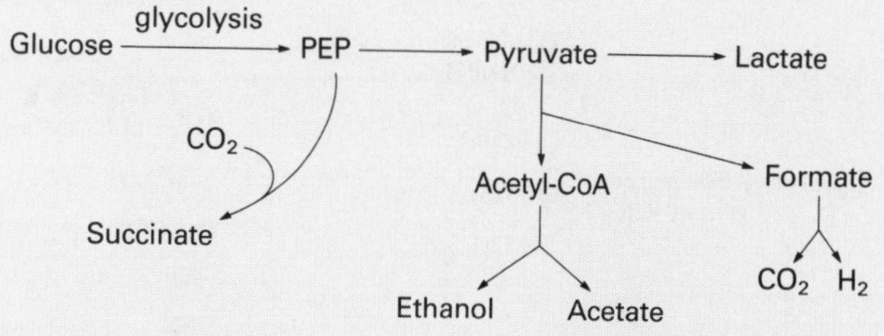

Glucose → (glycolysis) PEP → Pyruvate → Lactate
PEP + $CO_2$ → Succinate
Pyruvate → Acetyl-CoA → Ethanol, Acetate
Pyruvate → Formate → $CO_2$, $H_2$

(b) Butanediol

Glucose → (glycolysis) Pyruvate → Lactate
Pyruvate → Acetyl-CoA → Ethanol
Pyruvate → Formate → $CO_2$, $H_2$
Pyruvate → ($CO_2$) α-Acetolactate → ($CO_2$) Acetylmethyl carbinol (acetoin) → 2, 3-butanediol

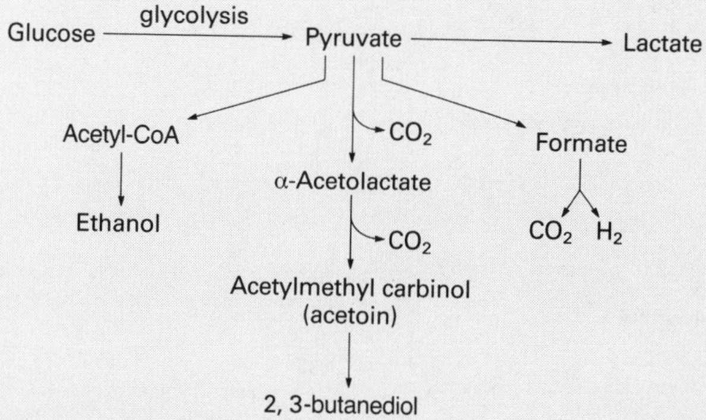

During the operation of the mixed-acid fermentation, four organic acids are produced: formic, acetic, lactic and succinic acids. Ethanol, $CO_2$ and hydrogen are also produced, but not 2,3-butanediol. When the organisms are cultured in a laboratory growth medium, the acids are produced in sufficient quantities to reduce the pH to a value of 4.5 or less. This can be detected by the use of the acid–base indicator methyl red.

Bacteria which use the butanediol fermentation produce mainly neutral end-products and, therefore, do not produce an acid response with methyl red. One of the intermediates of this second pathway is called acetylmethyl carbinol or acetoin, and this can be detected by the production of a red colour in the Voges–Proskauer test.

## Glucose breakdown by the glycolytic (Embden–Meyerhof–Parnas) or EMP pathway

The glycolytic pathway is widespread throughout nature and occurs in a wide range of microbes. During glycolysis, two molecules of $NADH_2$ are

**Figure 4.3** The glycolytic (Embden–Meyerhof–Parnas: EMP) pathway.

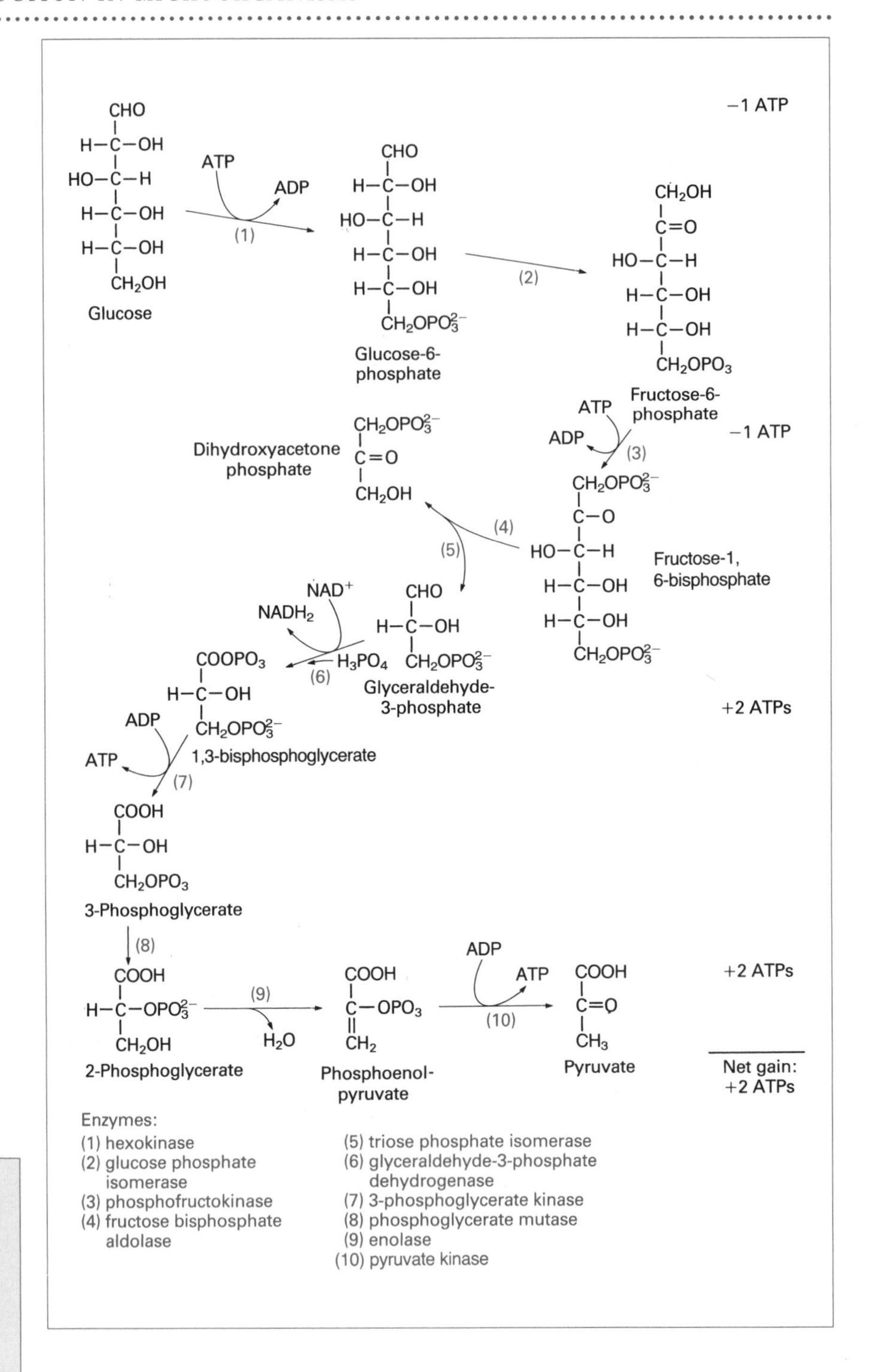


In appropriate enzyme-catalysed reactions, a bond in an organic substrate molecule possesses sufficient energy for the synthesis of an ATP molecule from ADP. This is **substrate level phosphorylation**.

produced and two molecules of ATP are generated by **substrate level phosphorylation**. The complete series of reactions making up the EMP pathway is shown in Figure 4.3.

A total of four molecules of ATP are synthesized for each molecule of glucose broken down to pyruvate. However, two molecules of ATP are used up in the initial reactions in which glucose is activated to produce fructose-1,6-bisphosphate. The net gain is therefore two molecules of ATP for each molecule of glucose catabolized.

### Fate of pyruvate: the regeneration of $NAD^+$

We now know that the oxidation of each molecule of glucose causes the reduction of two molecules of $NAD^+$ to $NADH_2$. However, the amount of $NAD^+$ in a cell is limited and, if it were to be used up completely, oxidation of further molecules of glucose would stop. This problem is overcome by additional reactions in which $NADH_2$ is used to reduce either pyruvate itself or one of its metabolic products. In this way, $NADH_2$ is oxidized to $NAD^+$, replenishing the supply of the oxidized coenzyme.

These reactions, in which $NAD^+$ is regenerated from $NADH_2$, lead to the production of characteristic end-products. The simplest is the **homolactic fermentation** found in most bacteria of the genus *Lactobacillus* and in all streptococci. This fermentation requires a single reaction catalysed by the enzyme lactic dehydrogenase and is called homolactic because, in contrast to some lactic fermentations which we shall examine shortly, lactic acid is the only end-product:

$$\underset{\text{Pyruvate}}{\begin{array}{c}CH_3\\|\\C=O\\|\\COOH\end{array}} \xrightarrow[NADH_2 \rightarrow NAD^+]{\text{lactic dehydrogenase}} \underset{\text{Lactic acid}}{\begin{array}{c}CH_3\\|\\CHOH\\|\\COOH\end{array}}$$

The best-known fermentation is the alcoholic fermentation in brewer's yeast (*Saccharomyces cerevisiae*), which under anaerobic conditions regenerates $NAD^+$ by reducing the acetaldehyde produced by the decarboxylation of pyruvic acid. Alcoholic fermentation in yeast is slightly more complex than homolactic fermentation and requires two enzymes:

$$\underset{\text{pyruvate}}{\begin{array}{c}CH_3\\|\\C=O\\|\\COOH\end{array}} \xrightarrow[CO_2]{\text{pyruvate decarboxylase}} \underset{\text{acetaldehyde}}{\begin{array}{c}CH_3\\|\\CHO\end{array}} \xrightarrow[NADH_2 \rightarrow NAD^+]{\text{alcohol dehydrogenase}} \underset{\text{ethanol}}{\begin{array}{c}CH_3\\|\\CH_2\\|\\OH\end{array}}$$

Notice that in both alcoholic and lactic fermentations, an internally generated organic molecule acts as a terminal electron acceptor. In both cases, the electrons are part of hydrogen atoms; pyruvate is the acceptor in the homolactic fermentation and acetaldehyde in the alcoholic fermentation.

During the fermentation of glucose using the glycolytic pathway, two molecules of ATP are synthesized for each molecule of glucose metabolized. This ATP can be used for cellular activities including biosynthesis, motility and solute transport.

## Alternative pathways for the fermentation of glucose

Although the glycolytic pathway is the most common method for the microbial breakdown of glucose, two alternatives with rather narrower distributions are also important.

The first of these is the **phosphoketolase pathway**. This is a lactic

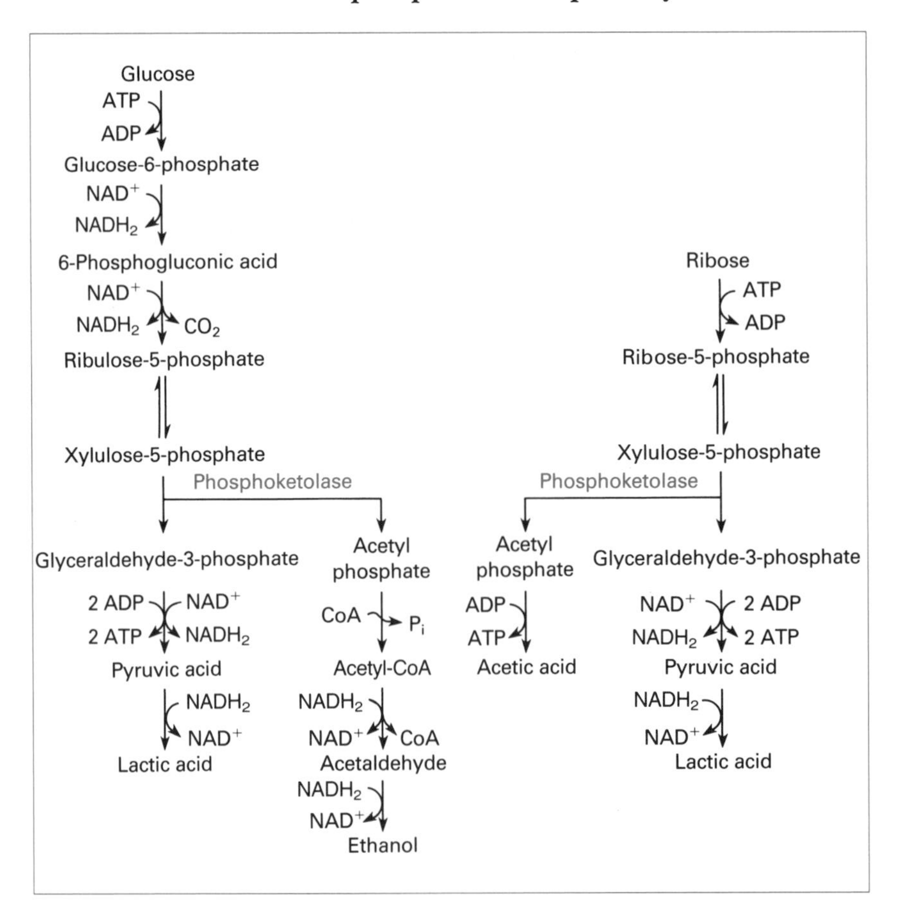


**Figure 4.4** The phosphoketolase pathway. This pathway is used by heterofermentative lactic acid bacteria such as *Leuconostoc mesenteroides*. In contrast to homofermenters, lactic acid and other major end-products are formed including $CO_2$ and a two-carbon compound. On ribose, acetic acid is produced, but on glucose, ethanol is produced.

fermentation in which lactic acid is produced together with other major end-products: $CO_2$ and either ethanol or acetate. To distinguish this lactic fermentation from that of most lactic acid bacteria (in which lactic acid is the only major end-product), it is referred to as a **heterolactic fermentation.**

The phosphoketolase pathway is found in the bacterium *Leuconostoc mesenteroides* when growing on glucose or on the five-carbon sugar, ribose. It is also used by some *Lactobacillus* species for the oxidation of various five-carbon sugars.

The phosphoketolase pathway is sometimes called the **hexose monophosphate pathway** to emphasize one of its major differences from the glycolytic pathway – the absence of fructose-1,6-biphosphate (or any other biphosphates) from its reaction sequence. The reactions of the phosphoketolase pathway are outlined in Figure 4.4 and you will see that, when glucose is the substrate, only one molecule of ATP is produced for each molecule of glucose metabolized. In the case of ribose, however, two molecules of ATP are produced for each ribose molecule oxidized.

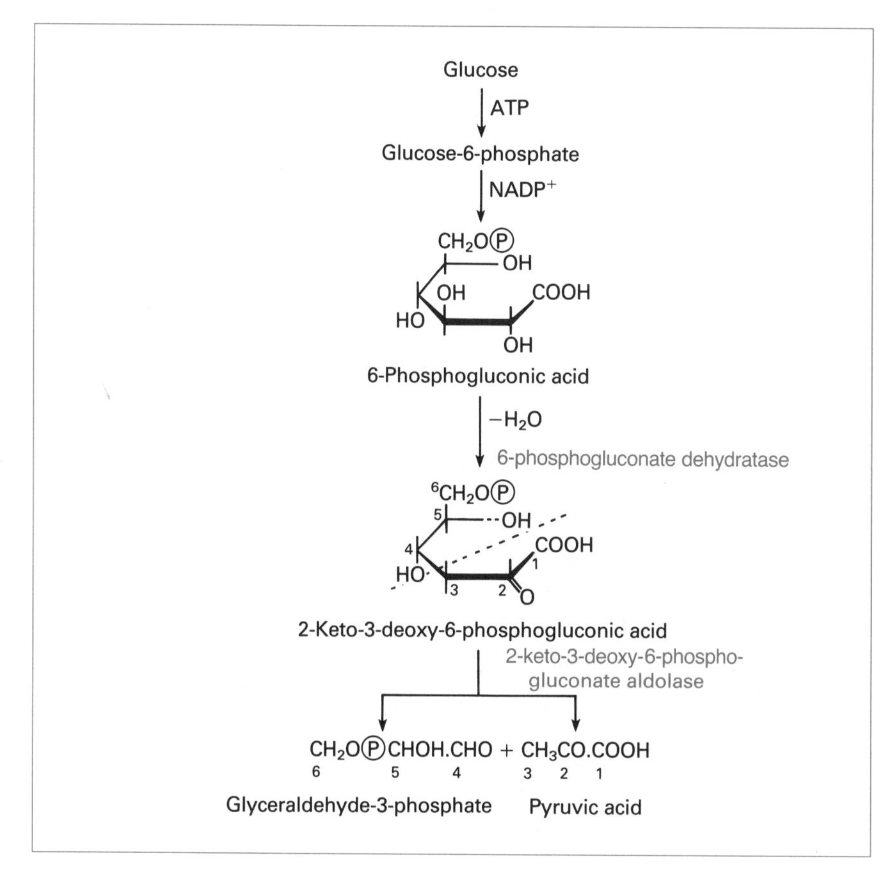


**Figure 4.5** The Entner–Doudoroff pathway for glucose degradation. This pathway is typically found in Gram-negative rods, some of which ferment glucose to $CO_2$ and ethanol. *Zymomonas mobilis* is an example. Alternatively, aerobic bacteria may completely oxidize glucose to $CO_2$ and $H_2O$, as in *Pseudomonas saccharophila*.

The second alternative to the glycolytic pathway for the fermentation of glucose is restricted to a number of Gram-negative, rod-shaped bacteria. The best known of these is *Zymomonas mobilis*, which ferments glucose to ethanol and carbon dioxide by the **Entner–Doudoroff pathway** (Figure 4.5). Again, this is a hexose monophosphate pathway and its operation in *Z. mobilis* may be established by demonstrating the presence of its key enzymes, a dehydratase and 2-keto-3-deoxy-6-phosphogluconate (KDPG) aldolase. The operation of different pathways for the fermentation of glucose to ethanol and carbon dioxide in *Saccharomyces cerevisiae* and *Z. mobilis* can also be demonstrated by the technique of radiorespirometry (Figure 4.6).

In South America, *Zymononas* is used to ferment cactus juice into an alcoholic drink called *pulque.*

In contrast to the glycolytic pathway, fermentation by the Entner–Doudoroff pathway yields only one molecule of ATP for each molecule of glucose fermented.

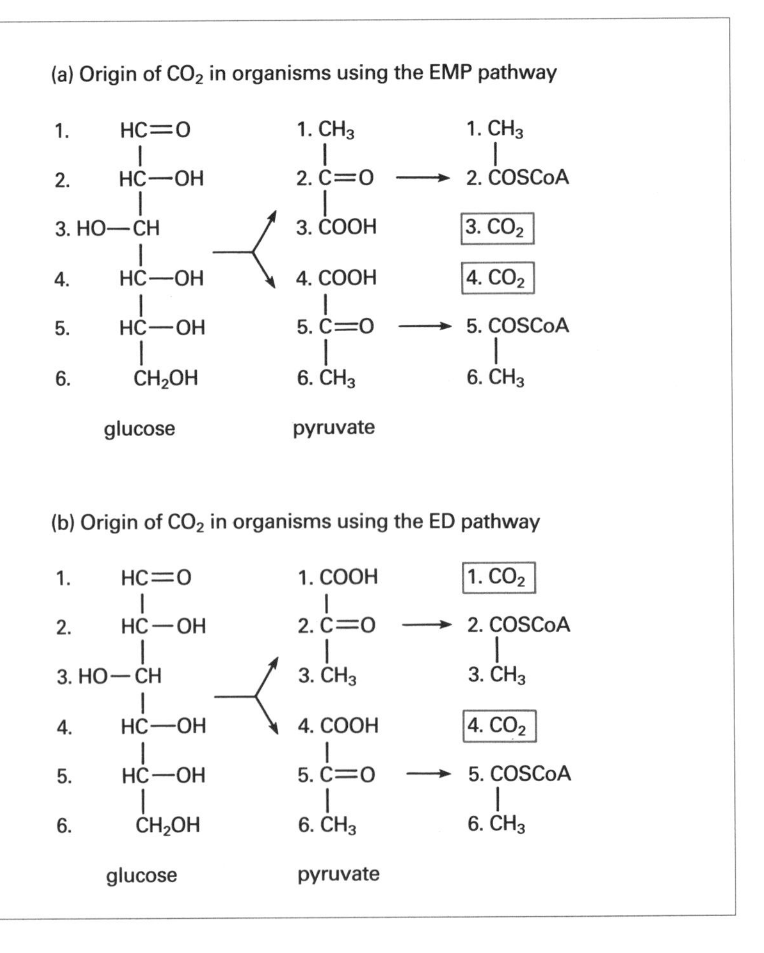


**Figure 4.6** The use of the technique of radiorespirometry to establish whether the glycolytic (EMP) or Entner–Doudoroff (ED) pathway operates in an organism. Cells of the organism are allowed to grow in the presence of glucose which is labelled with the radioisotope $^{14}C$ in one of the six carbon atoms, and the $CO_2$ evolved is examined for the presence of radioactivity. By repeating this procedure with different glucose molecules, each labelled at a different C atom, the origin of the $CO_2$ may be established. The figure shows that both the EMP and ED pathways produce $CO_2$ from C atom number 4 of glucose. However, organisms using the EMP pathway also produce $CO_2$ from C number 3 of glucose, whereas those using the ED pathway produce $CO_2$ from C number 1.

## The oxidative pentose phosphate cycle

Although the glycolytic pathway is widely distributed in nature, a second series of metabolic reactions using glucose as its substrate is usually found in the same cell. In fact, when an *Escherichia coli* cell is growing on glucose, 70% of the glucose is broken down by the glycolytic pathway and 30% by the so-called **oxidative pentose phosphate cycle** (Figure 4.7).

You may ask yourself, why the need for two quite distinct pathways for the breakdown of glucose in the same cell? We may find some clues if we examine the reaction sequence in Figure 4.8. If we ignore the details of the reactions, two general aspects strike us. First, the hydrogen atom acceptor for glycolysis is $NAD^+$ and for the oxidative pentose phosphate cycle is $NADP^+$. The significance of this difference is that $NADH_2$ is oxidized during operation of the electron transport chain and leads to ATP synthesis. Conversely, $NADPH_2$ is required as a source of reducing power for biosynthetic reactions.

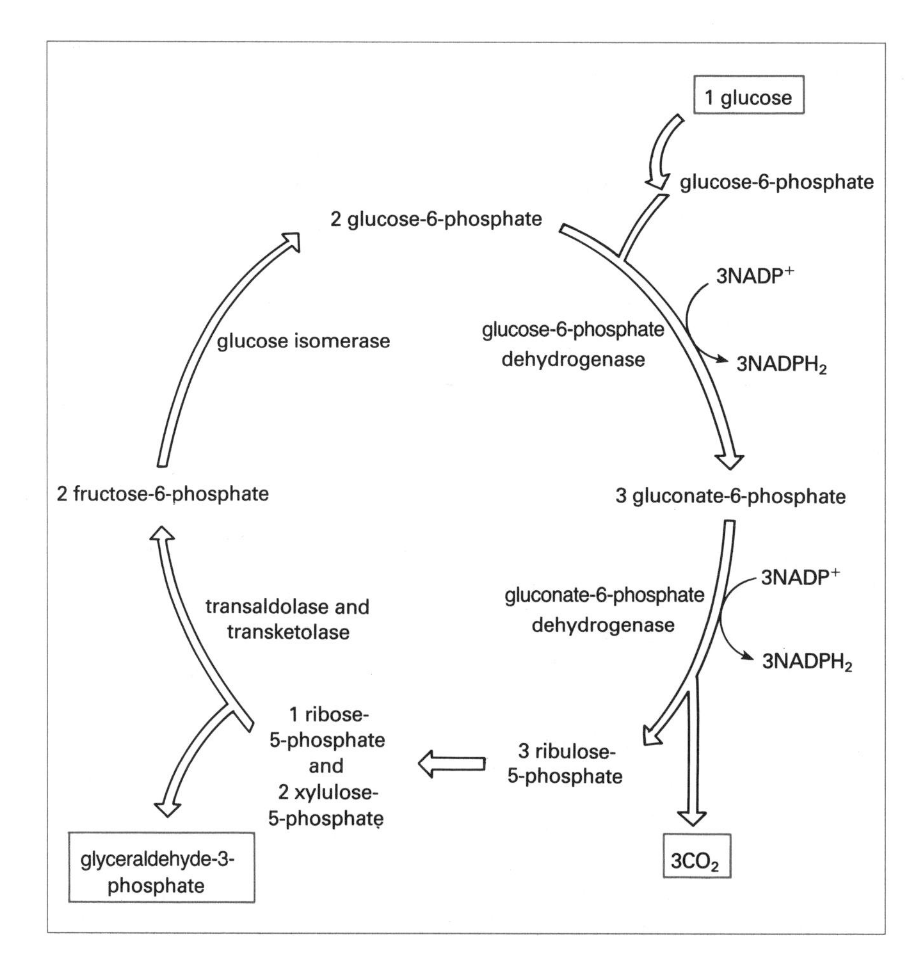


**Figure 4.7** The oxidative pentose phosphate cycle. Glyceraldehyde-3-phosphate is either completely oxidized by the TCA cycle to $CO_2$ and water or it re-enters the pentose phosphate cycle as half a molecule of glucose-6-phosphate.

The pathway for the breakdown of glucose, described here, is the *oxidative* pentose phosphate cycle. The pathway for autotrophic carbon dioxide fixation, usually called the Calvin cycle, is the *reductive* pentose phosphate cycle.

Because acetic acid bacteria (*Acetobacter* and *Gluconobacter*) lack a complete TCA cycle, the glyceraldehyde-3-phosphate produced by the pentose phosphate cycle is oxidized to acetic acid and excreted. This is the basis of the commercial production of vinegar.

Second, five-carbon sugar phosphates (pentose phosphates) are intermediates of the pentose phosphate cycle but not of glycolysis. Operation of the former pathway means two things. It permits an organism to use a pentose (such as ribose, ribulose, arabinose) as a substrate as well as a hexose (such as glucose). In addition, when growing on glucose, ribose phosphate, an intermediate of the pentose phosphate cycle, can be used for the formation of ribonucleotides and deoxyribonucleotides – the building blocks of RNA and DNA respectively.

## Respiration

As we have seen, during the fermentation of organic substrates the amount of energy released and ATP synthesized is small. There are two reasons for this. First, the carbon atoms of the organic substrate are only partially oxidized and second, the difference in reduction potential between the electron donor and electron acceptor is small.

Not all microbes are restricted to a fermentative way of life and many can completely oxidize organic substrates to carbon dioxide. Such complete oxidation or **respiration** requires an *external* electron acceptor. This is usually oxygen but may be a nitrate or a sulphate. If complete oxidation can be achieved, the amounts of energy released and ATP synthesized are far greater, and the limitations of fermentation are overcome. Complete oxidation of the substrate is accomplished and the difference in reduction potential between electron donor and acceptor is relatively large.

Let us look first at the pathway by which organic carbon is oxidized to carbon dioxide and then the mechanism of electron transport and ATP synthesis.

### Oxidation of organic substrates: the tricarboxylic acid (TCA) cycle

The initial stages of glucose oxidation lead to the production of pyruvate. In the absence of an external electron acceptor, pyruvate is fermented to products such as $CO_2$ and ethanol. In the presence of an external electron acceptor, however, pyruvate can be completely oxidized to $CO_2$ by means of the **tricarboxylic acid (TCA) cycle**.

The reactions of the TCA cycle, which is also known as the **Krebs cycle** after one of its discoverers (and the **citric acid cycle** after one of its characteristic intermediates), are shown in Figure 4.8.

Before entering the cycle, pyruvate is first **oxidatively decarboxylated** in a complex reaction catalysed by the enzyme complex, pyruvate dehydro-

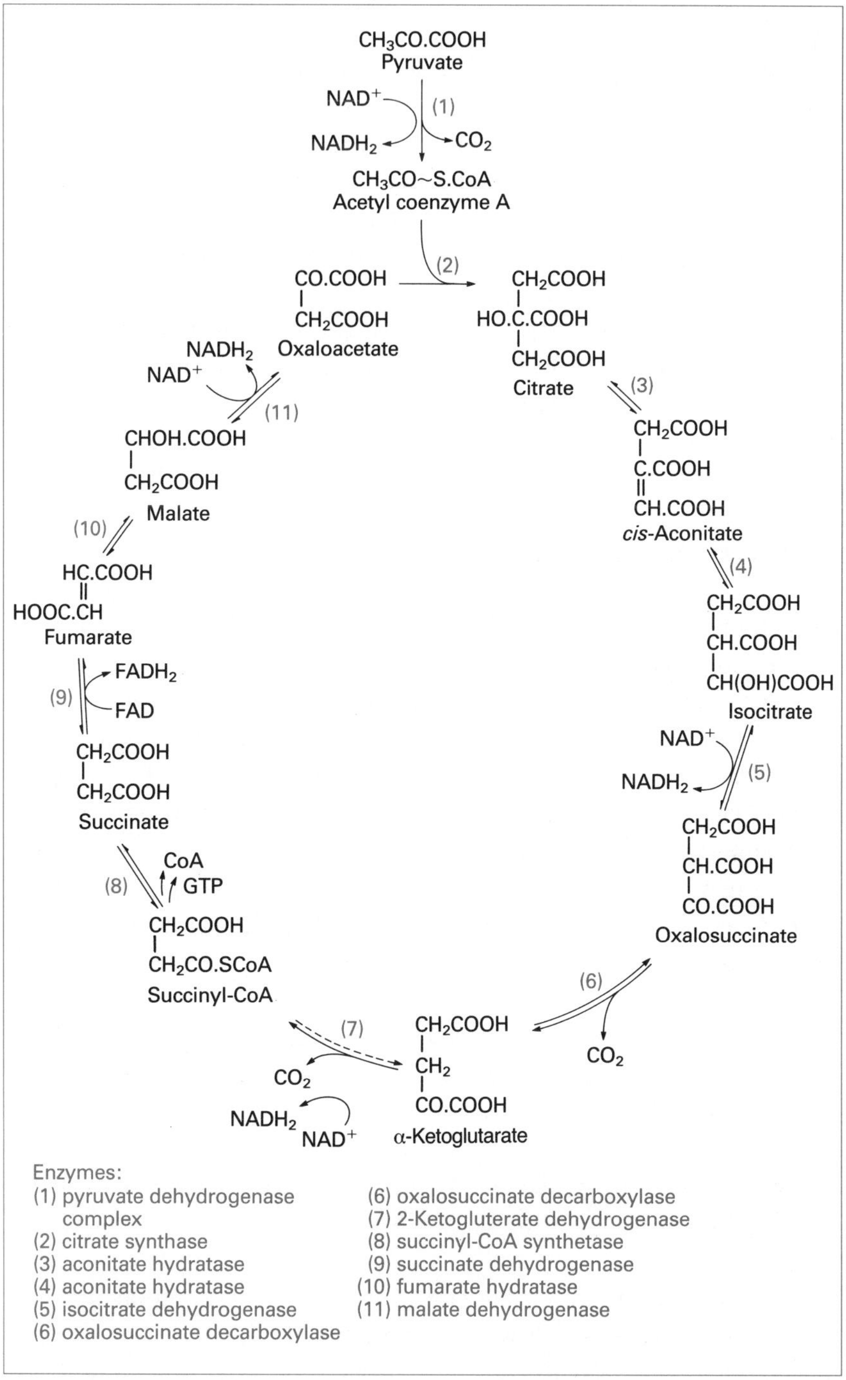

$CH_3CO.COOH$
Pyruvate
$NAD^+$ (1) $NADH_2$ $CO_2$
$CH_3CO{\sim}S.CoA$
Acetyl coenzyme A
(2)
$CO.COOH$ | $CH_2COOH$ Oxaloacetate
$CH_2COOH$ | $HO.C.COOH$ | $CH_2COOH$ Citrate
(3)
$CH_2COOH$ | $C.COOH$ || $CH.COOH$ *cis*-Aconitate
(4)
$CH_2COOH$ | $CH.COOH$ | $CH(OH)COOH$ Isocitrate
$NAD^+$ $NADH_2$ (5)
$CH_2COOH$ | $CH.COOH$ | $CO.COOH$ Oxalosuccinate
(6) $CO_2$
$CH_2COOH$ | $CH_2$ | $CO.COOH$ α-Ketoglutarate
(7) $CO_2$ $NADH_2$ $NAD^+$
$CH_2COOH$ | $CH_2CO.SCoA$ Succinyl-CoA
(8) CoA GTP
$CH_2COOH$ | $CH_2COOH$ Succinate
(9) $FADH_2$ FAD
$HC.COOH$ || $HOOC.CH$ Fumarate
(10)
$CHOH.COOH$ | $CH_2COOH$ Malate
(11) $NADH_2$ $NAD^+$

Enzymes:
(1) pyruvate dehydrogenase complex
(2) citrate synthase
(3) aconitate hydratase
(4) aconitate hydratase
(5) isocitrate dehydrogenase
(6) oxalosuccinate decarboxylase
(6) oxalosuccinate decarboxylase
(7) 2-Ketogluterate dehydrogenase
(8) succinyl-CoA synthetase
(9) succinate dehydrogenase
(10) fumarate hydratase
(11) malate dehydrogenase

**Figure 4.8** The tricarboxylic acid (TCA) cycle.

genase. During this reaction, carbon dioxide is released and $NAD^+$ reduced to $NADH_2$. The product of the reaction is an acetyl unit linked to coenzyme A (acetyl-CoA). The energy of acetyl-CoA is used to drive the condensation of the acetyl unit with a four-carbon acceptor molecule, oxaloacetate. The product of this reaction is the six-carbon tricarboxylic acid, citric acid.

As a result of further metabolism of citrate, the two carbon atoms of the acetyl group are released as carbon dioxide, and the hydrogen atoms reduce four further molecules of coenzyme. Through the cycle, oxaloacetate is eventually regenerated and can then act as an acceptor for the entry of further acetyl units into the cycle.

During operation of the TCA cycle, the three carbon atoms of pyruvate are completely oxidized to carbon dioxide, and its four hydrogen atoms used to reduce the coenzymes $NAD^+$ or flavin adenine dinucleotide (FAD). Reduced coenzymes are then oxidized by an **electron transport** or **respiratory chain** using an external terminal electron acceptor such as $O_2$, $NO_3^-$ or $SO_4^{2-}$. It is during this transport of electrons along a chain of carriers that ATP is synthesized.

## The electron transport chain

An electron transport chain is composed of a number of biochemical carriers located in cell membranes. Respiratory electron transport chains are located within the mitochondria (Chapter 2) of eukaryotic organisms or within the cytoplasmic membrane of bacteria. The role of an electron transport chain is twofold; first, since as we now know, the amount of a coenzyme such as $NAD^+$ in a cell is limited, the oxidation of reduced coenzymes by the electron transport chain regenerates oxidized coenzymes from the corresponding reduced compound. Second, some of the energy released during electron transport is used to synthesize ATP.

The electron transport chain has a number of components which fall into five general categories: (1) $NADH_2$ dehydrogenases; (2) flavoproteins; (3) cytochromes; (4) iron–sulphur proteins; and (5) quinones. We shall look at each of these briefly.

**(a) $NADH_2$ dehydrogenases.** These are enzymes that catalyse the transfer of hydrogen atoms from $NADH_2$ to flavoproteins.

**(b) Flavoproteins.** These contain a flavin prosthetic group which is reduced by accepting a hydrogen atom from $NADH_2$ and oxidized by loss of an electron. Two flavins are known, flavin mononucleotide (FMN) and flavin adenine dinucleotide (FAD).

**(c) Cytochromes.** These are proteins with iron-containing porphyrin rings attached to them (Figure 4.9). They may be alternatively reduced and

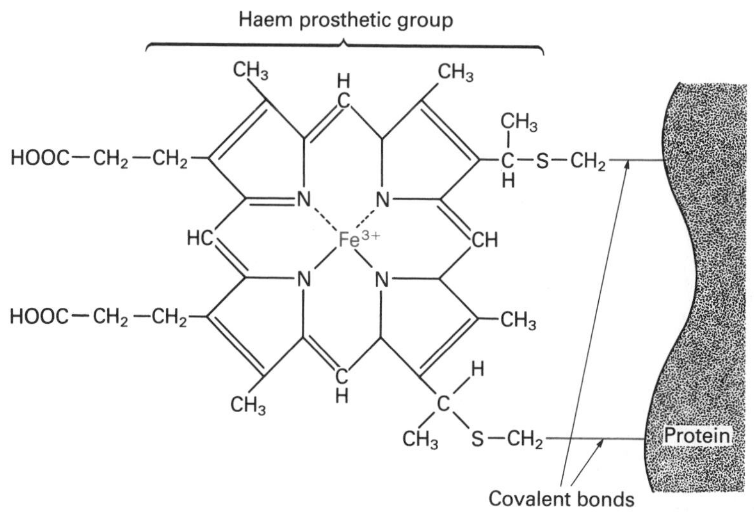

**Figure 4.9** Structure of cytochrome c. In cytochrome c, the haem prosthetic group is covalently linked to protein. In other cytochromes, non-covalent bonds are involved.

oxidized by the addition and loss of a *single electron* to the iron atom at the centre of the molecule:

$$\text{cytochrome-Fe}^{2+} \rightleftharpoons \text{cytochrome-Fe}^{3+} + e^{-}$$

Several classes of cytochrome are found in living organisms and, with different reduction potentials, electrons may be transferred from one cytochrome to another with a more positive reduction potential.

**(d) Iron–sulphur proteins**. In contrast to the cytochromes, the iron of this group of carriers is not linked to porphyrin rings. Instead, iron atoms are linked to free sulphur and to protein through the sulphur atoms of cysteine residues (Figure 4.10). Iron–sulphur proteins have a range of reduction potentials depending on the number and arrangements of the sulphur and iron atoms. Like the cytochromes, these carriers are electron carriers, not hydrogen carriers.

**(e) Quinones**. These are non-protein carriers and, because they are soluble in lipids, are able to diffuse through membranes transferring electrons from iron–sulphur proteins to cytochromes. Like the flavoproteins, quinones are acceptors of hydrogen atoms, not electron donors.

### ATP synthesis during respiratory electron transport

In contrast to fermentative processes, ATP synthesis during respiratory electron transport occurs by the process of **oxidative phosphorylation**. It

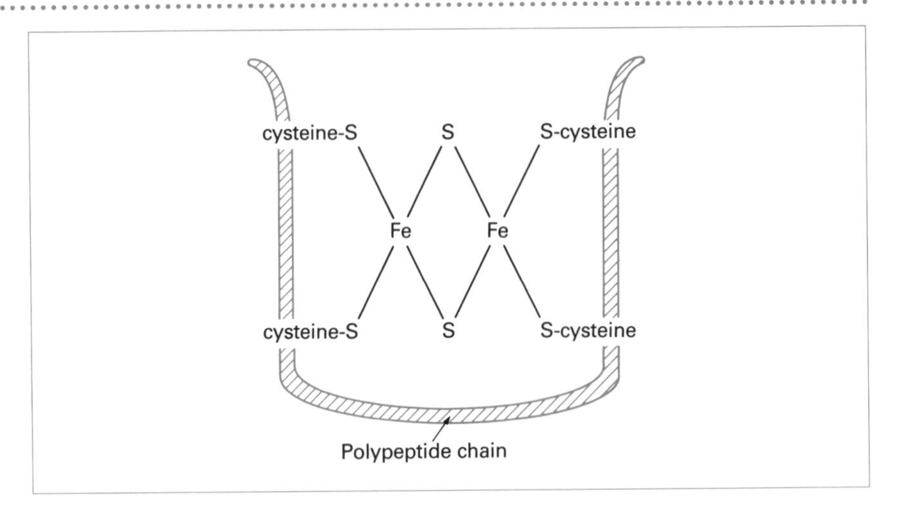


**Figure 4.10** The centre of an iron–sulphur protein. This consists of two Fe and two S atoms. Iron is bound to cysteine residues of the polypeptide chain and to sulphide.

Mitchell's chemiosmotic theory and its wide-reaching impact in the field of bioenergetics was recognized by the award of a Nobel prize in 1978.

was not until 1961 that the British biochemist Peter Mitchell put forward his **chemiosmotic theory** to explain the mechanism of oxidative phosphorylation. Since then, all available experimental evidence has supported Mitchell's idea.

The essence of Mitchell's theory is that protons ($H^+$ ions) are moved or translocated from one side of a membrane to the other. In this way a proton gradient is established across a membrane during electron transport reactions. The proton motive force (or **proticity**) generated can be likened to electron motive force (or electricity) and is used to do work, including the synthesis of ATP.

**Chemiosmosis**. An essential feature of the chemiosmotic process is the arrangement of components of the electron transport system within the inner mitochondrial membrane of eukaryotes or the cytoplasmic membrane of bacteria. The electron transport components are arranged *asymmetrically* so that some are accessible from one side of the membrane and the remainder are accessible from the other. As a result, protons are separated from electrons during the transportation process. A hydrogen atom removed from a hydrogen carrier such as $NADH_2$ is split into a proton and an electron inside the membrane. The electron is passed to the cytoplasmic side of the membrane by appropriate carriers and the proton is expelled or extruded to the cell's surroundings (Figure 4.11).

The final step in aerobic respiration is the reduction of oxygen to water. To achieve this, electrons are supplied by the electron transport chain and the protons are provided by the dissociation of water molecules present in the cell cytoplasm:

$$H_2O \rightleftharpoons OH^- + H^+$$

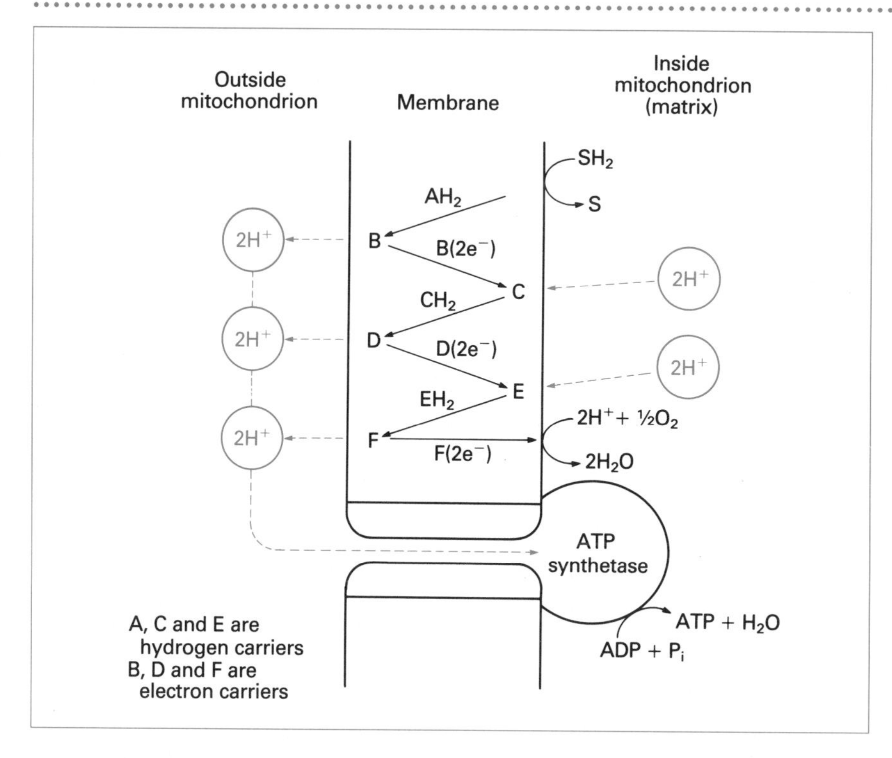


**Figure 4.11** Model for loop mechanism proposed by Mitchell to explain chemiosmosis. Hydrogen and electron carriers alternate to form three loops. Oxidation of substrate ($SH_2$) releases hydrogen atoms which are transferred by a hydrogen carrier, such as A. At the outer surface of the membrane an electron carrier (B) accepts electrons from $AH_2$ and protons ($H^+$) are released. Electrons are transferred back into the membrane and together with protons from the matrix side of the membrane reduce the hydrogen carrier, C, thus initiating a second loop. This continues until the terminal electron acceptor (for example $O_2$) is reached. Protons translocated to the outside of the membrane can re-enter it through a pore in the $F_0$ subunit of ATP synthetase and ATP is synthesized (Figure 4.14).

As a consequence of the need for protons to reduce oxygen, $OH^-$ ions accumulate on the inside of the membrane. In other words, $H^+$ ions (protons) accumulate on one side of the membrane and $OH^-$ ions on the opposite side. Since the membrane is impermeable to the free passage of $H^+$ and $OH^-$ ions, equilibrium cannot be restored spontaneously and conditions on opposite sides of a membrane vary in two important respects. First, there will be more protons on the outside of the membrane than the inside: this may be expressed in terms of a **pH gradient**. Second, since protons and $OH^-$ ions are also electrically charged, an **electrical potential** across the membrane will be produced with the inside being more electronegative and alkaline and the outside being more electropositive and acidic.

These two components, pH gradient and electrical potential, energize the membrane in a way which may be likened to an electrical battery. In the same way that the power (or electron motive force) of a battery is expressed in volts, the proton motive force of a membrane is also expressed in volts. The proton motive force generated during electron transport can be used to do work for living organisms including ion transport, bacterial flagella rotation and ATP synthesis.

Central to the mechanism of proton translocation is the asymmetrical or **vectorial arrangement** of certain carriers within the membrane. The

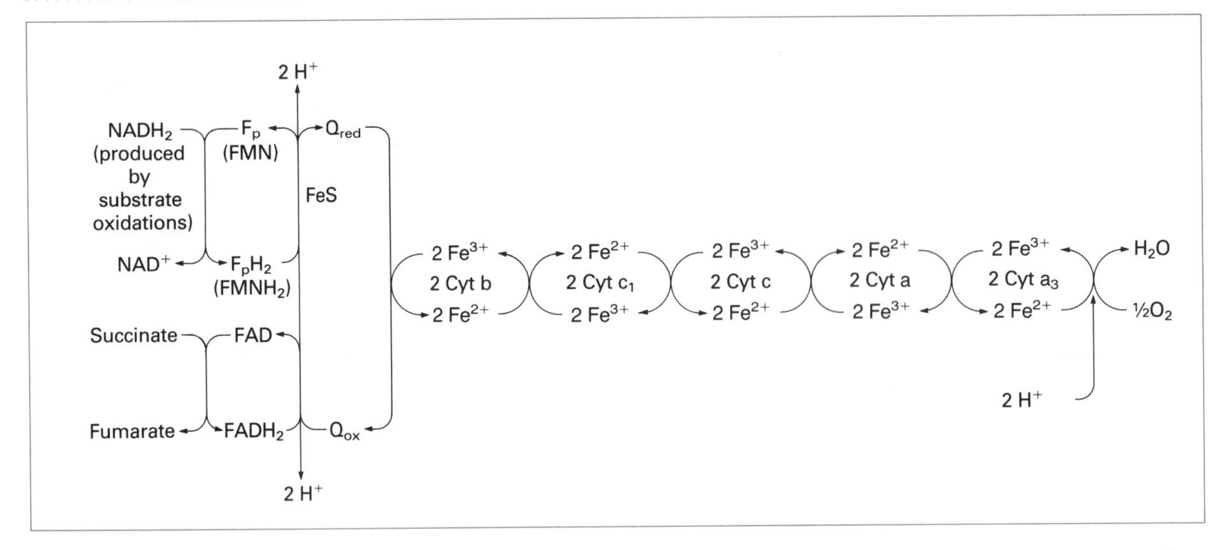


**Figure 4.12** The electron transport chain. Electrons from a variety of reduced coenzymes are transferred *via* a number of carriers, including cytochromes, FAD and a flavoprotein (Fp), to a terminal electron acceptor such as $O_2$. During transfer of electrons, protons are translocated and a proton motive force is generated.

flavin enzymes and the quinones, which are hydrogen atom acceptors but electron donors, are arranged in a membrane in such a way that protons are expelled to the outside of the membrane and electrons are transferred to the next carrier in the chain.

The components of the electron transport chain are shown in Figure 4.12, and an examination of their organization will show how proton translocation is accomplished. FMN is a hydrogen atom carrier and is reduced by the transfer of a pair of hydrogen atoms from $NADH_2$. The next carrier in the chain is an iron–sulphur protein which, you will remember, is an electron carrier only, so that when it is reduced by $FMNH_2$, the two protons cannot be passed to the iron–sulphur protein but are expelled into the surrounding environment instead. In the next step a hydrogen atom carrier, the quinone coenzyme Q, is reduced; the two electrons required to achieve this reduction may be donated by the preceding iron–sulphur protein but the two protons required must be taken up from the cytoplasm.

At this stage in electron transport, a **proton motive quinone cycle** or **Q cycle** accomplishes the extrusion of four protons (as two pairs of protons) during the course of two successive single electron transfers from the preceding iron–sulphur protein to cytochrome (Figure 4.13). Cytochrome oxidase then serves as a vectorial electron carrier from cytochrome c, to catalyse reaction with molecular oxygen during aerobic respiration. As we stated earlier, protons for this final reduction are supplied by the dissociation of water and give rise to a build-up of $OH^-$ ions on the inside of the membrane. The electron transport chain described above is found in the mitochondrion. The electron transport chains of bacteria differ in detail from that described. However, all electron transport chains translocate protons and generate a proton motive force.

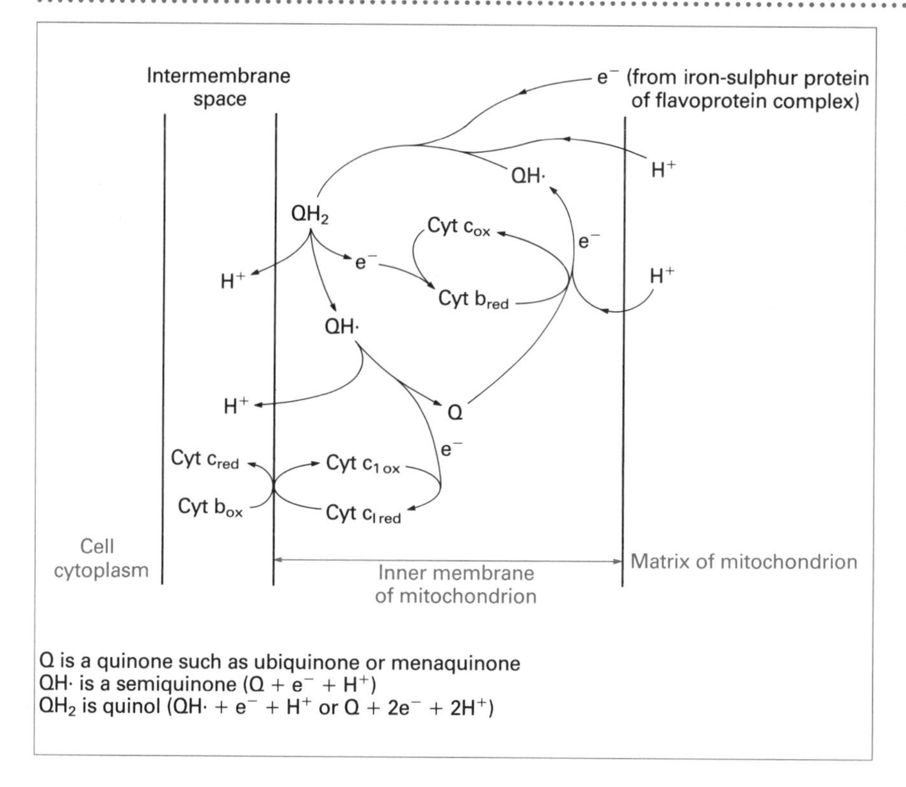


**Figure 4.13** The proton motive quinone cycle. This was proposed by Mitchell to account for proton translocation at those mitochondrial sites where hydrogen carriers and electron carriers do not alternate and the loop mechanism is, therefore, not possible.

## The mechanism of ATP synthesis during oxidative phosphorylation

As you may have guessed already, the use of a proton motive force to synthesize ATP requires the presence of a specific enzyme. The enzyme concerned was originally known as an ATPase because of its ability to hydrolyse ATP rather than for its capacity to synthesize it. So, when catalysing the formation of ATP, ATPase is known by the more appropriate name of **ATP synthetase**. The enzyme is complex and comprises two major parts, a multi-subunit, $F_0$, located on the inside of the membrane, and a proton-conducting tailpiece, $F_1$, which spans the membrane (Figure 4.14). Re-entry of protons from the outside of the membrane into the cytoplasm through the $F_0$ proton channel makes available a steady supply of energy to synthesize ATP.

In the reverse reaction, ATP is hydrolysed to ADP and inorganic phosphate with the release of three protons. In this way, the energy of ATP is converted into a different form of energy represented by a proton motive force. This is important because some cell processes, including flagella rotation and solute transport, require a membrane potential to drive them. Consequently, ATPase is also found in microorganisms, such as *Lactobacillus* and *Streptococcus*, which are unable to respire at all.

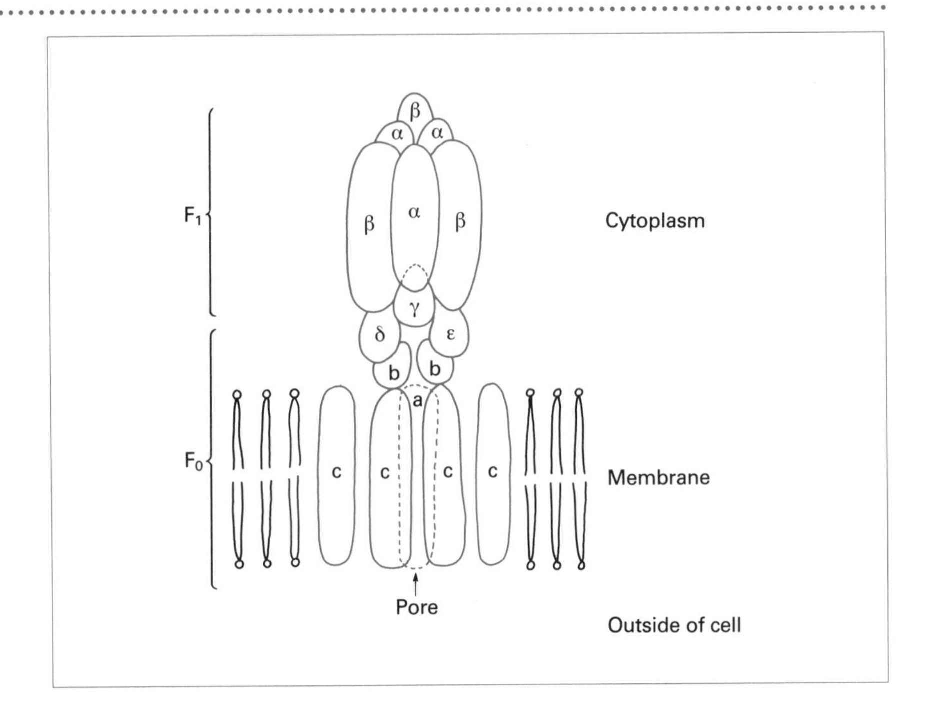


**Figure 4.14** Model of subunit organization of ATP synthetase from *Escherichia coli*.

## Energetics of respiration

You will recall that glucose is oxidized to two molecules of pyruvate by the reactions of the glycolytic pathway (page 74). During this process, there is a net synthesis of two molecules of ATP and two molecules of $NADH_2$.

During the operation of the TCA cycle, pyruvate is completely oxidized and the three carbon atoms of pyruvate are expelled as carbon dioxide. A maximum of 38 molecules of ATP are synthesized during the respiratory oxidation of one molecule of glucose by the glycolytic pathway, TCA cycle and electron transport chain (Table 4.5). This compares with a net gain of two molecules of ATP during the fermentative breakdown of glucose.

Since ATP contains an energy of approximately 28 kJ/mole, 1064 kJ (28 × 38 kJ) of energy can be converted to high-energy phosphate bonds in ATP by the complete oxidation of glucose. Compare this figure with the 2752 kJ/mole of free energy available from the complete oxidation of glucose by oxygen. Aerobic respiration is, then, approximately 39% efficient. The remaining 1688 kJ/mole are lost as heat. In contrast, the efficiency of lactic fermentation is almost 50%. Despite the fact that fermentation is energetically more thermodynamically efficient, the amount of energy conserved as ATP is considerably lower during fermentation than during respiration.

At this stage, we should remind you that in addition to its role in ATP synthesis, the TCA cycle also provides precursors for the biosynthesis of several amino acids, the porphyrins and fatty acids.

**Table 4.5** Net yield of ATP from the complete oxidation of glucose

| *Pathway* | *ATP synthesized per molecule of glucose* |
|---|---|
| **Glycolysis** | |
| Glucose → 2 pyruvate (ATP synthesized by substrate level phosphorylation) | 2 |
| Oxidation of 2 $NADH_2$ | 6 |
| **Activation of pyruvate to produce acetyl-CoA** | |
| Oxidation of 2 $NADH_2$ | 6 |
| **TCA cycle** | |
| Oxidation of 6 $NADH_2$ formed by the oxidation of two molecules each of isocitrate, α-ketoglutarate and malate | 18 |
| Oxidation of 2 $FADH_2$ formed in the oxidation of two succinates | 4 |
| Two succinyl-CoA → 2 succinate (ATP synthesized by substrate level phosphorylation) | 2 |
| Net yield | 38 ATP/glucose |

These values represent the theoretical maximum values in eukaryotes. Values for bacteria are smaller.

## Anaerobic respiration

In the absence of oxygen, a number of kinds of bacteria can use alternative terminal electron acceptors such as sulphate or nitrate. In some kinds of bacteria, called **facultative anaerobes**, this **anaerobic respiration** arises only if oxygen happens to be unavailable. Some other kinds of bacteria are actually unable to use oxygen and live as **obligate anaerobes**.

During anaerobic respiration the energy released from a particular electron donor is less than if the same compound were oxidized using oxygen as electron acceptor. The reason is that the reduction potential of the $O_2/H_2O$ couple is more positive than that of alternative electron acceptors. The reduction potentials for $Fe^{3+}$, $NO_3^-$ and $NO_2^-$ are near to that of oxygen but those of $CO_3^{2-}$ and $SO_4^{2-}$ are more electronegative.

**Nitrate respiration**. As an example of an anaerobic respiration, we shall describe the use of nitrates as terminal electron acceptors. Nitrates constitute some of the most common alternative electron acceptors in anaerobic respiration. The products formed from the reduction of nitrate are nitrous oxide ($N_2O$), nitric oxide (NO) and nitrogen gas ($N_2$). Because all these products are gases, they lead to the loss of nitrogen from an environment such as the soil. The process is called **denitrification**.

The first step in nitrate respiration is catalysed by the enzyme **nitrate reductase**. This unusual enzyme contains the element molybdenum. Synthesis of dissimilatory nitrate reductase occurs only under anaerobic conditions (dissimilation is the process of breakdown). Its formation is actually repressed by oxygen. The product of nitrate reductase activity is nitrite. Nitrite is excreted by most staphylococci and enterobacteria and not reduced any further.

Reduction of nitrite ($NO_2^-$) is catalysed by the enzyme **nitrite reductase**, with the eventual production of either ammonia or $N_2$. Reduction of nitrite to ammonia is carried out by several kinds of bacteria. If nitrate is reduced to nitrogen gas, two intermediates are involved: nitric oxide (NO) and nitrous oxide ($N_2O$). In some bacterial species, the reduction process proceeds only as far as nitrous oxide, but others produce $N_2$ as the gaseous product. The electron transport process involved in nitrate respiration in *Paracoccus denitrificans* is shown in Figure 4.15.

The biological breakdown of an organic substrate, such as glucose, is sometimes described as **dissimilation**. The synthesis of large molecules from small ones is **assimilation**. Because nitrate reduction can occur in either situation, it is important to indicate which process is meant. Nitrate respiration permits the anaerobic, respiratory breakdown of glucose and requires *dissimilatory* nitrate reductase. Nitrate is also used as a nitrogen source for the synthesis of proteins and nucleic acids and requires the reduction of nitrate: in this situation *assimilatory* nitrate reductase is used.

## Photosynthesis

Although many microbes oxidize organic or inorganic chemicals to obtain energy, a large number are able to convert the energy of sunlight into chemical energy and use it to synthesize ATP by a process known as **photophosphorylation**. In addition to green plants, a large number of microorganisms are able to photosynthesize (Table 4.6). These include both eukaryotic forms (the algae; Chapter 9) and some bacteria. Organisms that photosynthesize commonly have the capacity to use carbon dioxide as their

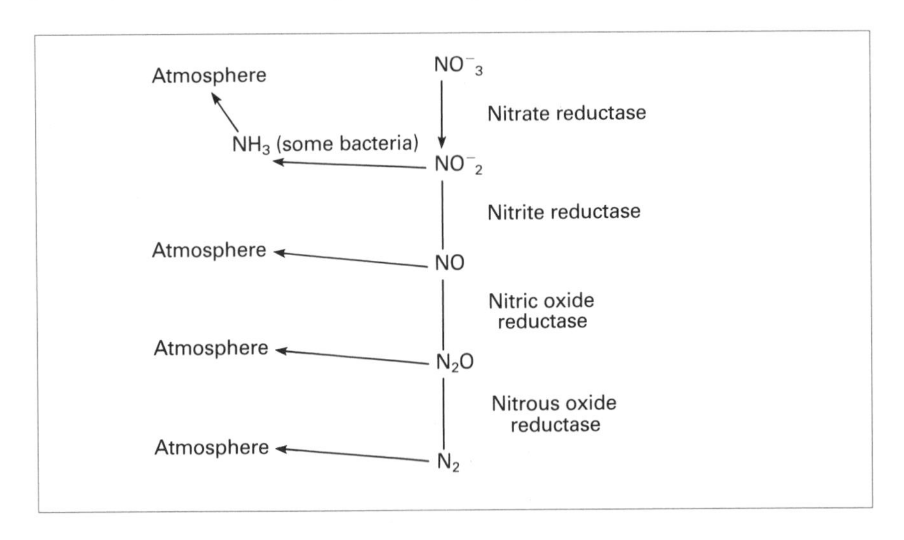


**Figure 4.15** Nitrate respiration. The pathway illustrated is provisional and some authorities believe that nitric oxide (NO) is not involved.

**Table 4.6** Photosynthetic microorganisms

| | *$O_2$ evolution* | *H-source* | *C-source* | *Classification* |
|---|---|---|---|---|
| Algae | Oxygenic | $H_2O$ | $CO_2$ | Photolithotrophs |
| Cyanobacteria (Blue-green bacteria) | Oxygenic | $H_2O$ | $CO_2$ | Photolithotrophs |
| Chlorobiaceae (Green S-bacteria) | Anoxygenic | $H_2$, $H_2S$, $S_2O_3^{2-}$ | $CO_2$ (acetate, butyrate) | Primarily photolithotrophs |
| Chromatiaceae (Thiorhodaceae, Purple S-bacteria) | Anoxygenic | $H_2$, $H_2S$, $S_2O_3^{2-}$ | $CO_2$ (acetate, butyrate) | Primarily photolithotrophs |
| Rhodospirillaceae (Athiorhodaceae, Purple non-S-bacteria | Anoxygenic | $H_2$, organic | Organic ($CO_2$) | Primarily photoorganotrophs but can grow as chemoorganotrophs aerobically in dark |
| Chloroflexaceae (Green gliding bacteria) | Anoxygenic | $H_2$, organic | Organic ($CO_2$) | Primarily photoorganotrophs but can grow as chemoorganotrophs aerobically in dark |
| Halobacteriaceae | NA | Organic | Organic | Photoorganotrophs |

NA, not applicable

only source of carbon. Such **autotrophic** organisms, as they are called, are able to synthesize all the organic molecules they need for growth and reproduction from carbon dioxide. Indeed, certain **blue-green bacteria** or **cyanobacteria** are also able to use nitrogen gas as their sole nitrogen source and have the most elemental nutritional requirements of any organism.

The photosynthetic process is a complex one and in photosynthetic autotrophs can be divided into two distinct sets of reactions: the **light reactions** in which light energy is converted into chemical energy and the **dark reactions** in which, for example, carbon dioxide is reduced to organic compounds using ATP synthesized in the light reactions. The reducing power required is provided by reduced coenzymes such as $NADPH_2$ which, in turn, are formed by the reduction of $NADP^+$ using a variety of electron donors.

Green plants, algae and cyanobacteria use water as an electron source, splitting it by a process called **photolysis** to produce the desired electrons, and oxygen as a waste product. Green and purple bacteria are unable to split water and use alternative sources for their electrons, including hydrogen gas and hydrogen sulphide. Others use organic molecules.

These bacterial forms of photosynthesis are anaerobic processes and oxygen is not produced as a waste product. The term **oxygenic** (oxygen-producing) is often used to describe the type of photosynthesis carried out by green plants and cyanobacteria, and the term **anoxygenic** (not oxygen-producing) to describe the type shown by green and purple photosynthetic bacteria.

### Chlorophylls and bacteriochlorophylls

With the exception of the photosynthetic archaebacteria, photosynthetic organisms contain some form of chlorophyll. Like the cytochromes, chlorophylls are porphyrins but contain a magnesium rather than an iron atom at the centre of the ring. In addition, various side-chains are bonded to the porphyrin ring. In particular, a long hydrophobic alcohol side-chain enables the chlorophyll molecule to associate with the lipid and hydrophobic proteins of the photosynthetic membranes.

The principal chlorophyll of higher plants, most algae and the cyanobacteria is **chlorophyll a** (Figure 4.16). Chlorophylls appear green because they *absorb* red and blue light (maximally at wavelengths of 680 nm and 430 nm respectively; Figure 4.17). However, a number of additional chlorophylls are known in nature which differ slightly from chlorophyll a in their chemical structures and in the wavelengths of light which are maximally absorbed. A summary of the types of chlorophylls and their absorption maxima is given in Table 4.7.

### Accessory pigments

In addition to chlorophylls and bacteriochlorophylls, a variety of additional pigments play a supporting role in the photosynthetic process. The most widespread of these accessory pigments are the **carotenoids**. These are membrane-located, water-soluble pigments with long, hydrocarbon chains of alternating C–C and C=C bonds. Carotenoids are yellow, red or green

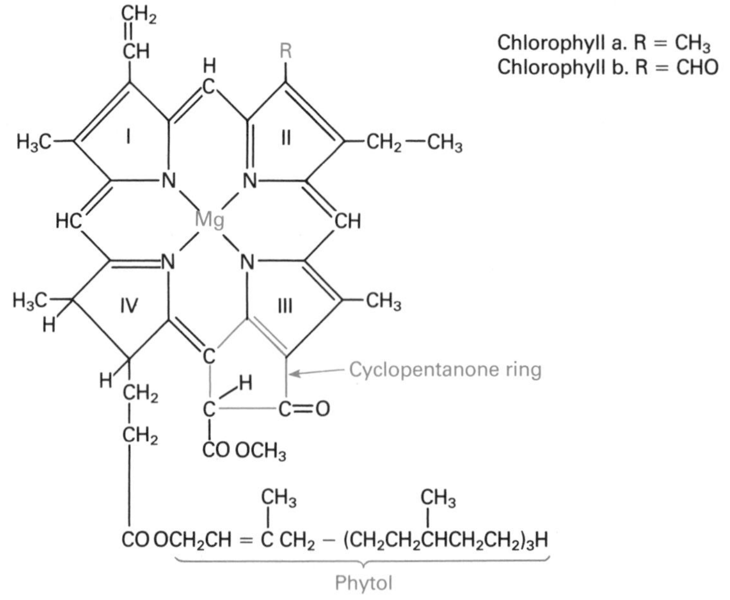

**Figure 4.16** The structure of the chlorophyll a and b molecules. Chlorophyll molecules, like cytochromes, are porphyrins but chlorophyll contains a magnesium atom instead of an iron atom at the centre of the porphyrin ring. In addition to the porphyrin ring, a cyclopentanone ring and a long hydrophobic alcohol (phytol) make up the completed chlorophyll molecule.

Chlorophyll a. R = $CH_3$
Chlorophyll b. R = CHO

Cyclopentanone ring

Phytol

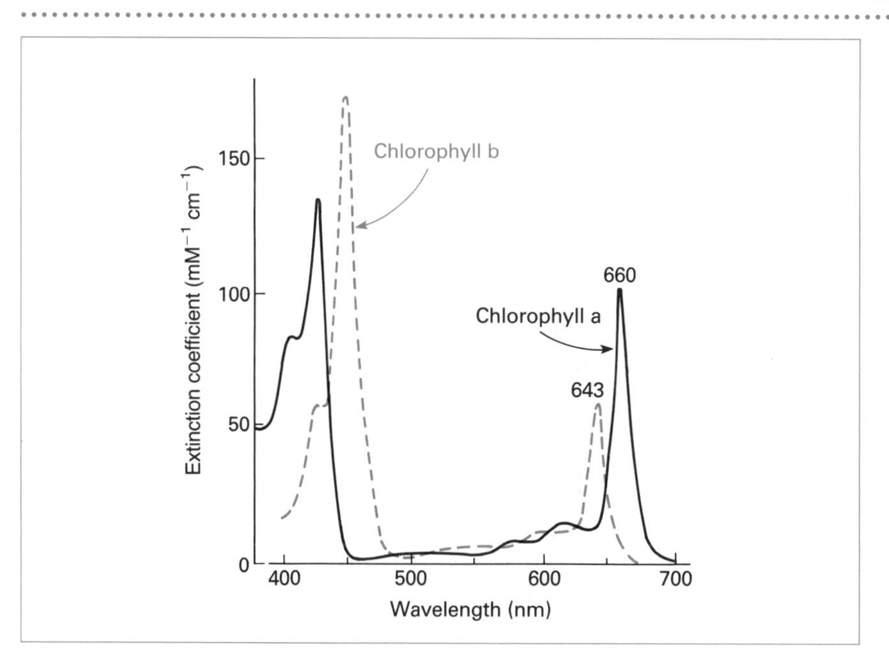


**Figure 4.17** The absorption spectra of chlorophylls a and b extracted in acetone.

in colour and absorb blue light, the energy of which they are able to transfer by fluorescence to chlorophyll molecules closely associated with them. Such transferred energy may then be used for ATP synthesis in the same way as that captured directly by chlorophyll.

Fluorescence is a process in which light energy absorbed by a molecule is re-emitted from the same molecule at a longer wavelength.

**Phycobiliproteins**, a second type of accessory pigment, are found in red algae and cyanobacteria. For example, phycoerythrin is red and absorbs light maximally at 550 nm; phycocyanin is blue and absorbs maximally at 620–640 nm. Phycobiliproteins are open-chain tetrapyrroles, called **phycobilins**, coupled to proteins. They occur as aggregates called **phycobilisomes**. These are closely linked to chlorophyll molecules and

**Table 4.7** Absorption maxima of plant and bacterial photosynthetic pigments

| *Pigment* | *Wavelength [nm]* | *Occurrence* |
|---|---|---|
| Chlorophyll a | 430, 670 | All green plants |
| Chlorophyll b | 455, 640 | Higher plants; green algae |
| Chlorophyll c | 445, 625 | Diatoms; brown algae |
| Bacteriochlorophyll | 365, 605, 770 | Purple and green bacteria |
| α-Carotene | 420, 440, 470 | Leaves; some algae |
| β-Carotene | 425, 450, 480 | Some plants |
| γ-Carotene | 440, 460, 495 | Some plants |
| Luteol | 425, 445, 475 | Green leaves; red and brown algae |
| Violaxanthol | 425, 450, 475 | Some leaves |
| Fucoxanthol | 425, 450, 475 | Diatoms; brown algae |
| Phycoerythrins | 490, 546, 576 | Red and blue-green algae |
| Phycocyanins | 618 | Red and blue-green algae |
| Allophycoxanthin | 654 | Red and blue-green algae |

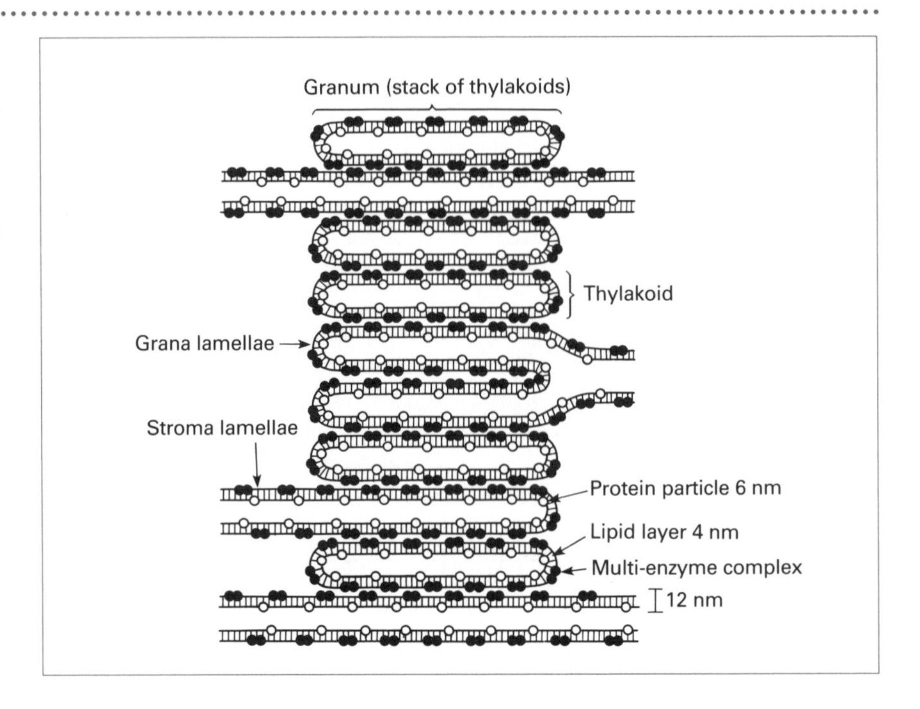


**Figure 4.18** Thylakoid structure of chloroplasts. Lamellae which contribute to the formation of **grana** (singular: **granum**) are called **grana lamellae**. Granum means 'a grain' and these structures appear as granules under the microscope. The parts of the lamellae between grana are called **stroma lamellae**.

facilitate almost complete energy transfer. Accessory pigments have a second function in that they prevent damage to the photosystem by bright light. They also increase the range of wavelengths which can be used for photosynthesis.

## Photosynthetic membranes

The photosynthetic apparatus is associated with highly developed photosynthetic membranes. In eukaryotic organisms, these membranes are located in bodies called **chloroplasts** (Chapter 2). Within the chloroplasts, the chlorophyll molecules are attached to membranes called **thylakoids** (Figure 4.18). Thylakoids divide the chloroplast into two separate regions, enabling protons to be translocated from one region to the other, generating the proton motive force necessary to synthesize ATP.

Within the chloroplasts, groups of between 200 and 300 chlorophyll molecules are grouped together on the thylakoid membranes. Of these, a small number participate directly in the conversion of light energy into ATP. These are the **reaction centre** chlorophyll molecules and they receive energy from the more numerous **light-harvesting** (or **antenna**) chlorophyll molecules (Figure 4.19).

Included among the important components of the photosynthetic apparatus are proteins which orientate the various chlorophyll molecules in the thylakoid membranes so that light energy absorbed by one chlorophyll molecule may be efficiently transferred to another.

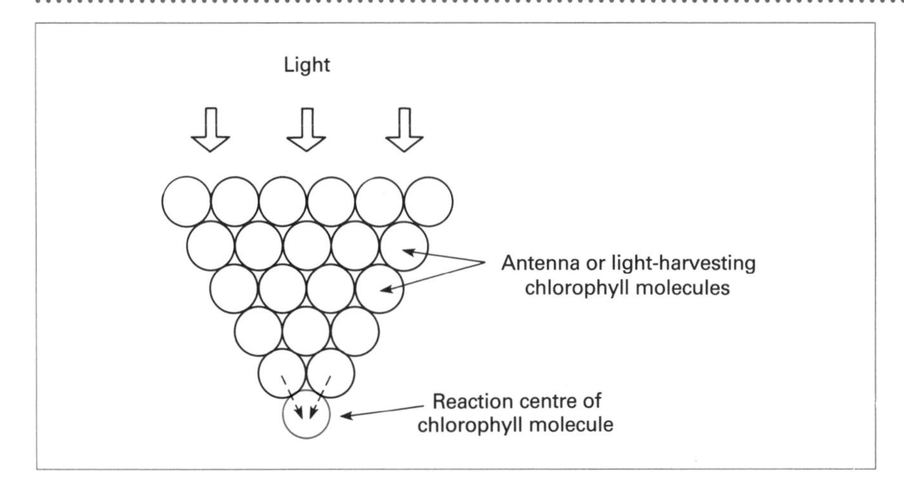


**Figure 4.19** The photosynthetic unit. **Antenna (light-harvesting) chlorophyll molecules** transfer their excitation energy to a specialized chlorophyll at the reaction centre, and it is this which takes part in further reactions involving electron transport. This funnelling of energy absorbed by a number of antenna molecules to a single reaction centre chlorophyll allows an organism to respond to different intensities of light.

Photosynthetic bacteria do not possess chloroplasts. Their chlorophyll molecules are located in a variety of internal membrane systems.

## Anoxygenic photosynthesis

The green and purple photosynthetic bacteria are unable to use water as a reductant in the photosynthetic process, and so do not evolve oxygen during photosynthesis. Bacteria which carry out anoxygenic photosynthesis are normally classified into four families: Chlorobiaceae, Chromatiaceae, Rhodospirillaceae and Chloroflexaceae. The first two groups contain bacteria which tend to grow as **photolithotrophs**. They use $CO_2$ as their sole carbon source and a variety of reductants, including hydrogen gas, and sulphur-containing inorganic and organic compounds (such as hydrogen sulphide, thiosulphate and sulphur).

Bacteria classified in the Rhodospirillaceae or Chloroflexaceae tend to grow as **photosynthetic heterotrophs** (or **photoorganotrophs**) and require organic molecules as reductants. Organisms belonging to these last two groups can, in the absence of light and under aerobic conditions, also live as **chemosynthetic heterotrophs (chemoorganotrophs**). Some of the properties of anoxygenic photosynthetic bacteria are summarized in Table 4.6.

The use of purple bacteria in particular as experimental organisms for the investigation of the light reactions of photosynthesis has a number of advantages over more complex organisms. Because they are unicellular, populations of purple photosynthetic bacteria can be mutated (that is their genetic material can be altered). The mutant cells produced will be abnormal in different parts of the photosynthetic process. By comparing the properties of such mutant cells with those of normal (non-mutated) cells, information can be produced about the details of the light reactions.

An analogy of this kind of approach to investigate biological phenomena would be to investigate the working of the internal combustion engine by removing different components (spark plugs, distributor and so on) and observing the result. For this reason, the extent of our knowledge concerning the light reactions in these organisms is more advanced than in others, and this is where we start our description of the light reactions.

The photosynthetic apparatus of purple photosynthetic bacteria is embedded in various kinds of intracytoplasmic membranes, including membrane vesicles or **chromatophores**. The photosynthetic apparatus consists of four membrane-bound pigment–protein complexes, plus a proton-translocating ATP synthetase for the synthesis of ATP. Two of the pigment–protein complexes are light-harvesting systems, the third is the reaction centre, and the fourth is a cytochrome b–c complex which is also used for respiratory electron flow when the bacteria are growing aerobically in the dark.

The reaction centre has been crystallized and its structure elucidated using X-ray diffraction techniques. Results from these investigations show the presence of:

1. A pair of bacteriochlorophyll a molecules called the **special pair**.
2. A further pair of bacteriochlorophyll a molecules which do not appear to participate in the photosynthetic process and which, for this reason, are called **'voyeur' bacteriochlorophylls**.
3. Two molecules of **bacteriophaeophytin** (bacteriochlorophyll without the magnesium atom) and two molecules of quinone.
4. In addition, a number (usually three) of polypeptide molecules bound to the photochemical complex.

## Electron flow during photosynthesis

The light reactions start when a quantum of light strikes the light-harvesting bacteriochlorophyll a molecules, which transfer their energy to bacteriochlorophyll molecules of the special pair. The primary photochemical event then occurs when the absorption of light energy converts the special pair into a **strong reductant** and the reduction potential of the reaction centre is reduced from approximately +0.5 volts to –0.7 volts. This reduction in potential represents work done on the system by light energy and drives the reduction of bacteriophaeophytin, which is an acceptor molecule of very low potential.

Once reduced, bacteriophaeophytin reduces a quinone molecule located towards the outer surface of the photosynthetic membrane. From quinone, electrons are transported in the membrane by a number of

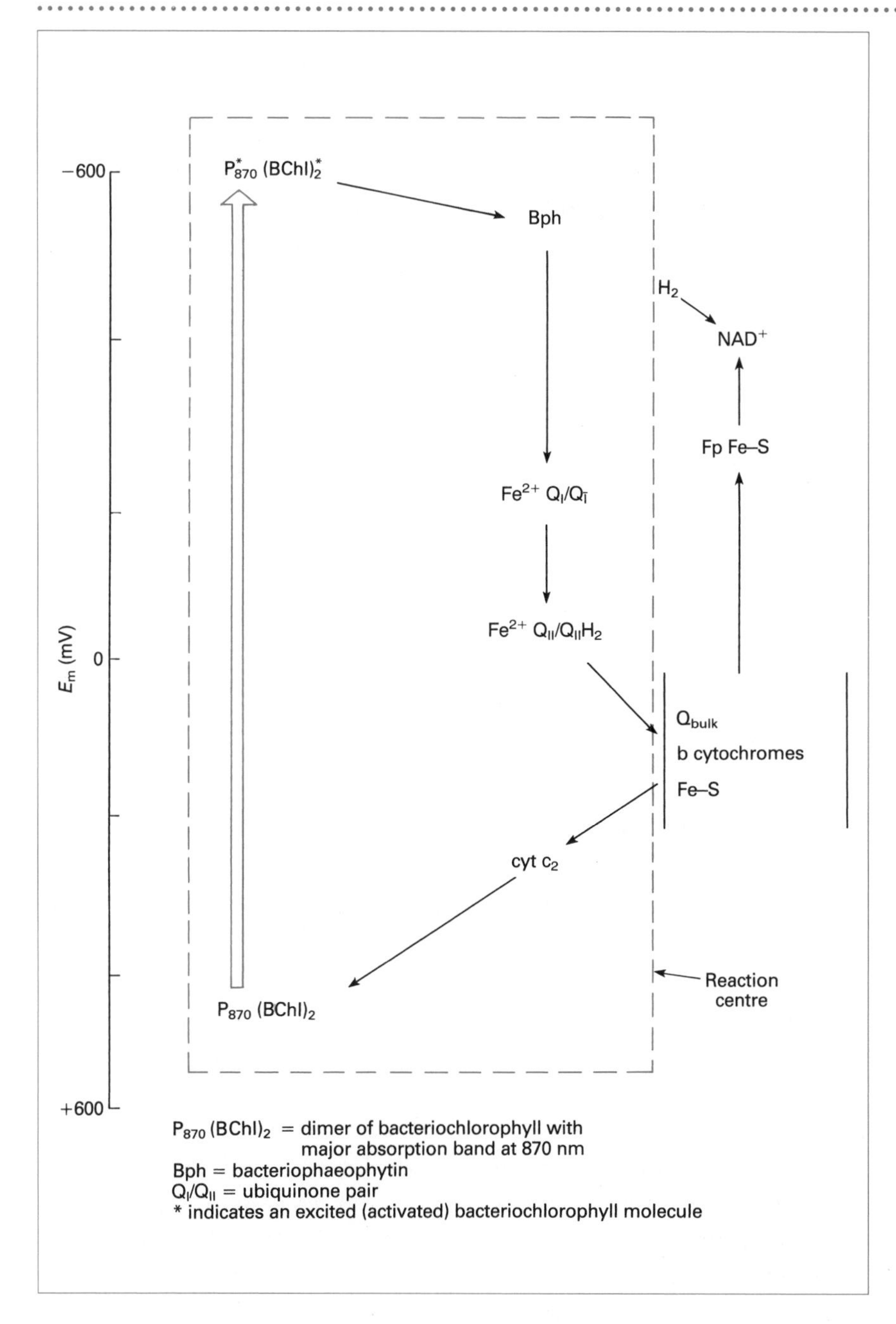


**Figure 4.20** Reaction centre of the purple photosynthetic bacterium, *Rhodopseudomonas spheroides*, and cyclic electron transfer.

iron–sulphur proteins and cytochromes, eventually returning to the same reaction centre (Figure 4.20).

## Photophosphorylation

The synthesis of ATP during the light reaction of photosynthesis occurs by a chemiosmotic process; that is to say, a proton motive force is generated by electron transport which can then be used by ATP synthetase for ATP synthesis. Electrons expelled from the reaction centre return to it by the electron transport chain and restore the bacteriochlorophyll of the special pair to its original reduction potential of +0.5 volts. In other words, electron flow during anoxygenic photosynthesis is a cyclic process and this method of ATP synthesis is called **cyclic photophosphorylation**. In contrast to respiration, there is no net change in the number of electrons present in the system.

As a result of electron transfer, a proton gradient is built up by the translocation of protons across the photosynthetic membrane and into the central space of the chromatophore. The transfer of one electron causes the translocation of three protons across the membrane. It is important to appreciate that the only contribution that light makes is to produce a strong reductant. The remaining reactions are not light-dependent but are thermodynamically favourable electron transfers.

## Reduction of $NADP^+$ during anoxygenic photosynthesis

Reduced coenzymes are required by green and purple photosynthetic bacteria for the reduction of carbon dioxide to produce cell material. You will remember that these bacteria use a variety of compounds to reduce $NADP^+$ to $NADPH_2$ (Table 4.6).

The way in which electrons from these compounds are used to reduce $NADP^+$ depends on their reduction potential. If the reduction potential of the external donor is more negative than the $NADP^+/NADPH_2$ couple (−0.32 volts), direct transfer of electrons can occur. This is the situation when hydrogen gas is an electron donor. Since its reduction potential (−0.42 volts) is lower than that of $NADP^+$, direct reduction is possible provided that the enzyme **hydrogenase** is also present.

If the external reductant has a reduction potential more positive than the $NADP^+/NADPH_2$ couple, such direct reduction is not possible. Under such conditions, electrons usually enter the photosynthetic electron transport chain at the level of the cytochromes. Energy, produced in the light reactions, is then used to force electrons backwards against the electrochemical gradient to reduce $NADP^+$ to $NADPH_2$. This method of producing $NADPH_2$ is called **energy-dependent reverse electron-flow** and is also used by chemosynthetic autotrophs for the same purpose.

## Oxygenic photosynthesis

The production of oxygen during the light reactions of photosynthesis is a consequence of using water as a reductant for carbon dioxide fixation.

The ability to use water as a source of electrons for the reduction of $NADP^+$ to $NADPH_2$ requires the operation of a second photosystem, called photosystem II, which is not found in organisms that exhibit anoxygenic photosynthesis. During oxygenic photosynthesis, light is responsible for the generation of both ATP and $NADPH_2$. Photosystem II catalyses the photolysis of water to produce electrons for the reduction of $NADP^+$ and oxygen is a waste product. The two photosystems are characterized by the presence of spectrally different chlorophyll a molecules in their reaction centres. Photosystem I contains a form of chlorophyll a (called P700 to indicate the wavelength of light which is maximally absorbed) and Photosystem II contains a type designated P680.

As we have seen with anoxygenic photosynthesis, oxygenic photosynthesis leads to ATP synthesis by a chemiosmotic process, which means that the photosynthetic apparatus of green plants and algae are located in chloroplast membranes, whereas those of the cyanobacteria are located within stacks of membranes within the cytoplasm. As we saw with our description of the organization of the reaction centre of the purple bacteria, proteins also play an important role in the organization of the photochemically active components of the reaction centre of oxygenic organisms.

The pathway of electron flow during oxygenic photosynthesis is shown in Figure 4.21. As you can see, the scheme as drawn looks rather like a letter Z turned on its side and for this reason a representation of electron flow during oxygenic electron flow is often called the **Z scheme**.

The reduction potential of P680 of Photosystem II is high (approximately +1 volt) and a little greater than that of the $O_2/H_2O$ couple (+0.82 volts). Following absorption of a quantum of light near 680 nm, an electron from water is passed to P680. As a consequence, P680 is converted to a fairly strong reductant which is able to reduce an acceptor molecule with a reduction potential of approximately +0.2 volts. The identity of this carrier is not known for certain but it may be **phaeophytin a** (chlorophyll a without its magnesium atom). From here, electrons are transferred along a chain of carriers, many of which are similar to those found in anoxygenic photosynthetic organisms, and include cytochromes, quinones and iron–sulphur proteins. In addition, the copper-containing protein **plastoquinone** donates electrons, originating in water, to Photosystem I.

Photosystem I has a role comparable with the photosystem of the green and purple bacteria. Following absorption of a quantum of light, the reaction centre chlorophyll of Photosystem II (P700) donates an electron to a primary acceptor with a very negative reduction potential of about –0.75 volts. Again, the identity of the primary electron acceptor of Photosystem I is uncertain but it may be an alternative form of chlorophyll a. This primary acceptor transfers an electron to the iron–sulphur protein, ferredoxin, which is then able to reduce $NADP^+$ to $NADPH_2$.

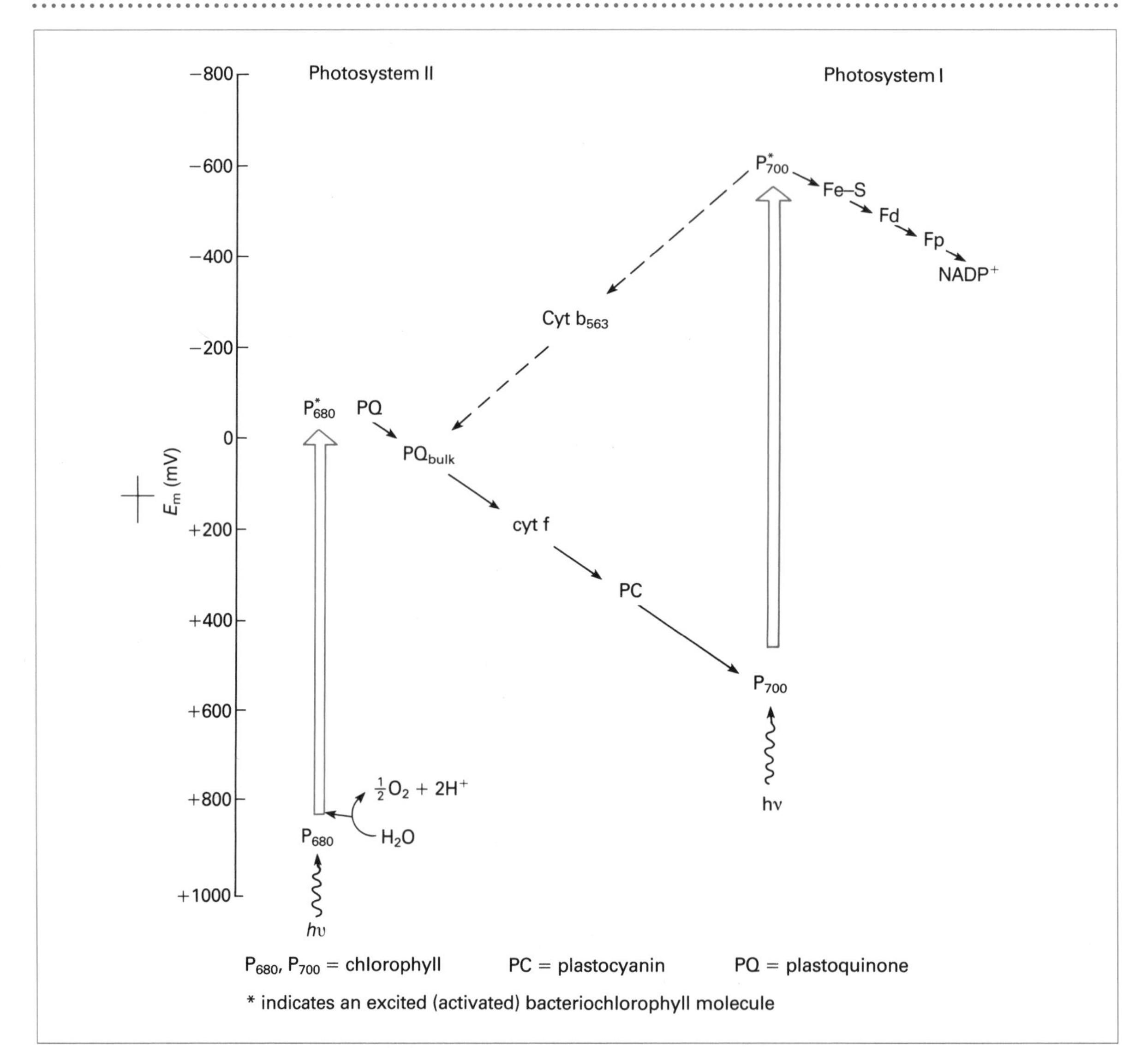


**Figure 4.21** Routes of photodependent electron transfer during oxygenic photosynthesis involving two photosystems and oxidation of $H_2O$.

## Photophosphorylation during oxygenic photosynthesis

During the transfer of electrons from the acceptor for Photosystem II to the reaction centre of Photosystem I, a proton gradient is generated which may be used to synthesize ATP. This type of ATP synthesis is called **non-cyclic photophosphorylation** because electrons expelled from Photosystem II *do not* return to it; instead they pass initially to Photosystem I and are eventually used to reduce $NADP^+$.

When sufficient $NADPH_2$ is present, electrons expelled from Photosystem I can return to it by way of the primary acceptor and cytochromes b and f. The path of electron flow is indicated by the dashed line in Figure 4.21. Under these conditions, the flow of electrons is cyclic, a proton gradient is produced, and ATP is synthesized by the process of **cyclic photophosphorylation**.

Under anaerobic conditions, many algae and some cyanobacteria are able to photosynthesize anoxygenically, using Photosystem I only, in a manner very similar to photosynthesis in green and purple bacteria. Under anaerobic conditions, algae and cyanobacteria require an external reductant to reduce $NADP^+$: this is often hydrogen gas in algae, and hydrogen sulphide in cyanobacteria.

The presence of a second photosystem in oxygenic photosynthetic organisms suggests that these may have evolved from anoxygenic organisms, and the evolutionary relationship between the groups of photosynthetic organisms is emphasized by the ability of certain algae and cyanobacteria to photosynthesize anoxygenically under appropriate environmental conditions.

### Autotrophic $CO_2$ fixation

In the preceding sections we have seen how photosynthetic organisms are able to use the energy of sunlight to synthesize ATP from ADP and inorganic phosphate during the operation of the light reactions of photosynthesis. We now move on to look at the **dark reactions** by which both chemosynthetic and photosynthetic autotrophs use $CO_2$ as a carbon source for the synthesis of cell material.

Autotrophic organisms are characterized by an ability to use $CO_2$ alone for the synthesis of all their organic molecules. The formation of organic material from $CO_2$ involves reduction of the carbon source. This overall process is often called **$CO_2$ fixation**. All reactions involved can occur in complete darkness using $NADPH_2$ and ATP formed either during the light reactions of photosynthesis or by the oxidation of inorganic compounds.

The most widespread pathway for $CO_2$ fixation appears to be the **Calvin cycle** (**reductive pentose cycle**). The technique that Calvin used to investigate the reactions involved in $CO_2$ fixation included the use of the newly available radioactive isotope of carbon, $^{14}C$, which he used to discover the nature of the first formed products of $CO_2$ fixation in a variety of unicellular algae including *Chlorella* and *Scenedesmus*.

## The Calvin cycle

Although the Calvin cycle contains a number of enzyme-catalysed reactions, only two of the enzymes concerned are peculiar to it. The rest of the

**Figure 4.22** The Calvin cycle of $CO_2$ fixation: key reactions.

(a) Phosphoribulokinase

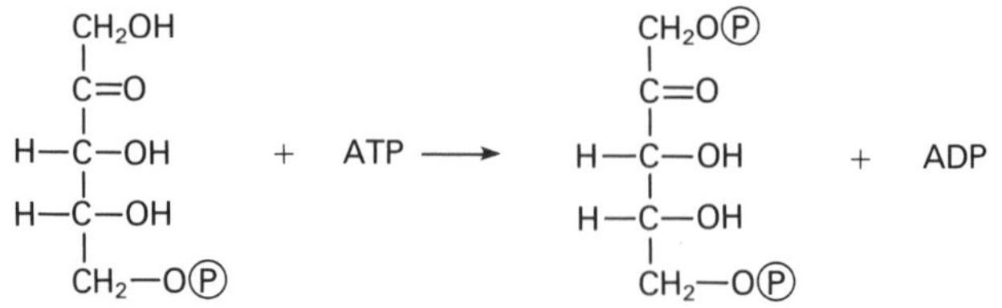

$$\begin{array}{c}CH_2OH\\|\\C{=}O\\|\\H{-}C{-}OH\\|\\H{-}C{-}OH\\|\\CH_2{-}O(P)\end{array} + ATP \longrightarrow \begin{array}{c}CH_2O(P)\\|\\C{=}O\\|\\H{-}C{-}OH\\|\\H{-}C{-}OH\\|\\CH_2{-}O(P)\end{array} + ADP$$

*Ribulose-5-(P)* *Ribulose-1,5-bis(P)*

(b) Carboxydismutase (ribulose bisphosphate carboxylase)

$$CO_2 + \begin{array}{c}CH_2O(P)\\|\\CO\\|\\H{-}C{-}OH\\|\\H{-}C{-}OH\\|\\CH_2O(P)\end{array} + H_2O \longrightarrow \begin{array}{c}CH_2O(P)\\|\\H{-}C{-}OH\\|\\COOH\\+\\COOH\\|\\H{-}C{-}OH\\|\\CH_2O(P)\end{array}$$

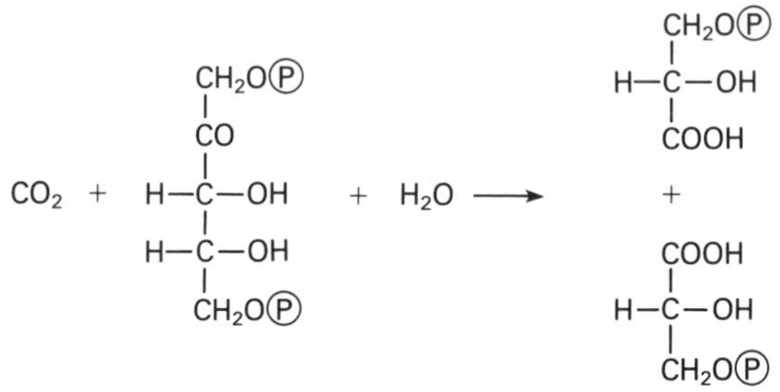

*Ribulose-1,5-bis(P)* *Glycerate-3-(P)*

Overall reactions:

Fixation 6 Ru-1,5-bis(P) + $6CO_2 + 6H_2O \rightarrow$ 12 PGA

Reduction 12 PGA + 12 ATP + 12 $NADPH_2$

12 glyceraldehyde-3-(P) + 12ADP + 12(P) + 12 $NADP^+$ + 12 $H_2O$

Regeneration 12 glyceraldehyde-3-(P) + 6 ATP$\rightarrow$

6 Ru-1,5-bis(P) + 6ADP + 5(P) + F-6-(P)

Sum $6CO_2 + 6H_2O$ + 18ATP + $12NADPH_2 \rightarrow$ F-6-(P) + 18ADP + $12NADP^+$ + 17(P)

enzymes are also involved in the operation of pathways such as the glycolytic pathway and the oxidative pentose phosphate pathway.

The first step in $CO_2$ fixation is a reaction catalysed by one of the two key enzymes of the Calvin cycle, **ribulose bisphosphate carboxylase (RuBisCo)**. In this cycle, $CO_2$ and ribulose bisphosphate react to form two molecules of 3-phosphoglyceric acid (PGA; Figure 4.22), the first identifiable product in the $CO_2$ reduction process. In further reactions, PGA is first activated by ATP and then reduced by $NADPH_2$ to produce glyceraldehyde-3-phosphate, which is reduced to the same extent as carbohydrates.

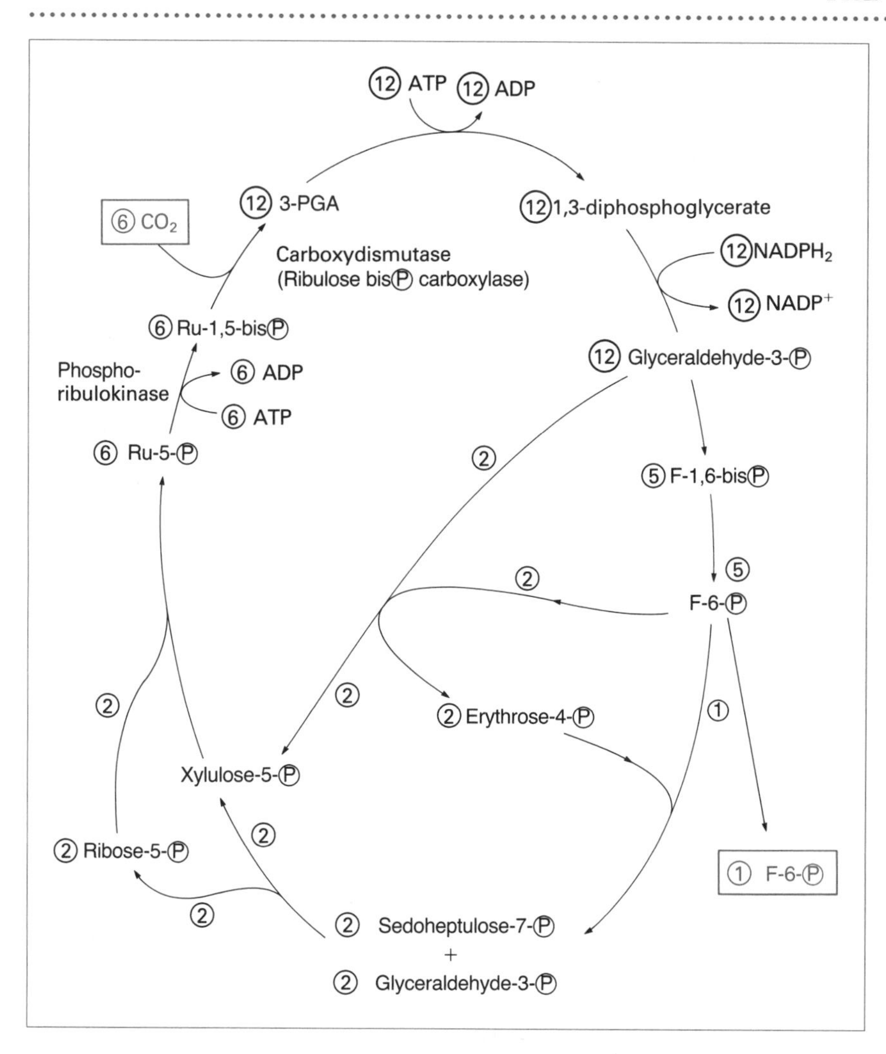


**Figure 4.23** The Calvin cycle of $CO_2$ fixation.

The second key enzyme of the Calvin cycle is **phosphoribulokinase**, which catalyses the synthesis of ribulose bisphosphate from ATP and ribulose-5-phosphate. The overall scheme of the Calvin cycle is shown in Figure 4.23, where fructose-6-phosphate is shown as the biosynthetic product of $CO_2$ fixation.

## The reductive TCA cycle

Although the Calvin cycle is widespread among autotrophic organisms, it does not appear to be responsible for the fixation of $CO_2$ in the green sulphur bacteria. This group of bacteria uses a completely different pathway, which is a reversal of the reactions of the TCA cycle and, for this reason, is known as the **reductive TCA cycle** (Figure 4.24).

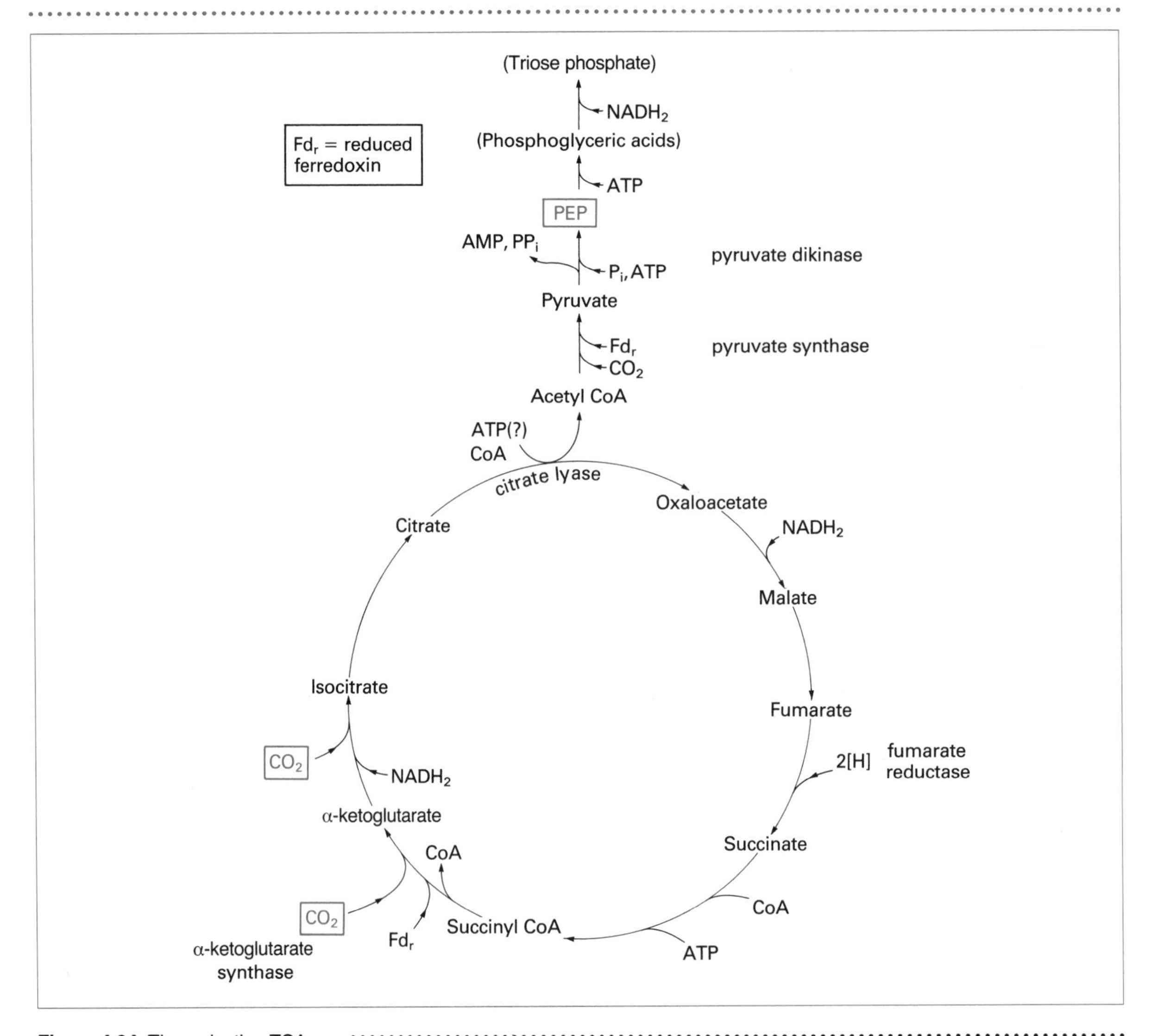


**Figure 4.24** The reductive TCA cycle. This is used for $CO_2$ fixation by a small number of green sulphur-bacteria, for example *Chlorobium thiosulphatophilum.*

## Archaebacterial photosynthesis

From our description of the light reactions of photosynthesis it will be apparent how important chlorophyll and bacteriochlorophyll are in harvesting light energy and converting it to chemical energy. However, one group of photosynthesizing bacteria does not contain either type of photosynthetic molecule. The extremely halophilic **archaebacteria** are a group of prokaryotes which grow in environments with high sodium chloride concentration. Typically these bacteria are able to grow in natural environments such as salt-lakes, or on the surfaces of certain artificially salted foods, where the concentration of sodium chloride may be as high as three or four molar.

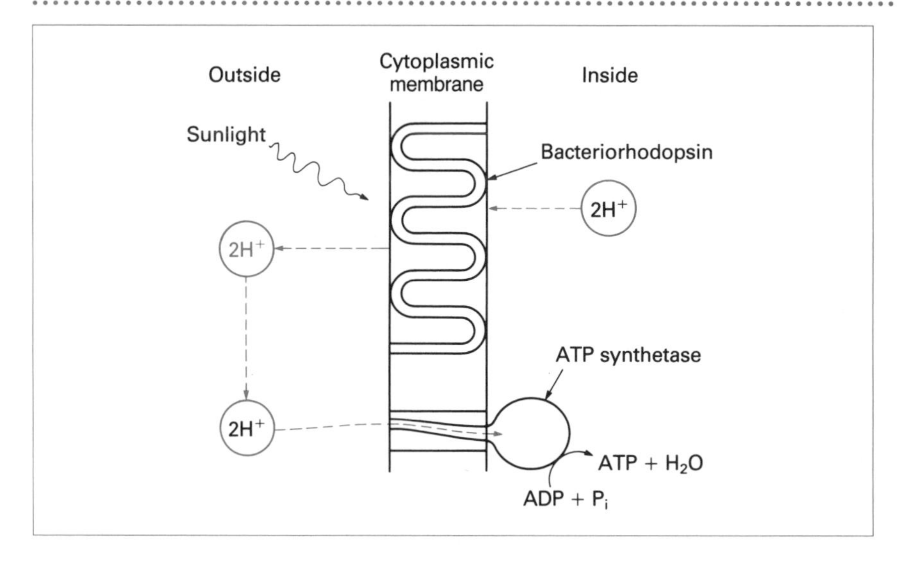


**Figure 4.25**
Photophosphorylation in *Halobacterium.* The absorption of one photon of light by bacteriorhodopsin leads to the translocation of one proton across the cytoplasmic membrane. The proton gradient drives the synthesis of ATP via the proton-translocating ATP synthetase. Two protons and therefore two photons are required to synthesize one molecule of ATP.

All halophilic bacteria are heterotrophic and most are obligate aerobes. They are able to oxidize organic substrates such as organic acids and amino acids and to carry out oxidative phosphorylation by a chemiosmotic process. Under conditions of low aeration, however, certain extreme halophiles such as *Halobacterium salinarium* synthesize the protein **bacteriorhodopsin**. This protein is purple in colour and is similar in structure and function to the visual pigment of the eye, **rhodopsin**. Bacteriorhodopsin is linked to a carotenoid-like molecule called **retinal**, which produces the purple coloration. Bacteriorhodopsin catalyses the translocation of protons across the cytoplasmic membrane to its outer surface. The accumulation of protons generates a proton motive force able to drive the synthesis of ATP by the action of a membrane-bound ATP synthetase. A model for the light-mediated bacteriorhodopsin proton pump of *Halobacterium* is shown in Figure 4.25.

## Chemolithotrophy: energy from the oxidation of inorganic electron donors

The ability to obtain energy from the oxidation of inorganic molecules is restricted to only a relatively small number of microorganisms, all of which are eubacteria. These bacteria typically obtain all of their carbon from $CO_2$ and, as we have seen, use the Calvin cycle to reduce $CO_2$ to organic molecules. The synthesis of ATP in these bacteria requires oxidation of the inorganic substrate and transfer of electrons along an electron transport chain containing carriers such as cytochromes. Oxidative phosphorylation involving a chemiosmotic mechanism produces ATP from ADP and inorganic phosphate in these organisms. Reduced coenzymes may be produced directly by the oxidation of the inorganic substrate, provided that it has a sufficiently low reduction potential. In most cases, however, reducing power requires ATP-dependent reverse electron flow.

### The groups of chemolithotrophic bacteria

The bacteria which live chemolithotrophically are specific for the inorganic substrate they are able to oxidize, and five different groups are recognized. The inorganic substrates which can be oxidized are as follows: (i) hydrogen; (ii) sulphur compounds; (iii) iron; (iv) ammonium; and (v) nitrite. Although some of these bacteria are obligately chemolithotrophic, others are facultative and can also grow as chemoorganotrophs using organic molecules as both energy and carbon sources. Many chemolithotrophs are aerobic and require oxygen as a terminal electron acceptor; some can grow anaerobically using alternative inorganic electron acceptors and carry out a type of anaerobic respiration.

## Chapter summary

In this chapter, we discussed the strategies adopted by different microbes to synthesize ATP. We saw that many microorganisms oxidize organic molecules by a variety of catabolic pathways, including the glycolytic, Entner–Doudoroff and phosphoketolase pathways. In some instances, oxidation of the organic substrate requires an internal electron acceptor and such pathways are described as fermentations. We saw that fermentations are important because they can lead either to the production of commercially valuable substances such as ethanol or to the production of characteristic end-products which may be used as aids in identification.

We also learned that complete oxidation of an organic substrate, such as glucose, can occur in the presence of an external electron acceptor. Under aerobic conditions, oxygen fulfils this role and the type of pathway used is described as respiration. We also found that respiration can occur in the absence of oxygen provided that alternative external acceptors, such as nitrate or sulphate, are available in the environment. In our review of respiratory processes, we saw the importance of the TCA cycle and electron transport chain in the generation of ATP by the process of oxidative phosphorylation.

Following our discussion of chemoorganotrophy, we looked at the ability of a number of microorganisms to use light energy to synthesize ATP. We emphasized the differences between the types of photosynthetic process and made a clear distinction between those which evolve oxygen (oxygenic) and those which do not (anoxygenic). The importance of chlorophyll as a photosynthetic pigment in most phototrophic microorganisms was emphasized although we noted that one group of photosynthetic archaebacteria, the extreme halophiles, used bacteriorhodopsin, and not chlorophyll, as their photosynthetic pigment.

Finally, we examined a relatively small number of bacterial species which are able to oxidize a variety of inorganic compounds and to synthesize ATP by a process of oxidative phosphorylation.

## Further reading

E.A. Dawes (1986) *Microbial Energetics*. Blackie and Sons Ltd., Glasgow, UK – a concise account of the subject.

G. Gottschalk (1986) *Bacterial Metabolism*. 2nd Edition. Springer Verlag, New York, USA – a general text which covers bacterial metabolism as a whole.

F.C. Neidhardt, J.L. Ingraham and M. Schaechter (1990) *Physiology of the Bacterial Cell – A Molecular Approach*. Sinauer Associates, Inc., Sunderland, MA, USA – sets microbial energetics in a physiological context.

D.G. Nicholls and S.J. Ferguson (1992) *Bioenergetics 2*. Academic Press Ltd., London – the friendly approach – containing cartoons!

## Questions

**1.** Multiple-choice questions:

(i) In which of the following groups of microorganisms does the EMP (glycolytic) pathway operate? (a) chemolithotrophs only, (b) anaerobic microorganisms only, (c) most cells, (d) photoorganotrophs only, (e) aerobic microorganisms only.

(ii) Which of the following microorganisms is a homolactic fermenter? (a) *Saccharomyces cerevisiae*, (b) *Lactobacillus casei*,

(c) *Leuconostoc mesenteroides*, (d) *Staphylococcus aureus*, (e) *Escherichia coli*.
(iii) Which of the following is the terminal electron acceptor in the alcoholic fermentation by *Saccharomyces cerevisiae*? (a) acetaldehyde, (b) acetic acid, (c) ethanol, (d) pyruvic acid, (e) acetyl-CoA.
(iv) Which of the following processes can lead to denitrification? (a) dissimilatory nitrate reduction, (b) assimilatory nitrate reduction, (c) oxidation of ammonia, (d) nitrogen fixation.
(v) Which of the following is usually involved in biological oxidation? (a) addition of oxygen, (b) loss of oxygen, (c) addition of electrons, (d) removal of electrons, (e) addition of H atoms.
(vi) In which of the following photosynthetic microorganisms is the reductive TCA cycle used to fix $CO_2$? (a) *Chlorella*, (b) *Chromatium*, (c) *Halobacterium*, (d) *Rhodopseudomonas*, (e) *Chlorobium*.

2. Short-answer questions:
(i) State the carbon and energy sources for each of the following microorganisms: (a) *Nitrobacter*, (b) *Chlorella*, (c) *Mucor*, (d) *Rhodopseudomonas*, (e) *Nostoc*.
(ii) Explain the biochemical basis for the differences between oxygenic and anoxygenic photosynthesis.
(iii) (a) Name **two** mechanisms which can lead to the generation of a proton motive force.
(b) Why is the generation of a proton motive force important to microorganisms?
(iv) State **two** functions of the TCA cycle.
(v) By which processes do chemolithotrophic bacteria (a) synthesize ATP, (b) produce $NAD(P)H_2$, (c) fix $CO_2$?
(vi) In what structures does photosynthesis occur in the following: (a) *Chlorella*, (b) *Halobacterium*, (c) *Rhodopseudomonas*, (d) *Chlorobium*?

3. Fill in the gaps:
(i) The energy which microorganisms derive from biochemical reactions is called __________ energy. This is the form of __________ that drives the __________ machinery of all organisms. From this point of view, heat is viewed as __________ energy because it __________ be used by __________ to do work.
(ii) In microorganisms free __________ is released through redox or oxidation- __________ reactions. Such reactions involve the transfer of an __________ or __________ from one molecule

to another. The donor molecule is the electron __________ and the accepting molecule is the electron __________. Together these two molecules constitute a redox ______________ or __________. In cells, redox reactions may involve __________ only, as in the cytochromes and iron __________ proteins or _____________ atoms as in the case of _____________ and _____________.

(iii) The standard __________ potential, $E^{0\prime}$ quantifies the ability of a __________ pair to __________ and accept __________ or hydrogen. A redox couple with a large __________ $E^{0\prime}$ such as $NAD^+$/ __________ will be readily oxidized by the $O_2$/ __________ couple with a larger __________ potential. In general, the direction of flow of __________ is from one couple with a more __________ $E^{0\prime}$ to another with a more __________ $E^{0\prime}$.

Choose from: positive (twice), negative (twice), donor, acceptor, wasted, biosynthetic, sulphur, hydrogen, quinones, redox, $H_2O$, $NAD^+$, $NADP^+$, $NADH_2$, pair, electron, electrons (four times), reduction (twice), cannot, organisms, couple, donate, free, energy (twice).

# 5 Biosynthesis

In order that microorganisms can grow and reproduce, new cell material must be produced from nutrients available in the environment. The production of new cell material is called **biosynthesis** or **assimilation**, and nutrients are the raw materials for this process. The first stage in the production of new cell material is the synthesis of small molecules, which are unavailable from the environment directly. These then serve as building blocks for the synthesis of large molecules.

In this chapter we shall first describe the reactions required for the synthesis of small molecules such as amino acids, purines and pyrimidines. We shall then examine the reactions which bring about the polymerization of these molecules to produce larger molecules or **macromolecules**.

Although the classes of precursors and macromolecules listed in Table 5.1 are found in all microorganisms and indeed in all living organisms, there is considerable variation in the chemical reactions which different microorganisms use to synthesize their cell material.

Such variation is a consequence of an organism's genetic make-up and it is this which determines the nature of the biosynthetic machinery. At one extreme, some representatives of one group of microorganisms, the **cyanobacteria** (blue-green bacteria) have minimal nutritional requirements in that they derive their energy from photosynthetic processes, their carbon from carbon dioxide and their nitrogen from nitrogen gas ($N_2$). As a consequence, such organisms possess the wide range of biosynthetic enzymes necessary to utilize such simple raw materials. At the other

**Table 5.1** Macromolecules and their low-molecular-weight precursors

| *Low-molecular-weight precursors* | *Macromolecular components* |
|---|---|
| Amino acids | Polypeptides and proteins |
| Deoxyribonucleotides | DNA |
| Ribonucleotides | RNA |
| Sugar phosphates | Polysaccharides |
| Fatty acids | Lipids |

extreme, lactobacilli and protozoa (Chapter 9) are poorly equipped with biosynthetic enzymes and can grow only provided that a large number of low-molecular-weight precursors are available in a preformed state.

The amount of biosynthesis (as reflected in enzyme content) required by the cyanobacteria at one extreme and the lactobacilli and protozoa at the other is the basis of the concept of **biosynthetic load**; that of the blue-green bacteria is heavy and that of the lactobacilli and protozoa is light.

## Anaplerosis

In Chapter 4 we discussed the processes by which microorganisms synthesize ATP and reduced coenzymes, such as $NAD(P)H_2$, using pathways such as the TCA cycle and glycolysis. The ATP and reducing power required for the biosynthetic reactions are described in the introduction to this chapter. Such energy-producing pathways perform another function in that their intermediates may be 'drained off' to be used as precursors for biosynthesis. The role of the TCA cycle in supplying precursors for biosynthesis is shown in Figure 5.1.

The diversion of such intermediates for biosynthesis means that unless steps are taken to replace the lost intermediates, there is a danger that an energy-yielding pathway may fail to operate at maximum efficiency or even

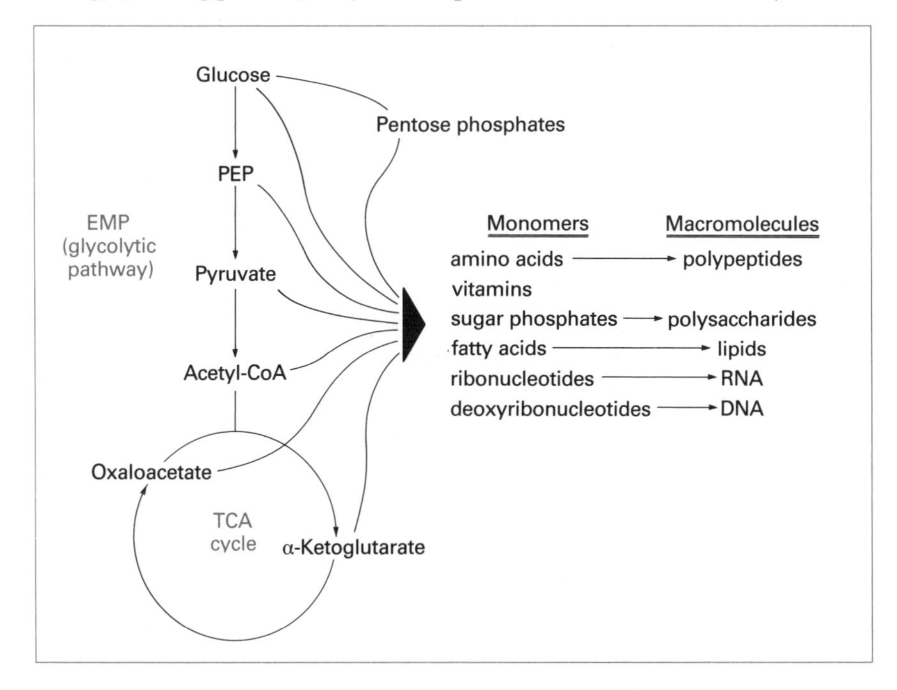


**Figure 5.1** The roles of the glycolytic pathway and TCA cycle in biosynthesis.

stop entirely. An organism may eventually die if the supply of ATP is insufficient to maintain its vital functions. This does not happen because microorganisms have evolved **anaplerotic reactions** whose function is to catalyse the synthesis of intermediates of energy-producing reaction sequences which may then replace those which have been removed and used for biosynthesis. In this way, energy production is able to proceed optimally with no loss of efficiency.

In heterotrophic bacteria, anaplerosis may be accomplished by a number of reactions (including reactions 1 and 2 below) in which a three-carbon compound such as pyruvate or phosphoenolpyruvate (PEP) reacts with carbon dioxide to produce an intermediate of the TCA cycle.

Reaction 1:

$$\text{PEP} + \text{HCO}_3^- \xrightarrow{\text{PEP carboxylase}} \text{oxaloacetate} + \text{P}_i$$

Reaction 1 is used to replenish oxaloacetate in enterobacteria, *Bacillus anthracis*, *Acetobacter xylinum*, *Thiobacillus novellus* and *Azotobacter vinelandii*.

Reaction 2:

$$\text{Pyruvate} + \text{HCO}_3^- + \text{ATP} \xrightleftharpoons{\text{Pyruvate carboxylase}} \text{oxaloacetate} + \text{ADP} + \text{P}_i$$

Reaction 2 is used to replenish oxaloacetate in a number of bacteria including *Pseudomonas aeruginosa*, *Arthrobacter globiformis*, *Bacillus coagulans* and *Acinetobacter calcoaceticus*.

Some bacteria can also utilize a two-carbon compound such as acetate as a carbon source, and anaplerosis under these conditions requires a different series of reactions known as the **glyoxylate shunt** or **bypass**. These reactions include a number of steps which also form part of the TCA cycle, but those reactions giving rise to $CO_2$ evolution are omitted to prevent loss of the carbon atoms of the acetate molecule, thereby ensuring their incorporation into TCA cycle intermediates. For this reason, the expression 'shunt' or 'bypass' is used. The glyoxylate shunt (Figure 5.2) does, however, include two key enzymes which do not participate in any other pathway: these are **isocitrate lyase** and **malate synthase**. Once again, this sequence of reactions produces TCA cycle intermediates which can replace intermediates drained off for biosynthesis.

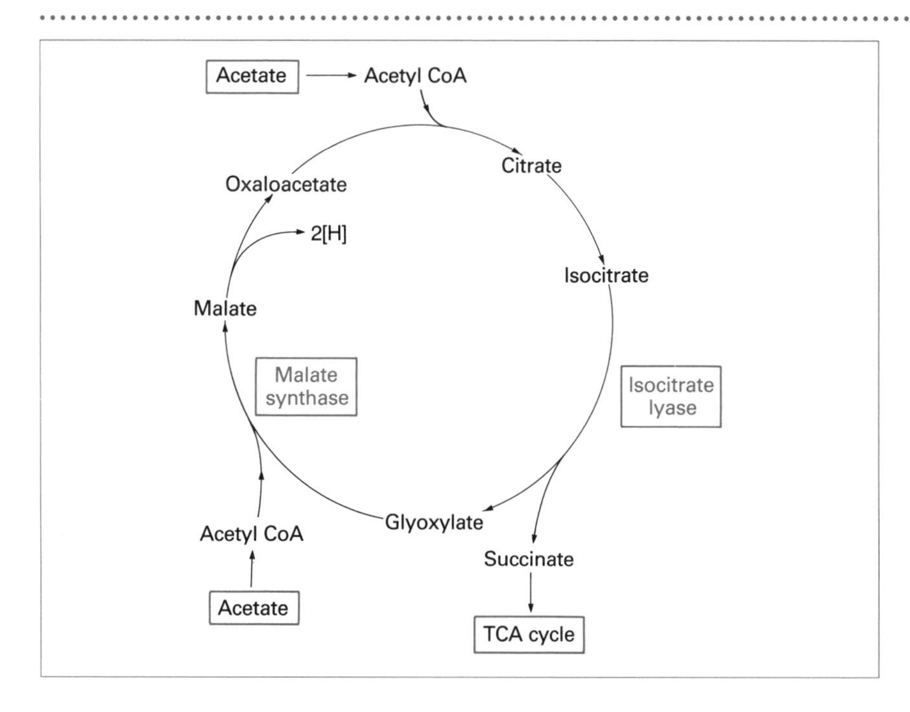


**Figure 5.2** Anaplerosis: the glyoxylate shunt (bypass).

## Biosynthesis of low-molecular-weight compounds

### Amino acids

Although there are twenty kinds of amino acid found in the polypeptides and proteins of living organisms, the number of pathways used for their synthesis is rather less than twenty. The reason for this is that groups or 'families' of amino acids exist in which members of the same family have a number of biosynthetic reactions in common and only a small number of specific reactions are required for one particular amino acid. The families of amino acids are shown in Figure 5.3. Furthermore, we have already met the precursors for amino acid biosynthesis as intermediates of either the glycolytic pathway or TCA cycle.

In the description of the biosynthesis of the protein amino acids, we can see two generalizations which apply to all of them. The first concerns the synthesis of the carbon skeleton which forms the backbone of each amino acid molecule and the second, the stage at which the amino group ($-NH_2$) is added to the carbon skeleton in an **amination reaction**. Taking the amination reaction first, there are two possibilities. First, ammonia reacts directly with the TCA cycle intermediate $\alpha$-ketoglutarate, to produce glutamic acid:

**Figure 5.3** The amino acid families.

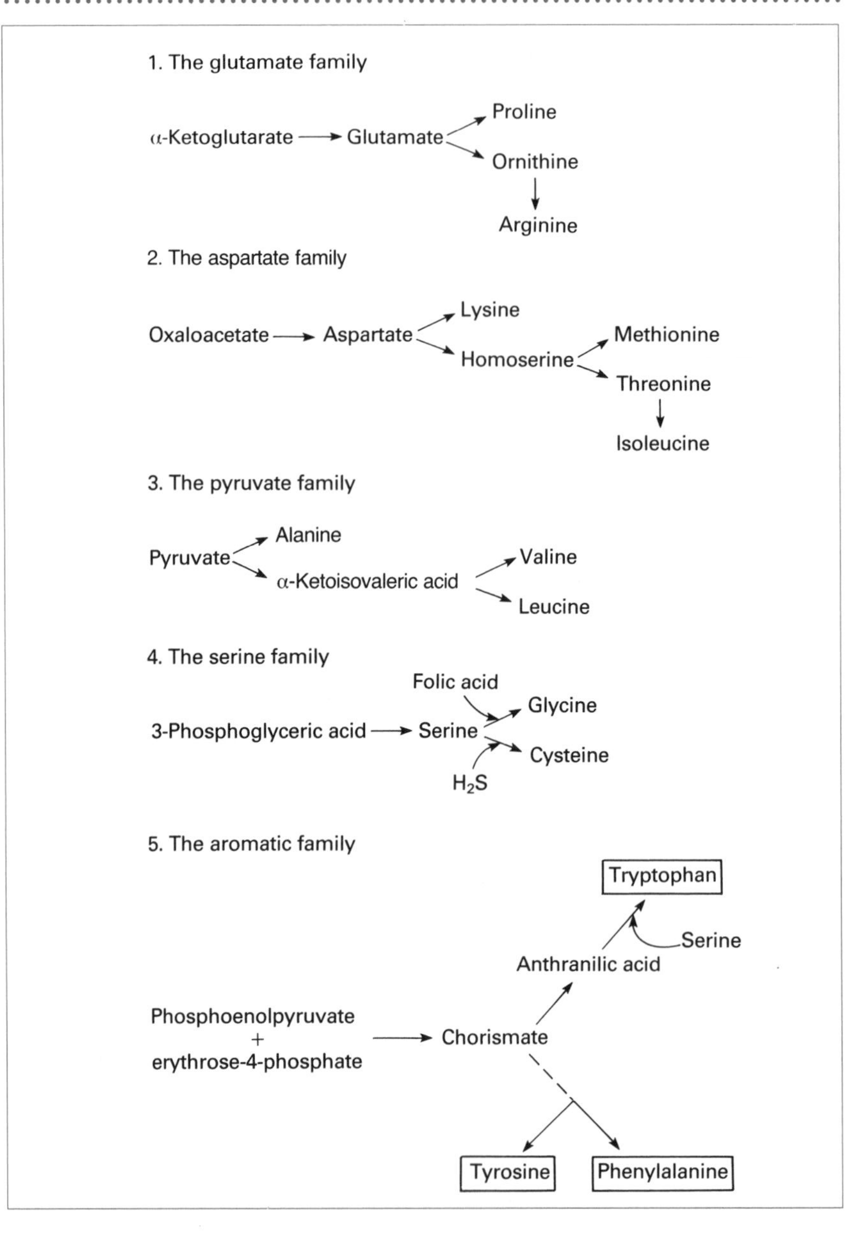


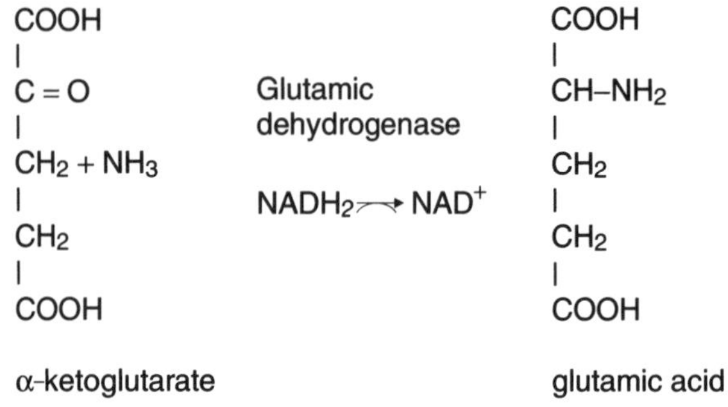

Alternatively, amination may occur at the very end of the biosynthetic pathway. In the former situation, the enzyme glutamic dehydrogenase catalyses the synthesis of glutamic acid. Glutamic acid can then act as a source of amino groups for the formation of further amino acids by the process of **transamination**:

| | | | | | |
|---|---|---|---|---|---|
| COOH | | COOH | | COOH | | COOH |
| \| | | \| | | \| | | \| |
| $CHNH_2$ | | C=O | transaminase $\rightleftharpoons$ | C=O | | $CHNH_2$ |
| \| | | \| | | \| | | \| |
| $CH_2$ | + | $CH_2$ | | $CH_2$ | + | $CH_2$ |
| \| | | \| | | \| | | \| |
| $CH_2$ | | COOH | | $CH_2$ | | COOH |
| \| | | | | \| | | |
| COOH | | | | COOH | | |
| glutamic acid | | oxaloacetic acid | | α-ketoglutarate | | aspartic acid |

Although glutamate dehydrogenase participates in the synthesis of a number of amino acids, alternative mechanisms are available making use of either different amino acid dehydrogenases or a completely different enzyme, glutamine synthetase. Carbon skeletons for amino acid biosynthesis are produced by the operation of energy-yielding pathways such as glycolysis or the TCA cycle. They are, therefore, ultimately derived from a sugar such as glucose in many heterotrophic microorganisms. This drain of intermediates is the reason why anaplerotic reactions are required.

## Purine and pyrimidine biosynthesis

In this section, we describe the synthesis of two classes of small molecules called **purines** and **pyrimidines**. These are nitrogen-containing compounds which are essential for the formation of nucleic acids. The nucleic

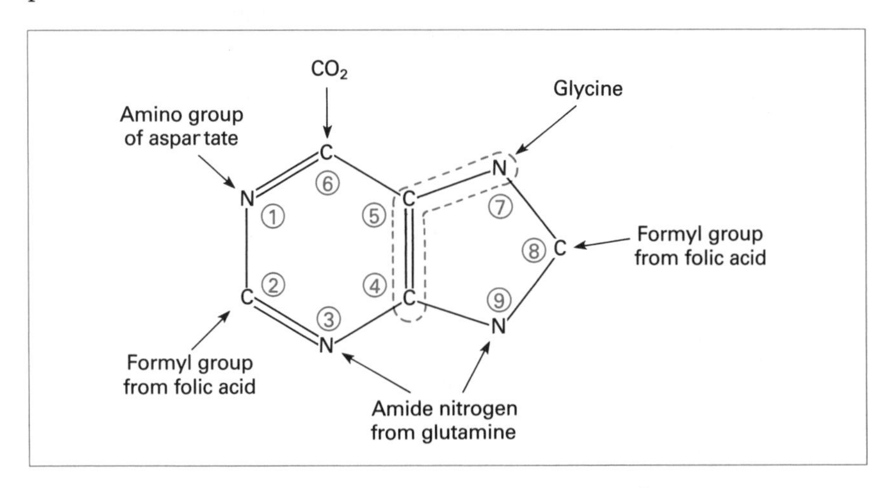


**Figure 5.4** Purine biosynthesis: the origins of the C and N atoms of the purine rings.

acids deoxyribonucleic acid (DNA) and ribonucleic acid (RNA) form the genetic material of all living organisms and a supply of purines and pyrimidines is essential if an individual's characteristics are to be passed from one generation to another. Nucleotides are also constituents of many vitamins and coenzymes such as adenosine triphosphate (ATP) and nicotinamide adenine dinucleotide ($NAD^+$), which we met in Chapter 4. The sequences

**BOX 5.1** Tetrahydrofolic acid and mode of action of the sulphonamides.

**Tetrahydrofolic acid and sulphonamide action**

Tetrahydrofolic acid (THFA) is a versatile carrier of activated one-carbon units required for the biosynthesis of a number of metabolites, including the purines, the pyrimidines and the amino acid methionine. Tetrahydrofolic acid consists of three groups: a substituted pteridine, *p*-aminobenzoate (PAB) and glutamate (see diagram). Because mammals are unable to synthesize the pteridine ring, they obtain THFA either from their diet or from microorganisms in their intestinal tracts. In contrast, most bacteria use THFA produced as a result of their own intracellular, biosynthetic activities.

Because of this difference, advantage may be taken of the selective toxicity of the sulphonamide drugs. These antibacterial agents are able to interfere with the synthesis of a bacterium's THFA, because they are structurally similar to PAB. Such agents are called **metabolic analogues** and are able to compete with PAB for the catalytic site of the enzyme which is responsible for incorporation of PAB into the THFA molecule. At sufficiently high concentrations of sulphonamide, no THFA is synthesized. The bacterium is deprived of its source of one-carbon units, and dies.

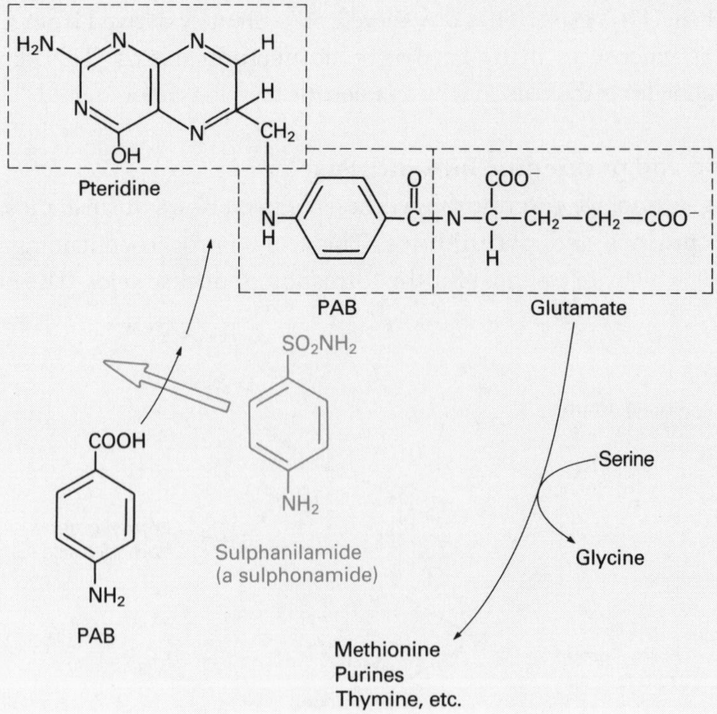

of reactions leading to the synthesis of purines and pyrimidines are complex but the broad outlines are given here.

The purines (adenine and guanine) are synthesized by addition of carbon and nitrogen atoms to the derivative of ribose phosphate known as phosphoribosyl pyrophosphate (PRPP). The carbon and nitrogen atoms come from a variety of amino acids, carbon dioxide and formyl groups (Figure 5.4). In cells, formyl groups are found in a coenzyme called **folic acid**. Folic acid is required in a large number of biochemical reactions in all kinds of living cells to attach part of a molecule containing a single carbon atom (a one-carbon fragment) to a carbon skeleton. The first purine formed is **inosine** and this serves as a precursor for the formation of adenine and guanine.

The biosynthesis of the pyrimidines differs from that of the purines in that the pyrimidine ring is first completely synthesized in the form of **orotic acid,** using carbon and hydrogen atoms from a number of sources (Figure 5.5) and it is only at a later stage that the addition of ribose and phosphate occurs. After the addition of these two components, thymine, uracil and cytosine are produced by further reactions.

## Synthesis of macromolecules

Small molecules, including those we discussed earlier in this chapter, serve as the building blocks for the synthesis of large polymeric compounds, which either form the substance of microbial cells (carbohydrates, lipids, proteins) or which direct (nucleic acids) or catalyse (polypeptides and proteins) the polymerization reactions required. In the following sections, we examine the steps needed for the synthesis of such molecules.

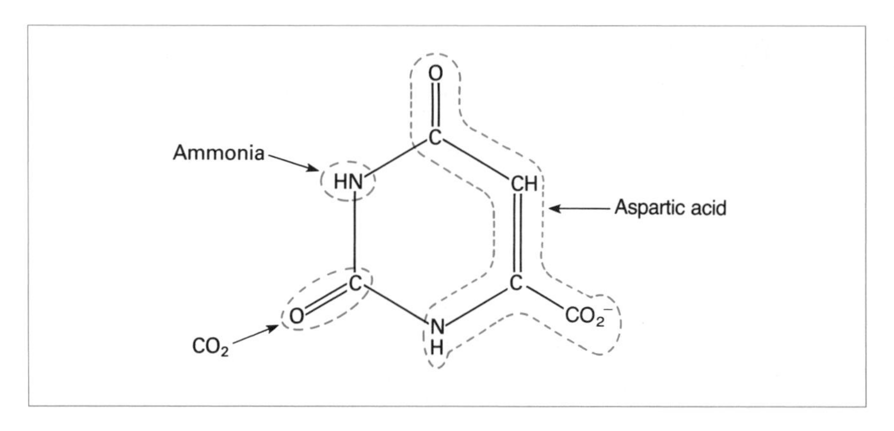


**Figure 5.5** Pyrimidine biosynthesis: the origins of the C and N atoms of orotic acid.

## Polypeptide synthesis

The proteins of all living organisms including microorganisms are usually described as either structural proteins or enzymatic proteins. In the former case, the protein is insoluble and forms part of the organism's structure; ribosomal proteins and cell membrane proteins are examples of this group. Enzyme proteins represent the major group of catalysts of all living organisms and speed up the rate of biochemical reactions. In spite of this distinction, the steps involved in the production of both types of protein are similar. Furthermore, although the strategies adopted by different groups of microorganisms are broadly the same, there are differences in detail between microorganisms belonging to the three primary kingdoms; that is to say, the two bacterial groups, eubacteria and archaebacteria, and the eukaryotic microorganisms (Chapter 9).

Investigation of protein synthesis requires a combination of the techniques of biochemistry and genetics and the knowledge gained forms one of the cornerstones of molecular biology. Research on a wide variety of microorganisms showed long ago that proteins differ from each other in the *sequence* in which individual amino acids are linked together by peptide bonds. The **configuration** of a protein molecule is a consequence of spontaneous folding determined by the sequence of amino acids (its so-called **primary structure**). The order of amino acids is determined by the sequence of nucleotides in one strand, called the coding strand, of a DNA molecule. So the ways in which polypeptide and protein molecules differ is ultimately determined by the nucleotide sequence of a section of the coding strand of a DNA molecule, which we refer to as a **gene** or **cistron**.

A sequence of amino acids polymerized to form a single chain is called a **polypeptide** and is the product of a single gene (called a **cistron**). Some proteins consist only of a single polypeptide, in which case the two terms are equivalent. In other examples, a protein consists of a number of different polypetide chains and a different gene is required for the synthesis of each polypeptide. A well-known example is haemoglobin, which consists of two $\alpha$-chains and two $\beta$-chains of polypeptide, and which requires two different genes for its synthesis.

The overall process of polypeptide synthesis is complex and, for simplicity, it is usual to divide the process into two main stages called **transcription** and **translation**. We shall look at transcription first.

### Transcription

The overall result of the transcriptional process is the synthesis of a messenger RNA (mRNA) molecule with a base sequence complementary to the coding strand of part of a DNA molecule which comprises a gene (cistron). In other words, the sequence of ribonucleotides in a mRNA molecule is dictated by the sequence of deoxyribonucleotides in DNA (Figure 5.6).

The building blocks for the synthesis of mRNA are the four ribonucleoside triphosphates, ATP, CTP, GTP and UTP. These are polymerized by the enzyme DNA-dependent RNA polymerase, which is usually known

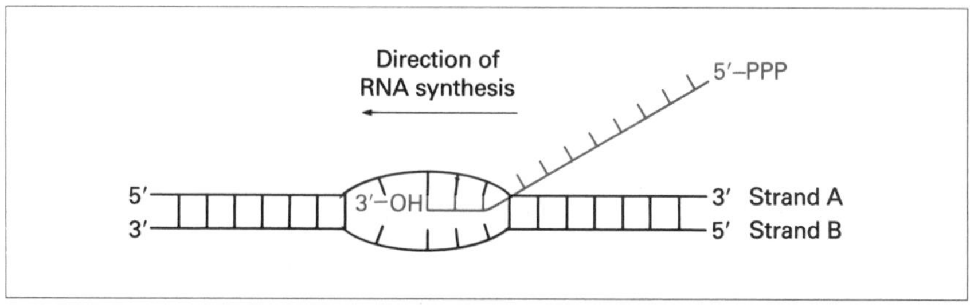


**Figure 5.6** The transcriptional process: the RNA molecule (blue) is copied from only one strand of DNA (strand A here) at a time. In other segments of the DNA strand B may be the coding strand and would therefore be transcribed. Notice that the growing RNA molecule is antiparallel to the DNA strand being copied and ends in a 5′-triphosphate at the non-growing end.

more simply as **RNA polymerase**. This enzyme functions by moving along a DNA strand and joining individual nucleotides together by linkages known as **phosphodiester bonds**. To do this, RNA polymerase must bind to the coding DNA strand at one end of the gene. This is possible because of the presence of so-called **recognition sequences** just before the DNA sequence which is to be transcribed into mRNA. Because the nature of the sequences varies somewhat, particularly between primary kingdoms, we shall describe the situation which exists in eubacteria such as *Escherichia coli*.

In *E. coli* there are two recognition sequences which together constitute a **promoter**: the – 35 sequence (so called because it is located 35 nucleotides before the first nucleotide to be transcribed), and the **Pribnow box** (5–10 nucleotides from the point at which transcription is initiated). All RNA polymerases are large molecules and that of *E. coli* is no exception; it consists of four types of subunit, designated $\alpha$, $\beta$, $\beta'$, and $\sigma$ and, since the $\alpha$ subunit is duplicated, the complete enzyme (or **holoenzyme**) is described as having a subunit structure $\alpha_2$, $\beta$, $\beta'$, $\sigma$.

Initiation of the transcription process is thought to occur by the $\sigma$ subunit binding to the –35 sequence, followed by interaction of the remainder of the molecule with the Pribnow box (Figure 5.7). As a consequence, the polymerase is aligned so that it can move along the DNA molecule in the correct direction. The Pribnow box determines whether the direction of transcription is from left to right or *vice versa*. This is very important because translation of a mRNA molecule starts before its synthesis is completed and the correct end of the mRNA molecule must be available for combination with a ribosome.

Having bound to the promoter with the correct orientation, the double helical form of DNA undergoes a **localized denaturation**; that is to say, the hydrogen bonds which hold the two strands of DNA together are broken to form an **open promoter complex**, thereby making the coding strand more accessible to the large RNA polymerase molecule. Once the open promoter complex has formed, the first nucleoside triphosphate to be incorporated into the new mRNA chain (almost always either ATP or GTP) becomes located in the initiation site of the polymerase opposite a

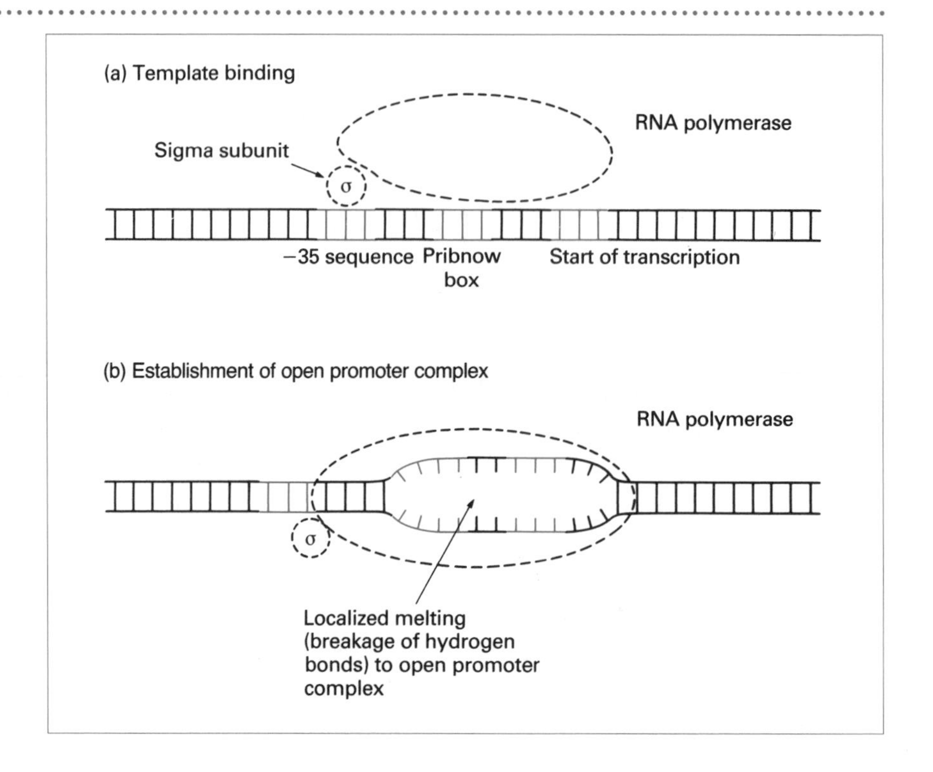


**Figure 5.7** Binding of RNA polymerase to promoter region showing initial binding of σ subunit to −35 sequence followed by orientation by the Pribnow box.

thymine or cytosine of the coding strand. Although the three phosphate groups of this initial nucleoside triphosphate remain intact, subsequent nucleotides are added to the 3′-OH of the initial ribose (Figure 5.8) with the simultaneous release of two of the three phosphate groups of each triphosphate. The growing mRNA molecules are, therefore, polymerized in a 5′-phosphate to 3′-hydroxyl direction. Once approximately eight nucleotides have been added to the chain, the σ subunit is lost from the polymerase and the so-called core enzyme (comprising $\alpha_2$, β, β′ subunits) completes the polymerization process following the steps described above.

To make sure that the polymerase ceases its activity at the end of a gene, a further recognition sequence, a **termination sequence**, is present at the end of a gene. The termination sequence for the genes concerned with tryptophan biosynthesis in *E. coli* has been well studied. It has a number of unusual features. It contains an inverted repeat base sequence with a central non-repeating sequence (Figure 5.9). This is significant in that either the DNA strand or the mRNA transcribed from it is capable of forming complementary base-pairs between bases located in the *same* polynucleotide stand; this is called **intrastrand base pairing**. This produces a **stem and loop** or **hairpin** configuration in the appropriate nucleic acid. A second region within the termination sequence is rich in guanine and cytosine residues and a third region (which is not always present) is rich in

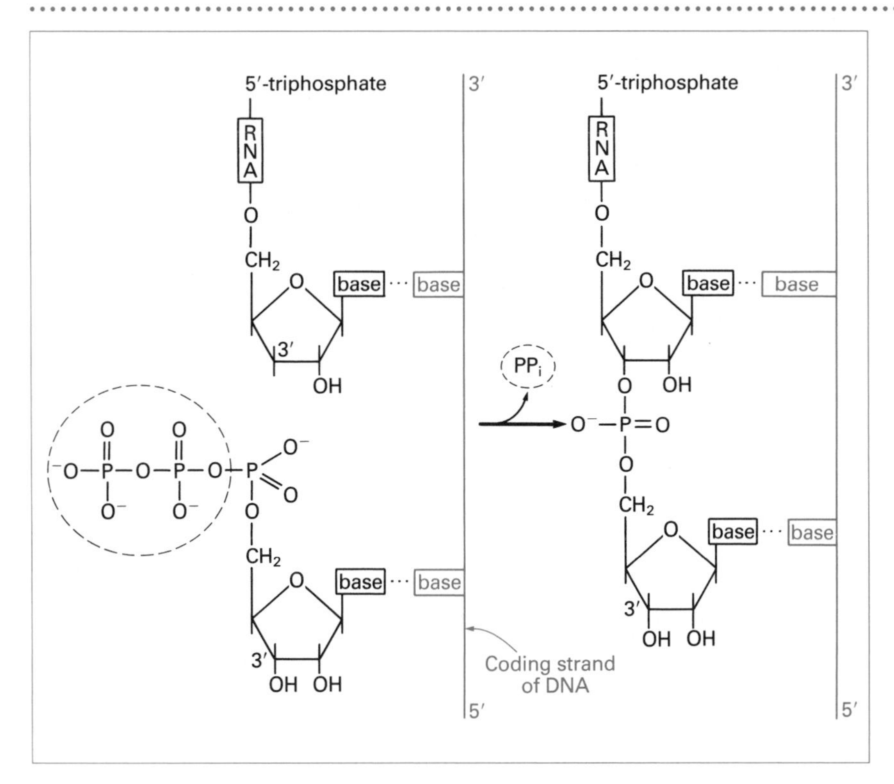


**Figure 5.8** Synthesis of RNA by RNA polymerase showing antiparallel orientation of the DNA coding strand and elongating RNA molecule. The direction of RNA polymerization is seen to be 5′- phosphate to 3′-OH.

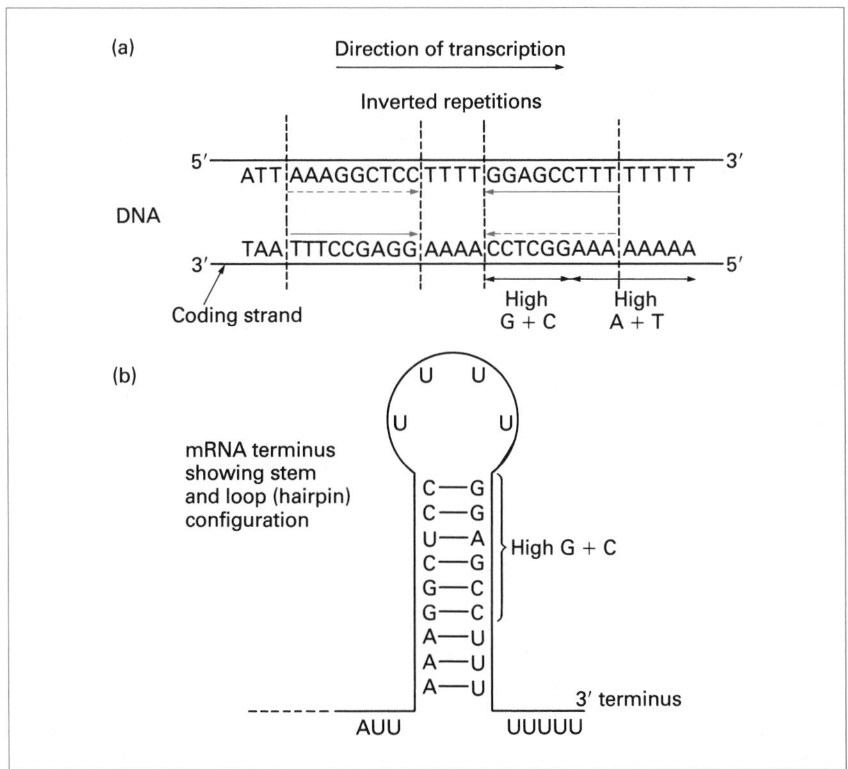


**Figure 5.9** Base sequence of (a) DNA of eubacteria showing a termination sequence and (b) the stem and loop (hairpin) configuration of the corresponding mRNA terminus.

adenine and thymine. In certain regulatory mechanisms an additional protein, **rho,** is required for termination to occur.

### Translation

In eukaryotic microorganisms, mRNA is synthesized in the nucleus (where the chromosomes are located) and passes into the surrounding cytoplasm to combine with a ribosome before translation of the polyribonucleotide code into the amino acid sequence of a polypeptide can occur. In other words, this class of RNA acts as a messenger between the genetic material, surrounded by a nuclear envelope, and the ribosomes located in the cell cytoplasm. Although there is no compartmentalization of DNA in prokaryotes, and both the ribosomes and DNA are located in the bacterium's cytoplasm, the designation 'messenger RNA' is retained for this functionally similar class of RNA molecule.

During the translation process, amino acids are polymerized to form polypeptide chains. There is no direct interaction between an mRNA molecule and the amino acids. Instead, a class of relatively small RNA molecules, known as **transfer RNA (tRNA)**, acts as an **adaptor molecule** by forming a chemical bond with a specific amino acid to form an aminoacyl-tRNA molecule. Each tRNA molecule can also align itself with the *corresponding* sequence of three bases in a mRNA molecule. In this fashion, sequences of three mRNA bases, called **codons**, determine the order in which individual amino acids are linked to each other by peptide linkages. As you would expect, the attachment of a particular amino acid to the corresponding tRNA molecule is catalysed by a specific enzyme known as an **aminoacyl-tRNA synthetase** or amino acid activating enzyme. Each of these enzymes has two specificities, one for an amino acid and one for the corresponding tRNA molecule. These enzymes are extremely accurate in attaching the correct amino acid to the correct tRNA. If they were not, the wrong amino acid would become incorporated at a particular position in a protein molecule with a corresponding decrease in functional activity.

The reactions catalysed by aminoacyl-tRNA synthetases occur in two stages. In the first stage, the amino acid is activated by reaction with ATP:

$$\text{amino acid} + \text{ATP} \rightleftharpoons \text{aminoacyl-AMP} + \text{PP}_i$$

The activated amino acid (aminoacyl-AMP) usually remains bound to the surface of the enzyme until reaction with the corresponding tRNA occurs. During this second stage, the amino acid is transferred to the appropriate tRNA molecule

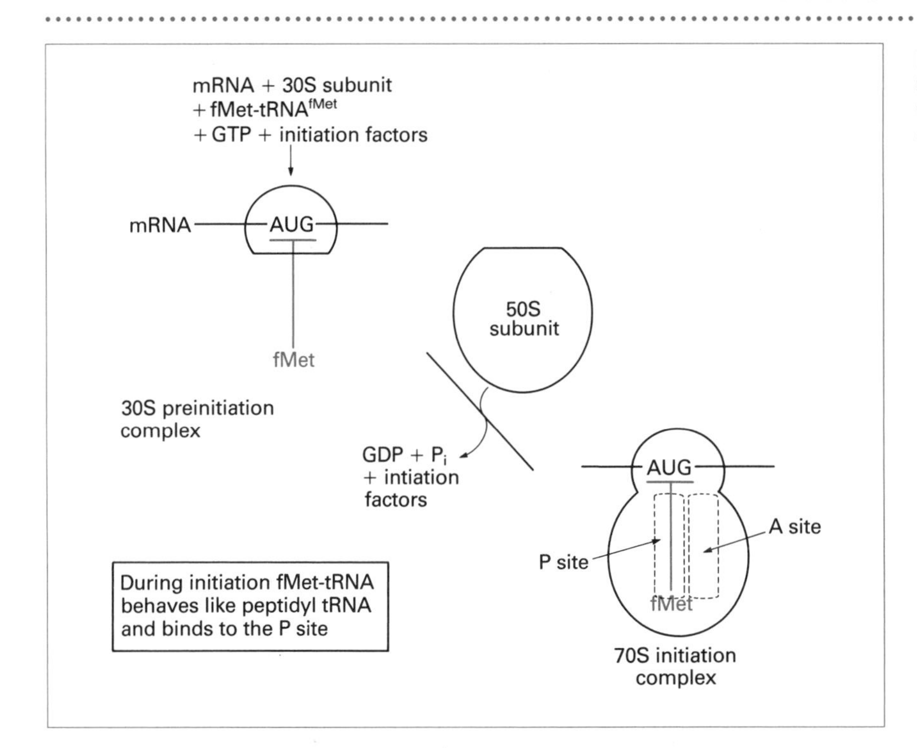


**Figure 5.10** Formation of the 30S preinitiation complex and the 70S initiation complex in eubacteria.

$$\text{aminoacyl-AMP} + \text{tRNA} \rightarrow \text{aminoacyl-tRNA} + \text{AMP}$$

The aminoacyl-tRNA molecules are then ready to act as precursors for the polymerization reaction directed by a mRNA molecule in association with a ribosome.

## Polymerization at the ribosome

The sequence of reactions which occurs at a ribosome is very complex and for this reason, the process is often divided into a number of steps to make discussion easier. In this account, we recognize four stages: initiation, elongation, termination–release and polypeptide folding. In addition to mRNA, tRNA and ribosomes, a number of additional protein factors, together with GTP, are also required.

In eubacteria, the polymerization of amino acids starts with the formation of an **initiation complex** consisting of a 30S ribosome subunit, mRNA, *N*-formylmethionyl-tRNA and initiation factors (Figure 5.10). A 50S ribosome subunit is then added to the initiation complex to produce an active 70S ribosome. In eubacteria, the ribosome attaches itself to the correct point of the mRNA molecule because of the presence of a recognition sequence (called the **Shine–Dalgarno sequence**) of between three and nine nucleotides located just before the first mRNA codon to be translated. The Shine–Dalgarno sequence is complementary to the nucleo-

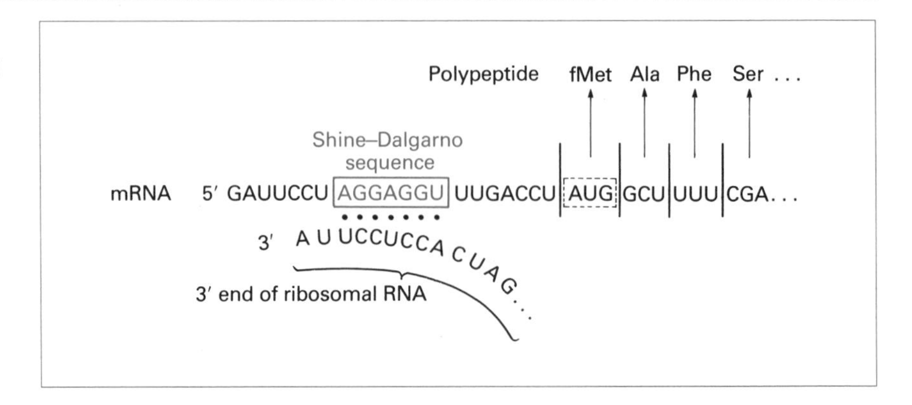


**Figure 5.11** Initiation of translation in eubacteria showing complementary base-pairing between the Shine–Dalgarno sequence of RNA and the complementary region near the 3′ terminus of 16S rRNA.

In eukaryotes and archaebacteria, methionyl-tRNA performs the same function as *N*-formylmethionyl-tRNA in eubacteria.

The size of sub-cellular particles, including ribosomes and ribosome subunits, can be determined by measuring the ease with which they sediment in an ultracentrifuge. Ease of sedimentation is expressed in Svedberg units (S units) after the Swedish scientist who developed the technique. Because ease of sedimentation is a reflection of a number of particle properties (including mass, shape and surface area), S values for the subunits of a particle such as a ribosome are not necessarily additive. For example, the 70S prokaryotic ribosome consists of one 50S subunit and one 30S subunit.

tide sequence at the 3′ end (Figure 5.11) of a ribosomal RNA molecule and the two **anneal** (bond together) by complementary base-pairing. The ribosome is now in a position to translate the message contained in a mRNA molecule in an ordered fashion to produce a polypeptide molecule with the correct sequence of amino acids.

During polypeptide synthesis, the mRNA is associated with the 30S eubacterial ribosomal subunit and aminoacyl-tRNA, and the growing polypeptide chain with the 50S subunit. The 50S subunit has two sites designated P (peptide) and A (acceptor) sites (Figure 5.12). The P site holds a tRNA molecule attached to a growing polypeptide chain and the A site receives an incoming aminoacyl-tRNA molecule. During peptide bond formation, the growing polypeptide chain moves to the A site and a new bond is formed. The tRNA molecule left behind in the P site is removed and the tRNA with polypeptide chain attached in the A site moves to the now empty P site (a process called **translocation**) leaving the A site empty and ready to receive the next incoming aminoacyl-tRNA molecule (Figure 5.12). This series of reactions is repeated until the ribosome reaches a **nonsense codon** (termination triplet). Such a codon has no complementary tRNA molecule and it marks the end of the mRNA sequence for one polypeptide molecule. At this stage, the eubacterial ribosome separates into 30S and 50S subunits and the polypeptide molecule is released from the ribosome.

## Secretory proteins

Although the ultimate destination of many proteins is within the confines of the cytoplasmic membrane others, called secretory proteins, can fulfil their roles only after they have been released into the surrounding environment. In prokaryotes, periplasmic and extracellular enzymes fall into this category. Secretory proteins are transported across the cytoplasmic membrane by means of an extra sequence of 15–20 amino acids called a **signal sequence**. The signal sequence contains predominantly hydrophobic

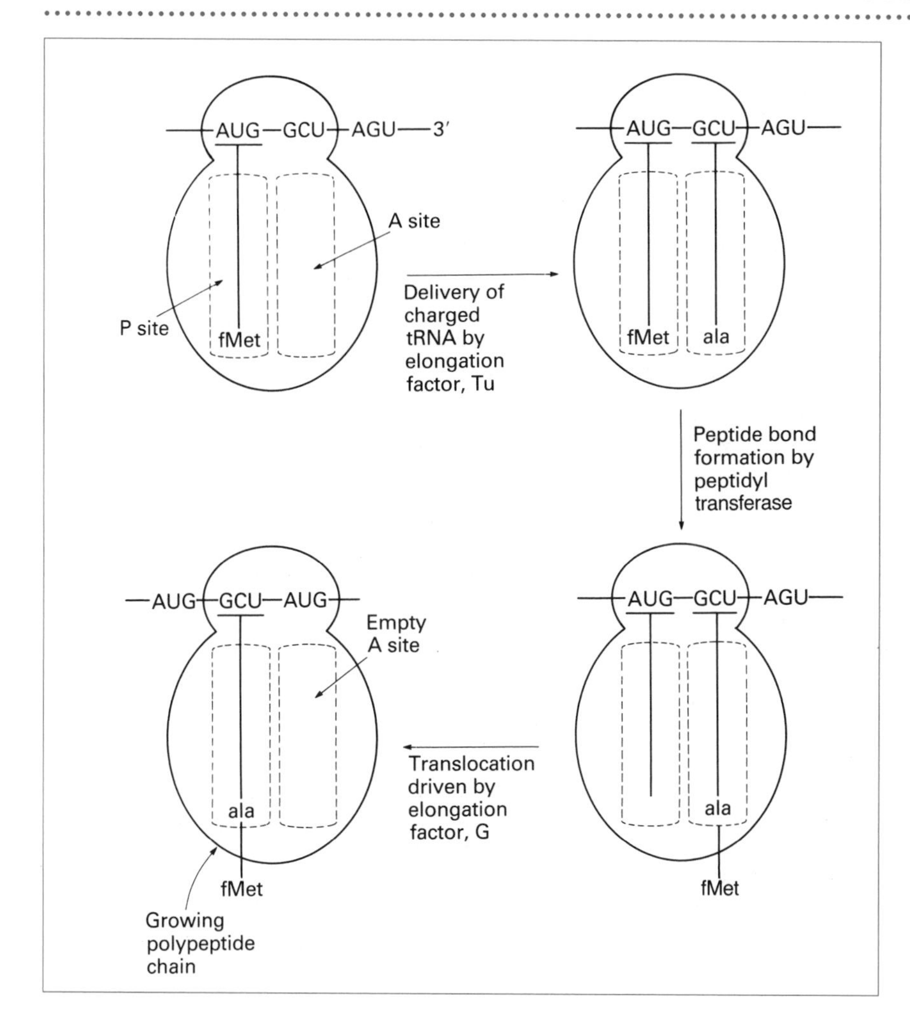


**Figure 5.12** Polypeptide synthesis: elongation process showing binding of aminoacyl-tRNA (charged tRNA), peptide bond formation and translocation.

amino acids which are believed to make it easier for the enzyme to cross the hydrophobic cytoplasmic membrane. In addition, the ribosomes responsible for the synthesis of secretory proteins are often bound to the cytoplasmic membrane, so synthesis and transport occur almost simultaneously. Once the protein has been secreted, the signal sequence can be removed by a peptidase enzyme, an example of **post-translational modification**.

The hypothesis just described is called the **signal hypothesis** and its consequences are potentially extremely important for the production of mammalian proteins by genetically engineered microbial cells. If a signal sequence can be added to a desired protein, it will be secreted into the surrounding medium and its isolation and purification then become relatively straightforward.

**BOX 5.2**

**Antibiotics which interfere with protein synthesis**

1. **Puromycin** binds to the part of the A site which is on the large subunit of the ribosome (Figure 5.12). It can accept the growing polypeptide chain (or *N*-formylmethionine) (fMet) from the P site. The problem with this is that puromycin is readily lost from the A site and consequently causes premature release of the growing polypeptide chain from either prokaryotic or eukaryotic ribosomes. Puromycin has a structure resembling the amino acid residue linked to the amino acid-accepting sequence of tRNA molecule.

2. **Tetracycline** binds to the small (30S) subunit of bacterial ribosomes and blocks the binding of aminoacyl-tRNA to the A site. The growing polypeptide chain, attached to tRNA, remains in the P site and further elongation is prevented.

3. **Chloramphenicol** binds to the large (50S) subunit of the bacterial ribosome and prevents transfer of the growing polypeptide chain from the P site to aminoacyl-tRNA in the A site; in other words the peptidyl transfer reaction is inhibited.

4. **Erythromycin** reacts with free ribosomes but not with polysomes. Once bound to the ribosome, it allows the formation of a very short polypeptide but prevents further chain elongation. The inhibited complex is unstable, ribosomes are released from the mRNA only to be attached to new mRNA where inhibition occurs once again.

5. **Streptomycin and related compounds**. Depending upon the concentration used, streptomycin has two effects on sensitive cells: at high concentrations, polypeptide synthesis is completely inhibited, whereas at lower concentrations, misreading of mRNA occurs. This difference in action depends upon whether the streptomycin molecule binds to a free ribosome or to one which is already engaged in polypeptide chain elongation. In the former case, the initiation complex is unable to participate in chain elongation. In the second case, less distortion of the ribosome occurs, allowing chain elongation, albeit with impaired accuracy.

When chain elongation is inhibited, fMet is the only amino acid bound to the ribosome and since such an initiation complex is unstable, the ribosomes are released from mRNA. Although they are inactive in polypeptide synthesis, released ribosomes may attach themselves to new mRNA where they again form a blocked initiation complex.

## Polynucleotide synthesis

### Polyribonucleotides

During our description of protein synthesis, we saw the parts played by the three types of RNA molecules: tRNA, mRNA and rRNA. We also saw how mRNA was synthesized by the enzyme RNA polymerase and how the sequence of nucleotides in it was complementary to the nucleotide sequence of the coding strand of a DNA molecule. The overall strategies for the synthesis of the other two classes of RNA – tRNA and rRNA – are essentially the same as that used for the synthesis of mRNA. In fact, in eubacteria it appears that a single RNA polymerase is responsible for the synthesis of all three classes of RNA. In eukaryotes the situation is different; there are three classes of polymerase, each responsible for the synthesis of a different class of RNA. The distribution and function of eukaryotic RNA polymerase is shown below.

| | *Class I* | *Class II* | *Class III* |
|---|---|---|---|
| Location | Nucleolus | Nucleoplasm | Nucleoplasm |
| Product | rRNA | mRNA | tRNA |

## RNA processing

Once the primary transcripts have been produced by the action of one or more RNA polymerase molecules, they often require modification or processing before they become functional. In prokaryotes, for example, tRNA and mRNA molecules are initially produced as long precursor molecules which are then cut by specific enzymes at a number of places to produce the final functional RNA.

Eukaryotic mRNA also requires processing because the coding regions of the genes (called **exons**) are separated from each other by sequences of non-coding DNA known as **intervening sequences** or **introns** (Figure 5.13). Since the primary mRNA transcripts contain ribonucleotide sequences corresponding to both exons and introns, the introns have to be removed and the remaining fragments joined together to form a functional mRNA molecule. Finally, mature eukaryotic mRNA molecules are modified at both the 3′- and 5′-ends. The 3′-end often has a number (between

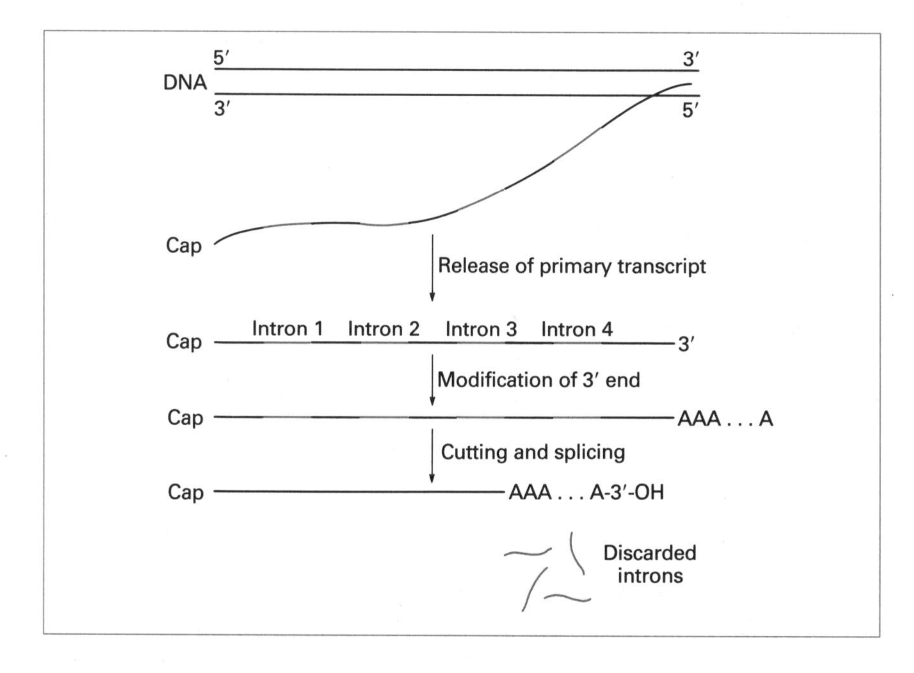


**Figure 5.13** Diagram showing transcription of eukaryotic mRNA. Capping occurs before mRNA is released and this is followed by addition of the polyA tail to the 3′ end. Finally the intervening sequences (introns) are excised.

20 and 200) of adenylic acid residues added to form what is called a **polyA tail** (Figure 5.13). Addition of adenylic acid residues is catalysed by an enzyme called poly(A) polymerase and occurs at one of a number of possible sites preceding the 3′-end of the primary transcript. This implies that the primary transcript is cut by an endonuclease enzyme at the point where adenylic acid residues are to be added. Although the function of the polyA tail has yet to be confirmed, mRNA molecules which differ in the form of their poly A tails may have different functions in the life-cycle of an organism. Specifically, such tails may increase the stability of eukaryotic mRNA and may be involved in attachment of mRNA to intracellular membranes.

At the opposite end (5′ end) of a eukaryotic mRNA molecule, a structure called a **cap** may be found. This consists of a 7-methylguanosine residue in an unusual 5′–5′ linkage to the 5′ terminal nucleotide of the primary transcript. Sometimes the sugars of the adjacent nucleotides may also be methylated. As with the polyA tail, the function of capping is not fully understood but it may protect mRNA from degradation by nuclease enzyme and provide a feature for recognition by the protein-synthesizing machinery.

It can be seen that synthesis of functional RNA molecules requires not only the polymerization of nucleotide precursors to produce a primary transcript but additionally a number of post-transcriptional modifications to produce a functional mRNA molecule.

### Ribozymes

For many years it was believed that all enzymes were proteins. However, with advances in studies of RNA synthesis it became apparent that some RNA molecules also possess catalytic activity. Such RNA catalysts are known as **ribozymes**, and RNaseP is one such example. RNaseP is a small, single-stranded molecule RNA of 300 to 400 nucleotides which removes short oligonucleotide segments from both the 3′- and 5′-ends of primary tRNA transcripts to produce active tRNA molecules.

### DNA synthesis

In order that each individual cell in a population of unicellular microorganisms, or each cell in a multicellular organism, receives a copy of the genetic material from its progenitor, replication of a cell's DNA complement must precede division of the cell itself. The replication of a cell's DNA presents problems because the amount of DNA present is very large and, if unfolded, the DNA would be many times longer than the length of the cell. An individual *Escherichia coli* cell is 2 μm in length, whereas its chromosome is between 1.1 mm and 1.4 mm in length – 600 times longer than the cell which contains it. Since a cell's DNA contains the blueprint for the structure and function of the whole cell, the replication must occur with extreme accuracy if the blueprint is to be faithfully transmitted from one generation to another. The replication process becomes even more amazing when the speed with which it occurs is taken into account. It has been calculated that synthesis of *E. coli* DNA requires the incorporation of 3300 nucleotides per second and that each strand of DNA increases in length by about 825 nucleotides per second!

Unwinding of the two strands of the original double helix must also occur in front of the advancing replication fork, and this is catalysed by the enzyme **DNA gyrase (topoisomerase II)**. Because the *E. coli* chromosome is a closed loop, often described as circular, unwinding of the double helix at the replication fork leads to overwinding of the non-replicating part of the chromosome and the replication process is brought to a halt. DNA gyrase is able to prevent overwinding by catalysing appropriate breaks and reunions of DNA at the replication fork.

An overall strategy for the replication of a DNA molecule was suggested by the Watson and Crick double helical structure in which nucleotides in one strand are complementary to those in the other. Specifically, only the pairings of adenosine (A) with thymine (T), and guanosine (G) with cytosine (C) are possible. It seemed likely, therefore, that each strand of the double helix could act as a template for the synthesis of a new complementary strand so that two double helices were produced from the original parent molecule, one for each daughter cell. It was further suggested that each of the two daughter DNA molecules contained one strand of the original (parental) molecule and one newly synthesized strand. This method of DNA replication is called **semi-conservative replication** (Figure 5.14) because each newly formed DNA double helix contains half (one strand) of the original. Experimental support for the semi-conservative mode of DNA replication has been provided by Meselson and Stahl.

Replication of the *E. coli* chromosome starts at a specific site called the **origin of replication**, which consists of about 300 bases recognized by specific initiation proteins. At the origin of replication, a **replication bubble** or **displacement loop** is first formed by localized breakage of hydrogen bonds, catalysed by a **helicase** enzyme, between complementary DNA strands. Each strand can then act as a template for the synthesis of a complementary DNA strand, the point at which polymerization occurs being called a **replication fork**. Once a replication fork has been formed, the first nucleotides to be laid down are *ribo*nucleotides and these form a **polyribonucleotide (RNA) primer**.

It may seem strange that the first nucleotides to be polymerized during

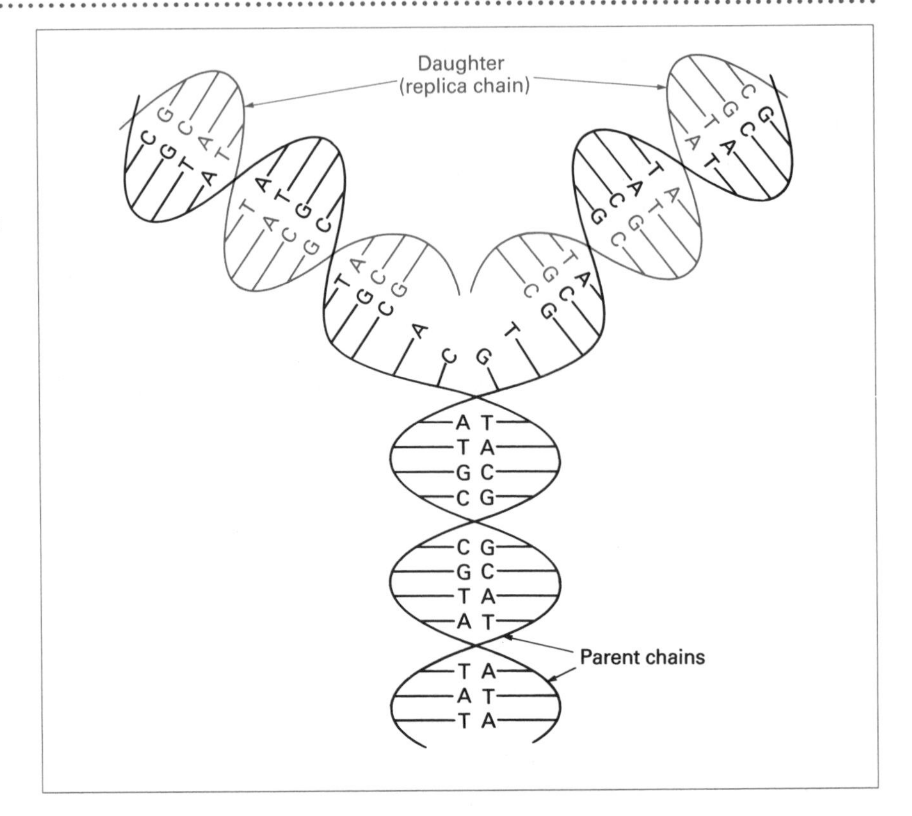


**Figure 5.14** Replication of DNA by a semi-conservative mechanism. The two newly formed replicas each consist of one parental strand (black) and one daughter strand (blue). Bases in the daughter strands are determined by a requirement to form a base-pair with the parental base.

DNA synthesis are in fact ribonucleotides. The reason is that the enzymes responsible for the polymerization of *deoxyribonucleotides* (DNA polymerases) can add nucleotides only to the 3′-OH end of a preceding nucleotide residue and at the start of a round of replication, no 3′-OH will be present. RNA polymerases have no such requirement for a free 3′-OH; they require only a DNA template. Once an RNA primer has been synthesized, further deoxyribonucleotides can be added, initially to the 3′-OH group of the last ribonucleotide in the growing chain (Figure 5.15).

The three *E. coli* DNA polymerases (DNA polymerase I, II and III) add nucleotides to 3′-OH groups; in other words, they can synthesize DNA only in a 5′ → 3′ direction. The strand which is being synthesized has the opposite polarity to the template strand: in other words they are *antiparallel* and if one strand runs in a 5′ → 3′ direction, the other will run in the opposite (3′ → 5′) direction. Furthermore the two strands in the original double helix are antiparallel, so that at a replication fork, synthesis of one of the two new strands has to occur in the *overall* direction of 3′ → 5′. This does not fit in with the observed action of DNA polymerases which can polymerize only in a 5′ → 3′ direction (Figure 5.16).

An answer to this paradox was provided by the experimental observations of Seiji Okasaki, who suggested that replication of the 5′ → 3′ parental

**BOX 5.3**

**Semi-conservative replication of DNA: the Meselson and Stahl experiment**

*E. coli* cells were cultured for approximately 14 generations in the presence of a nitrogen source ($NH_4Cl$) which contained the heavy isotope of N ($^{15}N$) until their DNA contained easily detectable quantities of $^{15}N$. This DNA was 'heavy' and could be separated from normal $^{14}N$-containing 'light' DNA by centrifuging it in a caesium chloride (CsCl) density gradient. Cells were then quickly transferred to a medium containing normal $NH_4Cl$. Samples were taken and the cells broken open by the addition of detergent. The cell contents were centrifuged in a CsCl gradient until the DNA had formed bands at its characteristic density. This procedure was repeated for a number of generations of bacteria, using both double-stranded DNA and DNA which had been denatured to produce single-stranded fragments. The results obtained were those which would be expected if, at each round of replication, one of the parental strands was retained in the newly formed double helix; that is to say, the results obtained were indicative of a semi-conservative mode of DNA replication. The results obtained confirm that the two parental DNA strands are **conserved** intact in one strand of each of the two daughter DNA molecules.

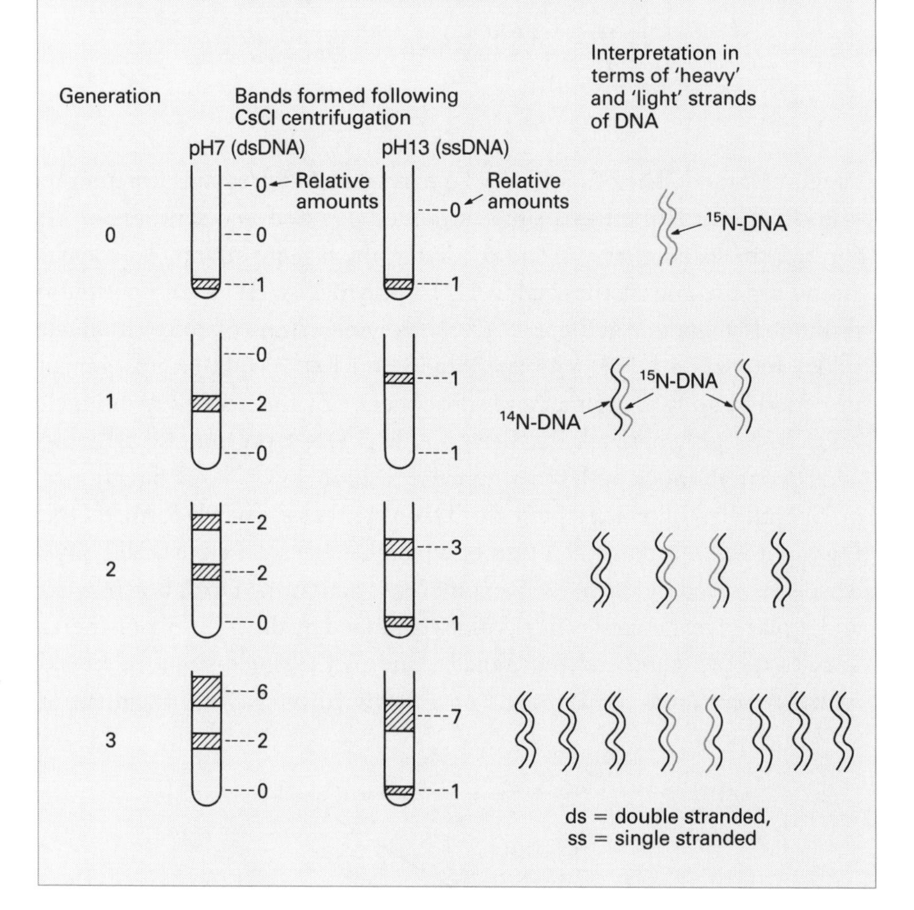


DNA strand requires **discontinuous synthesis**; that is, synthesis of the complementary 3′ → 5′ daughter strand is allowed to lag behind the replication of the other parental strand so that, for a period, one parental strand (3′ → 5′) may be present together with its newly formed (5′ → 3′)

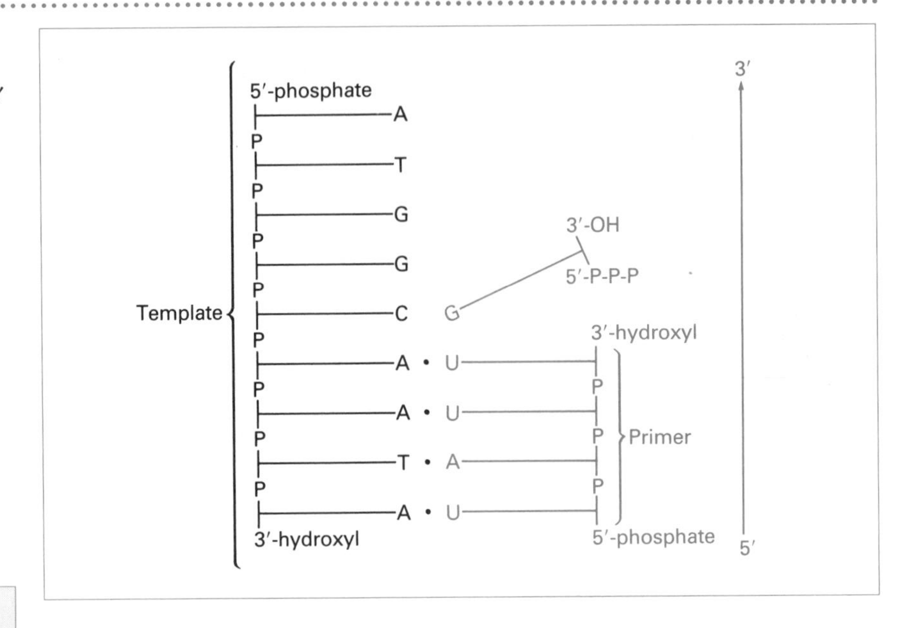


**Figure 5.15** Diagram to show the difference between template and primer and the meaning of 5′ → 3′ synthesis.

**Proofreading by DNA polymerases I and III**
Both DNA polymerases I and III of *E. coli* show exonuclease activity in the 3′ → 5′ direction (the opposite direction to that of chain elongation). As a result, DNA polymerase is able to remove non-complementary nucleotides incorporated during polymerization and replace them with the correct nucleotides. In other words, DNA polymerases 'proofread' the DNA chain and correct any errors in it.

daughter strand. The other (5′ → 3′) parental strand remains unreplicated – in other words there is a single-stranded gap between synthesis of the former, the leading strand, and the latter, the lagging strand. The lagging strand can then be synthesized by DNA polymerase III in a discontinuous fashion, as illustrated in Figure 5.17. So, polymerization of deoxyribonucleotides requires a DNA template strand and free 3′-OH groups, initially provided by an RNA primer and subsequently by the last deoxyribonucleotide to be polymerized. Furthermore, it is DNA polymerase III which adds deoxyribonucleotides in a stepwise fashion, in a 5′ → 3′ direction.

Eventually, following a period of DNA synthesis, a newly formed DNA fragment will meet an RNA primer which has been previously laid down. This is removed by the 5′ → 3′ exonuclease activity of DNA polymerase I and replaced by deoxyribonucleotides catalysed by the 5′ → 3′ polymerase activity of the same enzyme. Finally, any nicks (single-stranded breaks) which remain between the ends of a newly formed DNA fragment are

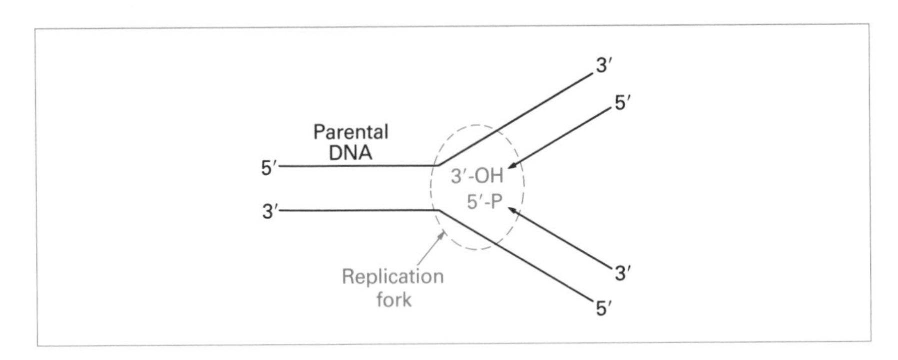


**Figure 5.16** Diagram to show the ends which would be present at a replication fork if both strands were to grow in the same overall direction. No known DNA polymerase can catalyse growth of a DNA strand in a 3′ → 5′ direction.

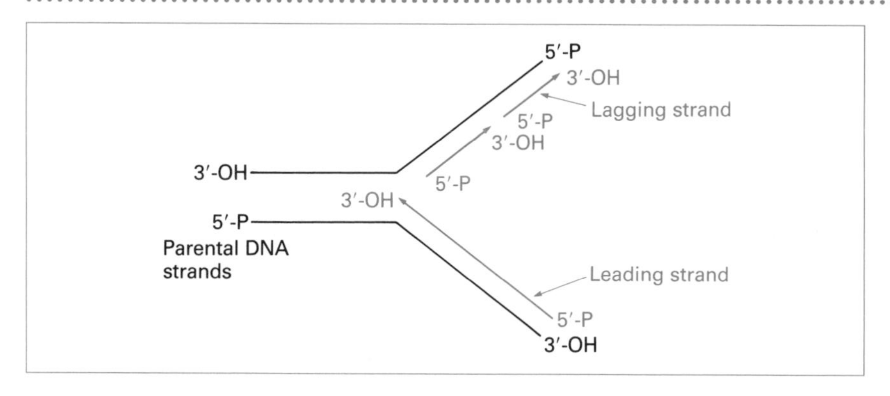


**Figure 5.17** A replication fork showing directions of growth of leading and lagging strands.

closed by the formation of phosphodiester linkages catalysed by the enzyme DNA ligase (Figure 5.18). The events which occur at a replication fork are summarized in Figure 5.19.

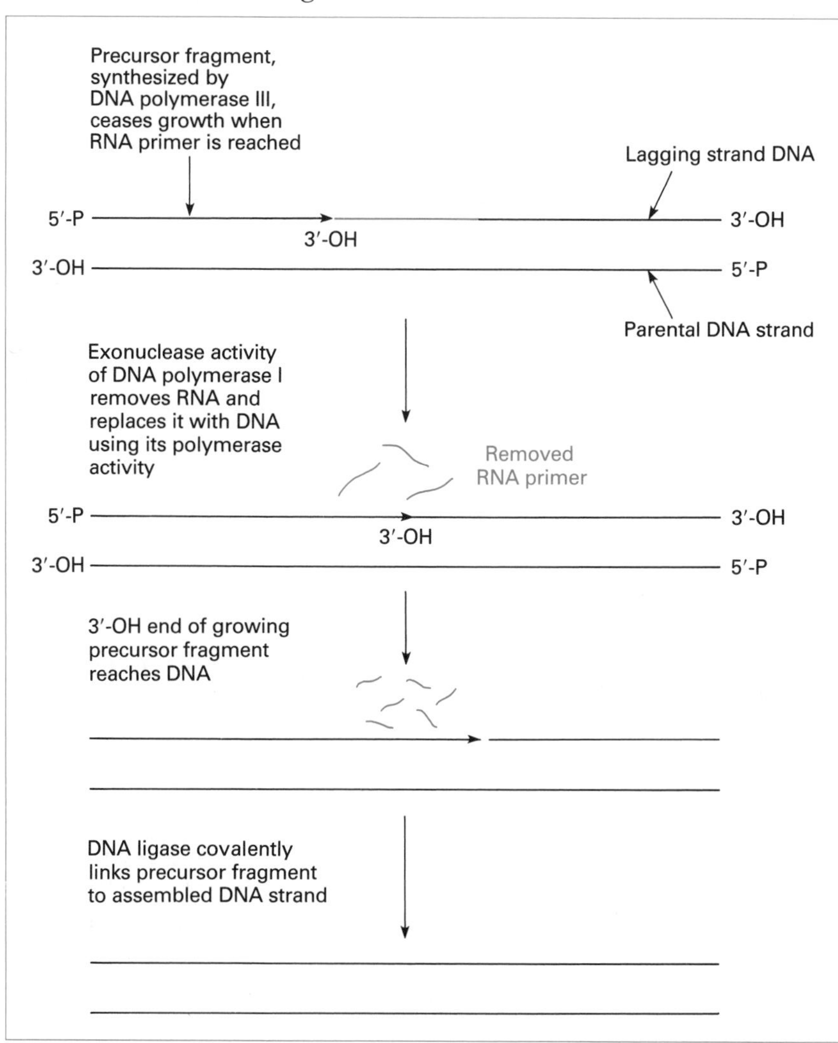


**Figure 5.18** Assembly of precursor fragments. RNA is in blue and the replication fork would be to the left of the diagram.

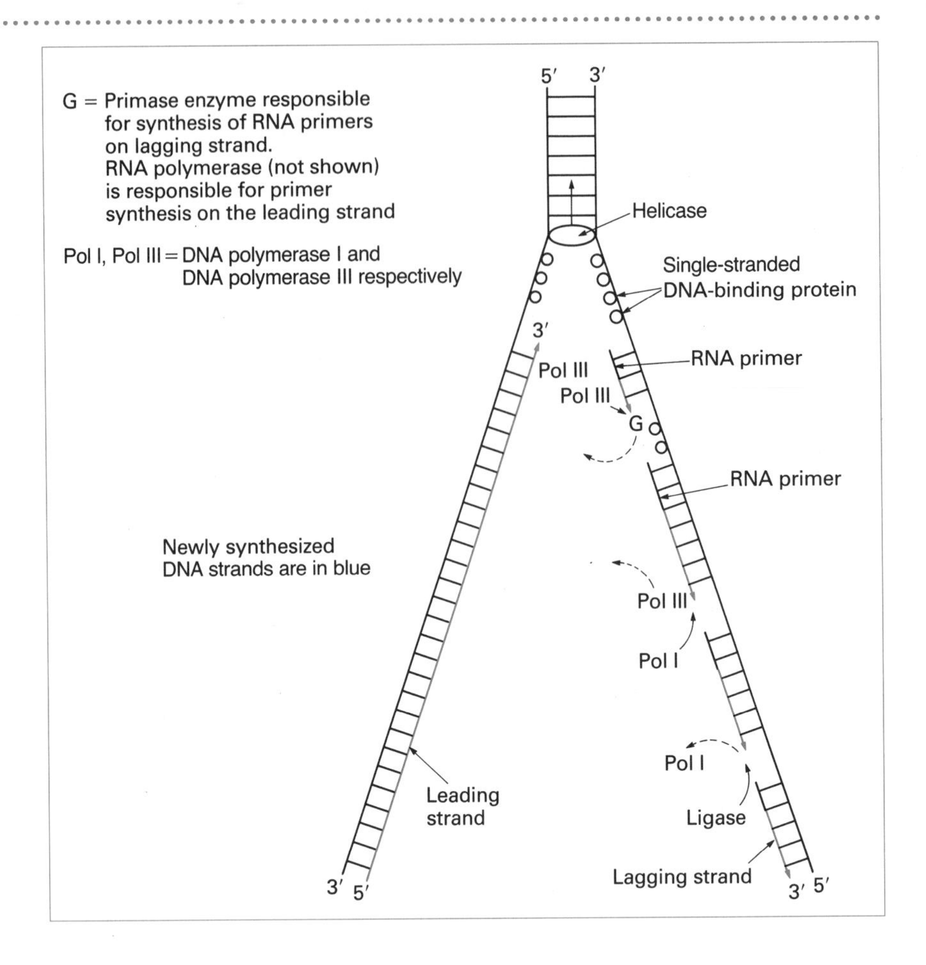


**Figure 5.19** Events at the DNA replication fork.

## The eubacterial cell wall

Almost all eubacteria possess a cell wall. The wall defines the shape of the cell and, perhaps more importantly, provides a rigid barrier between the cell and the surrounding environment. Without it, any influx of water would increase the volume of the cell cytoplasm and rupture the fragile cytoplasmic membrane.

The eubacterial cell wall contains chemical components found in no other living organism and its nature is a primary aid in distinguishing eubacteria from other groups of microorganisms. When discussing the structure of eubacterial cell walls, it is convenient to recognize two broad categories, depending upon the appearance of cells following staining with the so-called **Gram reagents**. The Gram reaction depends upon the propensity of bacterial cell walls to retain a crystal-violet–iodine complex. Bacteria which retain the complex are said to be **Gram-positive** and those which do not, **Gram-negative**. Experimental observations using the elec-

The structures of the cell walls of eukaryotic microbes are described in Chapter 9.

BOX 5.4

**The Gram staining reaction**

The Gram staining technique was developed in 1884 by the Danish bacteriologist Christian Gram to aid identification of bacteria in pathological specimens. The technique consists of staining a heat-fixed smear of the sample with a basic dye such as crystal violet, followed by **mordanting** (fixing the stain) with iodine. Smears are then washed with either ethanol or acetone and finally counterstained with a dye of a different colour such as safranin. Gram-positive bacteria retain the crystal violet–iodine complex and appear dark purple when viewed under the light microscope. Gram-negative bacteria are the colour of the counterstain (pink) because the crystal violet–iodine complex is washed out by the acetone or ethanol.

The structural basis for the reaction of different bacteria to the Gram staining procedure appears to lie in the structure of the cell wall. The relatively thick Gram-positive cell wall becomes dehydrated by treatment with ethanol or acetone and the crystal violet–iodine complex becomes trapped within it. Gram-negative eubacteria have a relatively thin cell wall high in lipid content from which the crystal violet–iodine complex is readily washed out.

Some bacteria which possess typical structural and compositional features of Gram-positive cells nevertheless show a weak or variable propensity to stain Gram-positive. This phenomenon is often seen in old cultures when **autolysis** of the cell wall has occurred.

Bacteria belonging to the genus *Mycobacterium*, including the causative organism of tuberculosis (*M. tuberculosis*), are surrounded by a waxy layer and, unless treated with heat or detergent, are unable to take up the crystal violet. Unless pretreated to disrupt the waxy layer, *Mycobacterium* will appear to be Gram-negative, in spite of possessing Gram-positive cytological properties.

The reaction of different bacteria to Gram staining reflects not only cell wall structure but also important biochemical, physiological and genetic differences. However, there are two groups of exceptions to this general rule: the mycoplasmas (Mollicutes), which lack cell walls and so do not respond to Gram staining, and the archaebacteria.

Archaebacteria are represented by a relatively small number of bacterial types. Archaebacteria inhabit extreme environments where there are high concentrations of salt (extreme halophiles) or hot, acid conditions (thermoacidophiles). The third archaebacterial group (methanogens) live in anaerobic conditions and produce methane gas as a metabolic waste product. The absence of peptidoglycan from the extreme halophilic, methanogenic and thermoacidophilic bacteria was a major factor in establishing the archaebacteria as a distinct taxonomic group.

tron microscope show that it is the overall structure of the cell wall which determines whether a bacterium is Gram-positive or Gram-negative. The former have a thick, single-layered wall, whereas the latter have a complex, multilayered wall which nevertheless is relatively thin.

## The peptidoglycan layer

As you will see from Figure 5.20, a peptidoglycan layer is present in the cell walls of both Gram-positive and Gram-negative eubacteria. In the former, the bulk of the single-layered cell wall is peptidoglycan, whereas in the latter it accounts for only the innermost layer and is of relatively modest dimensions.

Chemical analysis has shown that peptidoglycan contains a number of unique molecules. These include the amino sugar *N*-acetylmuramic acid,

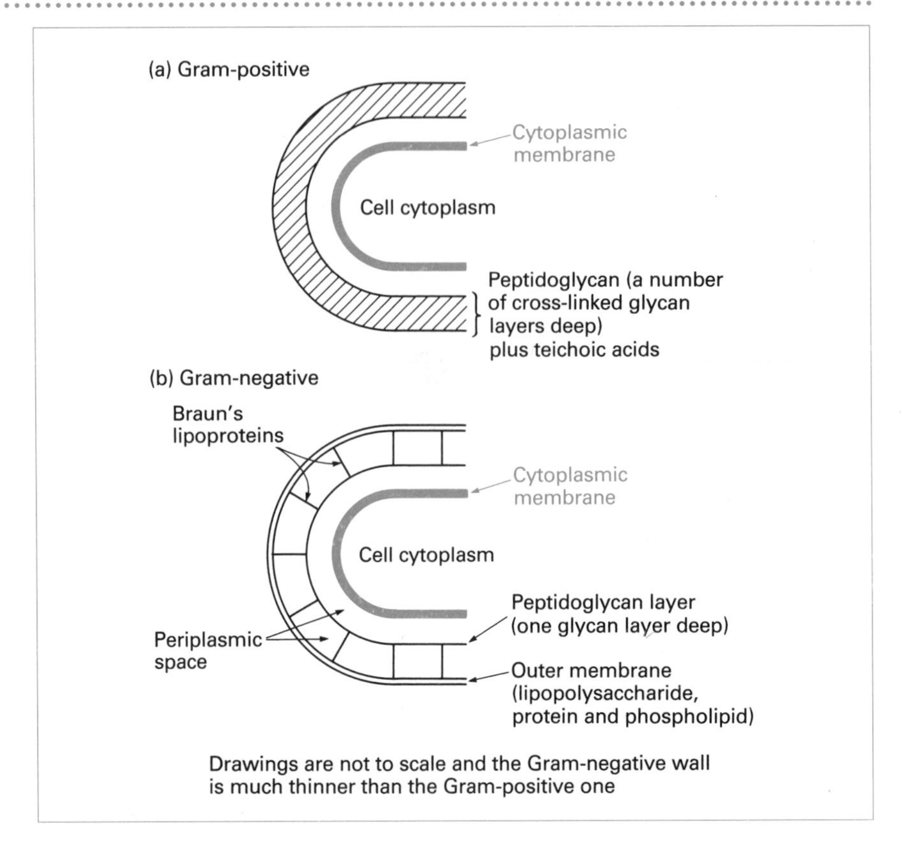


**Figure 5.20** Layers of Gram-positive and Gram-negative eubacterial cell walls.

and the amino acids D-alanine, diaminopimelic acid (DAP) and D-glutamic acid. Since these compounds are found only in peptidoglycan, their presence is evidence of the close taxonomic relationship between different microbial groups, including the eubacteria, the cyanobacteria and the actinomycetes.

Peptidoglycan forms a three-dimensional network which is sometimes likened to the reinforcing steel in concrete. It is likely that the peptidoglycan network is a single gigantic bag-shaped molecule (a **murein sacculus**) which surrounds the bacterial protoplast. The two amino sugars, *N*-acetylmuramic acid and *N*-acetylglucosamine, are linked by bonds described as β1–4 linkages, since carbon atoms 1 and 4 of adjacent residues are involved. These linkages are hydrolysed by the enzyme lysozyme, a component of tears and other body fluids (Figure 5.21). Linked in this way, the amino sugar residues make up long chains of alternating *N*-acetylglucosamine and *N*-acetylmuramic acid residues, usually referred to as **polysaccharide backbones** or **glycans**. Linked to the *N*-acetylmuramic acid residues of the polysaccharide backbones are short peptide chains of four amino acid residues. The amino acids which make up this chain are characteristic to each kind of bacterium but the terminal amino acid is

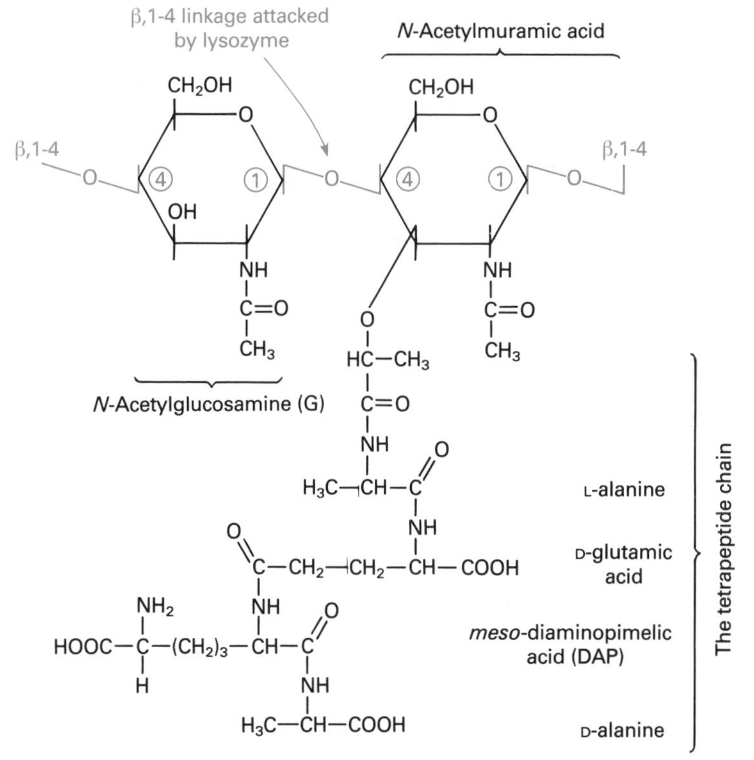

There may be some variation in the amino acids which comprise the tetrapeptide chain. *meso*-diaminopimelic acid (DAP) is a diamino acid (contains two amino groups) and may be replaced by different alternatives in other eubacteria

**Figure 5.21** The basic peptidoglycan subunit structure.

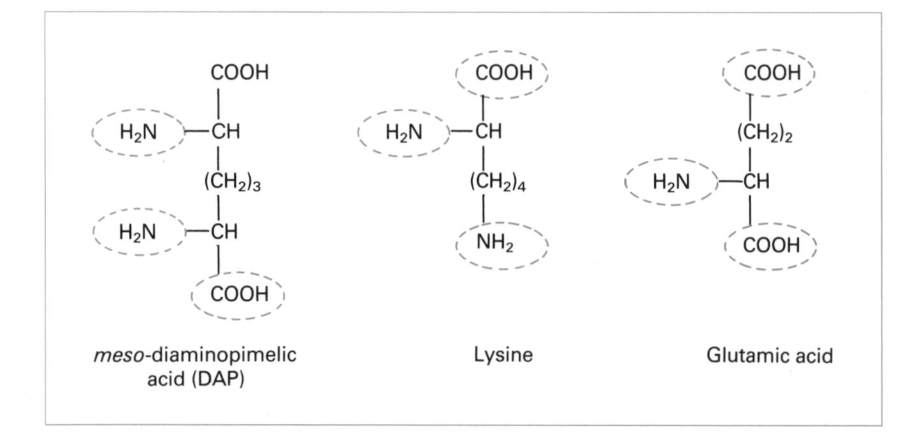


**Figure 5.22** The amino acid found at position three of the tetrapeptide chain is often required to participate in the formation of three peptide linkages: one of these is with the amino acid which precedes it and one with the amino acid which follows it in the tetrapeptide chain. The third is required for cross-bridge formation. Lysine and *meso*-diaminopimelic acid (DAP) each have two amino groups and these, together with a carboxyl group, allow formation of the three linkages. Glutamic acid is an acidic amino acid and because it has two carboxyl groups and one amino group can also form three peptide linkages.

usually D-alanine and the one next to it is usually a diamino acid such as lysine or diaminopimelic acid (DAP) (Figure 5.22).

**BOX 5.5**

**Lysozyme treatment of the Gram-positive rod, *Bacillus megaterium***

1. Intact rod-shaped cell

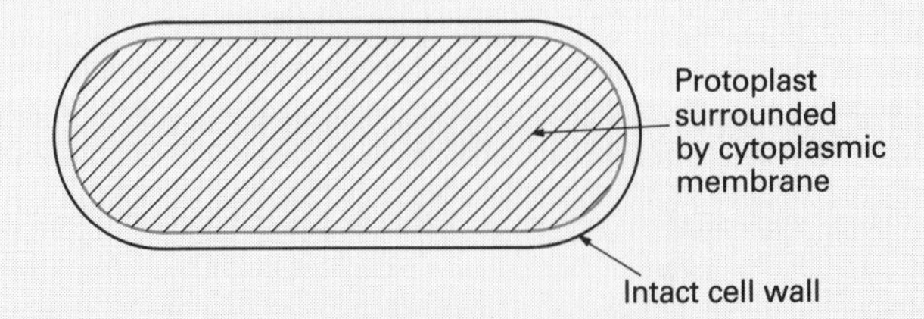


Lysozyme treatment of cells suspended in isotonic medium (for example 10% sucrose solution). Under these conditions the cell wall will be completely lost but because there will be no flow of water either into or out of the cell, the protoplast remains intact.

2. Intact spherical protoplast

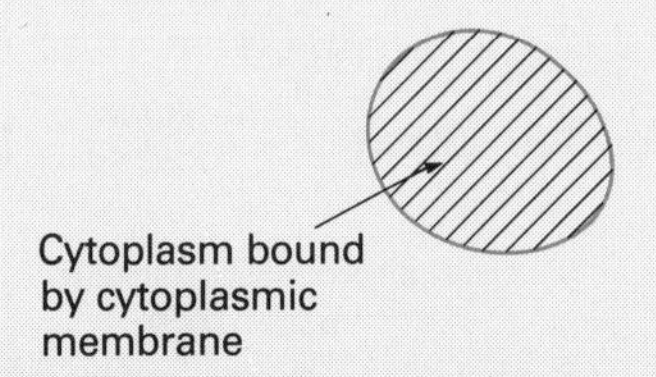


Since the protoplast is spherical, we can conclude that the cell wall is responsible for the shape of the intact cell.

Dilution of the suspending medium makes it hypotonic with respect to cytoplasm. As a result, water will flow into the protoplast and bursts the cell. We can conclude that the cell wall prevents the cell from bursting by resisting the increase in osmotic pressure from the influx of water.

3. Lysed protoplasts

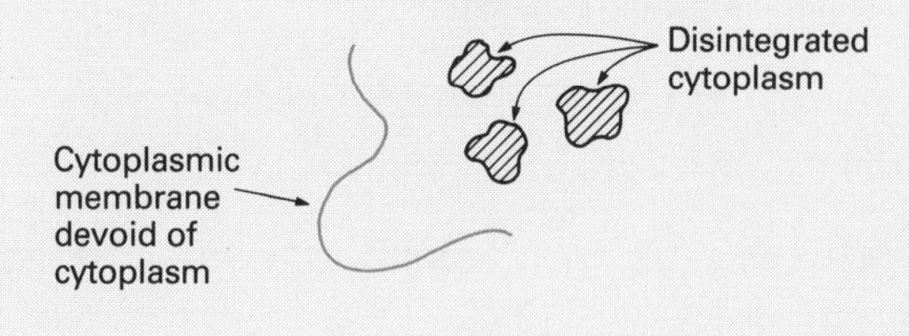


The peptidoglycan structure is completed by **interpeptide cross-bridges** which join together pairs of adjacent peptide chains. Again, these are variable in structure, but typically consist of one or more amino acid residues (Figure 5.23).

## Peptidoglycan biosynthesis

Peptidoglycan is a very large, insoluble molecule located on the outside of the cytoplasmic membrane. Peptidoglycan precursors are synthesized within the cytoplasm, transported through the cytoplasmic membrane, and then incorporated into the growing peptidoglycan molecule.

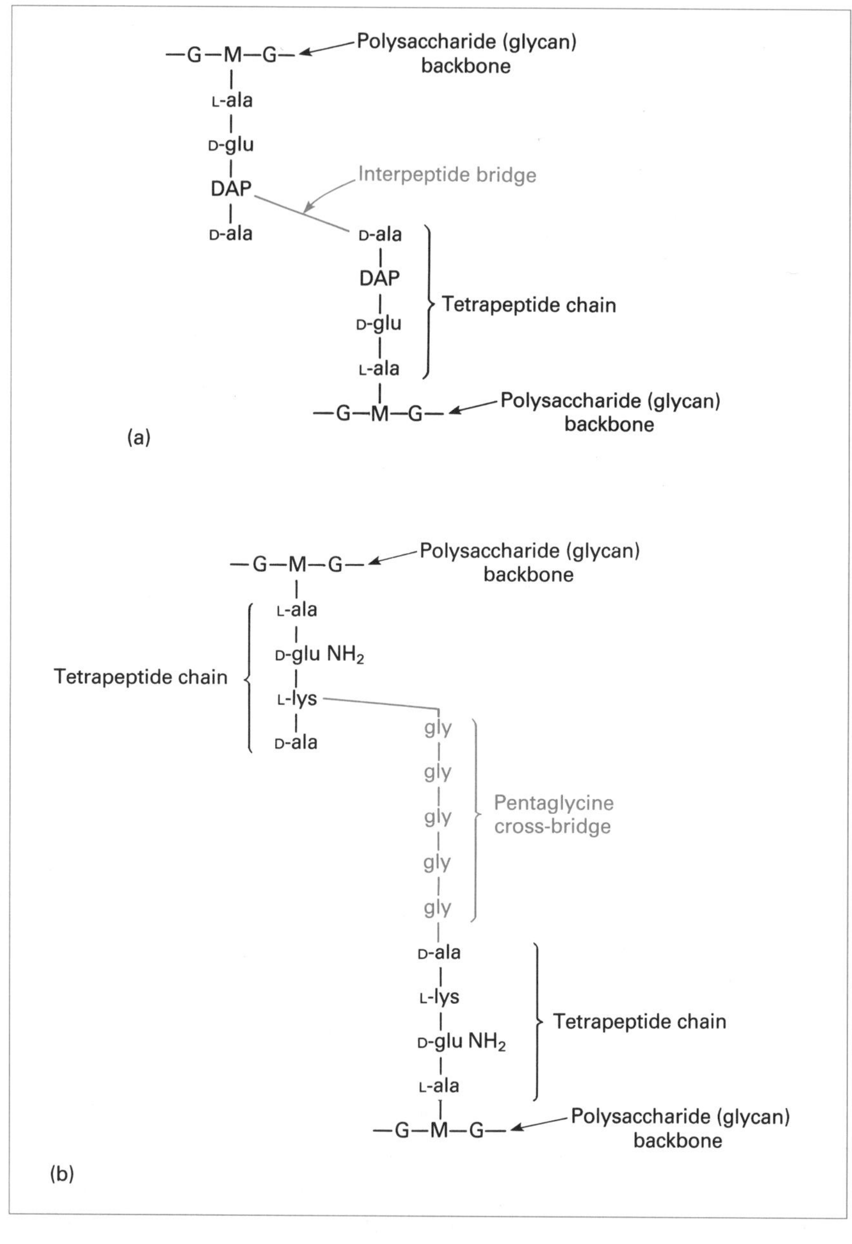


**Figure 5.23** Interpeptide cross-bridges in eubacteria. (a) This arrangement is seen in most Gram-negative eubacteria: a direct linkage joins the terminal D-alanine of one tetrapeptide chain to the DAP residue in an adjacent chain. Diaminopimelic acid (DAP), because it contains two amino groups, is able to make the three peptide bonds required. (b) This arrangement is seen in the Gram-positive eubacterium *Staphylococcus aureus*. Here the terminal D-alanine residue of one tetrapeptide chain is linked through a chain of five glycine residues to the lysine residue in an adjacent chain. Lysine contains two amino groups and is thus able to make up the three peptide bonds required.

Synthesis of precursors requires activation of *N*-acetylglucosamine with UTP to produce UDP-*N*-acetylglucosamine. Some of this is converted to UDP-*N*-acetylmuramic acid. Amino acids of the peptide chain (initially five residues long, not four as found in the completed structure)are added in a stepwise fashion to the UDP-*N*-acetylmuramic acid residue. Each addition requires a specific enzyme, together with ATP and either $Mg^{2+}$ or $Mn^{2+}$ ions as cofactors. The only exception is that two D-alanine residues are added together as a dipeptide unit to form the fourth and fifth members of

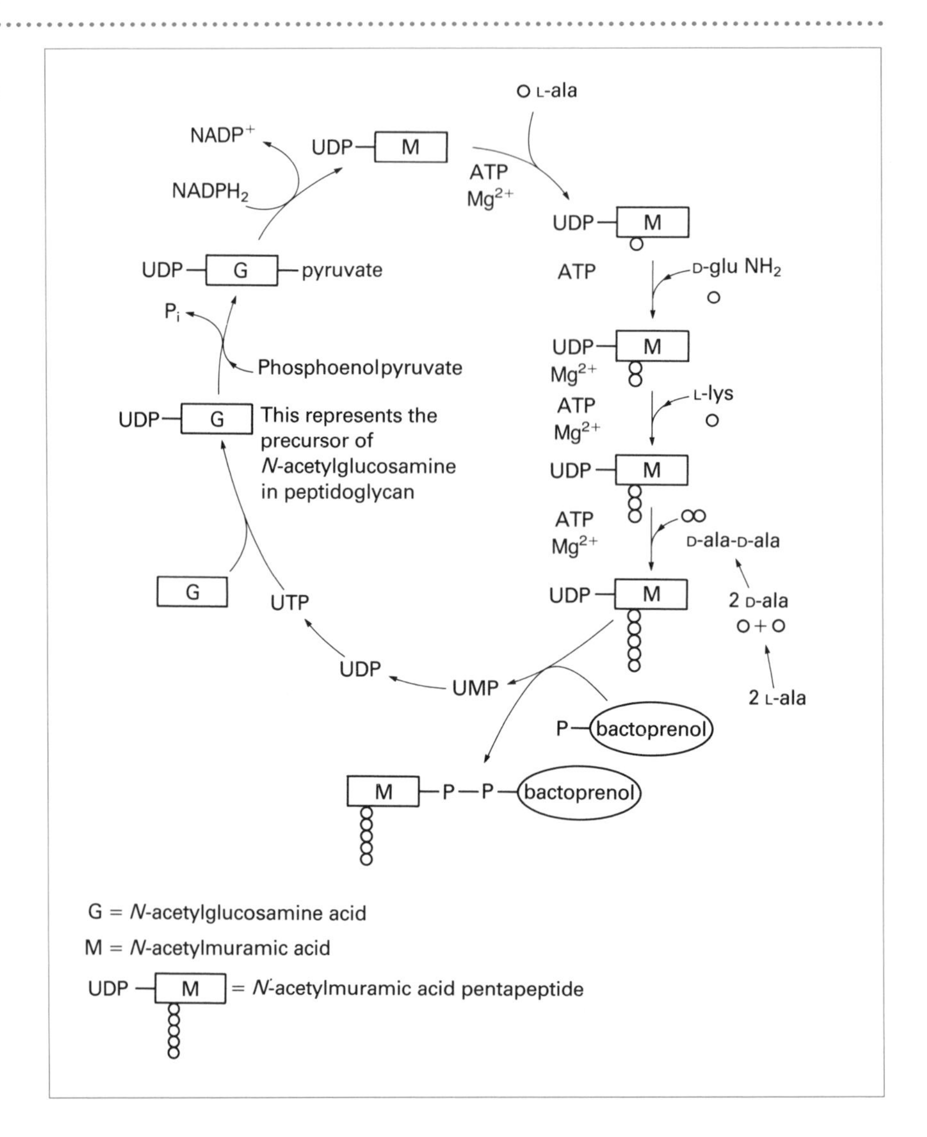


**Figure 5.24** Peptidoglycan synthesis: intracellular formation of nucleotide precursors.

the chain. At this stage the precursors **UDP-*N*-acetylglucosamine** and **UDP-*N*-acetylmuramic acid pentapeptide** have been produced (Figure 5.24). The two precursors are then linked to form a **disaccharide pentapeptide** unit. In this form, and linked to a lipid-soluble carrier called **bactoprenol**, the two precursors are carried across the cytoplasmic membrane (Figure 5.25).

Once on the outside of the cytoplasmic membrane, the disaccharide pentapeptide is incorporated into the growing peptidoglycan molecule (Figure 5.26), and the process is completed by the cross-linkage of adjacent pentapeptide chains. This final reaction typically makes use of a diamino acid in one peptide chain and the D-alanine residue at position four of the neighbouring chain. These two are linked by a peptide bond in a reaction

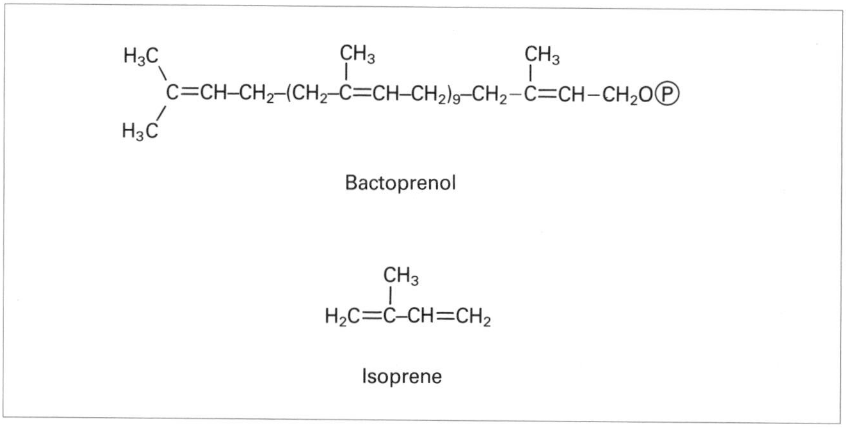

$$H_3C\backslash/H_3C>C{=}CH{-}CH_2{-}(CH_2{-}\overset{CH_3}{\overset{|}{C}}{=}CH{-}CH_2)_9{-}CH_2{-}\overset{CH_3}{\overset{|}{C}}{=}CH{-}CH_2O\text{Ⓟ}$$

Bactoprenol

$$H_2C{=}\overset{CH_3}{\overset{|}{C}}{-}CH{=}CH_2$$

Isoprene

**Figure 5.25** Bactoprenol (undecaprenyl phosphate). This carrier is required for the transport of precursors (*N*-acetylglucosamine and *N*-acetylmuramic acid pentapeptide) for peptidoglycan synthesis through the cytoplasmic membrane. It is also required for teichoic acid synthesis in Gram-positive eubacteria. Bactoprenol is formed from 11 (hence 'undeca') isoprenoid units. Isoprene is a branched five-carbon molecule and bactoprenol therefore contains a total of 55 carbon atoms.

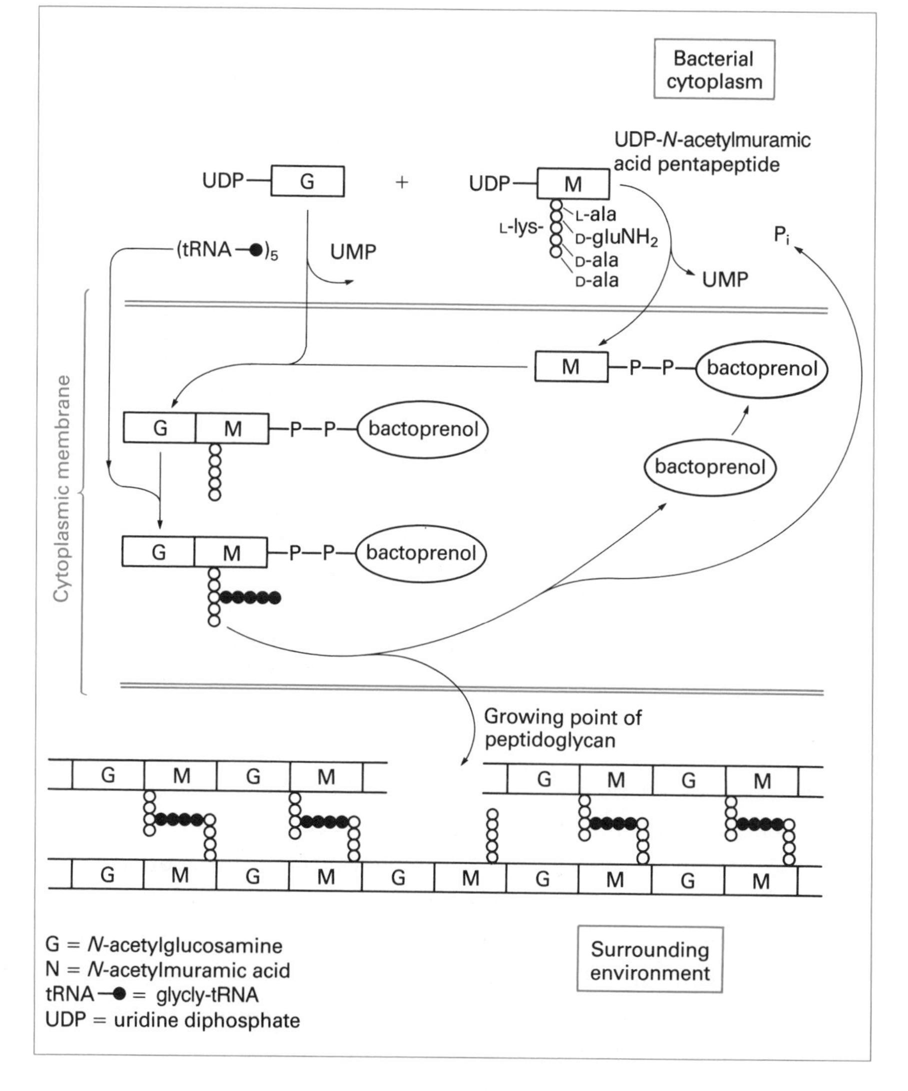


**Figure 5.26** Peptidoglycan synthesis: incorporation of disaccharide pentapeptide into the growing point of peptidoglycan.

**BOX 5.6**

**Penicillins produced by fungi of the genus *Penicillium* are important antibiotics**

Penicillin inhibits the final reaction (the bridge closure reaction) during peptidoglycan synthesis. This leads to the complete loss of the Gram-positive eubacterial cell wall or the production of a weakened Gram-negative cell wall. In both cases the cell **lyses** (bursts) and dies.

Because only eubacteria contain peptidoglycan, human and animal cells are unaffected by penicillin. It is this **selective toxicity** which makes treatment of bacterially induced diseases with antibiotics such as penicillin so effective. The industrial production of penicillin is described in Chapter 11.

Penicillin inhibits the bridge closure reaction because its shape is similar to the transition state of D-alanyl-D-alanine dipeptide (found at the end of the pentapeptide chain of peptidoglycan) bound to the surface of the transpeptidase enzyme responsible for the bridge closure reaction. Penicillin forms a covalent bond with a serine residue at the catalytic site of the transpeptidase and thereby excludes the correct substrate.

**Penicillin binding proteins (PBP)**

On the basis of their ability to bind to radioactive penicillin, a number of cell wall proteins have been identified and shown to have roles in cell wall synthesis and cell division. Examples of PBPs from *E. coli*, together with their properties, are shown in the table below.

| *PBP* | *Molecules per cell* | *Enzyme activities* | *Possible function* |
|---|---|---|---|
| 1A or 1B | 100 of each | Transglycosylase-transpeptidase (i.e. these are bifunctional enzymes) | Polymerization and simultaneous cross-linkage of peptido-glycan subunits |
| 2 | 20 | Transpeptidase | Cutting of pre-existing peptidoglycan before insertion of new sub-units (in conjunction with PBP4) |
| 3 | 50 | Transglycosylase-transpeptidase | Synthesizes peptidogly-can destined for septum formation before cell division |
| 4 | 110 | DD-endopeptidase, DD-carboxypeptidase | Hydrolyses cross-bridges during cell elongation |
| 5 | 1800 | DD-carboxypeptidase | Breakdown of unwanted pentapeptide |
| 6 | 600 | DD-carboxypeptidase | Breakdown of unwanted pentapeptide |

catalysed by an enzyme called a **transpeptidase**. Since this reaction is energy-dependent and takes place outside of the cell (where ATP is unavailable), the energy is provided by breaking the terminal peptide bond between the two terminal D-alanine residues. The final D-alanine is lost and the peptide chain originally five amino acids long is reduced to four. It is this last reaction which is inhibited by penicillin. Terminal D-alanine residues at the ends of peptide chains not involved in cross-bridge formation are removed by a different enzyme, **carboxypeptidase**.

**BOX 5.7**

**Penicillin enrichment technique**

Penicillin is effective only in cells where peptidoglycan synthesis is occurring. Cells which are not growing will not be killed. Because of this difference in action against growing and non-growing cells, penicillin can be used to increase the numbers of auxotrophic mutants in a mixed culture of mutant and non-mutant cells.

The technique depends upon using a medium containing penicillin and which allows growth of non-mutant individuals, but not that of auxotrophic cells. For example, mixtures of non-mutant cells and mutant auxotrophs which require the amino acid tryptophan for growth are incubated in a medium that lacks tryptophan. Under such conditions, non-mutant cells will be killed and mutant (auxotrophic) ones will survive. As a result, the number of mutant cells relative to non-mutant cells will increase.

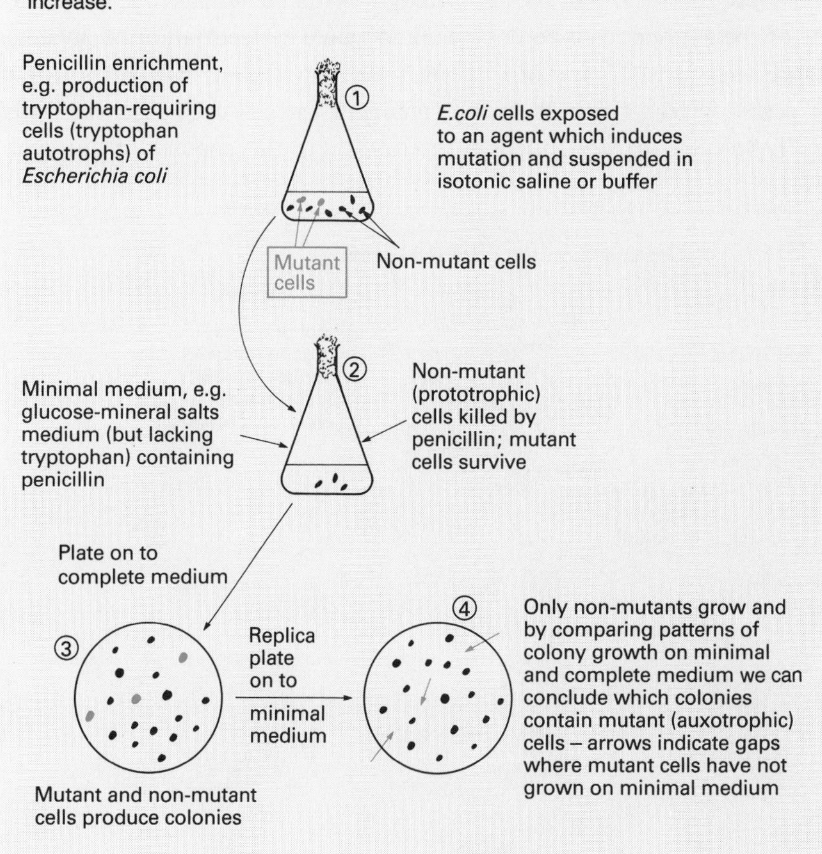


Recent work has shown that in *E. coli*, peptidoglycan is synthesized at approximately 200 sites in the cylindrical part of the rod-shaped cell. Each peptidoglycan chain is extended in a linear fashion around the cell circumference.

## Teichoic acids

This class of cell wall components is restricted to Gram-positive eubacteria. Teichoic acids are a mixed group of molecules with backbones of either

**polyribitol phosphate** or **polyglycerol phosphate** (they are **polyols**) but which contain a variety of other components including sugars, amino-sugars and amino acids (Figure 5.27). This group of cell wall components is often sub-divided, according to their location, into **membrane (lipo) teichoic acids** and **wall teichoic acids**. Although closely associated with the cytoplasmic membrane, the former group appear to extend through the cell wall to the external environment where they are able to react with specific antibodies. Because teichoic acids are negatively charged, they contribute to the overall negative charge of the bacterial cell surface and one of their functions is to ensure an adequate concentration of divalent cations such as $Mg^{2+}$ and $Mn^{2+}$ ions, to activate the enzymes responsible for peptidoglycan synthesis in the vicinity of the cell wall. Teichoic acids also have a regulatory function in controlling the activity of autolytic

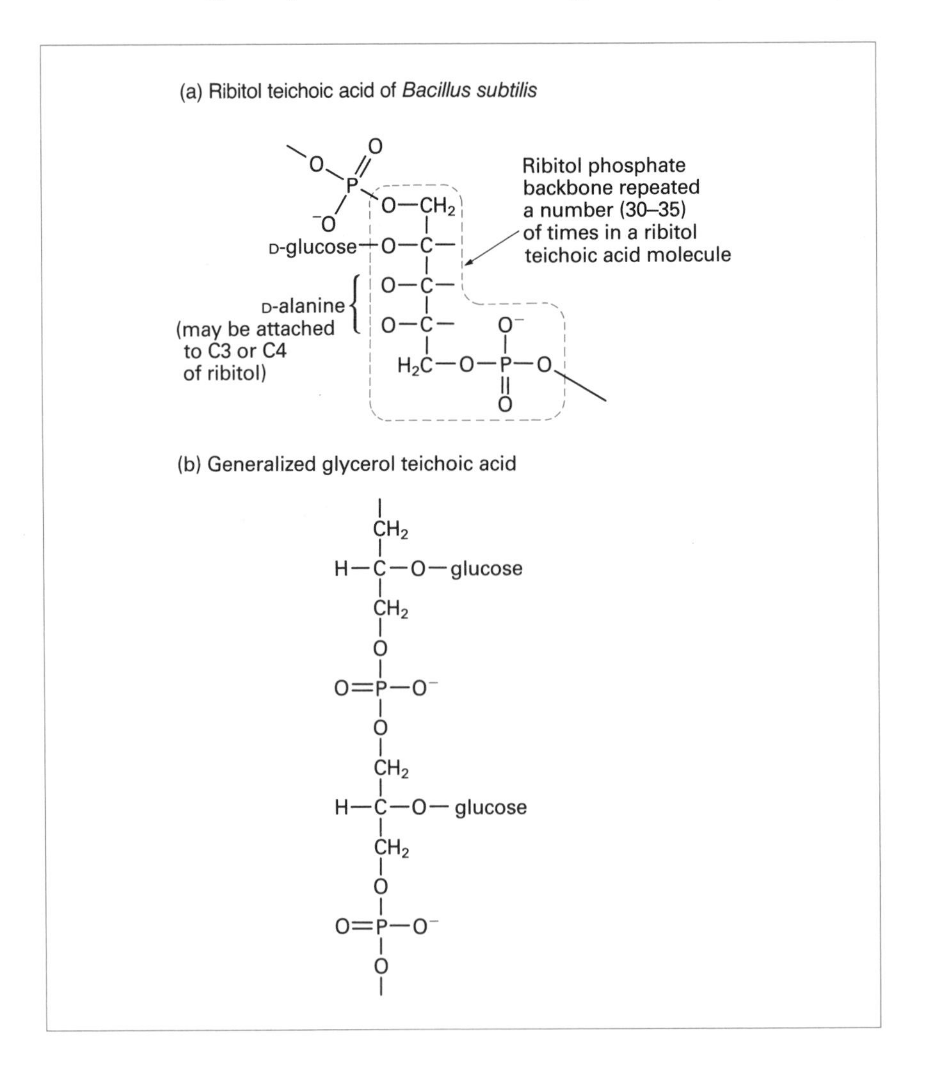


**Figure 5.27** Teichoic acids of Gram-positive eubacteria. (a) Ribitol teichoic acid of *Bacillus subtilis.* (b) Generalized glycerol teichoic acid.

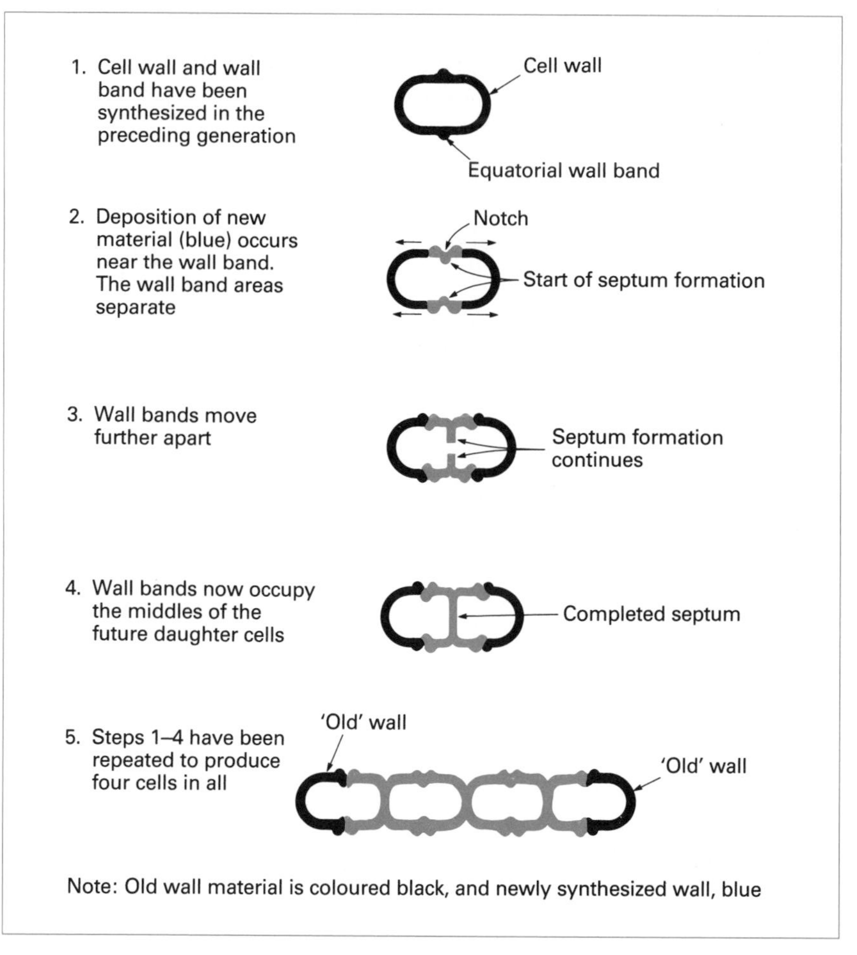


**Figure 5.28** Growth of the cell wall of *Streptococcus faecalis*.

enzymes which selectively cut the peptidoglycan macromolecule so that additional subunits can be added.

## Overall growth of the eubacterial cell wall

Synthesis of the cell wall of *Streptococcus faecalis* starts at the mid-line of the cell and proceeds outwards (Figure 5.28). Synthesis of cross-wall and peripheral wall occurs at the same time, so that the new halves of the daughter cells are formed in a back-to-back fashion. The old ends of the cocci remain intact but are gradually pushed further apart.

In contrast, the cylindrical portion of the cell wall of a rod-shaped bacterium such as *E. coli* is synthesized at many sites, so that new and old wall become interspersed. This observation is consistent with the finding that there are about 200 sites at which peptidoglycan is inserted into the cell wall. In contrast to the cylindrical part of the cell, it appears that insertion of peptidoglycan does not occur at the poles of rod-shaped bacteria. Here, peptidoglycan is conserved as a single unit.

**Table 5.2** The chemical composition of some archaebacterial cell walls

| *Characteristic cell wall component* | *Archaebacterial examples* |
|---|---|
| Pseudopeptidoglycan | Some methanogens including *Methanobacterium* |
| Thick Gram-positive wall made of polysaccharide | Some methanogens including *Methanosarcina* |
| Sulphated polysaccharides | *Halococcus* – an extreme halophile |
| Glycoprotein | Some methanogens, the extreme halophile *Halobacterium*, and the sulphur-dependent archaebacteria |
| Several proteins | Some methanogens such as *Methanococcus* and *Methanomicrobium* |
| No cell wall but a unique cytoplasmic membrane which resists powerful osmotic forces, an environmental temperature of 60°C, and a pH of 2. | *Thermoplasma* – a thermoacidophilic archaebacterium |

### The archaebacterial cell wall

Archaebacterial cell walls differ from those of eubacteria in that they do not contain peptidoglycan. Six types of cell wall are known and these are shown in Table 5.2.

## Chapter summary

In this chapter we began by emphasizing that microorganisms show considerable variation in the biosynthetic pathways used to synthesize the same end-products. Some blue-green bacteria produce all their cell material from carbon dioxide, nitrogen gas ($N_2$) and essential mineral salts. At the other end of the spectrum, protozoans and lactobacilli need many preformed amino acids, purines, pyrimidines and vitamins to be present in the environment because they lack the biosynthetic repertoire of the blue-green bacteria. We learned that the idea of biosynthetic load encapsulated this difference in biosynthetic capacity.

Next, we saw that the precursors for biosynthesis were often intermediates of metabolic pathways, such as glycolysis and the tricarboxylic acid cycle. We already know from Chapter 4 that these pathways are important in ATP synthesis. We learned that additional reactions, anaplerotic reactions, were needed to replenish the intermediates drained off for biosynthetic purposes.

We went on to describe the synthesis of small molecules such as amino acids, purines and pyrimidines which, we noted, would be used as building blocks for the synthesis of much larger, polymeric molecules, including polypeptide chains and nucleic acid molecules.

Following our consideration of the synthesis of small molecules, we turned to a description of the synthesis of some major macromolecules. Initially, we examined the production of polypeptide chains and noted that the amino acid sequences of individual polypeptides are ultimately determined by the sequence of bases in a DNA strand – in a gene. We learned that the overall strategy for polypeptide synthesis is the same for eukaryotes, eubacteria and archaebacteria, but there were variations in detail.

To facilitate our description of polypeptide synthesis we divided the process into two stages, transcription and translation. We saw that during transcription, the three types of RNA molecule (tRNA, mRNA and rRNA) were synthesized using one RNA polymerase enzyme in prokaryotic and three in eukaryotic organisms. The base sequence of a gene determined the base sequence of the complementary RNA molecule.

The translation stage is so called because a message in the form of the ribonucleotide sequence of an mRNA molecule is translated into the amino acid sequence of a polypeptide chain. We saw that this second stage of polypeptide synthesis is a complex process requiring ribosomes, tRNA molecules and a number of protein factors, in addition to mRNA. We concluded the section on polypeptide synthesis with a description of the signal hypothesis and its importance for the secretion of some proteins into the environment.

In the next section we examined the synthesis of the nucleic acids. We described the split genes of eukaryotic microbes and the way in which introns were removed from the primary transcript and the remaining exons spliced to produce a functional mRNA molecule.

We then learned about the steps required for the replication of DNA molecules. We noted the problems of duplicating a highly folded DNA molecule in the confines of an *Escherichia coli* cell and how these problems were overcome by enzymes such as helicase and DNA gyrase. We saw how the base sequences of daughter DNA strands were determined by the complementary base sequences of the parent strands; in other words, as Watson and Crick suggested, that DNA synthesis occurs by a semi-conservative mechanism. The requirement of an RNA primer was noted and the difference between continuous and discontinuous DNA synthesis examined. The contributions to the process of DNA polymerase and the DNA ligase enzymes were described.

Finally in this chapter we described the structures of bacterial cell walls and the synthesis of the major wall component, peptidoglycan. We examined the importance of bacterial cell walls in the context of penicillin action and the viability of bacteria.

## Further reading

D. Freifelder (1986) *Molecular Biology.* 2nd Edition. Jones & Bartlett. Boston – a well illustrated account of DNA replication, RNA and polypeptide synthesis.
A.G. Moat and J.W. Foster (1988) *Microbial Physiology.* 2nd Edition. John Wiley and Sons, New York – a concise treatment of microbial (not solely bacterial) physiology.
C. Smith and E.J. Wood (1992) *Molecular and Cell Biology.* Chapman & Hall, London – an up-to-date, well-written treatment.
C. Smith and E.J. Wood (1992) *Biosynthesis.* Chapman & Hall, London – a comprehensive account of biosynthesis taken from the biochemist's perspective.

## Questions

**1.** Multiple-choice questions:
(i) By reaction with which of the following does nitrogen become incorporated into the amino group of glutamic acid: (a) citrate, (b) fumarate, (c) α-ketoglutarate, (d) lactate, (e) pyruvate?
(ii) Which of the following is used to describe a reaction whose function is to replenish intermediates of an energy-producing pathway, which have been drained off for biosynthesis: (a) anaplerotic, (b) lysogenic, (c) anabolic, (d) catabolic?
(iii) From which of the following does the energy for the bridge closure reaction of peptidoglycan synthesis come: (a) ATP, (b) GTP, (c) acetyl-CoA, (d) the alanyl–alanine bond?
(iv) To which of the following is the disaccharide-pentapeptide precursor for peptidoglycan synthesis linked before it is transported across the cytoplasmic membrane and inserted into the growing peptidoglycan molecule: (a) lipopolysaccharide, (b) ATP, (c) UTP, (d) bactoprenol?

(v) Which of the following pathways is responsible for replenishing TCA cycle intermediates when *E. coli* is grown on acetate as a carbon and energy source: (a) glycerate, (b) glyoxylate, (c) glycolytic, (d) dicarboxylic acid?

**2.** Short-answer questions:

(i) Explain the meaning of the word anaplerosis.

(ii) Name a polymer found in Gram-positive eubacterial cell walls but not in Gram-negative ones.

(iii) List three classes of ribonucleic acid (RNA) required for polypeptide synthesis.

(iv) What are introns and exons?

(v) What is the function of DNA ligase?

**3.** Fill in the gaps:

The synthesis of one __________ of a DNA double __________ requires a number of __________. These include two DNA __________ (DNA polymerase I and __________). Both require __________ as substrates which are added to a __________ with a terminal __________ group which is available for reaction. Polymerization therefore consists of a reaction between a __________ group and the incoming __________. Following addition of the nucleotide, it supplies another free __________ group. Because each DNA __________ has a __________ terminus and a __________ terminus, the strand is polymerized in a __________ direction.

Choose from: $5' \rightarrow 3'$, primer, 3′-hydroxyl (four times) deoxyribonucleotides, DNA polymerase III, helix, enzymes, polymerases, strand (twice), 5′-phosphate, deoxyribonucleotide.

# 6 Coordination of metabolism

In this chapter we look at the ways in which an *individual* microorganism adapts its biochemical machinery to cope with changes in environmental conditions. Here we are not concerned with the long-term evolutionary changes which microorganisms undergo to produce new strains and species, but with physiological changes by means of which the individual survives when environmental conditions change. The kinds of environmental adaptations we have in mind include temperature, pH, osmotic pressure and nutrition.

The use of glucose as an energy source is described in Chapter 4.

What happens, for example, when a microorganism using glucose as a food finds that the glucose source has run out? It must find an alternative if it is to continue to grow and reproduce. This can happen to *Escherichia coli* living in your intestinal tract. If glucose becomes depleted but another sugar, say lactose, is available, the *E. coli* cell can switch its metabolism to use lactose instead. Such a switch requires enzymes which were not needed for growth on glucose. The *E. coli* cells in this example have *adapted* to lactose as a carbon and energy source, a change which requires the synthesis of different enzymes.

We have just said that a change from glucose to lactose requires different enzymes. What does this mean in molecular terms? In practice, physiological adaptation requires a change in either the *activity* or in the *amount* of one or more enzymes. Theoretically, the use of lactose could require either activation of enzymes previously present in an inactive form during growth on glucose, or the synthesis of new enzymes which were not present at all during growth on glucose. In bacteria at least, this kind of physiological adaptation can sometimes require changes in enzyme activity and sometimes changes in amounts of enzyme molecules produced. The latter mechanism accounts for adaptation to growth on lactose in *E. coli.*

## Control of enzyme activity

In this section we examine an example of the ways in which the activity of enzymes may change in response to changes in the nutritional content of the environment. Provided that appropriate nitrogen, sulphur and phosphorus sources are present, *E. coli* can produce all its cell material from a single carbon source such as glucose (Chapter 5). This is because *E. coli* possesses all the enzymes it needs to synthesize the proteins, nucleic acids, lipids and carbohydrates required for growth and reproduction. As an alternative, *E. coli* cells can take advantage of external sources of preformed amino acids, purines and pyrimidines and build them up further to produce similar cell materials.

If the starting materials for biosynthesis are available from the external environment, *Escherichia coli* is able to switch off its internal synthesis. Internal production of these substances requires a considerable investment of metabolic energy. It should, therefore, come as no surprise that control mechanisms have evolved to capitalize on the use of external, preformed precursor molecules.

The very large chemical components of living cells are built up in a stepwise manner from much smaller molecules. For example, polypeptides are assembled from twenty amino acids. For convenience, such low-molecular-weight building blocks are called **macromolecular precursors** or **precursors**.

Allosteric enzymes show sigmoid (S-shaped) kinetics when rection velocity (*V*) is plotted against substrate concentration ([S]).

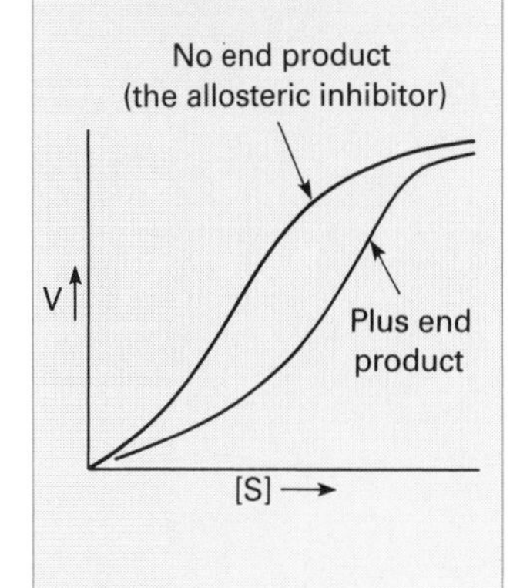


## Feedback inhibition

One way in which the throughput of a biosynthetic pathway can be reduced is by **feedback inhibition**. When a biosynthetic end-product exceeds a particular internal concentration, it binds to an earlier enzyme of the pathway concerned and inhibits its activity. The internal synthesis of the end-product decreases and may stop altogether. The external supply of the end-product can then be used instead. If the external supply becomes exhausted, inhibition is lifted and internal synthesis recommences.

**(a) Allosteric effects**. The way in which the end-product inhibits an earlier enzyme of the pathway is worth looking at further. The end-product binds to the enzyme at a site which is distinct from the active site of the enzyme. This second site is called the **allosteric site** and the end-product the **allosteric effector**. When the end-product binds to the allosteric site, the shape of the whole enzyme changes, including the active site. This change in conformation leads to a reduction in the enzyme's affinity for its substrate and the enzyme is inhibited (Figure 6.1). Allosteric control is a widespread phenomenon in microorganisms and plays an important part in the coordination and regulation of many biosynthetic and degradative pathways.

**(b) Covalent modification**. Although allosteric effects are very important as cellular regulatory mechanisms, the activity of some enzymes is controlled by a different mechanism. This second method is also used to inhibit an enzyme in the presence of excess end-product. The mechanism

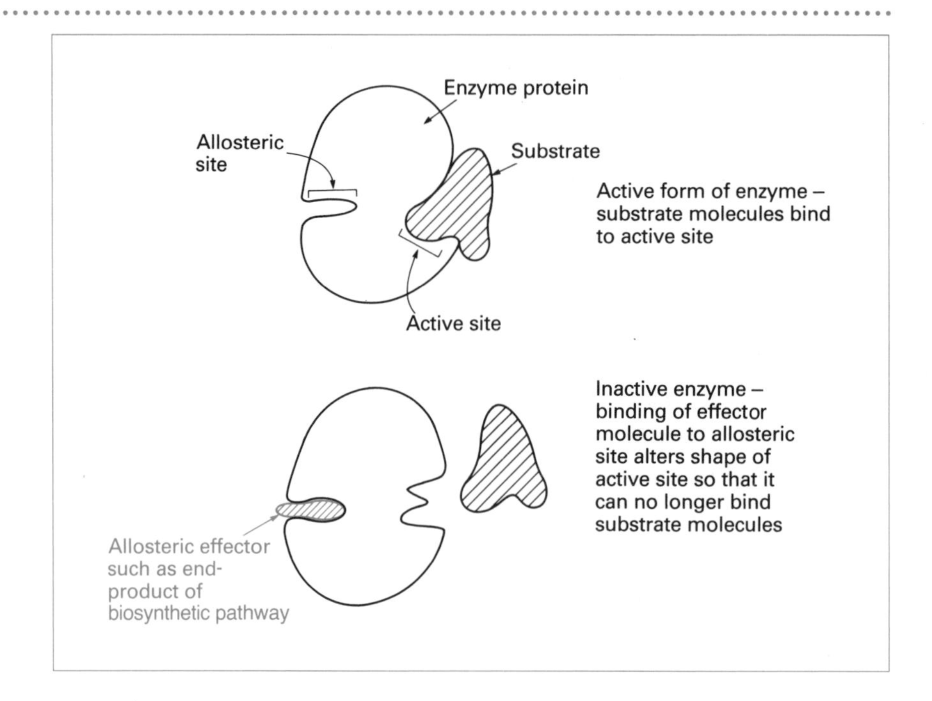


**Figure 6.1** Allosteric inhibition and feedback inhibition.

requires the modification of the relevant enzyme by linking it to a specific chemical group by a covalent bond. As we saw with allosteric inhibition, the addition of such groups changes the shape of the enzyme. The affinity for its substrate is reduced and the enzyme remains inhibited until the concentration of the end-product in the external environment has fallen. Table 6.1 lists a number of enzymes and the chemical groups used to modify them.

In microorganisms, a biosynthetic pathway may branch to produce two or more reaction sequences which lead to the formation of two or more end-products (Figure 6.2). From what you have already learned about the control of biosynthetic pathways, you will realize that branched pathways present a potential problem. If one end-product becomes present in excess, there is a danger that inhibition of enzyme activity (say enzyme E1 in Figure 6.2) would also prevent synthesis of a second end-product which may *not* be present in excess. If this happened, the organism might become starved of an essential metabolite and would eventually die. Microbes have evolved a variety of mechanisms to prevent such occurrences.

**(c) Concerted or multivalent feedback inhibition**. End-products separately have no effect on enzyme E1, but together they are efficient inhibitors. This situation is found in the regulation of the lysine–methionine–threonine pathway in *Rhodopseudomonascapsulatus*, *Bacillus polymyxa* and *Bacillus subtilis*.

**BOX 6.1** The control of lysine production in *Brevibacterium flavum.*

**The use of a mutant strain of a microorganism insensitive to feedback control to increase industrial production of the amino acid lysine**

The use of a microorganism to produce large quantities of a commercially valuable end-product is often limited by the organism's ability to regulate its own biosynthetic activity. **Regulatory mutants** which have lost one or more feedback control mechanisms by mutation synthesize such substances in much larger amounts and are much prized by microbial biotechnologists. Lysine is an example of such an end-product. It can be used to improve the nutritional value of cheap, lysine-deficient plant proteins. The usual source of lysine in the human diet is more expensive animal protein. By adding microbially produced lysine to a vegetable diet, it can be inexpensively given the nutritional value of meat.

Synthesis of the amino acid lysine is subject to concerted feedback inhibition by a combination of lysine and another amino acid, threonine, in *Brevibacterium flavum*, limiting the production of lysine. A strain of the bacterium has been isolated which contains a mutation in the structural gene for aspartokinase, the first enzyme of the pathway, such that the shape of its allosteric site is changed. This means that lysine (with threonine) can no longer bind to the altered allosteric site. Consequently, production of lysine occurs in an uncontrolled manner. From the industrial microbiologists' view, mutant *B. flavum* is a particularly valuable asset, because it produces enormous quantities of lysine. Industrial production of amino acids is described in greater detail in Chapter 11.

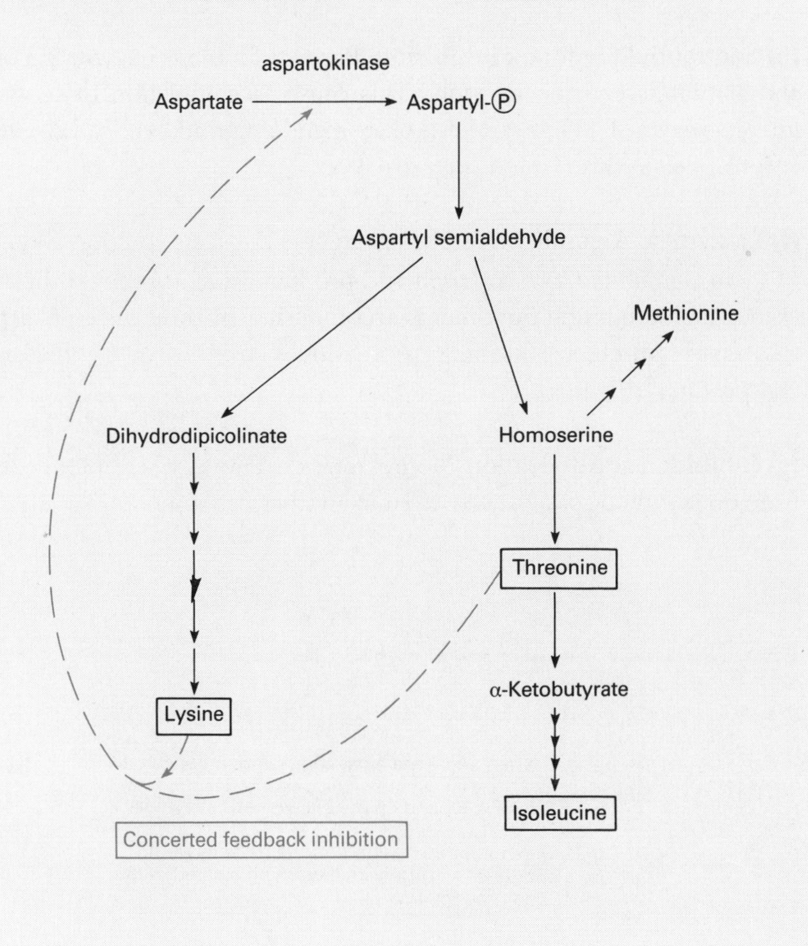

**Table 6.1** Examples of covalent modification of some bacterial enzymes

| *Enzyme* | *Organism* | *Modification* |
|---|---|---|
| Glutamine synthetase | *E. coli* and others | Addition of adenine residues |
| Isocitrate lyase | *E. coli* and others | Addition of phosphate groups |
| Citrate lyase | *Rhodopseudomonas gelatinosa* | Addition of acetyl groups |

**(d) Cumulative feedback inhibition.** A particular end-product inhibits enzyme E1 by a particular amount. The presence of a second end-product increases the extent of inhibition. If all end-products are present in excess, enzyme E1 is completely inhibited. Glutamine synthetase of *E. coli* possesses eight allosteric sites and is sensitive to cumulative feedback inhibition by eight compounds (Figure 6.3a).

**(e) Sequential feedback inhibition.** Product f inhibits enzyme E3 only, and g inhibits enzyme E5 only. This causes accumulation of c, which inhibits enzyme E1. The synthesis of aromatic amino acids in *Bacillus subtilis* is controlled in this fashion (Figure 6.3b).

**(f) Isozymes.** A number of enzymes catalyse the same reaction and each of them is inhibited by a different end-product. In *E. coli* the synthesis of lysine, methionine and threonine is accomplished by three different aspartokinases which are subject to feedback inhibition by different end-products (Figure 6.3c).

**(g) Inhibition and activation.** More complex pathways, where intermediates from one synthetic pathway are fed into another, are sometimes controlled

**Figure 6.2** A branched biosynthetic pathway.

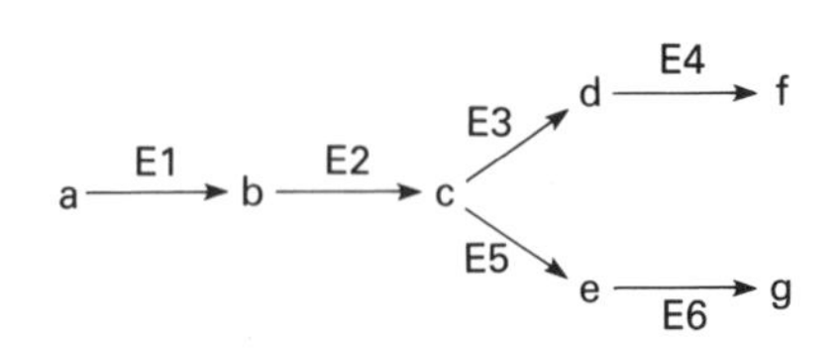


a, b, c, d, e, f and g are biosynthetic intermediates

E1, E2, E3, E4, E5 and E6 are biosynthetic enzymes

If E1 were inhibited by end-product g, there would be a danger of an organism becoming starved of the other end-product, f.

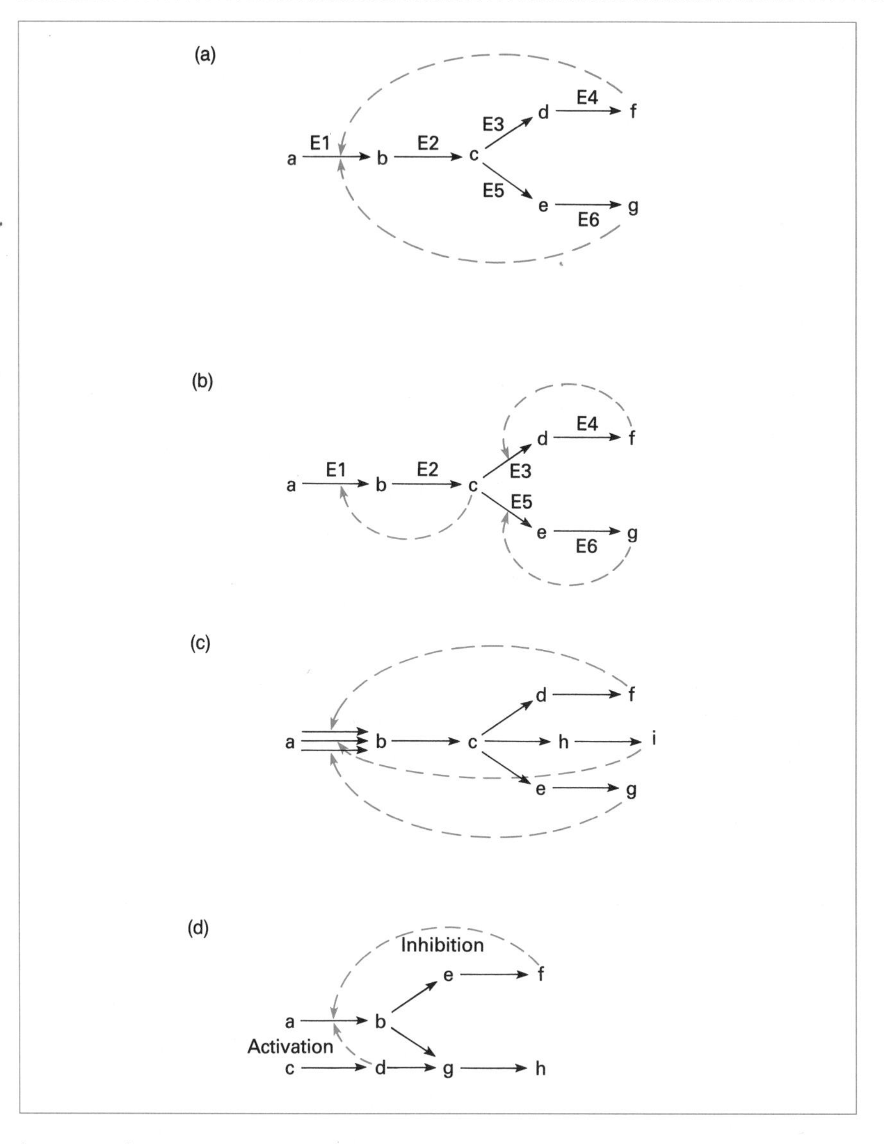


**Figure 6.3** Patterns of feedback inhibition found in branched biosynthetic pathways in bacteria.

by a combination of allosteric inhibition and activation. The end-product of one pathway is an inhibitor, and the last intermediate in the other pathway, before the two pathways join, is an activator (Figure 6.3d). The enzyme carbamoyl phosphate synthase in *E. coli* is regulated in this manner.

## Control of enzyme synthesis

So far we have described feedback inhibition as a method of changing the *activity* of an enzyme to accommodate to variations in environmental conditions. In this section, we look at a second type of regulatory mechanism in which the *amounts* of enzymes present in a cell can be made to fluctuate with changes in environmental conditions. Enzymes whose levels can be made to vary in this way are sometimes called **adaptive enzymes**. Adaptive enzymes are usually divided into two categories called **inducible enzymes** and **repressible enzymes**. In the following sections, we examine the importance of these two groups of enzymes.

**Allolactose as the inducer of the *lac* operon**
We now know that allolactose (an isomer of lactose) is the true inducer which binds to the *lac* repressor. Lactose does not possess this ability. Allolactose is produced by the small number of β-galactosidase molecules which are present in an uninduced cell. Once formed, allolactose molecules bind with the *lac* repressor to prevent its interaction with the *lac* operator. Transcription of the *lac* genes is initiated and metabolism of the remaining lactose molecules rapidly follows.

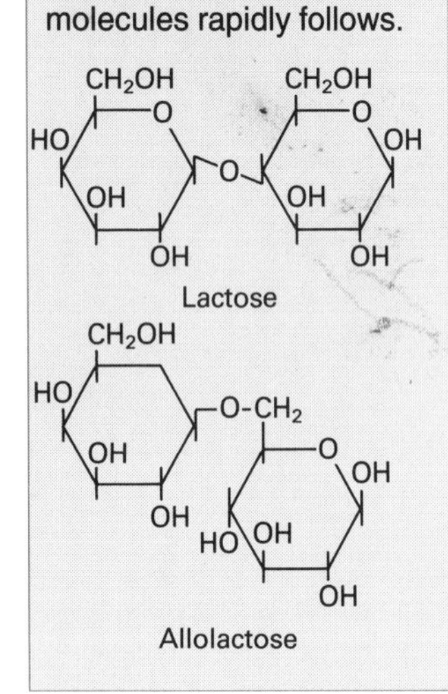


### Inducible enzymes

As we now know, the preferred carbon and energy source for *Escherichia coli* is glucose, though lactose may be utilized if glucose is unavailable. The enzymes required for growth on lactose are synthesized only if lactose is present in the environment and glucose is absent from it; in other words, they are synthesized only when needed. The lactose-utilizing enzymes are, then, examples of inducible enzymes because they are formed only in response to the presence of an **inducing substance** (lactose) in the environment. It would, of course, be possible for *E. coli* to synthesize the lactose enzymes (but not use them) even in the presence of glucose. In other words, they could be **constitutively** formed in much the same way as the glucose-utilizing enzymes are synthesized. The big advantage of producing inducible enzymes is that the organism does not waste energy synthesizing enzymes when they are not required.

The investigation of the molecular mechanisms underlying enzyme induction required a combination of genetic and biochemical approaches, and the experimental results obtained have provided considerable insight into the mechanisms of control of gene expression in bacteria. Information obtained from studies of the *lac* system of *E. coli* has formed the basis for the investigation of the control of gene activity in more complex organisms, including man, and may have a bearing on our understanding of diseases such as cancer, which are thought to be caused by a breakdown in the regulation of gene activity and cell division.

**(a) Enzymology of lactose catabolism**. Although both compounds are sugars, glucose is a monosaccharide and lactose is a disaccharide. Lactose consists of a glucose residue joined (by a β1–4 link) to a galactose residue (Figure 6.4a). For this reason, lactose is an example of a compound known as a β-galactoside.

The first step in the utilization of lactose by an *E. coli* cell is its uptake by the cell. Uptake is catalysed by the first of the *lac* enzymes, β-galactoside permease, which transports lactose across the cell envelope. Once inside the cell, a second *lac* enzyme, β-galactosidase, present at a basal level

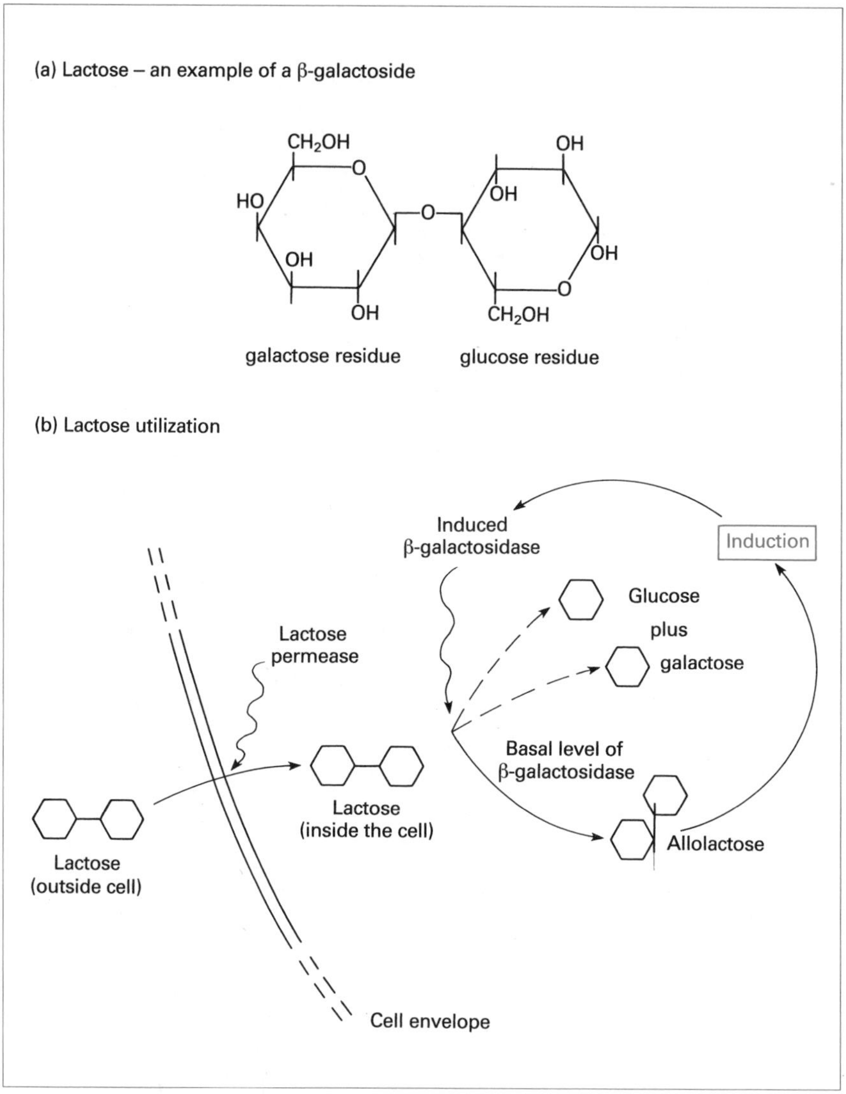

**Figure 6.4** Lactose and its utilization by *Escherichia coli.*

concentration, converts lactose to allolactose. The same enzyme at an induced level then hydrolyses further molecules of lactose to form one molecule of galactose and one of glucose: both of these molecules can then be further metabolized to provide energy (Figure 6.4b).

The third and final enzyme which makes up the *lac* enzyme system is β-galactoside transacetylase. A purified preparation of this enzyme can catalyse the transfer of an acetyl group to a β-galactoside. However, the role of the enzyme in nature is obscure since a mutant strain of *E. coli* which *lacks* this enzyme grows as well in the laboratory as a non-mutant strain. Some workers believe that, in the natural environment, the transacetylase forms part of a detoxification system.

**BOX 6.2**

**The naming of bacterial genes**

Individual bacterial genes are named according to a simple system, shown here by reference to *Escherichia coli.* Genes whose products have related functions (for example, enzymes of the same metabolic pathway) are shown as three letters set in italics, for example *lac.* To distinguish between genes of related function, the three italicized letters are followed by a capital letter. For example, *lacY* is a gene which belongs to the *lac* group of genes and which is responsible for the production of the enzyme β-galactoside permease. The normal (or **wild-type**) form of a particular gene is sometimes given a '+' superscript (for example $lacY^{+}$). Mutant forms are usually given a '–' superscript ($lacY^{-}$). For historical reasons, exceptions to this general rule are found: certain genes of the *lac* system designated $lacO^{c}$ and $lacI^{s}$ are both mutant *lac* genes.

The central feature of the *lac* system is the inducibility of the enzymes involved. Although levels of the three enzymes are very low in environments lacking lactose, cells growing in the presence of lactose contain thousands of times more of each of the three types of these enzyme molecules than do uninduced cells. The low levels of enzymes present in uninduced cells are known as the **basal levels**. It is these basal levels of β-galactoside permease and β-galactosidase which are responsible for the initial uptake of lactose (inducer) molecules from the environment and converting them to allolactose (Figure 6.4). Allolactose is then responsible for initiating the molecular processes which lead to increased rates of synthesis of the three *lac* enzymes.

Jacob and Monod measured the levels of *lac* enzymes present in *E. coli* cells growing under differing experimental conditions. Since their pioneering work, it has proved to be most convenient if levels of β-galactosidase are measured. For this purpose, a different β-galactoside, not normally present in the natural environment, is used as a substrate. This is *o*-nitrophenyl-β-D-thiogalactoside (ONPG for short!) which has the advantage of producing a yellow compound, *o*-nitrophenol (ONP), as one of the products of the reaction. This can be measured photometrically and used to quantify the amount of enzyme present in a cell at any given time.

For experimental investigation it has also proved convenient to separate the ability of a β-galactoside to act as an inducer from its role as an enzyme substrate, properties both possessed by the natural substrate, lactose. For this reason, a second β-galactoside, isopropyl-β-D-thiogalactoside (IPTG), is often used experimentally as an inducer. This compound has the advantage that, although it can induce synthesis of the *lac* enzymes, it is not a substrate for β-galactosidase. During investigations using different concentrations of inducer, the concentration of IPTG added to a laboratory culture would remain constant but the concentration of lactose would decrease. This is because lactose is also a substrate for the enzyme and would,

Unravelling the mechanism by which allolactose initiates increased synthesis of *lac* enzymes culminated in the award of the Nobel Prize to the French biologists Francois Jacob and Jacques Monod.

**BOX 6.3**

**Some synthetic analogues of lactose are used experimentally to study the *lac* operon**

(a) **Gratuitous inducers** bind to the *lac* (apo)repressor and initiate transcription of the *lac* structural genes. They are not substrates for any of the *lac* enzymes. The experimental advantage of using a gratuitous inducer is that once it is added to a bacterial culture, its concentration remains constant.
(b) **Chromogenic substrates** are useful because hydrolysis by β-galactosidase produces a coloured product. The intensity of the colour is proportional to the amount of enzyme present and can be quantified using a spectrophotometer.

| *Analogue* | *Chromogenic* | *Gratuitous inducer* |
|---|---|---|
| *o*-Nitrophenyl-β-galactoside (ONPG) | Yellow colour | No |
| 5-Bromo-4-chloro-3-indolyl-β-galactoside (X-gal) | Blue colour | No |
| Isopropyl-β-thiogalactoside (IPTG) | No | Yes |

Although lactose is a substrate and an inducer for the *lac* enzymes, its products are not coloured.

therefore, be degraded by β-galactosidase. A compound such as IPTG, which is an inducer but not a substrate, is often called a **gratuitous inducer**.

Once a convenient method for the determination of different levels of the *lac* enzyme had been developed, changes in levels could be conveniently followed under different experimental conditions. This paved the way for the use of genetic techniques for the production and investigation of mutant strains to elucidate the induction process at the molecular level.

## Induction of the *lac* operon

The observations which Jacob, Monod and their co-workers made concerning the induction of the lactose-utilizing enzymes of *Escherichia coli* led them to put forward the **operon model**. The operon model is the starting point for the investigation of the mechanisms of regulation of gene activity in many kinds of living organisms. As far as the *lac* system is concerned, the essential components are as follows:

1. *lacZ*, *lacY* and *lacA* are structural genes and determine the amino acid sequences of the *lac* enzymes, β-galactosidase, β-galactoside permease and β-galactoside transacetylase respectively.
2. *lacI* is a regulatory gene and specifies an allosteric (apo)repressor molecule.
3. *lacO* is the operator site. The (apo)repressor binds to the operator site (*lacO*) in the absence of lactose.

4. *lacP* is the promoter site. RNA polymerase binds to the *lacP* promoter site before synthesizing a complementary mRNA molecule.

We shall now see how these regulatory elements account for the observed facts about the induction of the *lac* enzymes.

In the absence of lactose, the (apo)repressor binds to the operator region and prevents transcription of the *lac* structural genes by RNA polymerase (Figure 6.5a). In the presence of an inducer such as lactose, the (apo)repressor forms a complex with an inducer molecule. As a result, the (apo)repressor can no longer bind to the operator. In the absence of the (apo)repressor, RNA polymerase binds to *lacP* and passes along the structural genes *lacZ*, *lacY* and *LacA*, transcribing them into the corresponding polycistronic (or polygenic) mRNA molecule. The *lac* messenger can then be translated by the ribosomes to produce the *lac* enzymes. The consequence of all this is that when lactose is the energy source, the *lac* enzymes are synthesized in large amounts (Figure 6.5b). In the absence of lactose (or any other β-galactoside) enzymes remain at basal levels.

**Figure 6.5** Negative control of the lactose (*lac*) operon of *Escherichia coli.*

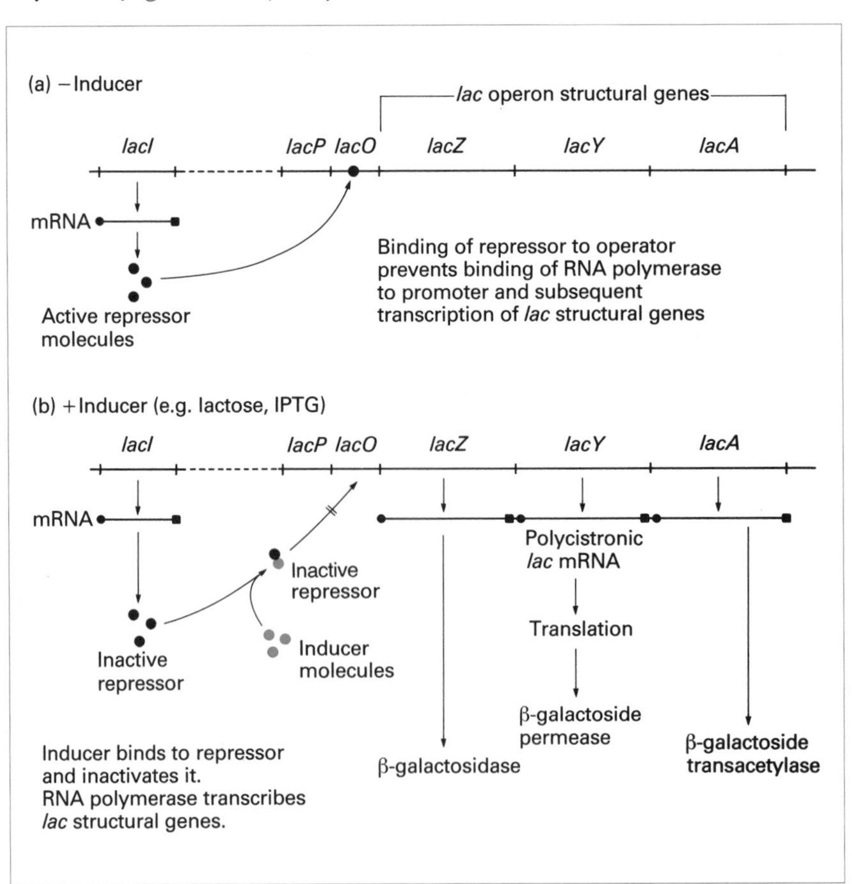

Remember that the *lac* enzymes are induced only when glucose is absent. From this, you may have concluded that glucose itself overrides the induction mechanism, and you would be right. This process is called **catabolite repression** (the **glucose effect** or **glucose repression** are alternative names for the same thing).

### Catabolite repression

In our examination of enzyme induction, we have been careful to emphasize that the enzymes needed specifically for lactose utilization are synthesized only provided that two conditions are satisfied. First, lactose or another β-galactoside must be present in the environment. Second, glucose must be absent. In other words, glucose is used preferentially as a carbon and energy source by many bacteria, including *Escherichia coli.*

Glucose overrides the induction of the enzymes of a number of different operons (including the *lac* operon) in the following fashion. If we continue to use the *lac* system as our example, two further regulatory elements are required in addition to those we have previously described:

1. *CRP* specifies a catabolite receptor protein which is able to bind to cyclic adenosine monophosphate (cAMP).
2. *cAMP–CRP* receptor site, which is thought to be part of the *lac* promoter (*lacP*) and which is able to bind to the cAMP–CRP complex.

When *E. coli* uses glucose as a carbon and energy source, the growth rate is high and, for reasons which are not fully understood, the level of cAMP is low. On lactose, the growth rate is lower but the intracellular concentration of cAMP is high. This means that a large quantity of cAMP–CRP complex will also be formed which can then bind to the cAMP–CRP receptor site. This interaction stimulates binding of RNA polymerase to *lacP*, and synthesis of the *lac* enzymes follows (Figure 6.6).

Notice that binding of the cAMP–CRP complex to its receptor site *activates* the induction process. Binding of inducer to operator *inhibits* the induction process. Because of these differences, induction of the *lac* operon is described as **negative control**, whereas catabolite repression is an example of **positive control**.

The importance of catabolite repression in the efficiency of microbial biotechnological processes, such as penicillin production, is described in Chapter 11.

### Isolation of the lac repressor

In an attempt to support the Jacob and Monod operon theory, Gilbert and Muller-Hill, working at Harvard University, attempted to isolate the protein *lac* (apo)repressor (thereby demonstrating the existence of such a molecule) and then to investigate some of its properties.

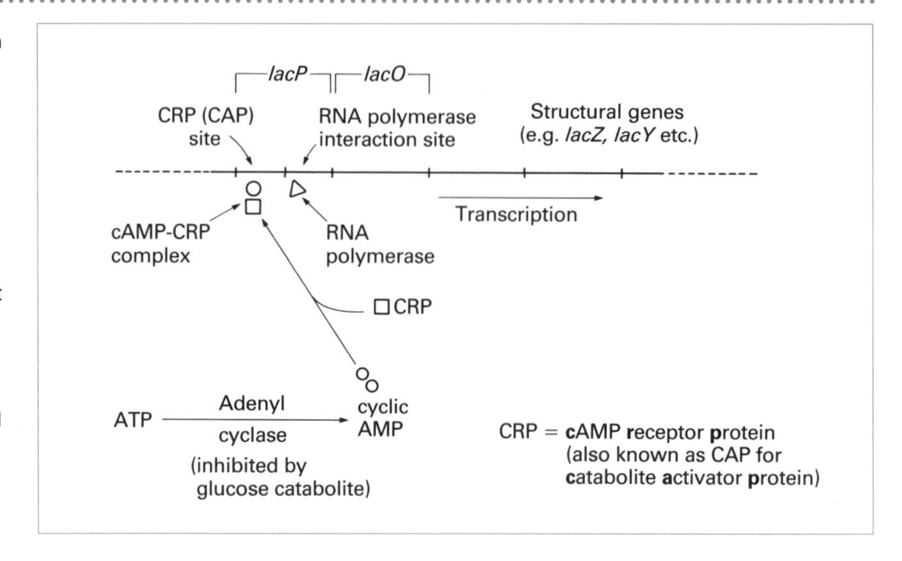


**Figure 6.6** Catabolite repression in *Escherichia coli.* When the organism is growing on glucose, the *lac* enzymes are not synthesized even if lactose (or any other *lac* inducer) is also present. This 'glucose effect' is mediated through cyclic AMP (cAMP). In the presence of glucose, the level of cAMP in an *E. coli* cell is low; in its absence it is high. In the absence of glucose, cAMP binds to a regulatory protein (CRP) to form a complex. The complex binds to a site within the *lac* promoter and enhances the binding of RNA polymerase to the RNA polymerase site. In the absence of glucose, therefore, transcription of the *lac* structural genes occurs, provided that an inducer is also present.

Since the number of repressor molecules in a wild-type *E. coli* cell is small (about ten) and represents only 0.001% of the total cell protein, a more plentiful source of repressor molecules was used as a starting point. This was in the form of a mutant strain of *E. coli* with more efficient synthesis of repressor. The mutant strain was also infected with a bacteriophage which carried the *lacI* gene and which could, therefore, further increase the number of repressor molecules produced by this strain to something like 20 000 molecules (2% total cell protein) per cell.

The isolated repressor molecule was shown to consist of four identical subunits, each with a molecular weight of 37 000 and each with a binding site for inducer. Kinetic studies showed that the repressor binds specifically to the *lac* operator region, which the repressor protein finds by diffusing along the DNA.

Once purified repressor became available, it was possible to isolate the operator region and to determine its base sequence. The *lac* operator was found to consist of 28 bp with a twofold axis of symmetry which, in the absence of an inducer, binds a molecule of *lac* (apo)repressor (Figure 6.7). Such symmetric DNA sequences are now known to be widely used by proteins to recognize specific DNA binding sites. An important feature of this region of the *lac* operon is that the operator and promoter regions overlap each other. This means that when the (apo)repressor binds to the operator, RNA polymerase is excluded from the promoter, and transcription cannot commence. The inducer–(apo)repressor complex is unable to bind to the operator, RNA polymerase can access the promoter, and transcription of the *lac* structural genes commences.

BOX 6.4

**Equilibrium dialysis and phosphocellulose binding for isolation of the *lac* (apo)repressor**

These two techniques were both used to demonstrate the presence of the *lac* (apo)repressor in a cell-free extract of *Escherichia coli* cells. In the first technique an experimental sample is placed in a special type of bag called a **dialysis bag**, suspended in a buffer solution. During dialysis, small molecules diffuse through holes in the bag into the surrounding buffer. Larger molecules, including proteins, are retained within the bag.

For equilibrium dialysis, a crude preparation of the *lac* (apo)repressor was placed in the dialysis bag and suspended in a solution of radioactive inducer ($^3$H-IPTG). You will remember that IPTG is a gratuitous inducer of the *lac* operon – the $^3$H-designation indicates that the molecule is radioactively labelled with tritium, an isotope of hydrogen.

During dialysis, free $^3$H-IPTG molecules diffuse into the bag until equilibrium is reached. If a bag contains (apo)repressor molecules, they will bind with $^3$H-IPTG molecules. If this happens, the total radioactivity of free and bound $^3$H-IPTG will be greater inside the dialysis bag than outside it.

More recent isolation techniques use phosphocellulose packed into glass tubes. Phosphocellulose has the ability to bind any protein which also binds to DNA. The *lac* (apo)repressor binds to this material and can be removed by washing it with sodium chloride solution.

## Repressible enzymes

As we have seen, inducible enzymes are found in biochemical pathways such as those concerned with energy production from sugar molecules. **Repressible enzymes**, on the other hand, are found in biosynthetic pathways concerned with the production of units for macromolecular synthesis (Chapter 5). In the context of biosynthetic pathways, we saw at the beginning of this chapter how feedback inhibition could be used to shut down the synthesis of a particular metabolite, if that metabolite became available in the external environment. In addition to feedback inhibition, the *amounts* of biosynthetic enzymes present may also be reduced. This can be achieved by reducing the rate of synthesis of all the enzymes specifically required to produce a particular end-product. Both feedback inhibition and repression of enzyme synthesis work together to regulate the synthesis of

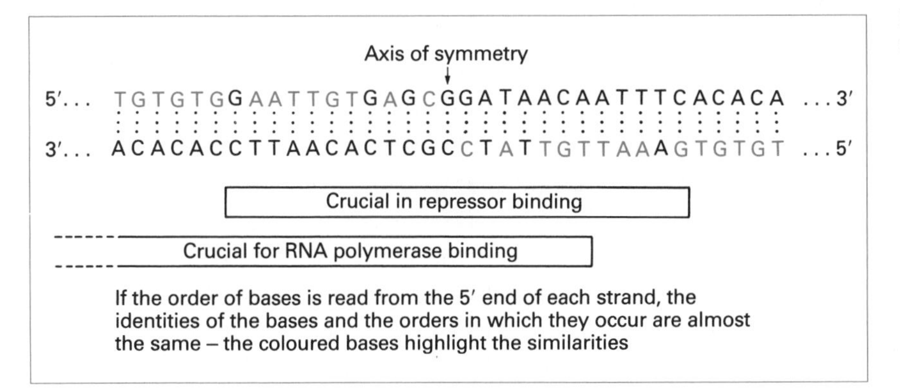


**Figure 6.7** DNA base sequence of the *lac* operon region of *E. coli.*

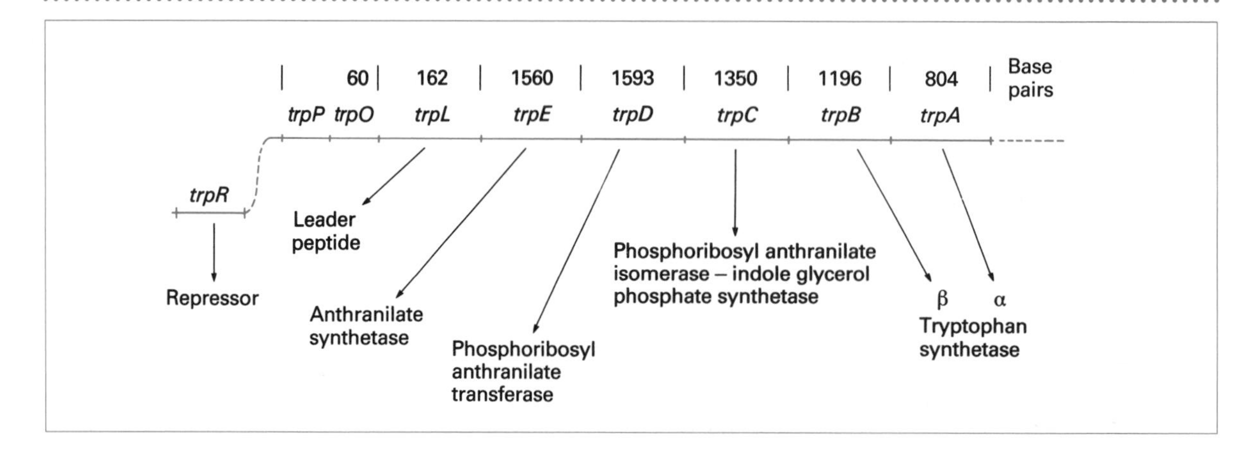


**Figure 6.8** The tryptophan operon of *Escherichia coli.* The structural genes, *trpE, trpD, trpC, trpB* and *trpA* are next to each other on the *E. coli* chromosome. *trpR* is the regulatory gene (c.f. *lacI*) and *trpP* and *trpO* are the promoter and operator sites respectively. *trpL* specifies the leader polypeptide sequence.

amino acids, purines and pyrimidines in line with the biosynthetic demands of individual cells. Whereas the effect of feedback inhibition is an immediate response to excess quantities of an end-product, repression of enzyme synthesis is a more drastic long-term change in physiology.

To illustrate the cellular mechanism by which the synthesis of biosynthetic enzymes can be controlled, we turn again to *Escherichia coli* and the synthesis of the amino acid tryptophan. In *E. coli*, the genes which specify the tryptophan biosynthetic enzymes are adjacent to each other and, as we saw with the *lac* system, together with the various regulatory elements, they constitute an operon called the **tryptophan** (or ***trp***) **operon** (Figure 6.8).

Further similarities between the *lac* and *trp* operons exist. Both have a regulatory gene (*lacI* and *trpR* respectively), both have operator regions (*lacO* and *trpO*), and both have promoter regions (*lacP* and *trpP*). These similarities suggest that the mechanism of induction of the *lac* enzymes and repression of the *trp* enzymes may have more in common.

The *trp* system has one feature which is not found in the *lac* system, however; there is an additional DNA sequence called *trpL*, which is transcribed and translated into what is known as a **leader polypeptide**. We shall see the importance of the leader sequence when we examine attenuation control of the *trp* operon later in this chapter. In fact, it turns out that the control of synthesis of the tryptophan enzymes really consists of *two* components. One mechanism, **repression control**, is essentially similar to the Jacob and Monod model and permits enzyme levels to fluctuate within one set of limits; the second mechanism, called **attenuation control**, allows levels to fluctuate within a different set of limits.

Observations on the levels of the tryptophan biosynthetic enzymes show that repression provides a 100-fold variation in the level of enzymes produced. The maximum level of enzymes produced in fully de-repressed cells can be increased a further 10-fold by removal of attenuation. So,

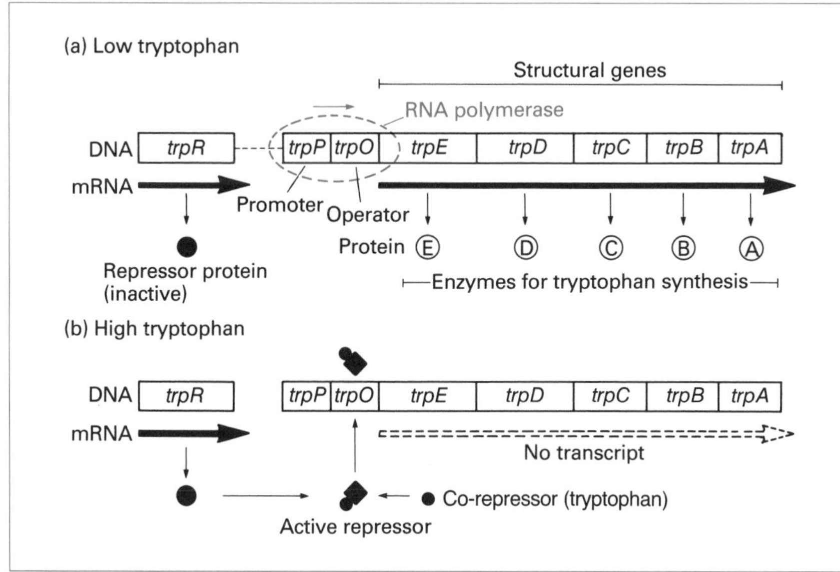


**Figure 6.9** Repression control of the *trp* operon of *Escherichia coli*. In the presence of a low tryptophan concentration, uncomplexed (apo)repressor protein is unable to bind to the operator. RNA polymerase is able to bind to the promoter and move to the structural genes which are transcribed into mRNA. This is translated into the corresponding enzymes. When the concentration of tryptophan is high, it can form a complex with (apo)repressor. Unlike uncomplexed aporepressor, the (apo)repressor–co-repressor (tryptophan) complex can bind to the operator (*trpO*). This prevents RNA polymerase from reaching the structural genes. In this situation, transcription of the structural genes cannot occur and the tryptophan biosynthetic enzymes are not produced.

together, repression control and attenuation control regulate the level of tryptophan enzyme synthesis over a 1000-fold range. Attenuation control of the *trp* operon in *E. coli* is less significant than repression control, but this may be an exceptional case because many biosynthetic operons are under attenuation control alone.

## Repression control of the tryptophan operon of *E. coli*

Repression control of the *trp* operon can be described using the Jacob and Monod model originally put forward to account for induction of the *lac* operon (Figure 6.9). The main difference between the two systems is that, in the *lac* system, the (apo)repressor–inducer complex is unable to bind to the operator, and transcription of the *lac* structural genes into the corresponding mRNA molecule can proceed. The mRNA can then be translated into the *lac* enzymes by the ribosomes.

In the *trp* system, the (apo)repressor–co-repressor (tryptophan) complex binds to the operator, thereby preventing transcription of the *trp* structural genes. The corresponding mRNA is not produced and translation to produce the tryptophan enzymes cannot occur. When the concentration of tryptophan is low or non-existent, the (apo)repressor cannot form a complex with tryptophan and the uncomplexed (apo)repressor is unable to bind to the operator. In these circumstances, transcription and translation occur and the tryptophan-forming enzymes are synthesized.

The mechanism we have described means that the levels of the *trp* enzymes are high if the intracellular concentration of tryptophan is low, and low if there is an abundance of tryptophan in the external environment.

> Attenuation operates through the standard components of the protein-synthesizing machinery. Special control elements, such as a regulatory gene or operator site, are not required for attenuation, only for repression of biosynthetic operons. This finding can be interpreted to mean that attenuation arose earlier in evolutionary history, whereas repression control had to await the evolution of the appropriate regulatory elements.

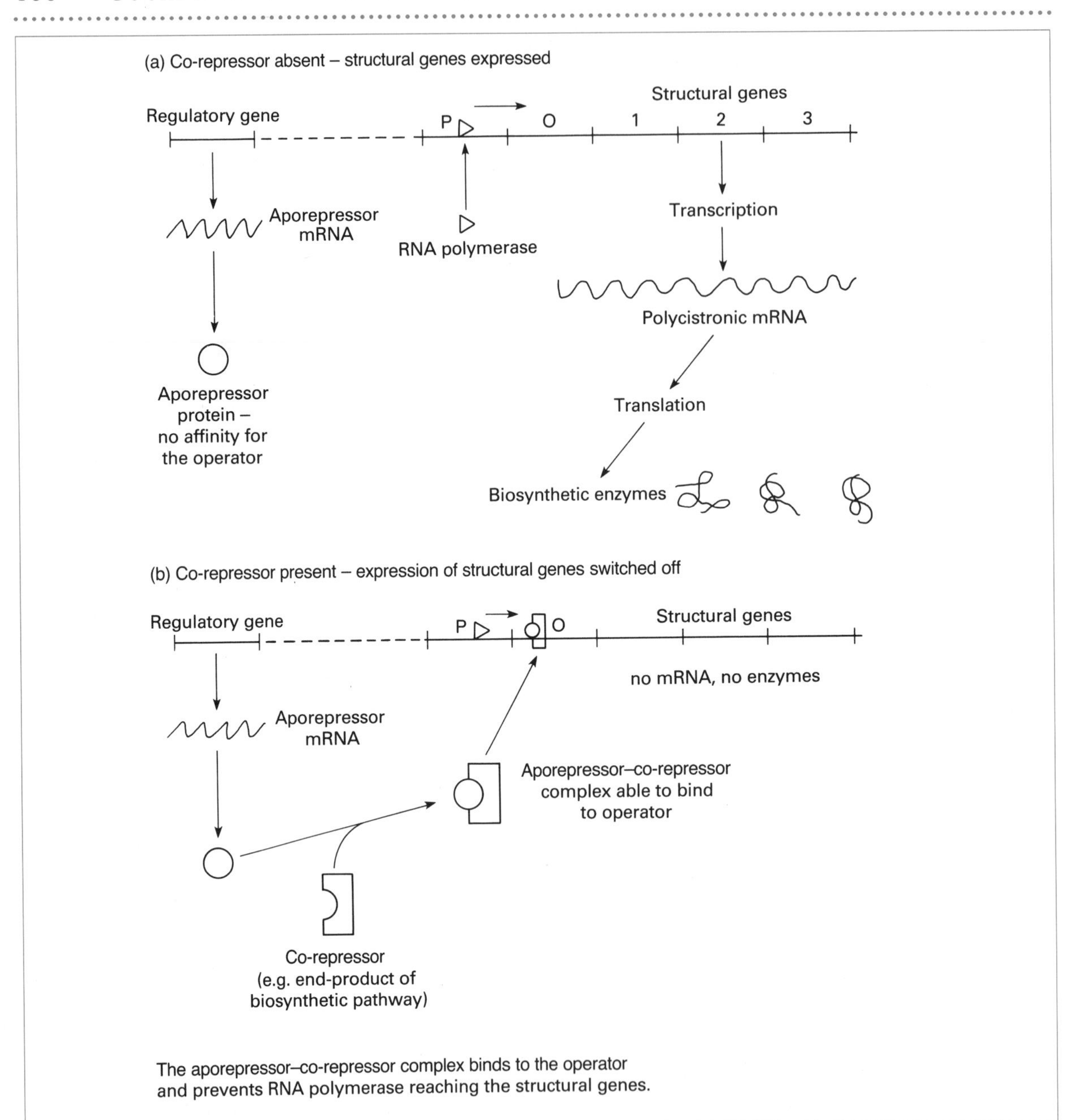


**Figure 6.10** Modification of the Jacob–Monod operon hypothesis to account for enzyme repression.

The parallels between *repression control* of the *trp* operon and *induction* of the *lac* operons are shown in Figure 6.10.

## Attenuation control

We have already mentioned the *trpL* region of the *trp* operon. This provides the key to understanding attenuation control. The *trpL* sequence is transcribed into mRNA and then translated into a leader polypeptide, 169

**Table 6.2** Leader peptides of attenuator-controlled operons containing genes for amino acid biosynthesis

| *Operon* | *Amino acid sequence of leader peptides* |
|---|---|
| Tryptophan | Met Lys Ala Ile Phe Val Leu Lys Gly [Trp Trp] Arg Thr Ser |
| Leucine | Met Ser His Ile Val Arg Phe Thr Gly [Leu Leu Leu Leu] Asn Ala |
| Phenylalanine | Met Lys His Ile Pro [Phe Phe Phe] Ala [Phe Phe Phe] Thr [Phe] Pro |
| Histidine | Met Thr Ala Val Gln Phe Lys [His His His His His His His] Pro Asp |

amino acid residues in length. In common with the leader sequences of other biosynthetic operons, it contains a high proportion of the amino acids whose synthesis the operon itself regulates (Table 6.2). In other words, the leader polypeptide of the *trp* operon contains more tryptophan than is found in *E. coli* proteins in general.

You will recall from Chapter 5 that mRNA determines the order in which polymerization of individual amino acids is carried out (the translation process) to produce a polypeptide chain. Furthermore, each amino acid reacts with a specific tRNA molecule to produce an aminoacyl-tRNA molecule. It is the anti-codon portion of the tRNA which aligns itself with the corresponding mRNA codon.

The two adjacent tryptophan codons in the *trpL* region of DNA (and its mRNA transcript) provide a mechanism for controlling synthesis of the *trp* biosynthetic enzymes. Because there are two *trp* codons in the leader mRNA, tryptophanyl-tRNA is required to complete synthesis of the leader polypeptide. This works well provided that there is sufficient tryptophan, but what happens if little or no tryptophan is available? Herein lies the answer to our problem! The mechanism of attenuation and de-attenuation is summarized in Figure 6.11.

## Repression control and attenuation in bacterial physiology

The existence of two control systems in *E. coli* cells, both concerned with tryptophan production, comes as a surprise. In normal circumstances, the levels of *trp* enzymes are regulated by repression alone. It is only when tryptophan is in very short supply that attenuation is adjusted. If a culture of *E. coli*, previously grown for some time in a tryptophan-rich medium, is transferred to a tryptophan-free medium, the following changes take place. The rate of synthesis of the *trp* enzymes increases 100-fold as a result of de-repression, and a further 10-fold increase can be attained by removing attenuation – a 1000-fold increase overall.

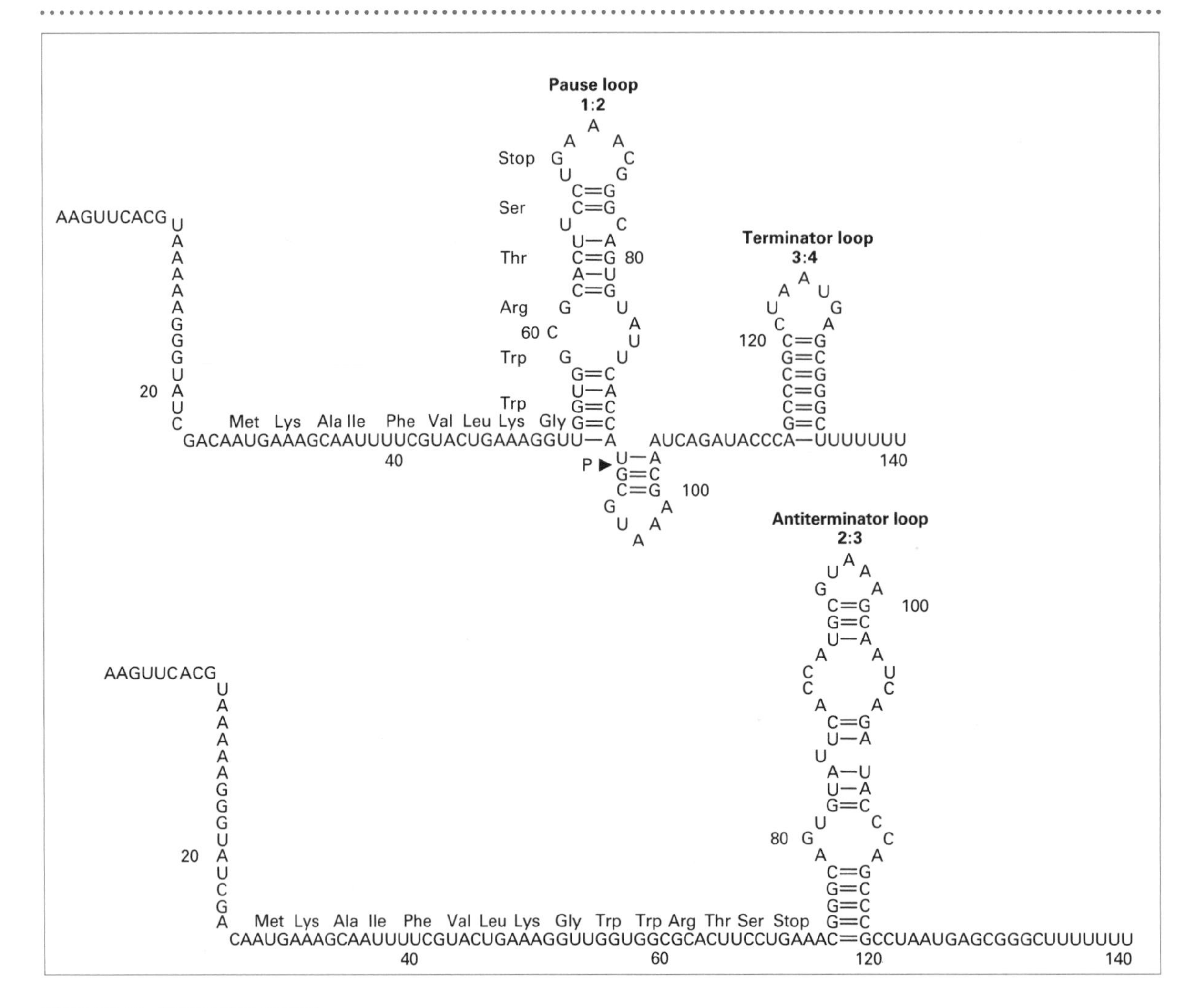


**Figure 6.11** Attenuation control of the *trp* operon of *Escherichia coli.*

Whether this provides a cell with any significant advantage over the 100-fold increase which de-repression alone provides is the subject of much debate. It may be that, in *E. coli*, attenuation evolved before repression and, in the *trp* system, attenuation is little more than an evolutionary relic and now of little physiological significance. This does not apply to a number of operons (including those listed in Table 6.2) which are regulated by attenuation alone.

Different regions of the leader mRNA can pair to form a variety of loop-like structures. The type of loop formed depends upon the concentration of tryptophan which is available. Different types of loops act as signals which either permit synthesis of the tryptophan enzymes or prevent it. For ease of description, the leader mRNA is usually divided into regions 1, 2, 3 and 4.

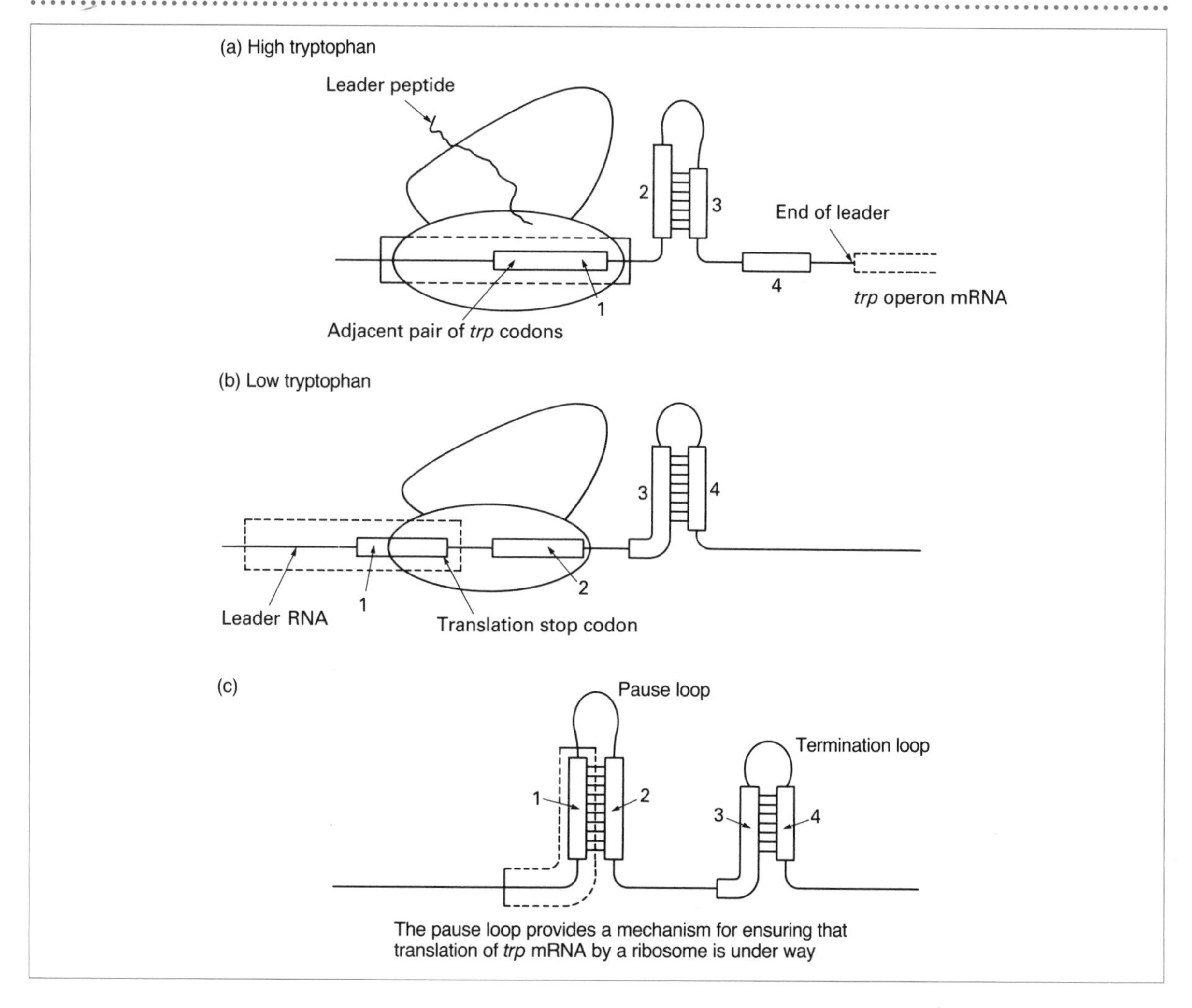


**Figure 6.11** *continued*

If the concentration of tryptophan is low (Figure 6.11a), the ribosome translating the leader mRNA sequence proceeds until it reaches the two *trp* codons in region 1. At this point the ribosome is unable to proceed further because of a lack of tryptophanyl-tRNA, and the ribosome is said to have 'stalled'. This is an analogy with the internal combustion engine under conditions of petrol shortage. When the ribosome stalls, a loop structure (regions 2 and 3) is produced which signals the transcription of the structural genes to continue. Under these conditions of tryptophan starvation, transcription and translation occur to produce *trp* biosynthetic enzymes to synthesize tryptophan for protein synthesis.

When tryptophan (and therefore tryptophanyl-tRNA) is present in abundance (Figure 6.11b), the ribosome does not stall at the *trp* codons and an alternative loop is produced (regions 3 and 4). This signal causes transcription to finish at the end of the leader mRNA sequence, and further transcripts are said to be attenuated.

Removal of attenuation can increase the maximum de-repressed level of *trp* enzymes by a factor of 10 to produce an overall 1000-fold increase. Think of loss of attenuation as 'supercharging' or 'turbocharging' the rate of synthesis of the *trp* enzymes. The mechanism of attenuation and de-attenuation is summarized in Figure 6.11.

## Alternative mechanisms of controlling protein synthesis

As we have seen, for both the *lac* and *trp* systems, control of synthesis of the corresponding enzyme proteins is exerted at the level of transcription. That is to say, it is the synthesis of mRNA molecules which fluctuates and which, in turn, determines the levels of corresponding enzyme proteins. Furthermore, the genes which determine the structure of the corresponding enzyme proteins (called **structural genes**) together with the corresponding regulatory elements (operator, regulatory gene and promoter) form a regulatory unit in which all the elements are side-by-side on the bacterial chromosome. This regulatory unit is called the operon (Table 6.3).

As you may have guessed, many variations on these basic themes are found in nature. In theory, any of the stages required to synthesize a protein molecule could be regulated and still produce the same overall effect on protein levels. As an example, let us look at the synthesis of ribosomal proteins (*r*-proteins) in *Escherichia coli*. Here, the regulation of synthesis of these molecules is exerted at the translation level. The way in which this mechanism works is as follows.

It appears that mRNA molecules contain nucleotide sequences similar to those found in rRNA. In both cases these sequences are recognition sites for the binding of protein molecules. Ribosomal proteins bind to appropriate sites of rRNA to ensure the correct assembly of the ribosomal

Ribosomal proteins are one of the structural components of ribosomes; ribosomal RNA (rRNA) is the other component. Details of ribosome structure and function can be found in Chapter 5.

**Table 6.3** Some prokaryotic operons

| *Operon* | *No. of enzymes* | *Function* |
|---|---|---|
| *lac* | 3 | Transport and splitting of lactose |
| *his* | 10 | Histidine synthesis |
| *gal* | 3 | Activation of galactose |
| *leu* | 4 | Leucine synthesis |
| *trp* | 4 | Tryptophan synthesis |

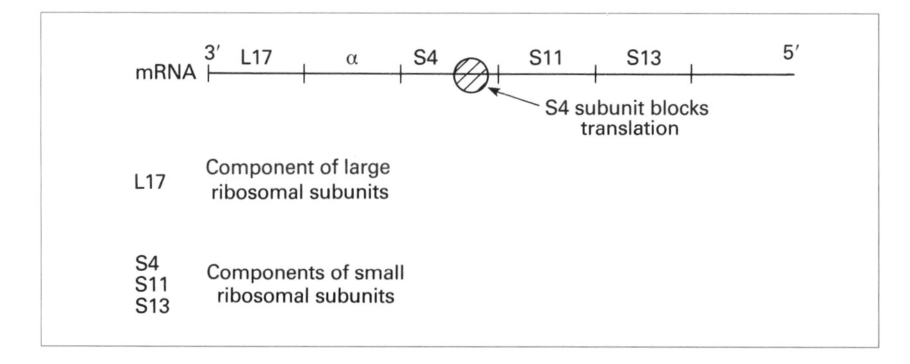


**Figure 6.12** A ribosomal protein (S4) represses translation of an mRNA which codes for components of the small and large ribosomal units and the α-subunit of RNA polymerase.

subunits to form a complete ribosome. The corresponding sequences of mRNA also bind *r*-proteins but, in this case, the proteins inhibit translation of the mRNA to prevent over-production of *r*-protein molecules. In this way, *r*-protein molecules which have not been incorporated into completed ribosomes inhibit the synthesis of yet more molecules of the same protein. This system means two things. First, the synthesis of *r*-proteins is matched to their rate of incorporation into ribosomes. Second, coordination of the production of different types of *r*-protein is possible. Because of their mode of action, *r*-proteins which act in this way are called **translational repressors** (Figure 6.12).

## Gene regulation in *Saccharomyces cerevisiae*, a eukaryotic microorganism

In contrast to bacteria, the genes of *S. cerevisiae* (Chapter 9) are not organized into operons. Genes with related functions may be close to each other on a yeast chromosome, but each of them is transcribed into a separate mRNA. Polycistronic (polygenic) mRNA molecules are not found in yeast and every yeast gene must, therefore, possess its own set of control elements. These include a promoter to which RNA polymerase binds, an upstream activating sequence (UAS) which turns the transcription process on, and an upstream repressing sequence (URS) which inhibits transcription of a gene.

The details of the mechanism by which UAS and URS sequences operate are still obscure, but because they are often located at points several hundred nucleotides away from the gene they control, the mechanism must differ significantly from that found in bacteria. It is clear, however, that the binding of a protein to a URS or UAS can cause an alteration in expression of the target gene. One novel feature of the proposed mechanism is that the DNA bends so that a distant URS or UAS site can interact with the corresponding promoter. In this way, the promoter may become more or less accessible for binding of RNA polymerase.

## Chapter summary

In this chapter, we have seen how microorganisms can adapt to changes in their environments using a variety of physiological processes. First, we saw that the activity of an enzyme could be changed in response to the presence of a nutrient in the environment. As far as biosynthetic pathways are concerned, we looked at two ways in which enzyme activity could be inhibited. During feedback inhibition, an early enzyme of a pathway is often subject to allosteric inhibition. This occurs when the end-product of the pathway binds to a site on an enzyme which is distinct from its active site, and changes its overall shape so that it can no longer efficiently bind substrate molecules. We then examined the inhibition of glutamine synthetase in *E. coli* by a process called covalent modification.

Second, we discussed the regulation of levels of enzymes by control of enzyme synthesis. To exemplify the principles involved, we described the *lac* operon of *E. coli*. We examined the negative control mechanism of enzyme induction and the positive control process of catabolite repression. From control of this catabolic enzyme sequence, we moved on to look at enzyme repression and attenuation of the *trp* operon of *E. coli*, an example of regulation of biosynthetic enzymes. Control of synthesis of ribosomal proteins emphasized that regulation of some systems occurs at the level of translation of mRNA.

## Further reading

T.A. Brown (1992) *Genetics: a Molecular Approach*, 2nd Edition. Chapman & Hall, London – a readable, concise treatment.
B. Lewin (1990) *Genes IV*. Oxford University Press, Oxford – a detailed account including chapters on gene regulation.
F.C. Neidhardt, J.L. Ingraham and M. Schaechter (1990) *Physiology of the Bacterial Cell*. Sinauer Associates Inc., Sunderland, Mass., USA. – coordination of metabolism in a physiological context.
J. Scaife, D. Leach and A. Galizzi (1985) *Genetics of Bacteria*. Academic Press, London – a more advanced treatment.

C.A. Smith and E.J. Wood (1991) *Molecular and Cell Biochemistry: Molecular Biology and Biotechnology*. Chapman & Hall, London – a very well presented, modern description of this important subject. A gentle introduction to the subject.

J.D. Watson, N.H. Hopkins, J.W. Roberts, J.A. Steitz and A.M. Weiner (1987) *Molecular Biology of the Gene*, 4th Edition. Benjamin-Cummings, Menlo Park, California, USA – a detailed, long treatise by a number of authors including J.D. Watson of Watson and Crick fame.

## Questions

1. Multiple-choice questions:
   (i) At which of the following levels does regulation of synthesis of the ribosomal proteins of *Escherichia coli* occur? (a) transcription, (b) translation, (c) post-translational modification, (d) all of these, (e) none of these.
   (ii) Which of the following enzymes are specified by genes of the *lac* operon? (a) β-galactosidase, (b) β-galactoside transacetylase, (c) lactonase, (d) β-galactoside permease, (e) lactic dehydrogenase.
   (iii) The intracellular level of which of the following is responsible for catabolite repression of the *lac* operon? (a) ATP, (b) AMP, (c) ADP, (d) cAMP, (e) UMP
2. Fill in the gaps:
   Regulation of the *trp* operon is controlled by two independent systems. One of these, __________, was first used to describe __________ of the *lac* operon. Tryptophan is a regulatory molecule called a __________ and lactose a regulatory molecule called an __________. In both instances, they bind to protein __________ molecules. This type of control mechanism is sometimes described as __________ control. The second system for control of the *trp* system is called __________. Operation of this system requires synthesis of a __________ peptide which contains a tandem pair of __________ residues. These are coded by corresponding __________ in mRNA. If the concentration of __________ is low, translation of mRNA by the ribosome will stall because __________ will be in short supply. If this happens a corresponding __________ and __________ structure is formed

and this signals the __________ of the *trp* __________ genes to continue. If the level of tryptophan is high, stalling does not occur and a different __________ and __________ structure is formed which signals a halt to the transcription process.

Choose from: attenuation codons, co-repressor, inducer, induction, leader, loop (twice), negative, repression control, (apo)repressor, stem (twice), structural, transcription, tryptophan (twice), tryptophanyl-tRNA.

3. True or false?
Which of the following statements are true?
(i) Feedback inhibition is a form of competitive inhibition.
(ii) Feedback inhibition operates on all of the enzymes of a biosynthetic pathway.
(iii) Sigmoid kinetics is characteristic of enzymes subject to feedback inhibition.
(iv) Covalent modification of an enzyme protein is never used as a regulatory mechanism.
(v) *Escherichia coli* synthesizes three aspartokinase enzymes which catalyse the same reaction.
(vi) Operons are equally frequent in eubacteria and eukaryotes.
(vii) Polycistronic mRNA is synthesized during transcription of the *lac* structural genes.
(viii) Polycistronic mRNA is common in eubacteria and eukaryotes.
(ix) Induction of the *lac* operon is repressed in *E. coli* cells grown on glucose.
(x) RNA polymerase binds to the operator region of an operon.
(xi) The *lac* repressor binds to the promoter.
4. Short-answer questions:
(i) What selective advantages do the possession of regulatory mechanisms give to a microorganism?
(ii) Which regulatory mechanisms control the synthesis of tryptophan in *E. coli?*
(iii) What is a regulon?
(iv) Explain the significance of allostery in the regulation of enzyme reactions.

# The microbial genome: organization, mutation and repair

# 7

In the previous chapters we described a number of structural and biochemical features of the microbial cell. Such characteristics remain more or less constant for a given microorganism and are controlled by the genes it possesses. If you like, you can think of the genes as holding the information necessary for the day-to-day operation of a microorganism in much the same way as a computer program stores information.

In this chapter we start off by looking at the chemical nature of the genetic material and the way it is organized in both eukaryotic and prokaryotic microbes. After this, we shall spend some time describing how genetic material can **mutate** (change). Mutation is a crucially important process for a number of reasons. First, the differences between individuals of the same species, which we call **variation**, are partly caused by differences in the genetic composition of individuals. These differences can arise by gene mutation and are important because they provide the raw material on which the forces of **natural selection** can operate and thus lead to the evolution of new forms.

Second, mutant strains of microbes are extremely powerful tools for investigating biological processes. The approach is to cause a malfunction in a biological process by mutation of a gene whose unimpaired activity is required for normal function. As a simple analogy, if you wanted to understand how a motor car works, you could try removing various parts one at a time (carburetter, distributor and so on) and note the effect on the car's operation. In this way it is possible to infer the role of each component.

The final part of this chapter is concerned with another source of variation called **recombination**. Recombination is usually associated with a sexual method of reproduction and is a consequence of shuffling the genes from each of the parents together and passing onto each individual of the next generation its own unique set of genes. So recombination is a second source of genetic variation. Whereas mutation produces completely new genes, recombination produces only new combinations of genes. Sexual reproduction in eukaryotic microorganisms is described in Chapter 9.

## Advantages of using microorganisms for genetic studies

The importance of the contributions of microbial genetics in our understanding of genetic processes cannot be overstated. It is from studies of bacteria in particular that much of our current understanding of gene structure and function stems. The concept of molecular biology arose out of microbial genetics and all the techniques involved in genetic engineering were first developed using bacteria. The enormous impact of microbial genetics on the field of genetics as a whole is more easily understood when we look at the advantages which microorganisms offer for this kind of experimental investigation.

1. **Microorganisms are very small**. Prokaryotic cells rarely exceed a width of 1 μm and a length of 5 μm. Consequently, large populations may be studied in a small volume. For example, 10 ml of a nutrient broth culture of *E. coli* would typically contain $5 \times 10^9$ individuals, roughly the human population of the whole world.
2. **Microorganisms have a short generation time**. The speed with which they grow and reproduce means that large populations may be established quickly. The population doubling time of *E. coli* under optimum conditions is about 20 minutes. Even a slow-growing bacterium such as *Nitrobacter* can double its population size in 18 hours.
3. **Microorganisms reproduce by simple cell division**. This ensures that if a population is grown from a single cell all the members of the population (with the exception of a very small number which have undergone spontaneous mutations) will be identical. When the population is close to genetic homogeneity it is *easy to select mutant cells which differ from the population even by a single characteristic.*
4. **Microorganisms grow on simple, chemically defined media**. Manipulation of such simple media (for example adding or omitting a single amino acid) can *make the selection of mutants a relatively simple process.*
5. **In a microorganism, every gene present in an individual is reflected in some kind of observable or measurable characteristic**. In higher plants and animals things are not so straightforward because each cell (except the sperms and eggs) has two copies of each gene. One copy of a gene can sometimes mask the presence of the other, thus presenting a false picture.

## DNA is the genetic material

In these modern times of gene replacement therapy and DNA fingerprinting we take it for granted that DNA is the genetic material – the molecule of heredity. It was not always believed to be so. In the late 1800s and early 1900s there was much controversy over whether nucleic acids or proteins were the genetic material. Although it was known that the chromosomes were responsible for the inheritance of characteristics, it was also known that they contained protein (called **histones**) as well as nucleic acid.

Bacterial transformation, described by Griffith in 1928, was the first major step in establishing that DNA was the genetic material. He demonstrated that it was possible to transfer heritable traits from one population of bacteria to another. We shall review Griffith's experiments.

Griffith worked with the bacterium *Streptococcus* (also called *Diplococcus*) *pneumoniae*, which was almost invariably fatal to mice injected with a suspension of the living bacterium. Virulent *S. pneumoniae* cells possess a polysaccharide capsule which makes their colonies appear smooth and shiny. He isolated different strains of smooth *S. pneumoniae* which differed in the polysaccharide component of their capsules. These different strains were referred to as IIS, IIIS and so on. Occasionally mutants of the smooth strains arose which lacked a capsule altogether, producing colonies with a rough appearance. These were given designations IIR, IIIR, and so on, according to the smooth strain from which they had evolved. When these rough strains were injected into mice, the mice survived (Figure 7.1). In other words, the rough strains were not virulent. Occasionally, a rough strain would mutate back to the smooth strain. If it did, the polysaccharide was always the same as that of the original S strain. That is, a IIR bacterium would back-mutate only to a IIS strain and never to a IIIS strain for example.

Griffith found that if he injected killed smooth bacteria into a mouse, the mouse would survive. But if he injected heat-killed IIIS cells *and* living rough IIR cells into a mouse, it died, and *live IIIS bacteria could be recovered from its body*. Griffith knew these could not be mutant IIR cells because they could only mutate to IIS cells. He concluded that a substance from the dead IIIS bacteria had converted the IIR cells to IIIS cells. He called this process **transformation** and the substance responsible, the **transforming principle**. Additionally, he demonstrated that the IIIS characteristic was passed from generation to generation. Although Griffith did not establish the chemical identity of his transforming principle, he did show that the inheritance of a characteristic has a physical foundation.

**Some explanations**
**Haploid**: microbes, and the sex cells (sperms and eggs) of higher animals and plants contain only *one copy* of each gene: this is called the **haploid state**.
**Diploid**: each cell of a higher plant or animal (but not the sex cells) and some stages in the life-cycles of fungi, protozoa and algae contain *two copies* of each gene: this is called the **diploid state**.
**Allele**: each of the two copies of a gene in a diploid cell is called an allele.
**Recessive allele**: in a diploid cell, the presence of one copy of a pair of alleles may be hidden or masked by the other. The allele whose presence is hidden is described as **recessive**.
**Dominant allele**: the allele which masks the presence of a recessive allele is the **dominant allele**. It is the characteristic determined by a dominant allele which is seen in a diploid cell.

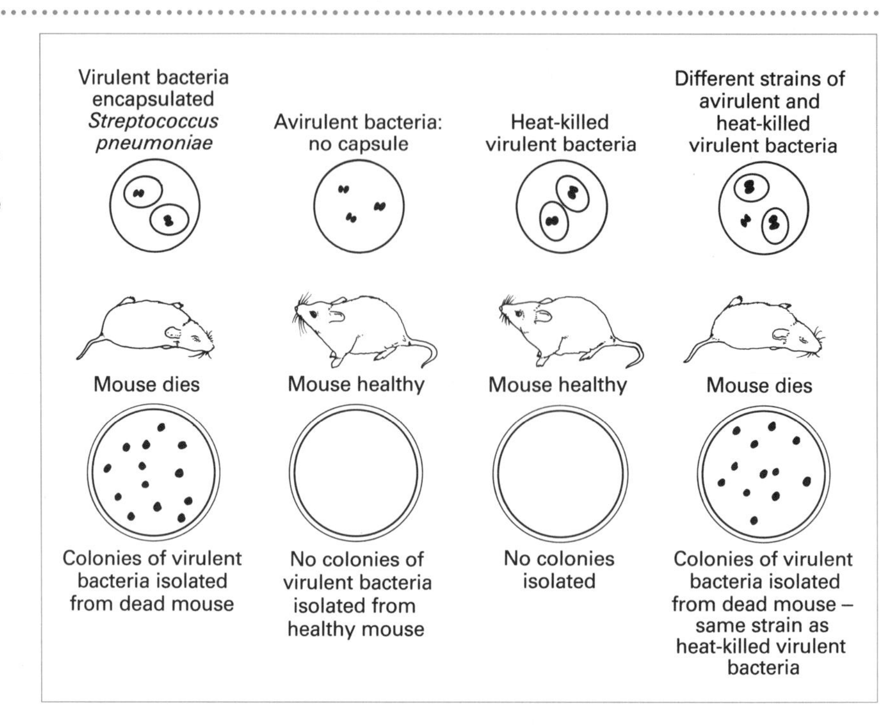


**Figure 7.1** Bacterial transformation in *Streptococcus (Diplococcus) pneumoniae*. When a mouse is injected with a mixture of dead, virulent (smooth colonies) and living, non-virulent (rough colonies) cells, the mouse is killed. Furthermore, living, virulent bacterial cells can be isolated from the dead mouse. As a result of this experimental observation, Griffith (1928) concluded that something (the transforming principle) had passed from dead virulent cells to living non-virulent ones. The latter cells were transformed into living virulent cells. The transforming principle was later identified as DNA (Avery, McLeod and McCarty, 1944).

Further progress on the identity of the transforming principle had to wait until 1944 when a group of United States workers succeeded in identifying the substance responsible. Avery, MacLeod and McCarty subjected samples of transforming principle to the activity of a number of degradative enzymes. They then investigated whether the treated sample was still capable of carrying out transformation. The enzymes they used and their effects on transforming activity are given here:

| *Enzyme* | *Substrate* | *Effect on transforming activity* |
|---|---|---|
| RNase | RNA | Retained |
| DNase | DNA | Lost |
| Protease | Proteins | Retained |

Under appropriate experimental conditions, DNA from any source (including human beings) can be introduced into a living bacterial cell using the transformation technique. This is of enormous importance to genetic engineers wishing to produce human proteins (such as hormones) from engineered bacterial cells.

On the basis of these results, what is the identity of Griffith's transforming principle? Yes, like Avery, MacLeod and McCarty, you would be right in concluding that DNA is the essential component of the transforming principle. This investigation of transformation of *Streptococcus pneumoniae* provided one of the first pieces of evidence that DNA (not protein or RNA) determines the nature of inherited characteristics.

Subsequently it has been shown that transformation occurs in a variety of bacteria, including *Bacillus*, *Haemophilus* and *Neisseria*. Furthermore, other genetically determined characteristics, such as sugar fermentation and

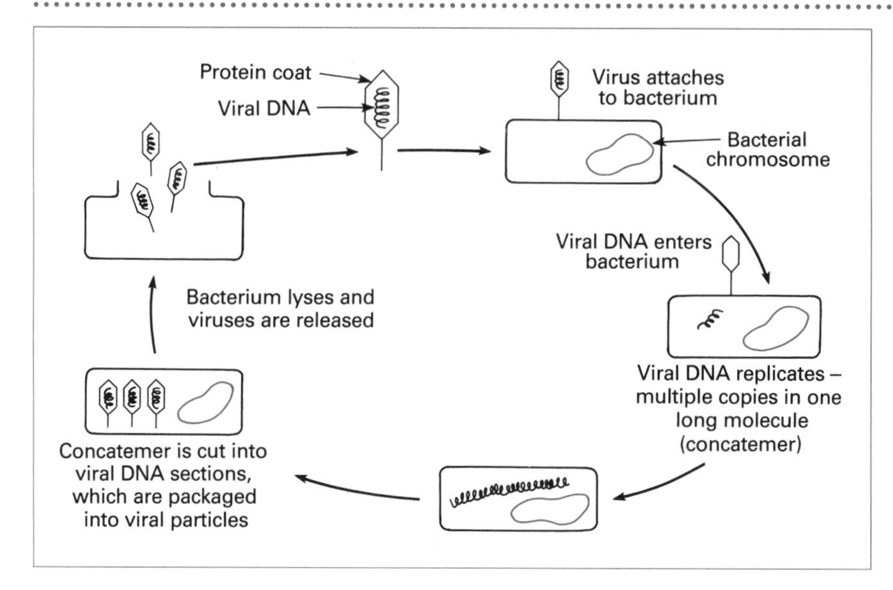


**Figure 7.2** The lytic cycle of a bacteriophage. Some phages exhibit only a lytic cycle of replication and therefore always lyse and kill the host bacterial cell. Such phages are called virulent bacteriophages: examples are T2 and T4 of *Escherichia coli.*

antibiotic resistance, can be transmitted by the transformation process, suggesting a more general role for DNA as hereditary material.

The second major piece of experimental evidence that DNA is the genetic material was published by Hershey and Chase in 1952. They studied the reproduction of bacteriophage T2 (normally abbreviated to phage T2) in *E. coli.* The life-cycle of a typical phage is shown in Figure 7.2. The phage has a protein coat (called the head) containing DNA, and a contractile protein tail by which it attaches itself to the *E. coli* cell wall. It then injects its DNA into the cell. In the host cell, replication of phage DNA occurs first, followed by synthesis of the protein coats and tails. The complete phages are assembled and the bacterium breaks open, releasing the mature phage particles.

Hershey and Chase investigated these events using radioactively labelled phage T2 particles. Proteins contain sulphur but not phosphorus, whereas DNA contains phosphorus but not sulphur. They took advantage of this difference to distinguish the protein coat of the phage from its nucleic acid. By culturing phage in *E. coli* which had been grown either on a medium containing $^{32}$P or $^{35}$S (radioactive isotopes of phosphorus and sulphur) they obtained two populations of T2. One population had its DNA labelled with $^{32}$P and the other population had its protein coats labelled with $^{35}$S (Figure 7.3). The labelled phages were used to infect bacteria grown on a non-radioactive medium. Immediately following infection, the bacteria were subjected to severe shear forces in a blender, stripping the phage coats (which no longer contained phage DNA) from the bacterial cell walls. The bacterial cells and the phage coats were then separated by centrifugation.

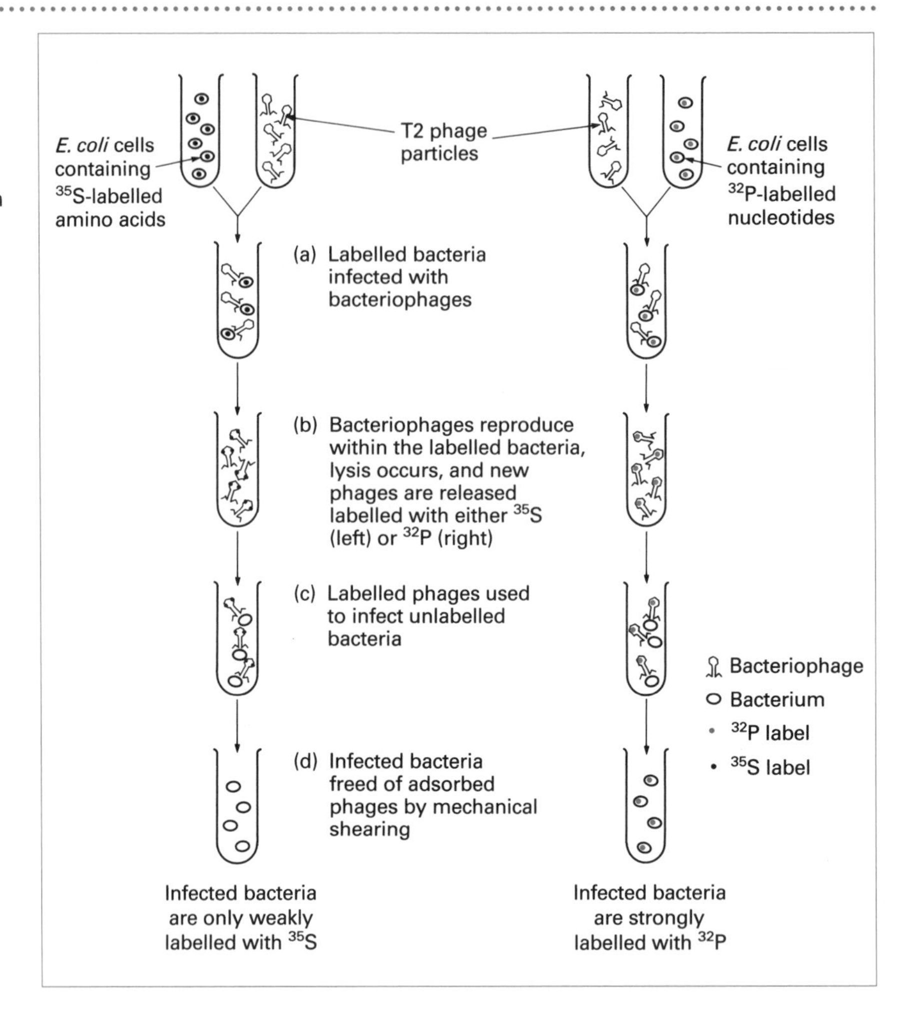


**Figure 7.3** The Hershey and Chase experiment. In this experiment, infection of *E. coli* cells proceeds with phages in which either protein is radioactively labelled with $^{35}S$ or DNA with $^{32}P$. Following infection with $^{32}P$-labelled phage particles only, *E. coli* cells become radioactively labelled. This does not happen following infection with $^{35}S$-labelled phages. It was concluded that DNA, not protein, carries the information which directs host cells to produce phage offspring.

Hershey and Chase found more than 80% of the $^{32}P$ but only about 1% of the $^{35}S$ in the bacteria, although viable phages were produced from both cultures. They reasoned that since the phage coat remained outside the cell it could not be involved in the production of new phage particles. On the other hand, phage DNA enters the cell and must be responsible for directing phage multiplication. Thus, in phage T2, DNA and not protein is the genetic material.

In 1953 Watson and Crick provided the theoretical framework to explain how the structure of DNA enabled it to act as a carrier of hereditary information. These early transformation and phage replication experiments, together with Watson and Crick's proposals, were the acorns from which the great oaks of molecular biology have grown.

## Organization of the hereditary material

Microbes range from subcellular particles (viruses), through the prokaryotes (bacteria), to the eukaryotes (yeasts, filamentous fungi, algae and protozoa). It would be impossible to describe in detail the organization of the genetic material of all these groups. But it is important to know the major characteristics of prokaryotes and eukaryotes.

### The prokaryotic genome

Bacteria possess a single chromosome which carries all the genes essential for survival under normal conditions. The chromosome of *E. coli* is about 1.3 mm long and consists of about 4000 kb ($4 \times 10^6$ base-pairs) of DNA. *E. coli* is only about 2 μm long and 1 μm wide, so the DNA has to be very tightly packed into the cell. This is possible because *E. coli* DNA is *supercoiled*: that is, the helix is coiled on itself. Additionally it is associated with positively charged molecules to maintain its highly condensed state.

The chromosome is repeatedly folded or looped around a protein core (Figure 7.4). In *E. coli* there are roughly 100 loops, each comprising 40 kb.

The word **genome** is used to denote the total genetic complement of an individual cell. This is usually its DNA content, but some viruses possess RNA as their genetic material.

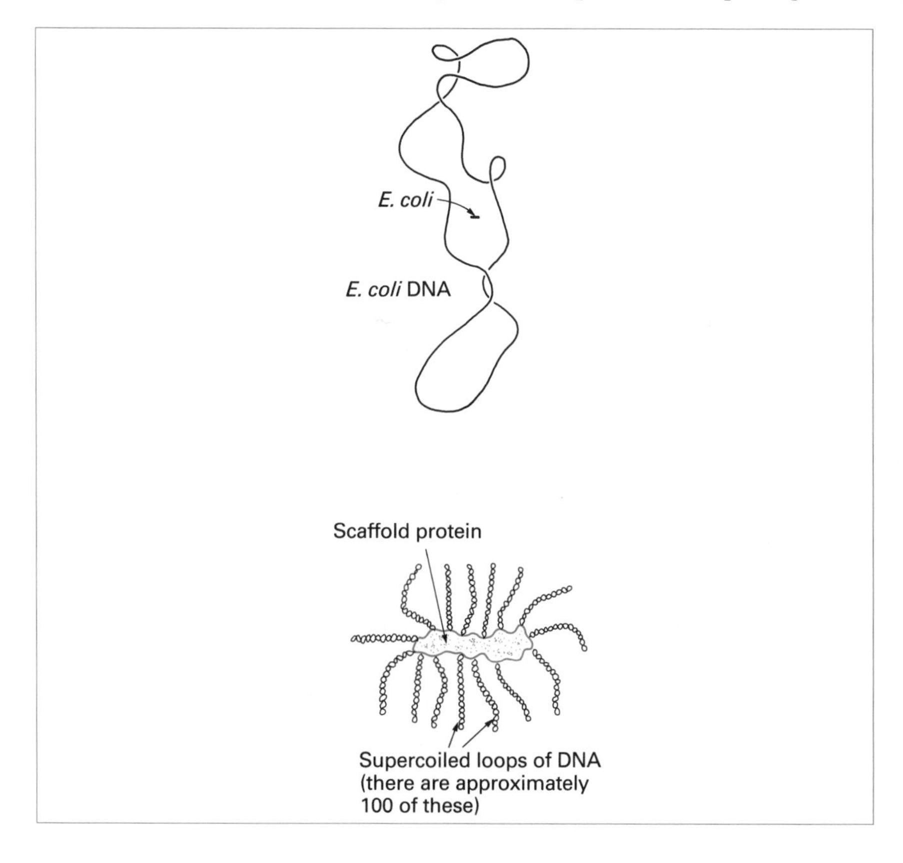


**Figure 7.4** The *Escherichia coli* chromosome. (a) Schematic diagram showing the relative lengths of the *E. coli* cell and its chromosome. The chromosome is approximately 600 times the length of the cell. To accommodate the chromosome, it is tightly coiled and condensed within the cell confines. (b) The *E. coli* chromosome occurs as a number of loops attached to **scaffold protein** and this constitutes the **nucleoid** (c.f. eukaryotic **nucleus**).

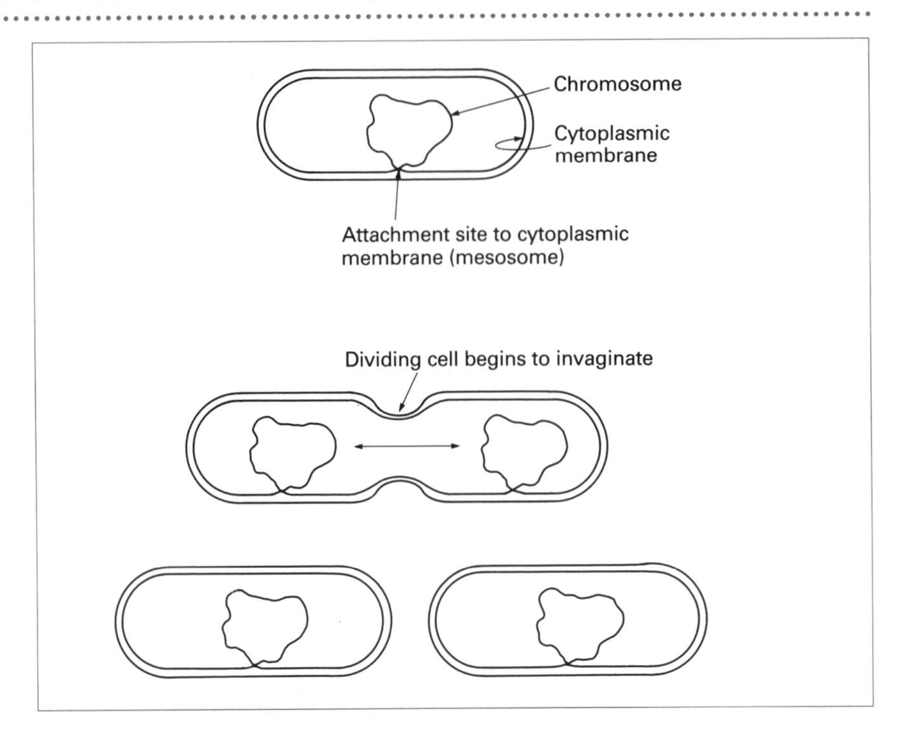


**Figure 7.5** Role of membrane attachment site (mesosome) in nucleoid partitioning at cell division.

The word **nucleoid** is used because no membrane surrounds the bacterial chromosome. The term **nucleus** is reserved for the eukaryotic structure, in which chromosomes are surrounded by a nuclear membrane.

The length of each loop is between 2 and 2.5 µm. If fully extended, 40 kb would measure 13.6 µm, suggesting that, in the cell, DNA is condensed about sixfold. The chromosome is restricted to a region in the cell called the **nucleoid**, which takes up about one third of the cell's volume. The chromosome is attached to an invaginated part of the cell's inner membrane called the **mesosome**. Attachment to the mesosome is important in ensuring that a copy of the chromosome goes to each daughter cell at cell division (Figure 7.5).

In contrast to eukaryotes, virtually the whole of the bacterial genome is functional; either it determines the amino acid sequence of polypeptide chains or it has a regulatory role in the same process. Eukaryotic genomes, on the other hand, contain considerable amounts of 'junk' DNA, which appears to have no function at all. Bacterial genes are typically present as single copies which are often clustered to form groups of genes called **operons** (Chapter 5) and which are subject to a common regulatory mechanism. Of those *E. coli* genes which have so far been identified, about one quarter are organized into operons.

Related bacteria (for example *E. coli*, *Salmonella*, *Shigella* and *Klebsiella*) have quite similar distributions of genes along their chromosomes. Indeed, the genomes of *E. coli* and *Salmonella typhimurium* are virtually identical with the exception of a region (about 10% of the chromosome) which is the other way round in *S. typhimurium*. Both the gene order and gene content vary widely in distantly related bacteria. The maps for *Pseudomonas aeruginosa*, *Bacillus subtilis* and *Streptomyces coelicolor*, for example, differ greatly from each other, and from that of *E. coli*.

## The eukaryotic genome

Eukaryotic microorganisms include fungi (yeasts are fungi too), algae and protozoa. In eukaryotes, the genetic material is confined to a membrane-bound region of the cell called the nucleus (Chapter 9). The genetic material is divided between several DNA molecules, associated with proteins called **histones**, to form a chromosome. The number of chromosomes varies between species but, in normal individuals, is constant for a given species.

The distribution of genes along a chromosome can be shown in the form of a **gene map**. The information required to produce such maps can be obtained from appropriate experiments described in Chapter 8.

The yeast *Saccharomyces cerevisiae* has 17 chromosomes, the alga *Chlamydomonas reinhardii* has 16 and *Aspergillus nidulans* has 8. Although eukaryotic chromosomes are linear, each of them contains a continuous loop of DNA. Eukaryotic microorganisms are diploid for at least part of their life-cycle (though this can be very brief). The genes in eukaryotes are not arranged in operons, nor are they under the common control of a promoter (page 120). Rather, each eukaryotic gene has its own regulatory element (an **enhancer**) which may be at some distance from the gene. Eukaryotic genes themselves are interrupted by inert regions called **introns** and are, therefore, called **split genes** (Figure 7.6). A full description of the organization of viral genomes is given in Chapter 10.

A promoter is a short segment of DNA which is recognized by the enzyme RNA polymerase. This enzyme is responsible for the transcription of a gene into a complementary mRNA molecule. The promoter is, therefore, a recognition site for RNA polymerase to initiate the transcription process.

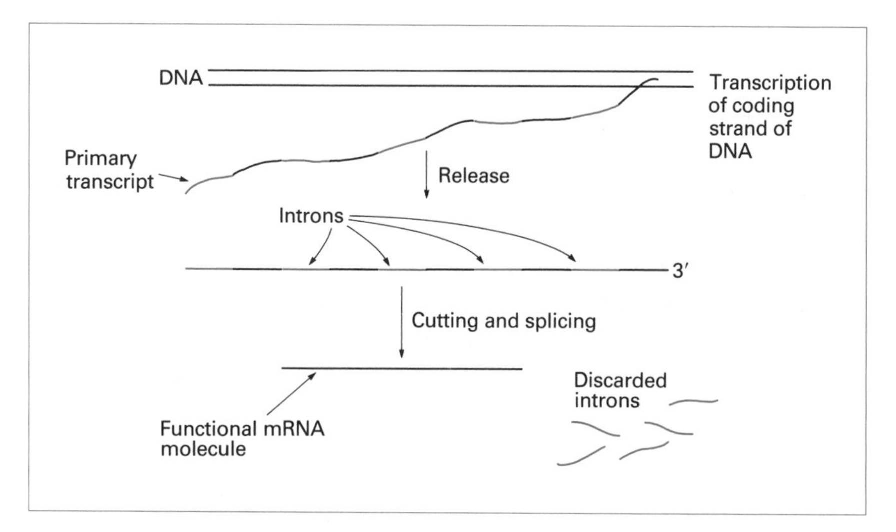


**Figure 7.6** Split genes of eukaryotic organisms.

The nature of an individual's genetic material is its **genotype**. The way in which the genotype expresses itself is an individual's **phenotype**. Production of a bacterial capsule, virulence, growth factor requirements and colour are all examples of phenotypic characteristics; they are all determined by genes but they are observable or measurable.

## Mutation

The word **mutation** simply means change, so genetic mutation is used to describe a change in an organism's genetic make-up – a change in its genome. For the sake of convenience, **chromosomal mutations** and **nucleic acid mutations** are recognized. A chromosomal mutation refers to major changes, such as the loss of part or all of a chromosome. A nucleic acid mutation is a more subtle change at the DNA level. In either case, the change in genetic make-up may be reflected in some measurable or observable change in a characteristic of the organism.

Mutations which arise in essential genes may be lethal. Some mutations, called **conditional lethal mutants**, will kill the cell only under certain environmental conditions. For example, temperature-sensitive mutants of *E. coli* will grow normally at a temperature of 30°C but will die at 37°C. Similarly, mutants which require a particular growth factor will die only if that growth factor is absent from the medium.

It is now widely accepted that mutations arise at random and are selected (or counter-selected) by environmental pressures. This was not always believed to be so. There was a school of thought which suggested that mutations were directed. That is to say, they arose in response to challenges in the environment, such as the presence of antibiotics.

The basis of this belief was that when bacteria sensitive to antibiotics were spread onto agar containing an antibiotic, some colonies arose which were resistant to the antibiotic and this resistance was heritable. An (incorrect) interpretation of this type of experiment was that the antibiotic had induced the bacteria to adapt to the presence of the antibiotic.

### Types of mutation

Nucleic acid mutations are changes in the base sequence of genetic material. Mutations which cause major rearrangements of the organism's genome are called **multi-site mutations**. These are summarized in Figure 7.7. A mutation that affects one, or a very few, nucleotides is called a **point mutation**. Base substitutions are point mutations in which one nucleotide replaces another without an overall change in number. If this involves the substitution of one purine for another (or a pyrimidine for a pyrimidine) it is known as a **transition mutation**. Those in which a purine replaces a pyrimidine (or *vice versa*) are called **transversion mutations**. The base substitutions which give rise to transitions and transversions are shown in Figure 7.8. To understand the effect of mutations on protein structure, it is important that you understand how the DNA base sequence determines

BOX 7.1

**The random nature of bacterial mutations**

Until the mid-1900s, it was widely believed that bacterial mutations arose only in response to environmental factors. For example, it was thought that antibiotic resistance arose only in response to the presence of an antibiotic in the environment.

An alternative hypothesis (and the one now known to be true) is that antibiotic resistance (or any other mutant characteristic) occurs at random: antibiotic-resistant mutants arise whether the antibiotic is present or not. The presence of an antibiotic in the environment simply selects for antibiotic-resistant mutants. Antibiotic-resistant cells can grow in the presence of the corresponding antibiotic; antibiotic-sensitive cells cannot.

**(a) Random origin of antibiotic-resistant mutants**

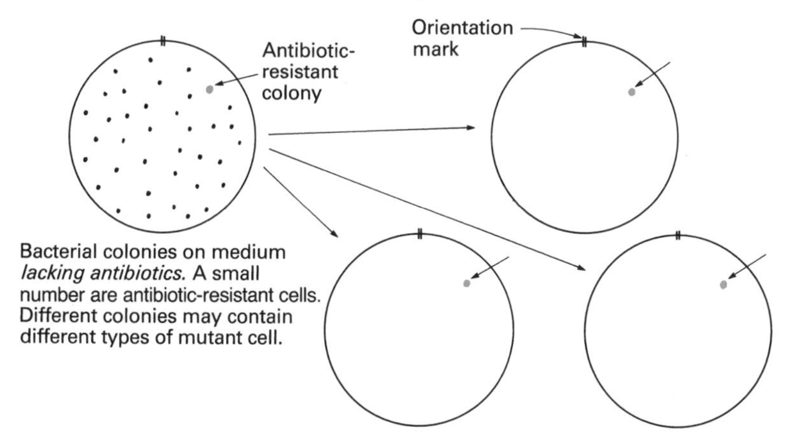


When samples of colonies are transferred to identical positions on a number of *antibiotic-containing* plates (replica plating), resistant colonies *always arise in the same position.*

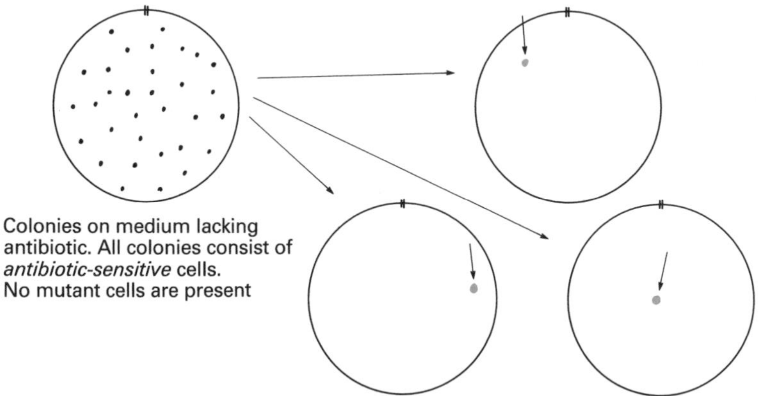


Cells transferred to plates containing antibiotic may mutate to antibiotic resistance once they have come into contact with the antibiotic.
Each cell has the same small chance of mutating and colonies of resistant cells appear *at different locations on different plates.* Since this does *not* happen in practice, we can conclude that bacterial mutation is not a directed process and that, in common with mutations in other organisms, bacterial mutations occur randomly.

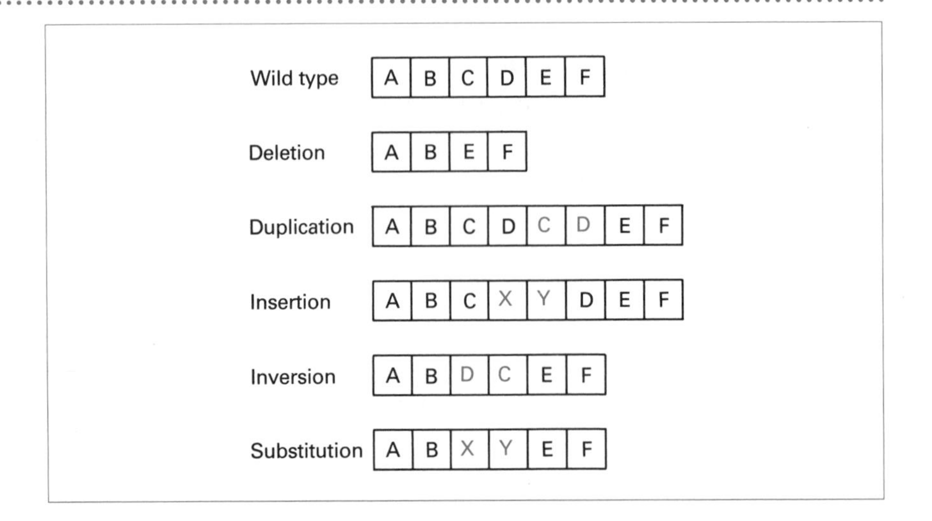


**Figure 7.7** Gross chromosome rearrangements (chromosome mutations).

the amino acid sequence of a polypeptide chain. This is described in detail in Chapter 5. There are four types of point mutation.

**(a) Nonsense mutations**. There are 64 possible codons in the genetic code (Figure 7.9). Of these, 61 code for amino acids. The other three are not translated but are 'stop' signals which terminate polypeptide synthesis. Stop codons are known as **nonsense codons** because they do not code for an amino acid. Single-base substitutions which give rise to nonsense codons may be transitions or transversions (Figure 7.10). Nonsense codons give rise to truncated polypeptides.

**(b) Same-sense mutations**. Of the 20 amino acids found in proteins only two (methionine and tryptophan) are coded by a single codon (AUG and UGG respectively). The remaining amino acids can be coded by two or more codons. This lack of codon specificity, or **degeneracy** of the genetic

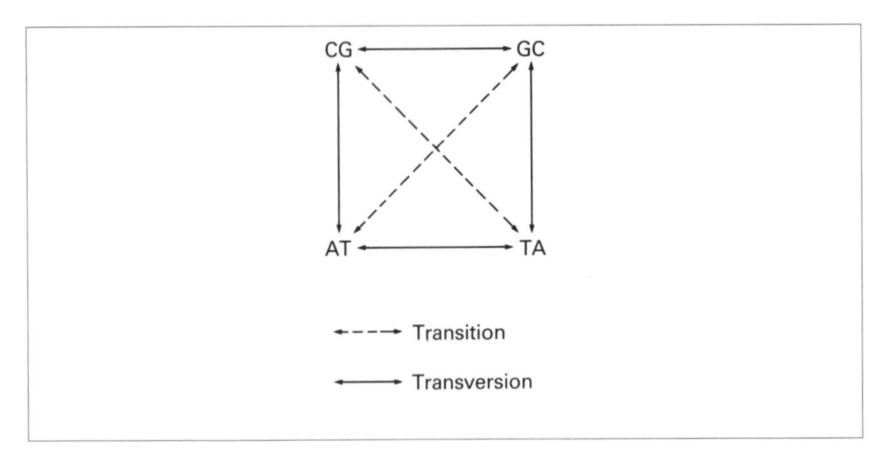


**Figure 7.8** Base substitutions which give rise to transition and transversion mutation.

| First letter | Second letter U | | Second letter C | | Second letter A | | Second letter G | | Third letter |
|---|---|---|---|---|---|---|---|---|---|
| U | UUU | Phe | UCU | Ser | UAU | Tyr | UGU | Cys | U |
| | UUC | | UCC | | UAC | | UGC | | C |
| | UUA | Leu | UCA | | UAA | STOP | UGA | STOP | A |
| | UUG | | UCG | | UAG | | UGG | Trp | G |
| C | CUU | Leu | CCU | Pro | CAU | His | CGU | Arg | U |
| | CUC | | CCC | | CAC | | CGC | | C |
| | CUA | | CCA | | CAA | Glu | CGA | | A |
| | CUG | | CCG | | CAG | | CGC | | G |
| A | AUU | Ile | ACU | Thr | AAU | Asn | AGU | Ser | U |
| | AUC | | ACC | | AAC | | AGC | | C |
| | AUA | | ACA | | AAA | Lys | AGA | Arg | A |
| | AUG | Met | ACG | | AAG | | AGG | | G |
| G | GUU | Val | GCU | Ala | GAU | Asp | GGU | Gly | U |
| | GUC | | GCC | | GAC | | GGC | | C |
| | GUA | | GCA | | GAA | Glu | GGA | | A |
| | GUG | | GCG | | GAG | | GGG | | G |

**Figure 7.9** The genetic code. The triplets UAA, UAG and UGA act as signals for terminating polypeptide chains. Although the genetic code is sometimes said to be universal, that of mitochondria and chloroplasts shows some differences to that shown here.

code, means that some base changes can occur with no corresponding alteration of the amino acids in the polypeptide (Figure 7.11). This is known as a same-sense mutation because the code is translated into the same amino acid sequence in the mutant as in the wild (or normal) type.

**(c) Mis-sense mutations.** A simple nucleotide substitution causes the incorporation of a different amino acid from the one in the wild-type polypeptide (Figure 7.12). A mis-sense mutation may produce no change in the individual's phenotype, it may produce a slight change, or it may produce a drastic change. It will depend upon whether the amino acid change occurs at an essential or non-essential position in the polypeptide chain. If an amino acid required for the integrity of the active site of an enzyme is changed, a drastic change in phenotype is likely to ensue.

**(d) Frameshift mutations.** The genetic code is read as blocks of three nucleotides (codons), starting at a fixed point. The addition or deletion of one or two bases alters the translational reading-frame of the mRNA carrying the mutation (Figure 7.13). If three bases are added or deleted, the

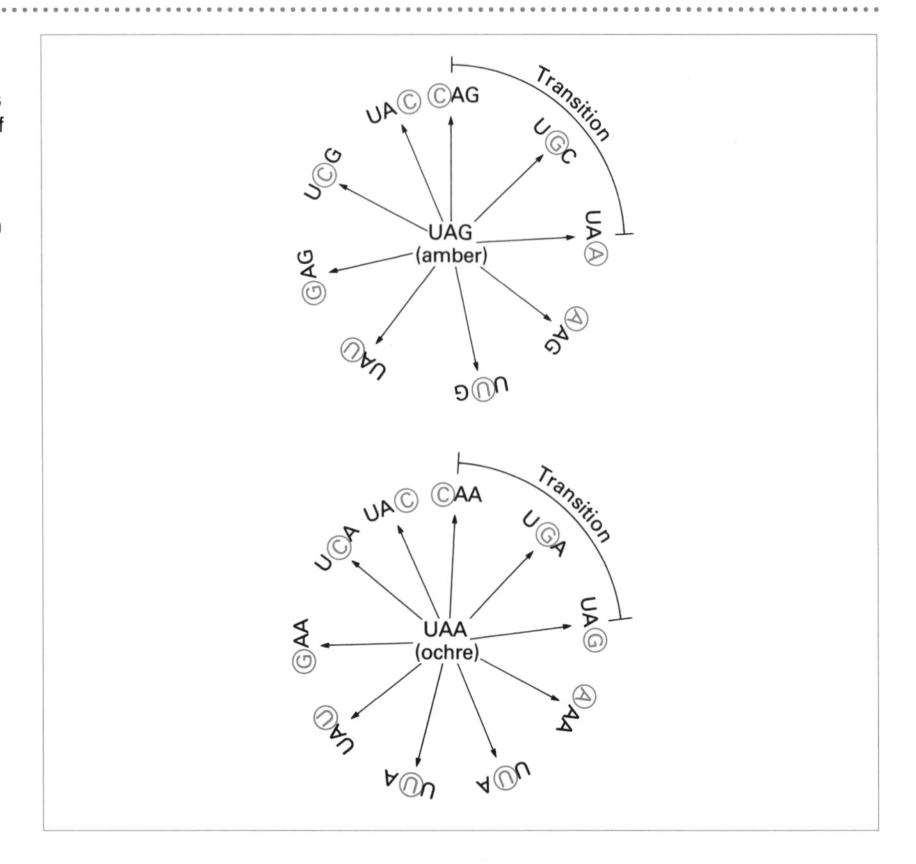


**Figure 7.10** Single base changes in sense codons which give rise to the nonsense codons amber and ochre. The majority of base substitutions involve transversions. In each case only three of the substitutions are transitions. These are marked on the outside of the ring.

reading frame would remain unaltered, but an amino acid residue would be added to, or deleted from, the polypeptide.

Mutations can arise spontaneously in a variety of ways. The fidelity of DNA replication relies on complementary pairing of the bases. That is,

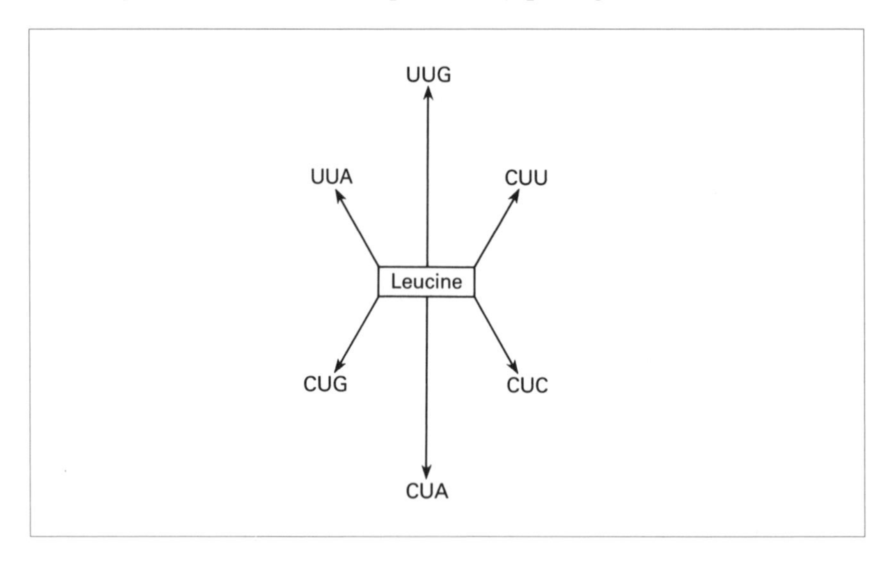


**Figure 7.11** Degeneracy of the genetic code. Of the 20 amino acids found in proteins only methionine and tryptophan are assigned to single codons: AUG and UGG respectively. All the other amino acids can be assigned to at least two codons. Leucine has six codons.

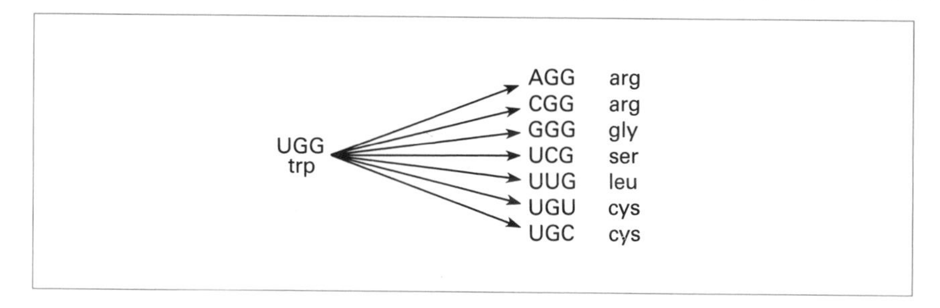


**Figure 7.12** Single base mutations in the tryptophan codon which give rise to mis-sense mutations. Single base mutations in the tryptophan codon give rise to seven possible sense codons which could result in one of five possible amino acid substitutions in the polypeptide. Note that two other base substitutions are possible but these would result in the nonsense codons UGA (opal) and UAG (amber).

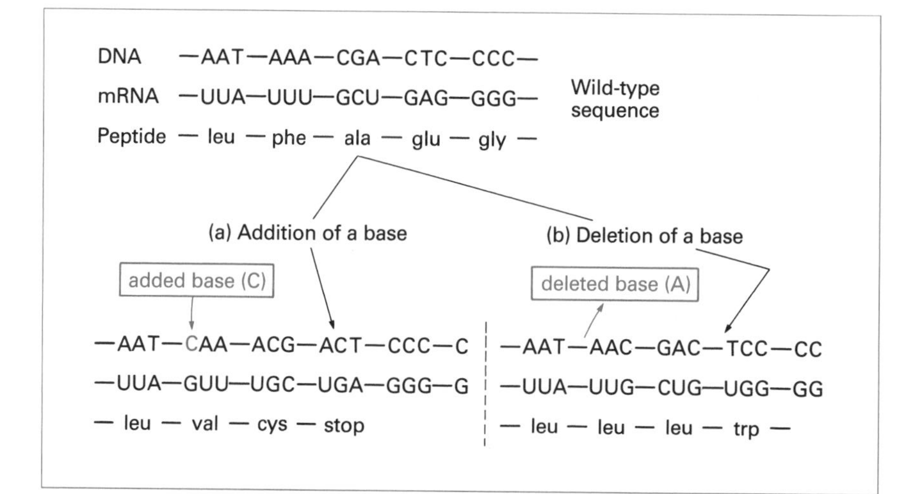


**Figure 7.13** Frameshift mutations. These can arise from either the addition or the deletion of a base pair in DNA. The addition of a base causes the reading frame to move one base backwards. The deletion of a base causes the reading frame to be advanced by a base. A frameshift mutation can result in a dramatic change to the structure of a polypeptide chain: it may be truncated or a number of amino acids may differ from those of the wild type. If a major change occurs, this will be reflected in a mutant phenotype.

adenine (A) pairs with thymine (T), and guanine (G) pairs with cytosine (C). **Tautomerism** of the nucleotide bases can cause incorrect base-pairing which can produce a spontaneous mutation.

All the bases which occur normally in DNA have rare tautomers (Table 7.1). Adenine, for example, exists predominantly in the amino form in which it pairs with thymine. In its rare imino form it will pair with cytosine (Figure 7.14). The overall effect of this will be the mutation from an A:T pair to a G:C pair (Figure 7.15).

**Table 7.1** Tautomeric bases of DNA

| *Base* | *Natural (common) form* | *Tautomeric form (rare)* | *H-bonding characteristics of tautomer* | *Mutation incurred* |
|---|---|---|---|---|
| Adenine | Amino | Imino | $A^*_{(imino)}$–$C_{(amino)}$ | AT → GC |
| Cytosine | Amino | Imino | $C^*_{(imino)}$–$A_{(amino)}$ | GC → TA |
| Guanine | Keto | Enol | $G^*_{(enol)}$–$T_{(keto)}$ | GC → AT |
| Thymine | Keto | Enol | $T^*_{(enol)}$–$G_{(keto)}$ | TA → GC |

**Figure 7.14** Tautomeric changes in adenine which cause altered base-pairing. The alteration in charge distribution following migration of the proton from the amino group allows hydrogen bonding to occur with cytosine. Normal (*amino*) adenine pairs with thymine. In this example an A:T pair would mutate to a G:C pair.

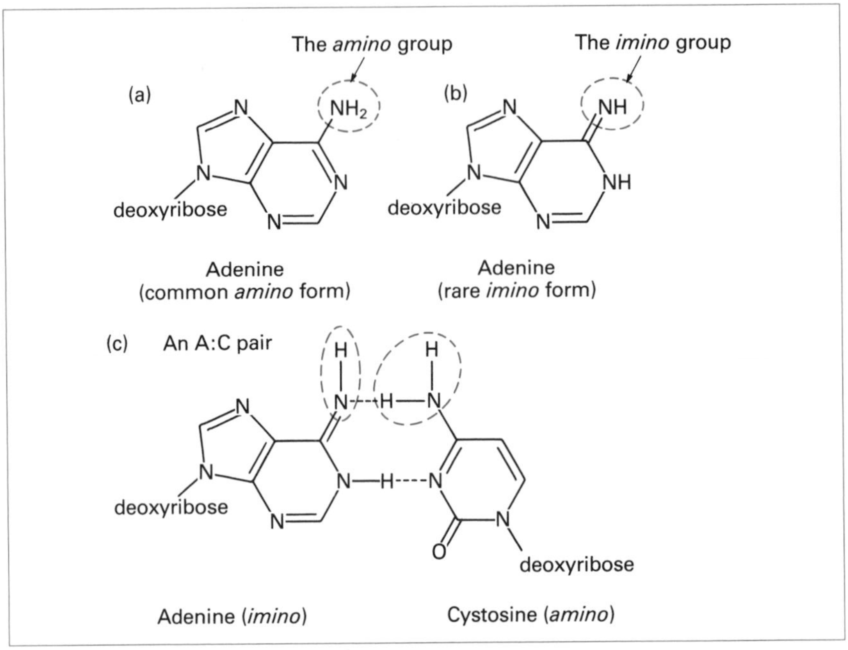

The *amino* group
The *imino* group
(a)
(b)
deoxyribose
Adenine (common *amino* form)
Adenine (rare *imino* form)
(c) An A:C pair
deoxyribose
Adenine (*imino*)
Cystosine (*amino*)

Although spontaneous mutation was described earlier as a random process, this is not strictly the case. The frequency at which individual genes mutate varies from $10^{-4}$ to $10^{-11}$; that is to say, one mutation in 10 000 to 100 000 000 000 replications. Some gene sites are more susceptible to mutation than others. These are known as **hot-spots** and were first

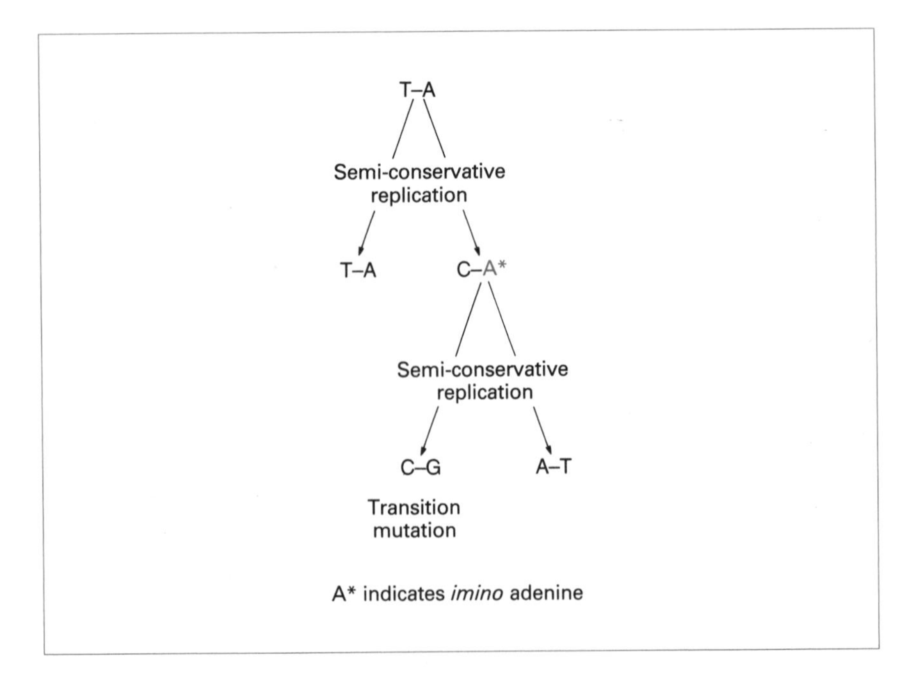


**Figure 7.15** Spontaneous mutation arising from the tautomerism of adenine to its rare *imino* tautomer.

demonstrated in the rII region of phage T4 by Benzer in 1961. He found that almost half of all mutations occurred in one or two sites. As a general rule, a mutation will spontaneously arise within a gene on one occasion in approximately one million individuals. That is, in a population of a million cells, in which each cell contains one copy of a specific gene, only one gene (or individual) is likely to be mutated. More succinctly, one can express this mutation rate as $10^{-6}$. By the same token, a nucleotide base-pair mutates with a frequency of approximately $10^{-9}$, or, to put it another way, one given base-pair in a given gene will mutate on only one occasion in a billion base-pairs. This is because the average gene comprises approximately one thousand base pairs.

**Tautomers**: each of the bases in a DNA molecule can exist in two forms (tautomers). Each form has the same component atoms but they differ in the location of a single hydrogen atom. For example, adenine can exist as two tautomers which differ in the presence of either an *amino* group or an *imino* group. Thymine tautomers possess either an *enol* or a *keto* group.

Adenine

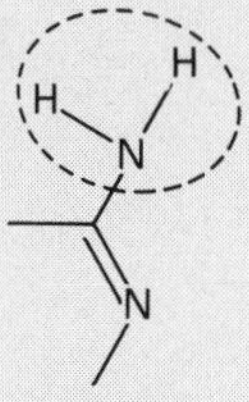

*amino* form

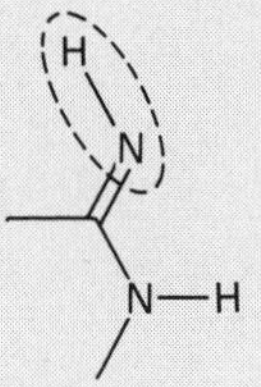

*imino* form

Thymine

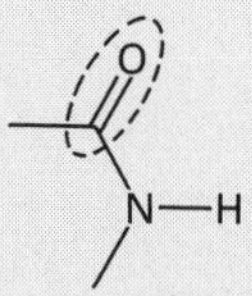

*keto* form

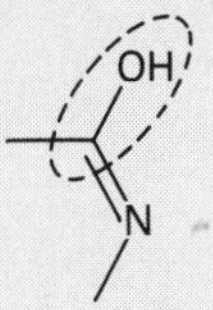

*enol* form

## Mutagenesis: making mutations happen

Mutations are rare events. However, the frequency with which they arise can be increased by treating cells with chemical or physical agents known as **mutagens**. Such mutations are referred to as **induced mutations**.

### Physical mutagens

Ultraviolet (UV) light, X-rays and other ionizing radiations have mutagenic effects on microorganisms. The action of UV light is the best understood. There is close agreement between the absorption spectrum of UV light and its mutagenic action. Purines and pyrimidines absorb UV maximally at 260 nm, which corresponds to the wavelength which is most efficient at inducing mutations in bacteria. Although UV rays possess insufficient energy to induce ionizations, their energy is absorbed by purines and pyrimidines. When this occurs, they become more reactive and enter an **excited state**.

The primary effect of UV irradiation is the production of **pyrimidine dimers** – most often **thymine dimers** (Figure 7.16). A thymine dimer is produced when two adjacent thymine residues in the same DNA strand become linked by additional chemical bonds absent from untreated DNA molecules. However, it is not the formation of the dimer itself which causes the mutation, but one of the cell's DNA repair mechanisms. If sufficient DNA damage is caused, an error-prone repair mechanism, called the **SOS repair system**, is induced. The SOS repair system frequently causes the insertion of incorrect bases where a thymine dimer occurs (see later in this chapter). Transition or transversion mutations may also arise following induction of the SOS system.

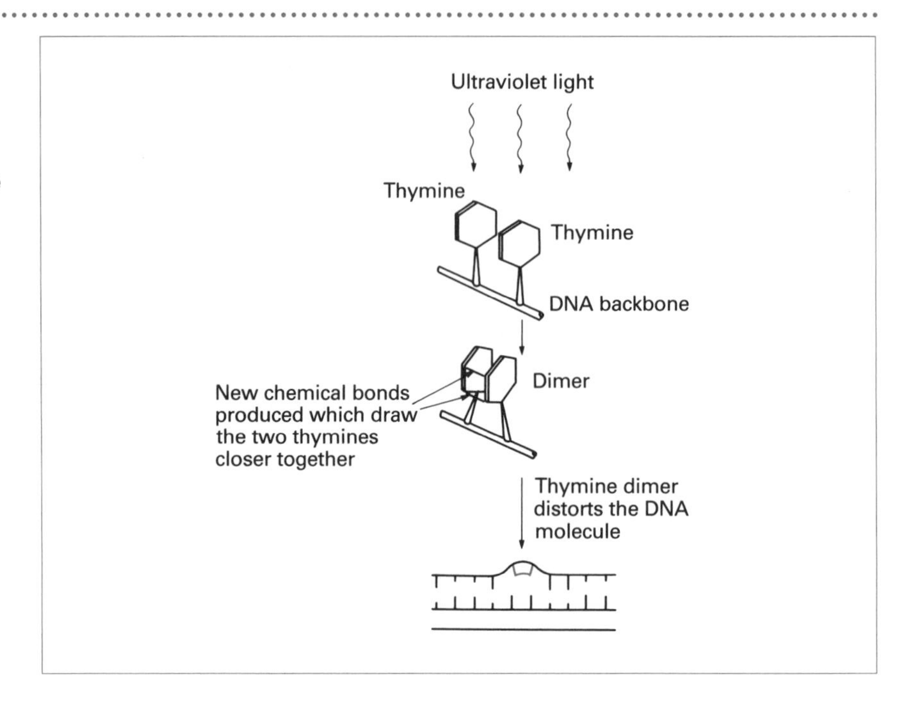


**Figure 7.16** Ultraviolet light is mutagenic. Following exposure to UV light, new chemical bonds form between some pyrimidine residues (normally thymines) situated next to each other in the same DNA strand. Thymine dimers can be removed by the SOS repair mechanism: this is an error-prone mechanism and mutations commonly result.

## Chemical mutagens

There are several types of chemical mutagens. Some cause damage during DNA replication, whereas others modify the DNA at any time, altering the normal bases so that they form abnormal base-pairs.

**(a) Base analogues**. Two examples are 5-bromouracil (5BU) and 2-aminopurine (2AP); these are similar in structure to the naturally occurring bases thymine and adenine respectively. During replication, the base analogue is incorporated into the DNA molecule in place of the naturally occurring base. It is not this event itself which causes the mutation. Base analogues undergo tautomeric shifts in the same way that naturally occurring bases do, but with much higher frequencies.

The alternative tautomeric form of the base analogue in one DNA strand forms an abnormal pairing with a base in the other strand. As a result, an incorrect base is incorporated opposite the analogue in the complementary strand of DNA. For example, 5BU in its usual keto state mimics thymine and pairs with adenine. However, its enol tautomer mimics cytosine and pairs with guanine (Figure 7.17). Similarly, whereas the amino form of 2AP pairs with thymine, the rare imino tautomer pairs with cytosine. In both, an AT pair is replaced by a GC pair.

**(b) Intercalating agents**. These are flat, three-ringed compounds whose dimensions correspond roughly to those of a purine–pyrimidine nucleotide

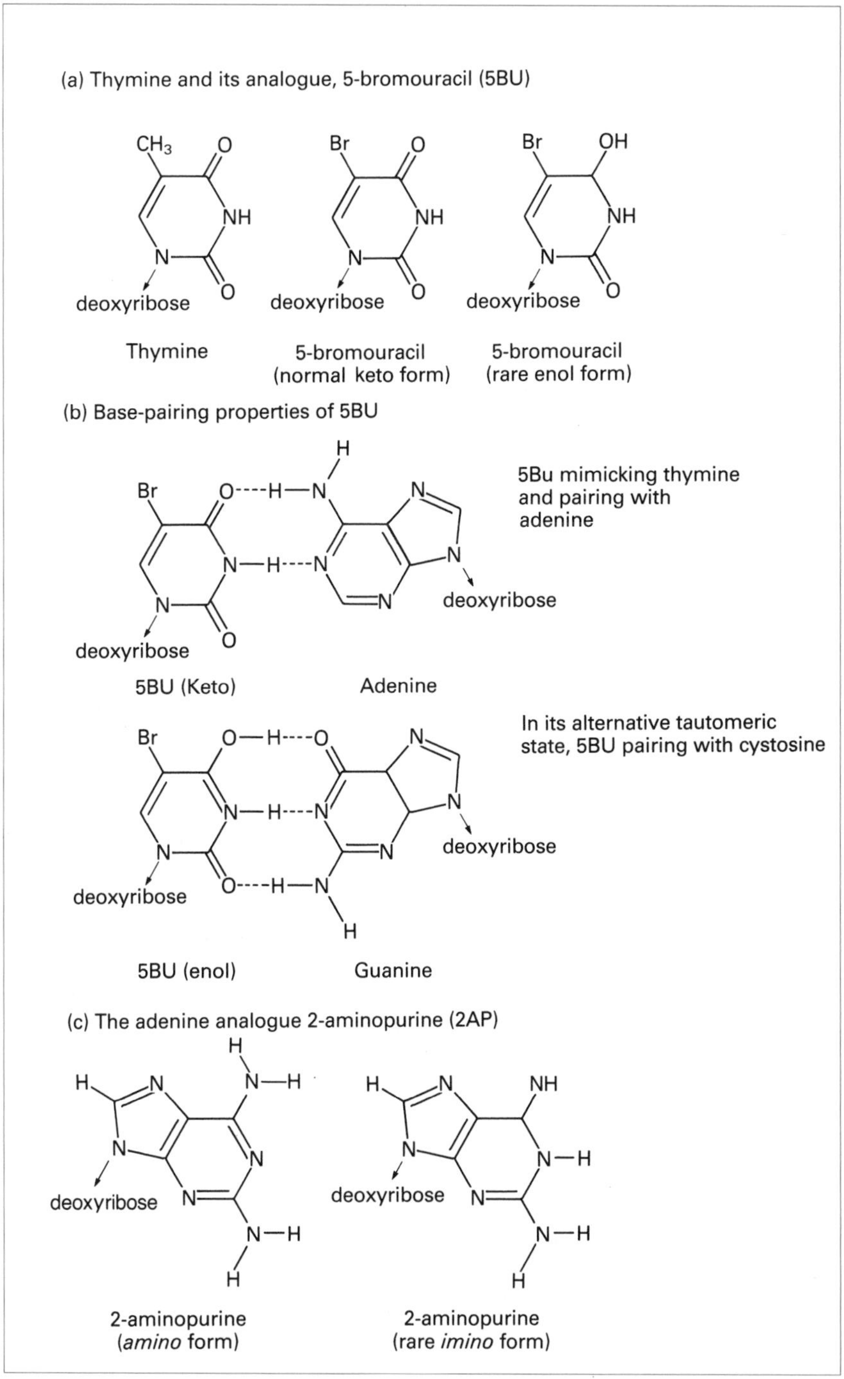

(a) Thymine and its analogue, 5-bromouracil (5BU)

Thymine — 5-bromouracil (normal keto form) — 5-bromouracil (rare enol form)

(b) Base-pairing properties of 5BU

5BU (Keto) — Adenine: 5Bu mimicking thymine and pairing with adenine

5BU (enol) — Guanine: In its alternative tautomeric state, 5BU pairing with cystosine

(c) The adenine analogue 2-aminopurine (2AP)

2-aminopurine (*amino* form) — 2-aminopurine (rare *imino* form)

**Figure 7.17** Base analogues.

base-pair. Some typical intercalating agents are shown in Figure 7.18. These compounds are thought to exert their effects by slotting (intercalating) between adjacent base-pairs in the double helix. The distortion of the helix

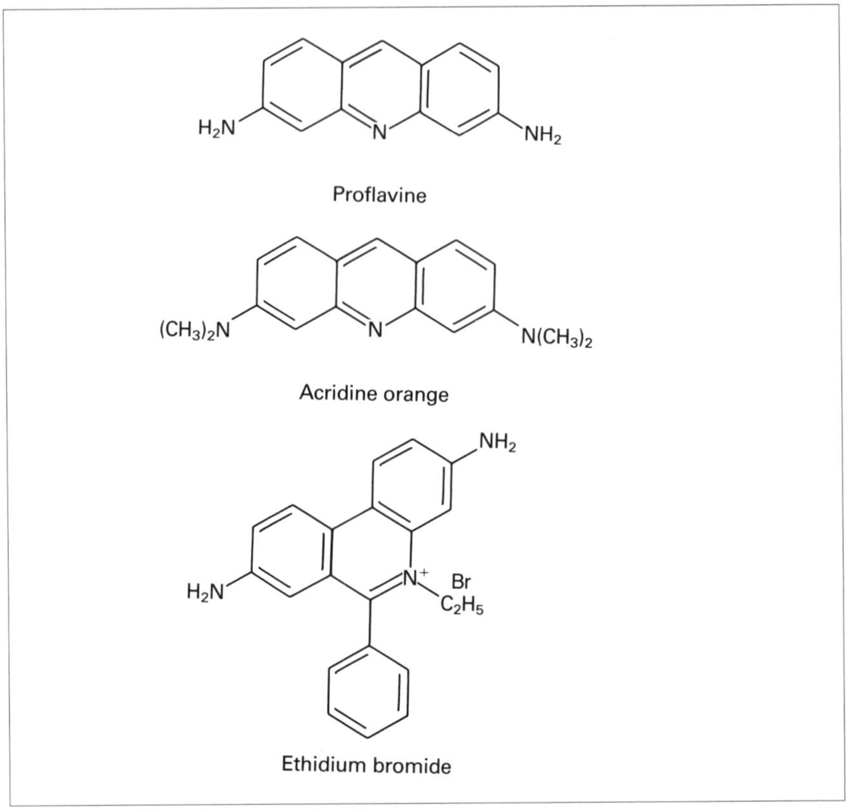

**Figure 7.18** Examples of mutagens which act as intercalating agents.

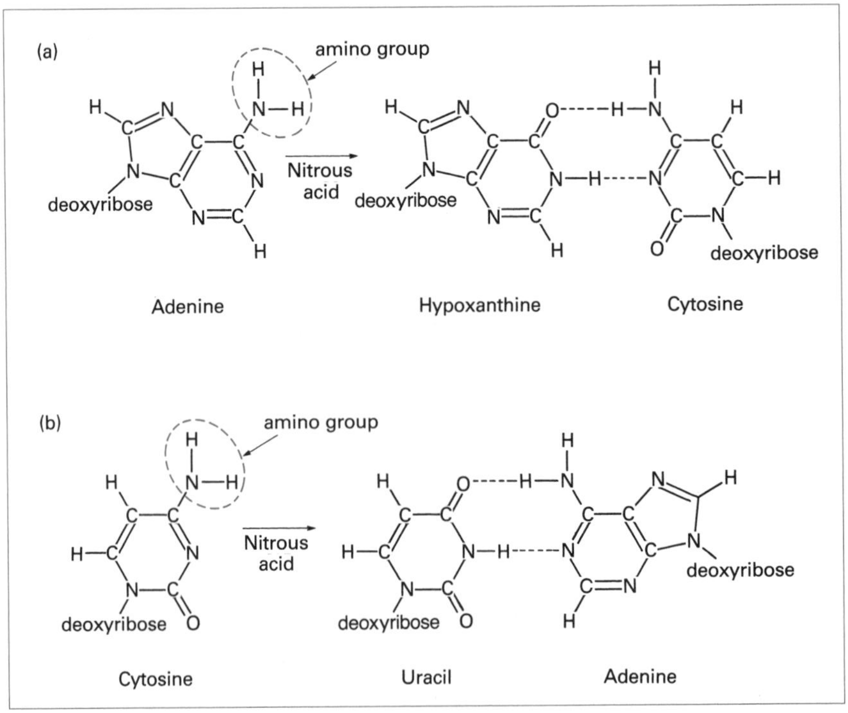

$H_2N$ N $NH_2$

Proflavine

$(CH_3)_2N$ N $N(CH_3)_2$

Acridine orange

$NH_2$ $H_2N$ $N^+$ Br $C_2H_5$

Ethidium bromide

**Figure 7.19** The mutagenic action of nitrous acid. The amino groups of adenine and cytosine are oxidized to carbonyl (>C=O) groups and produce hypoxanthine (H) and uracil (U) respectively. These modified bases have altered hydrogen-bonding specificities: H pairs with C and U pairs with A. Transition mutations result. (Guanine can be changed to xanthine but its base-pairing characteristics remain unaltered.)

(a) amino group

Nitrous acid

deoxyribose

Adenine Hypoxanthine Cytosine

(b) amino group

Nitrous acid

deoxyribose

Cytosine Uracil Adenine

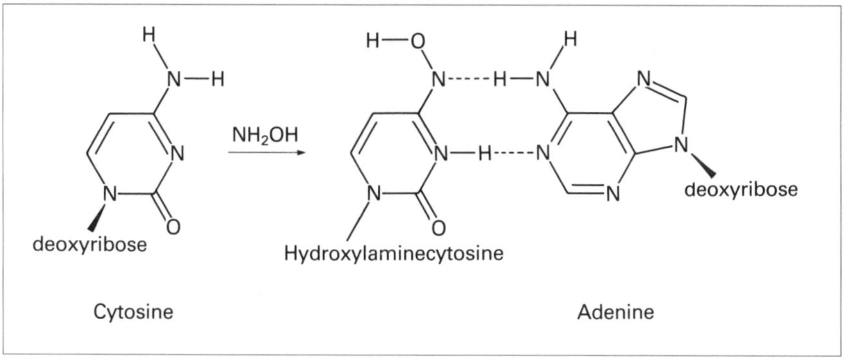

**Figure 7.20** The action of hydroxylamine on cytosine.

caused by insertion of the intercalating agent leads to the addition or deletion of a base, depending on whether the intercalating agent is inserted into the template strand or the newly copied strand. What word is used to describe the type of mutation produced? Yes, a frameshift mutation would occur.

DNA-modifying chemicals alter a base that is already incorporated into the DNA so that mis-pairing during replication occurs. Nitrous acid ($HNO_2$) removes amino groups from bases and replaces them with keto groups (Figure 7.19). Cytosine is converted to uracil, which pairs with adenine. Adenine is converted to hypoxanthine, which pairs with cytosine. The overall result is GC to AT and AT to GC transition respectively.

Hydroxylamine ($HONH_2$) reacts only with cytosine giving hydroxylaminocytosine (Figure 7.20). This modified base pairs with adenine. The result is a one-way transition mutation of GC to AT.

## Selection of mutants

Even in the presence of a mutagen, a mutation in a specific gene is a rare event. To study a particular mutant cell it is necessary, therefore, to select it from a large population of randomly mutated and non-mutant cells. Direct selection is possible for mutants which are resistant to particular agents (for example antibiotics, bacteriophage or heavy metals). Direct selection uses a mutant characteristic as a basis for separation. For example, bacteria resistant to streptomycin will be the only survivors from a population of bacteria treated with a concentration of the antibiotic known to kill the sensitive parent population.

An example of **indirect selection** is that used for the isolation of **auxotrophic mutants**. Mutation to auxotrophy gives rise to bacteria which are unable to grow on a minimal medium.

**BOX 7.2**

**The replica-plating technique for identification of auxotrophic mutants of bacteria**

This technique is one by which bacteria from colonies on one agar plate (the master plate) can be transferred to identical positions on a second agar plate (the replica). For the identification of auxotrophic mutants, the growth medium used for the master plate is one on which both mutant and non-mutant cells can grow. If the replica lacks an amino acid, such as methionine, methionine auxotrophs can be identified and isolated because they grow on the master plate but not on the replica.

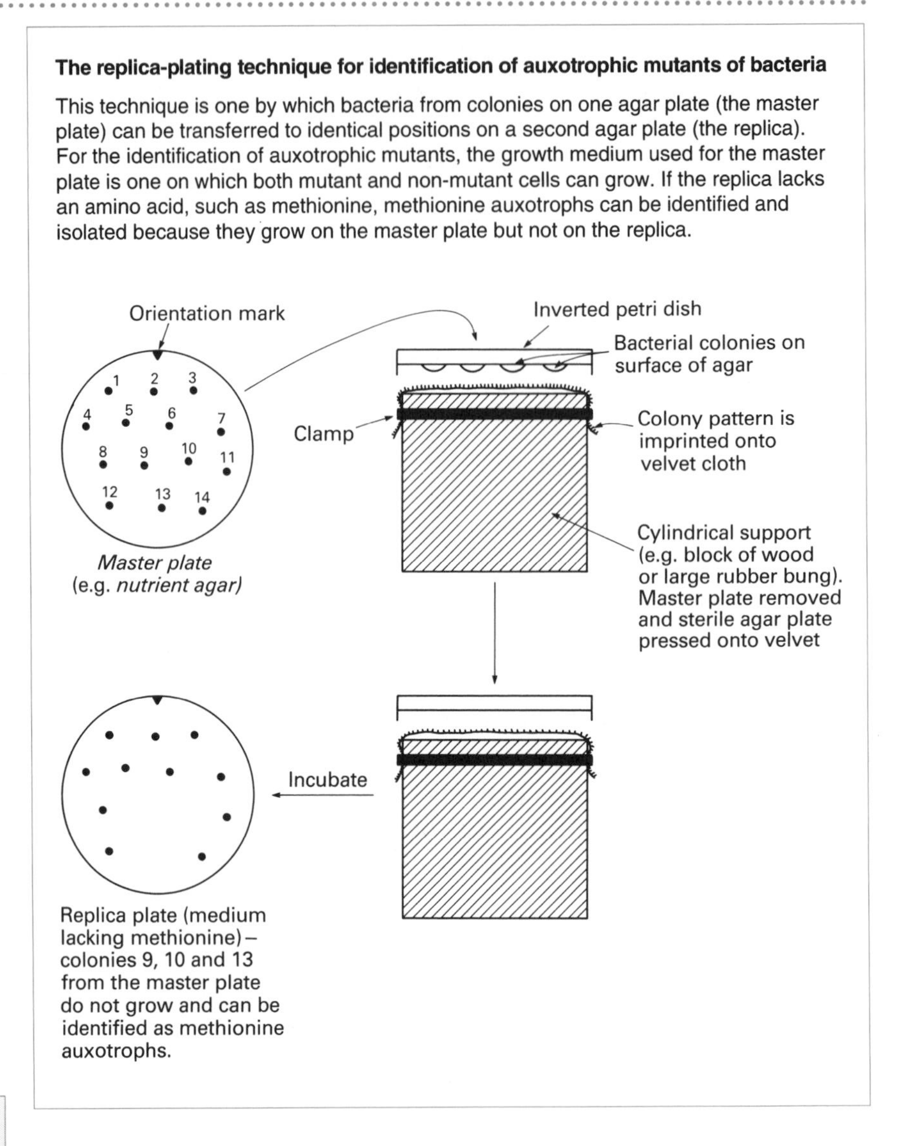


A **minimal medium** is a culture medium which contains the minimum nutritional requirements necessary to support the growth of a non-mutant (wild-type) strain of a microorganism. An auxotrophic mutant will require the addition of one or more growth factors to the minimal medium. The use of culture media is described in Chapter 3.

To distinguish auxotrophs from **prototrophic bacteria** (those which do not require a growth factor in the medium), the **replica plating technique** can be used. After the population has been treated with the mutagen it is diluted and plated onto nutrient agar (NA). Nutrient agar supports growth of prototrophic parent and mutant cells. Plates which have well-spaced colonies on them are selected and a velvet pad used to transfer a replica of the colony distribution to plates containing minimal agar (MA). Following incubation, the MA plates are examined for growth and the pattern of colonies noted. When compared with the original nutrient agar plates (master plates) it is evident that some colonies present on the NA plates have not grown on MA. These are the auxotrophic mutants. The colonies on the NA plates corresponding to these gaps can be grown on for further study.

## Reversal of mutation

The effects of a mutation can be reversed in two ways. The genome of a microorganism carrying a point mutation can **back-mutate** (revert) to the original base sequence. The genome of the microorganism carrying the back-mutation and that of the original (wild type) are indistinguishable.

This mechanism of back-mutation is quite distinct from a second mechanism of restoring the wild-type phenotype, called **suppression**.

**BOX 7.3** Typical results of an Ames test experiment.

### The Ames test for carcinogenicity

The Ames test uses auxotrophic strains of bacteria to test the potential carcinogenic effect of agents which are found in the environment. These include food additives, preservatives and many other kinds of substances.

The test was originally developed using histidine-requiring mutants of *Salmonella typhimurium*, but any auxotrophic mutant of any bacterium or yeast can be used in the same way.

The test makes the assumption that carcinogens are also mutagens, and depends upon reversion of auxotrophic cells to prototrophic ones. Although reversion can happen spontaneously and without exposure to a mutagen (carcinogen), such occurrences are very rare. The number of revertants to prototrophy will be very much increased in the presence of a mutagen.

The test is carried out by spreading a sample of auxotrophic bacteria on the surface of solid medium which lacks the growth factor required by the test organism. The test substance (such as a food additive) is incorporated into the medium and, if a large number of revertant colonies grow, the test substance is identified as a potential carcinogen.

The test may be made more sensitive by using a strain of the test organism which has increased permeability to external substances and lacks mechanisms for the repair of mutations. Finally, because non-carcinogenic compounds can be converted to carcinogens by enzymes which are found in mammals but not in bacteria, the test substance may be given a preliminary treatment in which it is incubated with a preparation of rat liver enzymes.

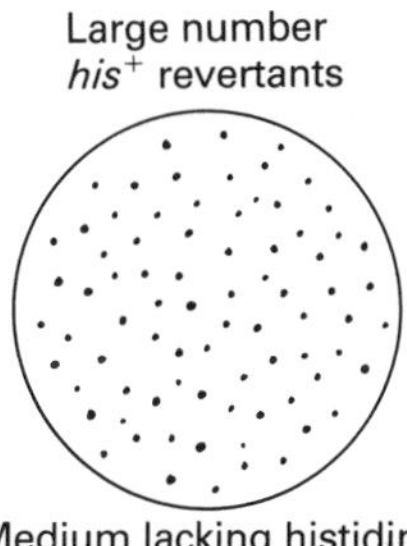

Suppression occurs when a second mutation, a **suppressor mutation,** arises in the genome which partially or wholly restores the wild-type phenotype. These mutants carry both the initial mutation and the suppressor mutation. Therefore, the wild-type microorganism and the one carrying the suppressor mutation are genetically distinct. There are two categories of suppressor mutation. If the second mutation occurs in the same gene as the first, we call it **intragenic suppression.** If the original and suppressor mutations occur in different genes, the description **intergenic suppression** is used.

## Intragenic suppression

Intragenic suppression may be effected in several ways. Where the original mutation is a nonsense mutation, the mutant terminator codon can be altered to a codon that specifies an amino acid. This is the case in the example given in Figure 7.21. The suppressed mutant has a phenotype similar to that of the wild-type individual. That is, its amino acid sequence differs from that of the wild type by a single amino acid occurring at the site of the original mutation. The codon specifying the altered amino acid contains two altered bases.

Where the original mutation is a mis-sense mutation, its effect may be suppressed by base substitution in a non-mutant codon. For example, a primary mutation in the tryptophan synthetase A gene of *E. coli* brings

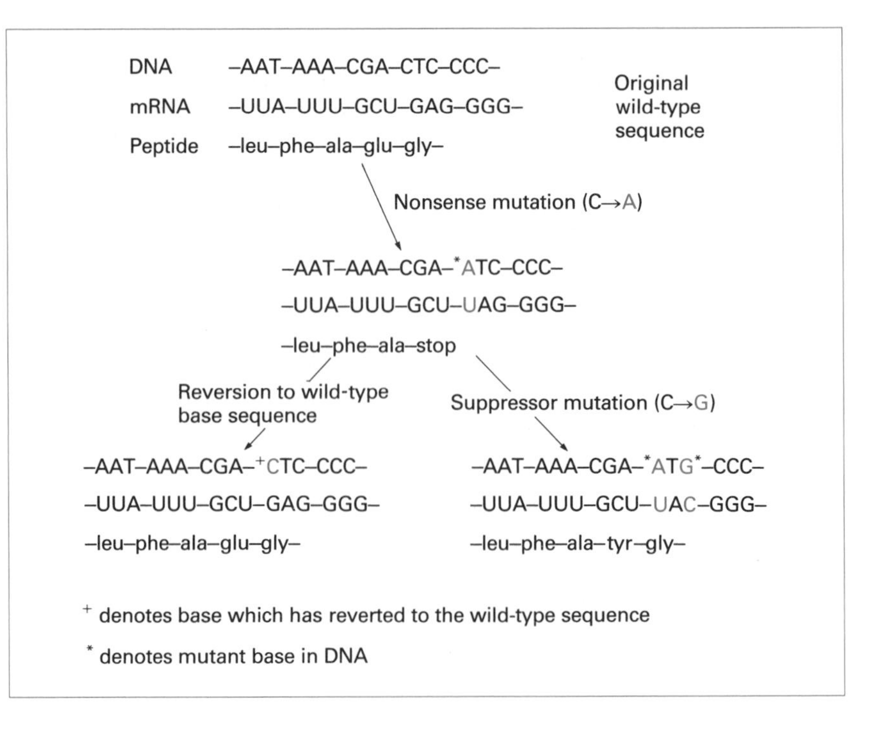


**Figure 7.21** Reversion and suppressor mutations. The penultimate amino acid in the sequence shown is glutamic acid. The original mutation alters the codon for glutamic acid to a chain-terminating sequence (amber). In the case of the revertant, the original mutation is reversed directly. The suppressor mutation changes the amber codon to a tyrosine residue. The overall effect of the suppressor mutation is that the whole peptide is translated but a tyrosine residue replaces the original glutamic acid residue.

about the substitution of a glutamic acid residue by glycine, which inactivates the enzyme. This inactivation may be corrected by a second mutation, 36 codons from the first, which replaces a cysteine residue with a tyrosine residue. The primary mutation alters the configuration of the enzyme which is, at least partially, corrected by the second (suppressor) mutation.

Frameshift mutations can be reversed by the addition of a base to compensate for a deletion or removal of a base to correct an addition. In the example shown (Figure 7.22), the original mutation is the insertion of a cytosine base. This insertion causes the mRNA to be transcribed as –leu–val–cys–stop in place of the normal peptide (–leu–phe–ala–glu–gly–). A suppressor mutation, the deletion of cytosine in another codon, restores the transcriptional machinery to the original reading frame. A complete polypeptide is produced but the amino acids between the two mutant sites are different from the wild-type polypeptide (val and cys have replaced phe and ala respectively). Where frameshift mutations are involved, the closer the primary and secondary mutant sites are, the more effective suppression is likely to be.

## Intergenic suppression

Intergenic suppression may be direct or indirect. Suppression is indirect if the second mutation alters a metabolic pathway which ameliorates the

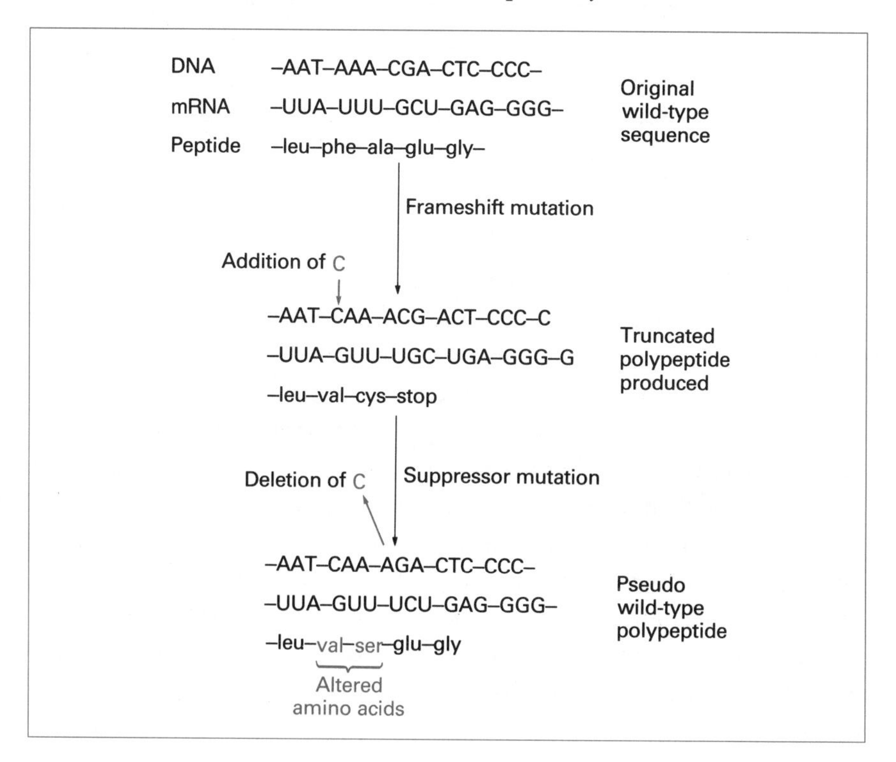


**Figure 7.22** Intragenic suppression of frameshift mutations.

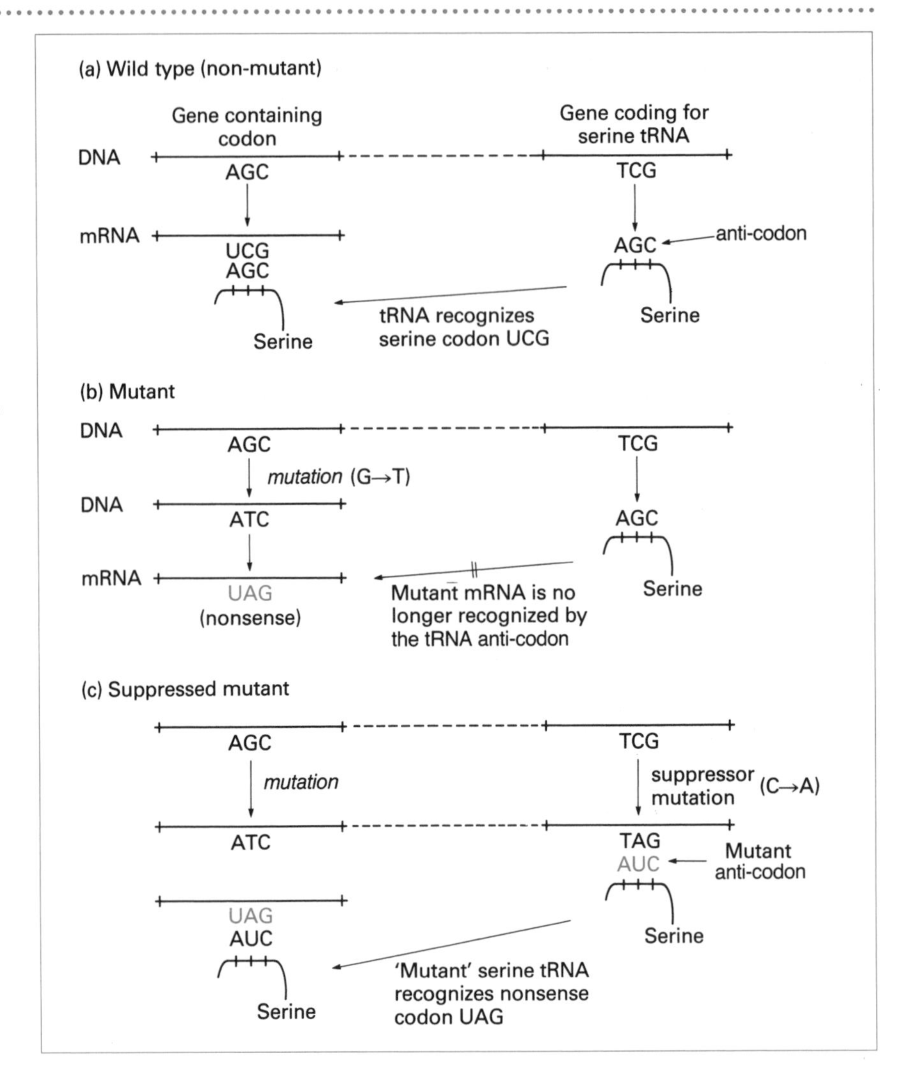


**Figure 7.23** Intergenic suppression of a nonsense mutation. (a) In the wild type, the serine codon (UCG) is recognized by the anti-codon (AGC) of a charged tRNA molecule (that is a tRNA with an amino acid residue attached). (b) The nonsense mutation (UCG → UAG) causes the peptide to terminate prematurely: the anti-codon cannot interact with the modified codon. (c) A mutation in the gene for the serine tRNA alters the anti-codon to AUC. The charged tRNA can now interact with the nonsense codon as if it were a sense codon: translation progresses through the 'stop' signal.

influence of the original mutation. *Neurospora* mutants have been isolated which are sensitive to high intracellular concentrations of the amino acid arginine. These can undergo a second mutation in the arginine biosynthetic pathway which decreases the rate of arginine synthesis in the cell.

Direct intergenic suppression takes place when the suppressor mutation occurs in a base which determines the anti-codon of a tRNA molecule. The mutant tRNA charged with an amino acid 'mistakes' the nonsense codon for a sense codon and incorporates an amino acid into the growing peptide rather than terminating it prematurely (Figure 7.23). The amino acid inserted at the site of the suppressed nonsense codon need not necessarily be the same as the original (wild-type) amino acid. However, in the majority of cases a single amino acid substitution is unlikely to affect the activity of the polypeptide.

*Neurospora* is an ascomycete fungus (Chapter 9).

## Recombination

Recombination is another mechanism for generating genetic variety in cells. Whereas mutation generates altered nucleotide sequences, recombination is a rearrangement of existing polynucleotides.

**(a) General (non-homologous) recombination**. This occurs between sections of DNA which have identical, or at least very similar, base sequences. Several models for general recombination have been proposed, most of them sharing several features with the so-called **Holliday model**. We shall consider a revised version of the Holliday model here. General recombination is an enzyme-mediated process which involves the coordinated breakage and reunion of single strands of DNA.

**(b) Site-specific recombination**. In some cases, recombination requires the presence of specific sites in the DNA. Site-specific recombination is exemplified by phage lambda (Chapter 10). These recombination events, which usually involve sequences of fewer than 25 base-pairs, are described elsewhere as part of the life-cycle of the phage (Chapter 8).

**BOX 7.4**

**Recombination between two homologous, linear, double-stranded DNA molecules**

(a) Two double-stranded DNA molecules, which have a region in which the nucleotide sequences are identical, or at least very similar, line up adjacent to one another.

A 5′ 3′ — 3′ 5′ B

a 3′ 5′ — 5′ 3′ b

(b) Endonuclease enzyme breaks each DNA molecule in equivalent places in identical nucleotides.

A — B

a — b

(c) The DNA strands containing the breaks are displaced and a cross-over branch is formed. A **heteroduplex** is produced in which nucleotides of each of the broken strands are shared between the two double helices.

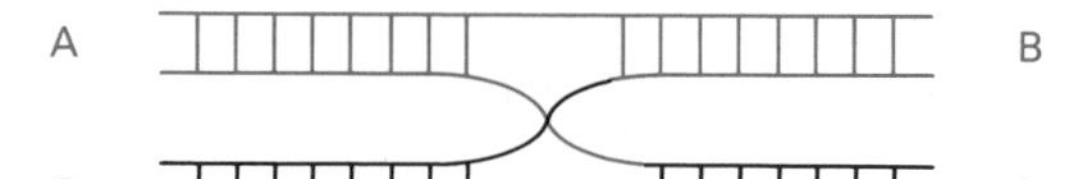

**BOX 7.4** *continued*

(d) Single-strand breaks are sealed by the enzyme DNA ligase and the cross-over point moves (or migrates) along the two molecules. The polynucleotides which have been exchanged can form base-pairs with the complementary strands in either DNA molecule. This is because the original DNA molecules are homologous.

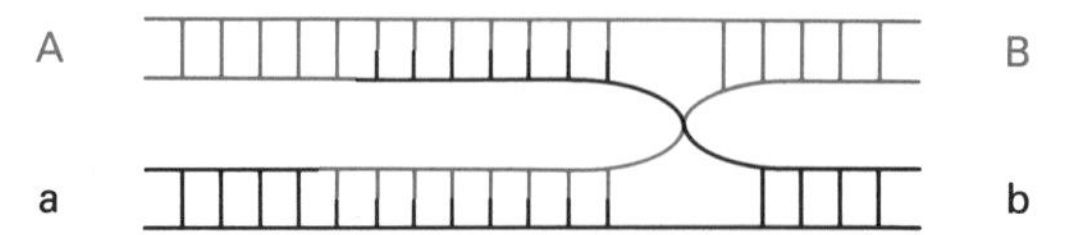


This structure is sometimes called the *Holliday* structure after Robin Holliday who first proposed its existence.

(e) The structure shown in (d) can also be drawn as a cross-like structure called the **chi form** after the Greek letter $\chi$ (pronounced ki, to rhyme with eye). The existence of this structure has been confirmed by electron microscope observations of recombining DNA molecules.

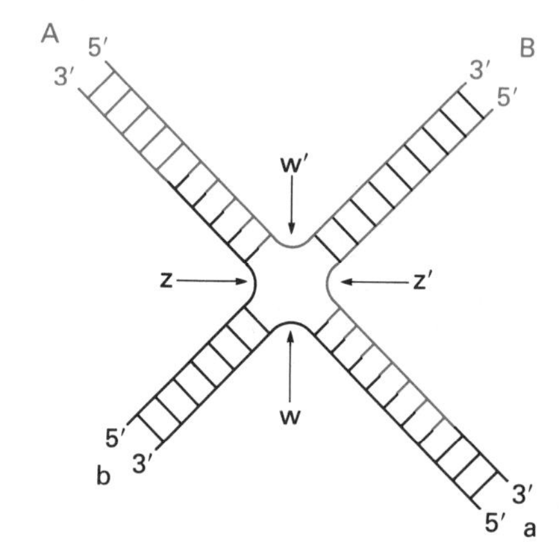


In order to once again produce separate double-stranded DNA molecules, breaks are made in the chi structure in two alternative ways: either at points w and w′ or at z and z′.

(f) If breaks occur at z and z′, the molecules produced after ligation have exchanged only single-stranded segments from their centres. Because the exchanged segments are homologous, they will have very little effect on the genetic constitution of each molecule.

A B

a b

(g) On the other hand, if breaks occur at w and w′, significantly different double helices are produced in which the two ends of the same molecule come from different homologues. The effect on the genetic composition of each molecule will be much more serious.

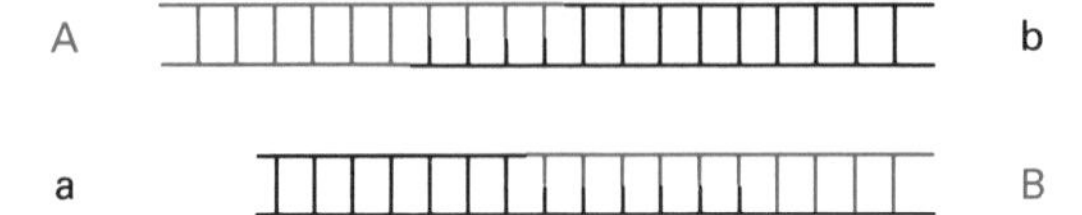

## DNA repair systems

Although DNA damage can be caused by a number of factors, most information has come from a study of processes which can remove thymine dimers from DNA. These are formed when adjacent thymine residues on the same DNA strand become covalently linked. The dimers distort the helix so that it can no longer form hydrogen bonds with adenine residues on the opposite strand.

Mechanisms which remove thymine dimers can be divided into a light-induced mechanism (**photoreactivation**), and three light-independent (**dark repair**) mechanisms: **excision repair**, **recombination repair** and **SOS repair**. The accuracy of DNA replication afforded by these three systems is different. Photoreactivation, excision repair and recombination repair are very accurate, error-free repair processes. In contrast, the SOS system causes frequent mutations and is known, therefore, as an error-prone repair mechanism.

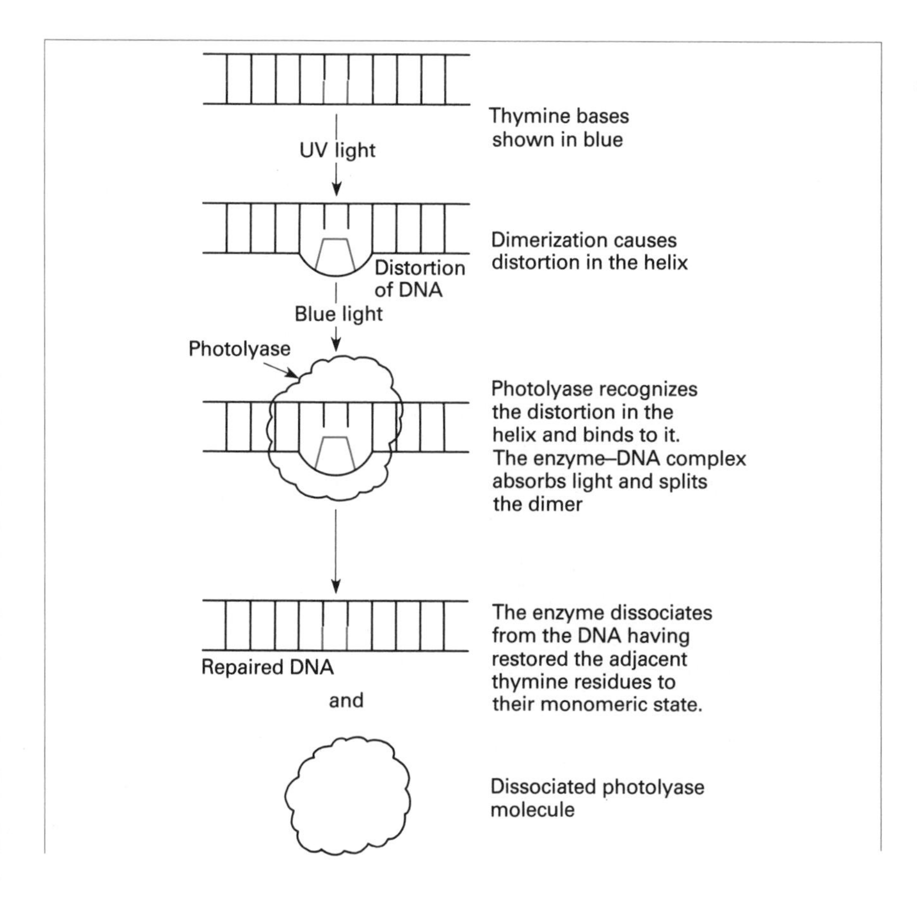


**Figure 7.24** Reversal of DNA damage by photoreactivation (light repair).

**(a) Photoreactivation**. This is mediated by enzymes and requires the presence of light. It is an intriguing twist of evolution that those elements, such as strong sunlight, which in nature are most likely to cause a cell to suffer DNA damage, have been harnessed to a mechanism which removes damage. The enzyme **photolyase** 'recognizes' the DNA damage, probably because of distortion of the helix caused by the dimer, and binds to the site (Figure 7.24). The enzyme–DNA complex absorbs visible light (300–600 nm) and uses the energy from it to split the covalent bonds which form the thymine dimer. Once the dimer has been removed, the helical shape is restored and the enzyme is released from the DNA.

**(b) Excision repair**. This is the major repair pathway in *E. coli*. This mechanism excises short sections of less than 20 kb of single-stranded DNA (ssDNA for short) from the genome. Analogous systems are probably the main way of repairing DNA damage in the majority of prokaryotic and eukaryotic cells (Figure 7.25).

In *E. coli* the four genes involved in excision repair were identified because their presence increased the cell's resistance to ultraviolet light. They are called the *uvr* genes (for ultraviolet resistance) and are known as *uvrA*, *uvrB*, *uvrC* and *uvrD*. The polypeptides produced from these genes form an enzyme complex called the **endonuclease complex**. If the products from *uvrA* and *uvrB* are combined, the enzyme complex is active, but its activity is further increased by the presence of the *uvrC* gene product. The role of the *uvrD* gene is still unknown.

**(c) Recombination repair**. This is a post-replication repair process which occurs after the dimer has been bypassed during replication (Figure 7.26). Cessation of replication on one side of the dimer and recommencement on the other side is known as **post-dimer synthesis**. Once the replication fork has bypassed the dimer (Figure 7.26), there are four DNA strands: two intact DNA strands (one parental and one daughter), a daughter strand containing a gap where the dimer occurred in the defective parental strand (the third strand), and the defective parental strand (the fourth strand).

In recombination repair the DNA strands are exchanged so that each of the two intact strands has a defective strand opposite it. This is called **sister strand exchange**. Thus, in each duplex there is a region of DNA which can act as a template against which intact DNA can be polymerized. Where the gap occurs this is straightforward. The gap is filled by the enzyme DNA polymerase I, which synthesizes a short stretch of DNA using the intact strand as a template. The dimer in the other strand can be removed by excision repair during the next round of DNA replication. In the event

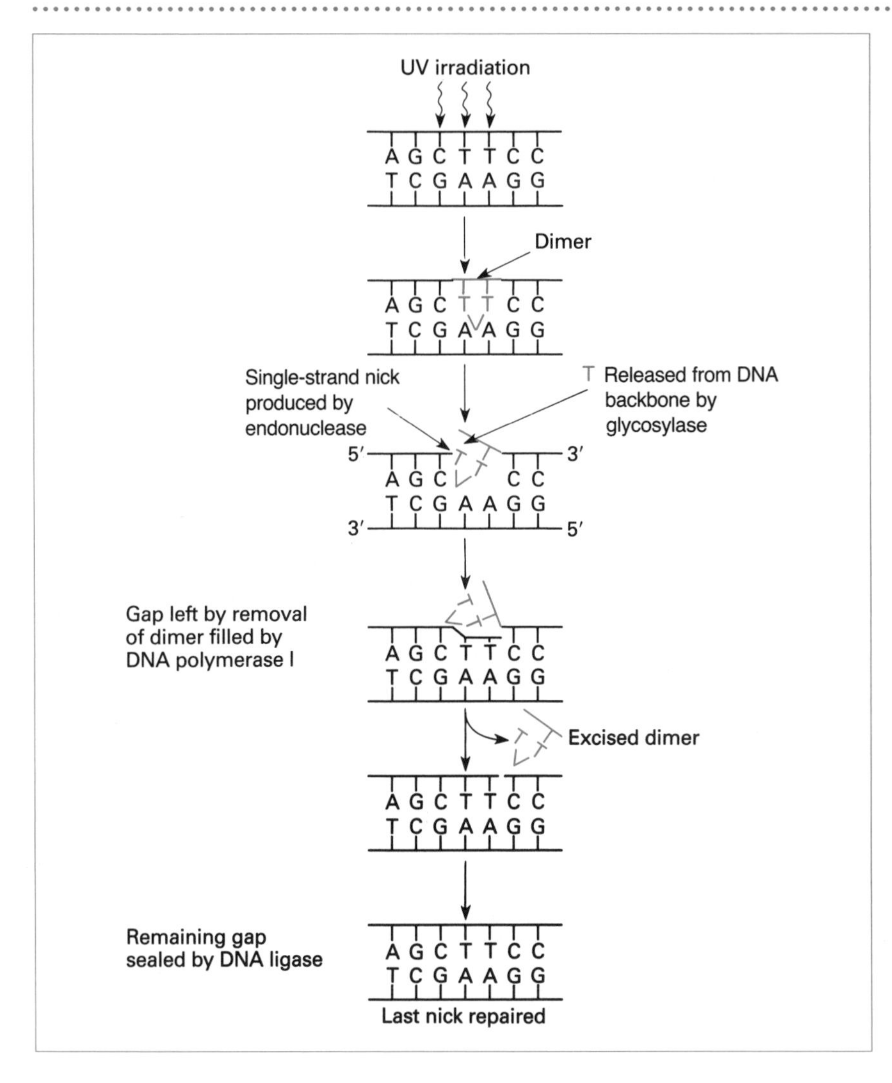


**Figure 7.25** Excision repair: an example of dark repair of ultraviolet light-induced thymine dimers. During this process the thymine dimer is removed (excised) from the mutant molecule by the combined action of four enzymes: glycosylase, endonuclease, DNA polymerase I and DNA ligase. Although details of excision repair vary from organism to organism, the process described here is thought to occur in the bacterium *Micrococcus luteus*.

that the dimer is not removed, recombination repair will occur again, effectively restricting the mutation to one strand of the DNA.

The genetic make-up of the mutant *E. coli* cell is important in recombination repair. The process relies on the general recombination system which occurs in the cell. It is important, therefore, that the cell possesses the recombinase gene, *recA*. It is this dependence on general recombination which gives recombination repair its name. However, since only the gaps opposite dimers are repaired, and not the dimers themselves, the term **daughter-strand gap repair** is sometimes used.

**(d) The SOS repair system**. This is perhaps the most significant repair system in terms of mutation and variation of the genome because it can lead to the incorporation of incorrect bases at the site of the original mutation.

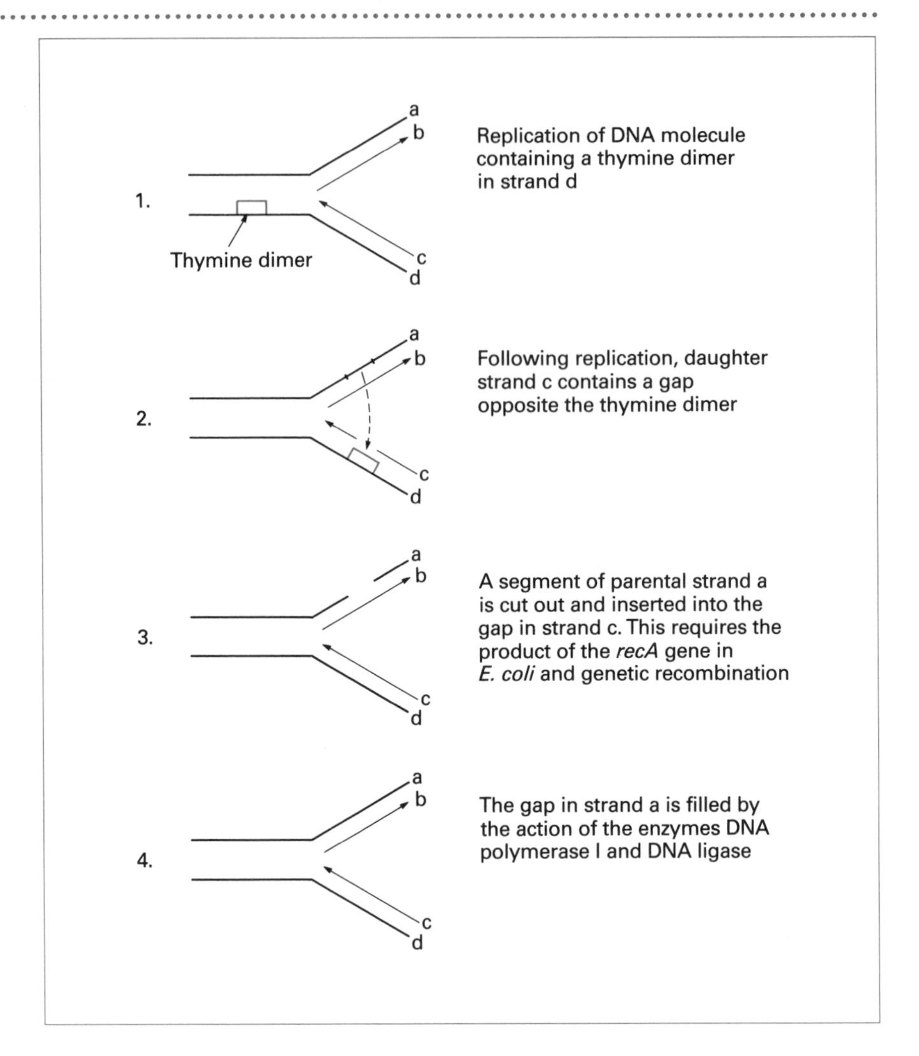


**Figure 7.26** Recombination repair.

Another feature which sets the SOS system apart is that it is **inducible**. That is to say, the enzymes involved in SOS repair are synthesized *only in response to DNA damage*. The products of DNA damage which induce the SOS repair system occur only at comparatively high levels when other repair systems cannot cope. This occurs only when DNA damage is severe. The SOS system is a final attempt by the cell to repair the DNA damage before the cell becomes seriously harmed.

Regulation of the SOS system is complex, relying on the interaction of several proteins. The SOS system is under the control of the *recA* and *lexA* genes. Interaction of the products of these two genes induces the transcription of at least 14 damage-inducible (*din*) genes. These include *recA, lexA, uvrA, uvrB, uvrC, uvrD, umuC* and *umuD*. Induction of these SOS genes gives rise to two recognizable error-prone repair events:

1. **Transdimer synthesis**: normally DNA polymerase (Pol III) incorporates bases into DNA accurately because it has a proof-reading capacity whereby it recognizes bases it has incorporated incorrectly, removes them, and incorporates the correct base. When the SOS system is induced, the proof-reading capacity of Pol III is relaxed and it can insert bases opposite a dimer and give rise to many mistakes.
2. **Long patch repair** is similar to excision repair but is dedicated to the excision of a stretch of around 1500 nucleotides. Gap filling is error-prone, possibly because the Pol III proof-reading function is relaxed.

## Chapter summary

In this chapter we have emphasized the value of bacteria and other microorganisms as experimental organisms for the investigation of genetic phenomena of much broader import and, to underline this thesis, we described a number of properties which make microorganisms particularly well-suited for genetic studies.

We then examined the key experiments on bacterial transformation using *Streptococcus pneumoniae* and those using phage T2, which provided convincing evidence that DNA was the hereditary material.

We looked at gene mutation and recombination and its importance in producing variation in populations of microbes. Events at the DNA level were described in some detail. Spontaneous mutations and the induction of mutations by a variety of physical and chemical agents were also discussed. We saw that mutation in a forward direction produced individuals with mutant phenotypes but these mutant individuals could also revert to the wild type, either by back-mutation at the original site or by means of a suppressor mutation at a different site.

By particular reference to ultraviolet light-induced pyrimidine dimers, we learned that cells contain a number of light and dark repair mechanisms which they can use to repair nucleic acid mutations.

## Further reading

T.A. Brown (1992) *Genetics: a Molecular Approach*. 2nd Edition. Chapman & Hall, London – a well-written and readable text.

J.W. Dale (1989) *Molecular Genetics of Bacteria.* John Wiley and Sons Ltd, Chichester, UK – a basic introduction for the beginner.
D.M. Freifelder (1987) *Microbial Genetics.* Jones and Bartlett Publishers, Inc., Boston, MA – an advanced text for the enthusiast.
C.A. Smith and E.J. Wood (1991) *Molecular Biology and Biotechnology.* Chapman & Hall, London – a well-presented marriage of molecular biology and biotechnology.
F. Stahl (1987) Genetic Recombination. *Scientific American*, **256** (2), 53–63 – a sympathetic treatment of this difficult subject.

## Questions

1. Fill in the gaps:
The inherited characteristics of an individual microorganism are the consequences of two genetic processes, __________ and __________ . The former usually arises as a result of a __________ mode of reproduction and requires an __________ of chromosome segments. The latter process may arise __________ or may be induced by a variety of agents, called __________ . Some of these, such as __________ __________ , are physical agents but others such as base __________ and __________ __________ are chemicals.
Mutant individuals arise when the __________ sequence of its __________ is changed. A __________ __________ can lead to the incorporation of an incorrect __________ __________ into the __________ chain whose sequence is specified by a __________ . This is referred to as a __________ - __________ mutation. Alternatively a __________ mutation arises and because the mutant __________ is a __________ termination __________ , a __________ polypeptide is produced.
When a base is added to or __________ from the DNA strand, a __________ - __________ mutation arises and because the __________ __________ is changed, the mutant polypeptide contains many different __________ __________ .
The effect which a mutation has on a mutant organism's __________ varies. If the mutant polypeptide contains only a single amino acid change, there may be little difference unless it

determines the shape of the active site of an ___________ molecule for example and then a loss of ___________ ___________ may be seen. If a polypeptide is produced by a nonsense mutation, the effect is likely to be more serious; the alteration in several amino acids caused by a ___________ - ___________ mutation will also be more dramatic.

Choose from: mutagens, base, exchange, spontaneously, analogues, mutation, sexual, recombination, ultraviolet light, nitrous acid, DNA, amino acid, substitution mutation, chain, mis-sense, polypeptide, gene, codon, nonsense, triplet, shortened, base sequence, enzyme, added, phenotype, catalytic activity, deleted, frameshift (twice), amino acids.

**2.** Multiple-choice questions:

(i) A frameshift mutation may be corrected by a second mutation in the same gene. This is called (a) intergenic suppression, (b) intragenic suppression, (c) gene repression, (d) mutation repression.

(ii) A nonsense mutation can be suppressed by a second mutation which changes the (a) anti-codon of a tRNA molecule, (b) amino acid recognition site of a tRNA, (c) the reaction between a tRNA and a ribosome, (d) the ribosome structure.

(iii) SOS repair of DNA may give rise to (a) many mutations, (b) no mutations, (c) few mutations, (d) early repair.

(iv) Light repair of DNA requires (a) photosynthesis, (b) photolyase, (c) isocitrate lyase, (d) photophosphorylation.

(v) To cause mutations, a base analogue must be (a) present during DNA replication, (b) absent during DNA replication, (c) added to undividing cells, (d) used in the light.

(vi) The chromosome of *Escherichia coli* is (a) found within a nuclear membrane, (b) visible under a light microscope, (c) much shorter than the *E. coli* cell, (d) a large, double-stranded molecule of DNA with the ends joined.

**3.** True or false?

(i) The *E. coli* chromosome is approximately 600 times longer than an *E. coli* cell.

(ii) Eukaryotic chromosomes contain DNA and protein.

(iii) Microorganisms are diploid for most of their life-cycles.

(iv) The Ames test is a test for carcinogenicity.

(v) Recombination is never required for DNA repair.
(vi) UV light produces adenosine dimers in DNA.
(vii) Acridine dyes are intercalating agents.
(viii) An auxotrophic mutant is an example of a conditional lethal mutant.

# Bacterial gene transfer and genetic engineering

# 8

A common feature in the processes of gene transfer and genetic engineering is the participation of genetic elements *other than chromosomes:* these are the so-called extrachromosomal genetic elements. In the previous chapter, we examined genetic variations which arise through changes in the genome. Here, we shall see how extrachromosomal genetics can introduce variability into microbes and influence their fitness to compete and survive in their environments.

Extrachromosomal elements have been exploited by man to investigate the genetics of microorganisms, particularly of bacteria. Although study of these genetic elements has provided much interesting information about their own molecular biology and genetics, their greatest value is in genetic engineering. Our ability to employ some extrachromosomal elements in the cloning and maintenance of genes has led to an explosion in our understanding of the expression and control of biological processes.

In this chapter we describe some of the techniques of genetic engineering; in Chapter 11 we look in detail at the production of insulin and hepatitis B vaccine using these techniques. But first, what *are* these extrachromosomal genetic elements?

## Bacterial plasmids

Plasmids are, typically, circular pieces of double-stranded DNA (called dsDNA for short) which exist independently of the bacterial chromosome. They are generally small in comparison with the bacterial chromosome and rarely account for more than 2% of the cellular DNA.

Plasmids normally carry one or more genes which confer some advantage to the host cell under certain conditions. For example, genes on a plasmid might confer on the bacterial cell resistance to antibiotics or toxic

Some Japanese workers have isolated linear plasmids, and some *Bacillus* plasmids produce single-stranded DNA (called ssDNA) at intermediate stages during replication.

**Table 8.1** Properties of some bacterial plasmids

| *Bacterial host* | *Plasmid* | *Phenotype* | *Size* (kb) | *Copy number* | *Incompatibility group* |
|---|---|---|---|---|---|
| *Escherichia coli* | F | Fertility, the ability to transfer genes by conjugation | 93 | 1–2 | F |
| *Pseudomonas* | pWWO | Degradative, the ability to catabolize toluene-related aromatic compounds | 117 | ? | ? |
| *Salmonella paratyphi* | R1 | Antibiotic resistance | 87 | | F11 |
| *Escherichia coli* | ColE1 | Colicin production | 63 | 15 | ? |
| *Agrobacterium tumefaciens* | Ti plasmids | Induction of crown gall tumours in dicotyledonous plants | 144 to 240 | ? | ? |
| *Staphylococcus* | pT127 | Tetracycline resistance | 4.4 | | |
| *Bacillus subtilis* | pBD6 | Kanamycin and streptomycin resistance | 8.7 | | |

metals (Table 8.1). Hence, the possession of the plasmid enables the bacterium to survive and grow in environments containing these toxic substances. However, environments containing toxic metals or antibiotics are comparatively few, so the majority of bacteria would rarely make use of genes conferring resistance to these substances. This highlights a major feature of plasmid-borne genes. Although plasmids can be invaluable under some growth conditions, they can be dispensed with in most circumstances. Plasmid genes are said, therefore, to be **non-essential genes**.

## Plasmid classification

There are several ways of grouping plasmids. One way is to classify them according to their function. This gives rise to five main types:

1. **Fertility plasmids** (called F plasmids), which confer the ability to transfer genes by a sexual process called **conjugation**.
2. **Col plasmids**, which carry genes for the production of low-molecular-weight proteins (called **colicins**) which kill other bacteria. The plasmid ColE1 of *E. coli* is an example of a 'col' plasmid.
3. **Resistance plasmids**, which confer the ability to grow (or sometimes merely survive) in the presence of antimicrobial agents such as antibiotics, some metal ions or colicins. For example, plasmid R1 carries genes which confer resistance to the antimicrobial agents ampicillin, streptomycin, sulphanilamide, chloramphenicol and kanamycin.

4. **Virulence plasmids** confer on the bacteria which carry them the ability to more successfully colonize a host organism such as an animal or human being. Some virulence plasmids enable a bacterium to adhere more strongly to an internal body surface from which it would otherwise be dislodged and expelled.
5. **Degradative plasmids** confer on the host bacterium the ability to degrade (catabolize) some unusual substance not often encountered in the environment. Degradative plasmids are common in species of *Pseudomonas.* Perhaps the best studied example is the TOL plasmid of *P. putida* (which has been redesignated pWWO), which carries genes for the catabolism of toluene and related aromatic compounds.

The disadvantage of this type of classification is that, although plasmids are grouped according to function, they are not necessarily related at the molecular or ancestral level. Additionally, some may carry out more than one function. For example, plasmid R455 from *Proteus morganii* confers resistance to mercury in addition to several antibiotics, and the tellurite resistance plasmid of *Salmonella ohio* also carries genes which enable it to use sucrose as a source of food. Finally, although a large number of plasmids have a function which can be categorized in the ways we have just described, many others have been identified which carry genes for other functions. And some plasmids, called **cryptic plasmids**, appear to have no recognizable function at all.

Another method of classifying plasmids, reflecting how closely they are related, is based on their ability to coexist. Closely related plasmids cannot stably coexist in the same cell, possibly because of competition for a membrane attachment site at cell division. They are said to be **incompatible** and are placed in the same incompatibility group (Inc group). For example, plasmid TP125 (a drug-resistance plasmid carrying genes for resistance to streptomycin, sulphanilamide, chloramphenicol and tetracycline), and plasmid TP113 (which carries a gene for resistance to kanamycin) cannot coexist in the same cell and both belong to incompatibility group B (Inc B). However, either of these can coexist stably with plasmid R1, which belongs to the incompatibility group FII (IncFII). All incompatibility groups are designated by a letter (for example, IncC, IncB, IncF). Some incompatibility groups are subdivided by numbers (for example IncFI, IncFII, IncP1, IncP2, and so on).

Although the basis for incompatibility is still largely a matter for conjecture, two theories have been suggested in an attempt to understand the phenomenon. The first proposes the existence of a repressor molecule which limits the number of plasmid molecules to a characteristic value.

Replication of incompatible plasmids is inhibited by the same repressor molecule. The alternative theory suggests that incompatible plasmids compete for the same attachment binding site in the cell membrane. If one type of plasmid is excluded from the attachment site by a second type, the unattached plasmid will not be replicated and will be lost from the bacterial population.

## Conjugation and the F factor

The first plasmid to be discovered was the F, or fertility, factor of *Escherichia coli.* The F factor plays a central role in the transfer of genes from a **donor** *E. coli* cell to a **recipient** *E. coli* cell. In this section we describe this **conjugation** process in detail.

The ability of *Escherichia coli* to transfer genetic material from a donor cell to a recipient cell when the two cells are in physical contact is called **conjugation**. Conjugation between bacteria was first reported by Joshua Ledeberg and Edward Tatum in 1946, although at that time the mechanism was not fully understood. The discovery was of immense significance though, because it was the first report of an apparently sexual process in bacteria, a process which led to the reassortment (recombination) of genes between two parental cells. However, it is not a sexual process as we understand it from reduction division (meiosis) and fusion of gametes in eukaryotes (Chapter 9). Rather, in bacteria, there is a unidirectional transfer of genes from donor to recipient cell and recombination occurs only in the recipient cell. It is more accurate to refer to bacterial conjugation as a **pseudo-sexual process**.

The Lederberg and Tatum experiment was relatively simple (Figure 8.1). They used mutant bacteria which required several amino acids to be provided in a minimal growth medium (glucose–mineral salts medium) before they could grow. Bacteria which require such growth factors (which can include purines, pyrimidines and vitamins) are known as **auxototrophs**, or, where several growth factors are required, **multiple auxotrophs**. The auxotrophic requirements of the two strains employed were complementary: that is to say, if one strain required methionine for growth (*met*$^-$) but not thiamine (*thi*$^+$) the other strain had the opposite characteristic. It did not require methionine to be supplied in the growth medium (*met*$^+$) but required thiamine (*thi*$^-$) instead.

Bacteria which require no additional growth factors are known as **prototrophs**. (Refer also to page 196 in Chapter 7.)

Lederberg and Tatum demonstrated that, by growing both strains together in a complete medium and subsequently plating samples onto a

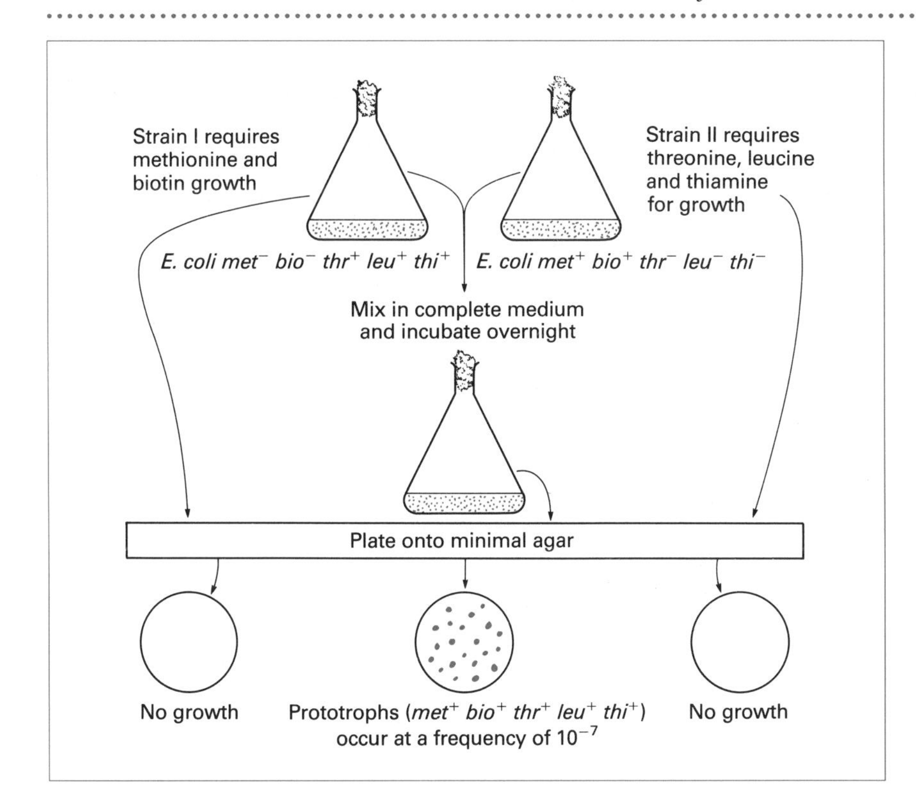


**Figure 8.1** The Lederberg and Tatum protocol for demonstrating gene transfer by conjugation.

glucose–mineral salts medium, they could isolate prototrophs at a frequency of approximately one in every 10 million bacteria (1 in $10^7$). No prototrophic bacteria were isolated from controls in which each of the strains was plated separately onto the same medium. Both the frequency of occurrence of prototrophs and the results of the control experiment indicated that the mutants had not back-mutated (discussed in Chapter 7). This result indicated that genes had been transferred between the strains. Furthermore, Lederberg and Tatum demonstrated that if they added the filtrate of one of the strains to the cells of another, no prototrophs could be isolated. This finding suggested that cell-to-cell contact was required for transfer of genes to occur.

The American bacteriologist, Bernard Davis, cleverly exploited Lederberg and Tatum's work by repeating it in a glass U-tube which was separated into two sections by a bacterial filter (Figure 8.2). Growth medium could pass through the filter but bacteria could not. One auxotroph was introduced on one side of the filter and the second on the other side. The growth medium could be mixed by applying pressure or suction to the tube. No prototrophs were isolated from these experiments, confirming that contact was required between the two strains before gene transfer could occur.

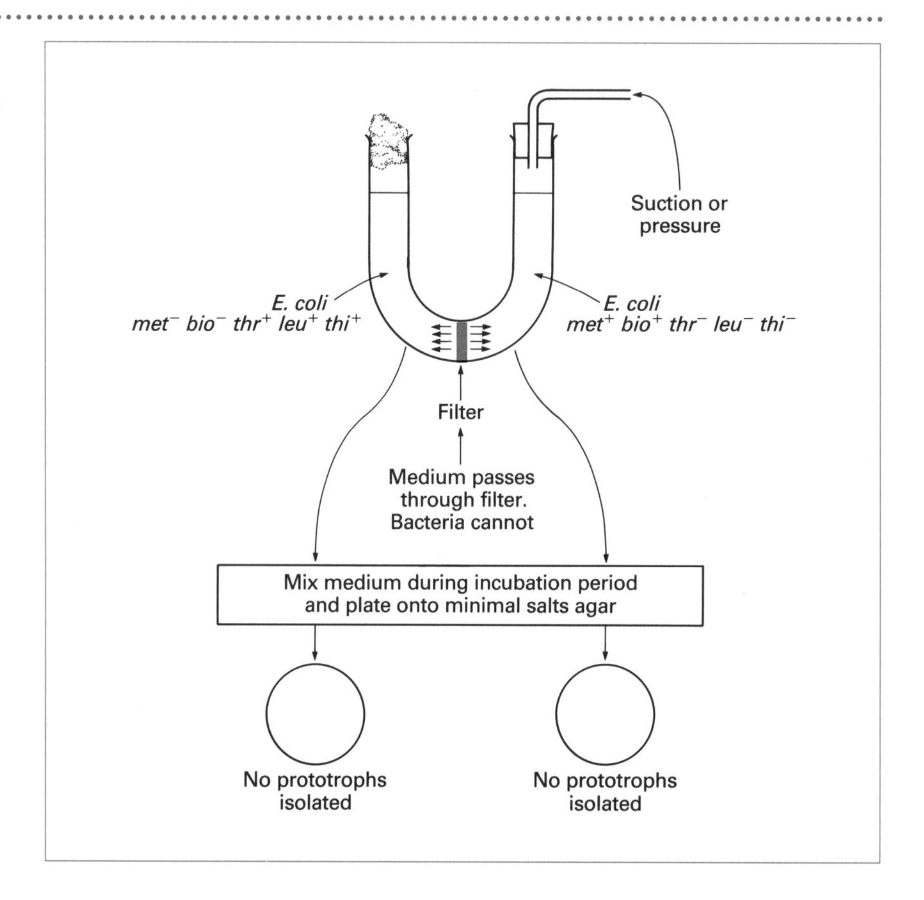


**Figure 8.2** The Davis U-tube experiment to demonstrate the requirement for physical contact in conjugation.

Following his observation in 1952 that transfer of genetic material was unidirectional – that is, it is always transferred from a donor population of cells to a recipient population – it was Bill Hayes in London who first suggested that gene transfer was mediated by an infective agent which we now call the **fertility factor**, F. The donor cell ($F^+$) serves as the 'male' and the recipient ($F^-$), the 'female'. The F factor can be lost from cells, rendering them $F^-$ (recipients). In matings between $F^+$ and $F^-$ populations, nearly all the recipients become $F^+$. The F factor subsequently was shown to be a circular, double-stranded DNA molecule – in other words, a plasmid. It usually exists independently of the bacterial chromosome (Figure 8.3).

The presence of the F factor causes a cell to produce a number of hair-like appendages called **pili** (singular, **pilus**). It also confers on a cell the ability to form a conjugation tube between $F^+$ and $F^-$ cells through which genetic material can be transferred. The conjugation tube is a long narrow pilus through which it has been suggested the F factor passes during conjugation (Figure 8.4). However, the presence of DNA in the pilus has never been demonstrated. An alternative proposal is that the pili may simply serve to stabilize cell-to-cell contact while DNA is transferred by membrane contact.

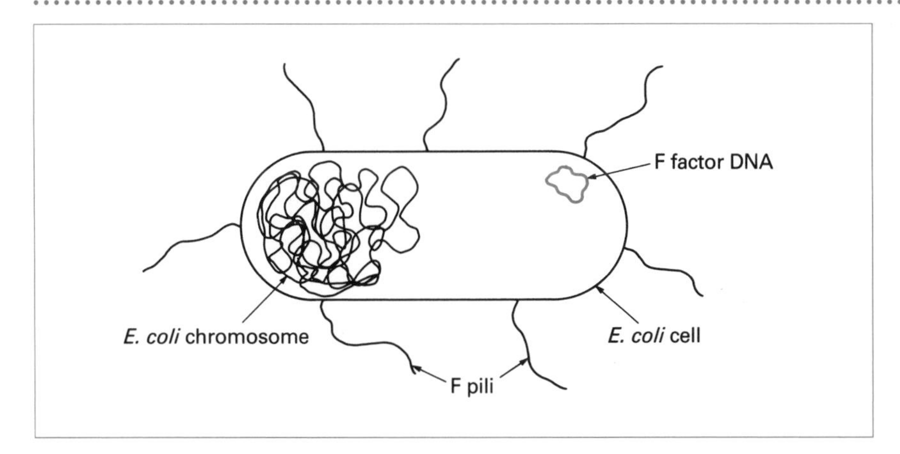


**Figure 8.3** The F factor of *E. coli*.

This description does not explain how bacterial genes are transferred during conjugation, only how the F factor transfers itself. It was Cavalli-Sforza and Hayes who first demonstrated that, in every $F^+$ population of *E. coli*, there was a sub-population of cells which could transfer chromosomal genes at a high frequency. These were called **high frequency recombinant** (Hfr) cells.

In crosses between Hfr and $F^-$ populations, Cavalli-Sforza and Hayes observed that some chromosomal genes were always transferred at a high

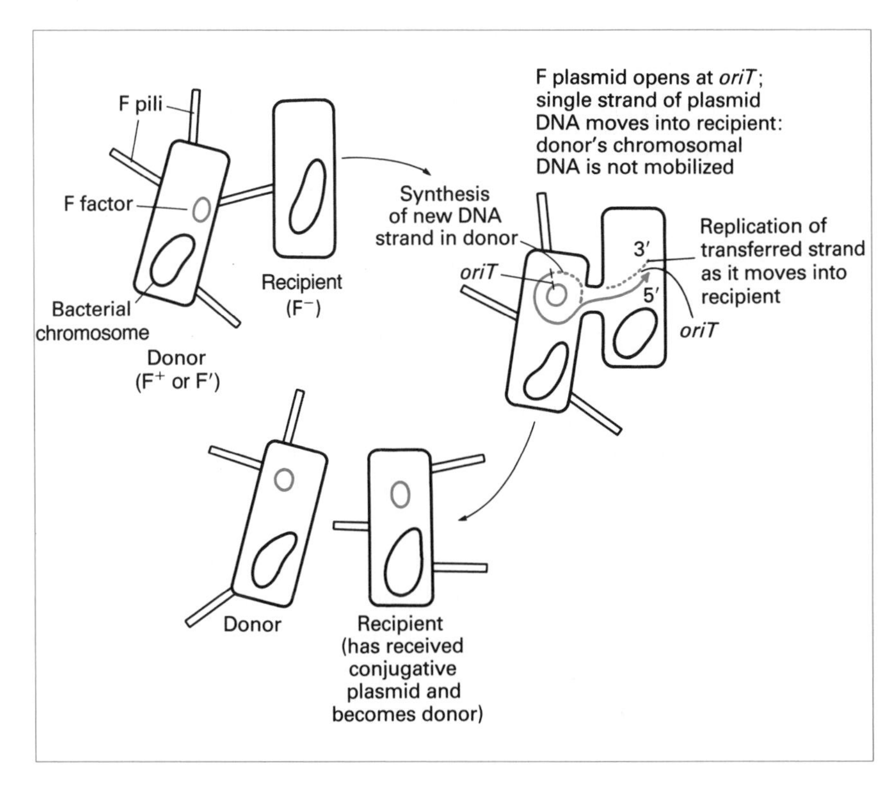


**Figure 8.4** Conjugation involving the F factor. Transfer of the F factor during conjugation involves one strand of the F factor being transferred through the conjugation pilus. A complementary strand is synthesized in the recipient to produce a dsDNA molecule. In the donor, the complementary strand is synthesized continuously against the circular ssDNA template.

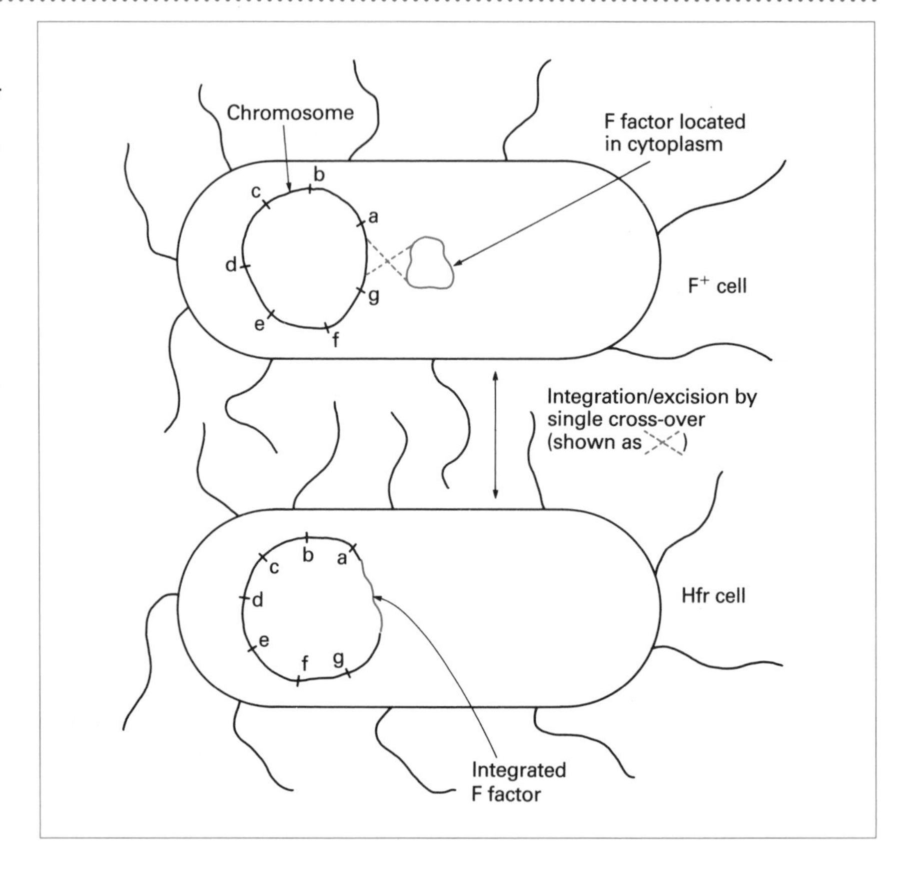


**Figure 8.5** Formation of a Hfr strain by integration of F into the *E. coli* chromosome. The F factor recombines with the *E. coli* chromosome to form a Hfr strain. Plasmids which can integrate into the chromosome are sometimes called **episomes**. In the case shown here, the F factor has integrated between genes a and g.

frequency while others were rarely or never transferred. In fact, in the classical **interrupted mating experiment**, conjugating bacteria were mechanically separated after specific time intervals and samples plated to determine which Hfr genes had been transferred to F$^-$ cells.

The results of the experiment showed that each Hfr gene was always transferred after a characteristic period of conjugation. This was interpreted as meaning that Hfr genes are transferred in a linear fashion and in a fixed order or sequence. Additionally, in Hfr × F$^-$ crosses, recipients remain F$^-$, unlike F$^+$ × F$^-$ crosses where nearly all recipients become F$^+$. This polarity of transfer is explained by the integration of the F factor into the chromosome of Hfr cells but not into F$^+$ cell chromosomes (Figure 8.5).

When the integrated F factor is transferred, it takes bacterial chromosomal genes with it. However, the conjugation tubes often break, so genes which are close to the origin of transfer (*oriT*) of the F factor and which first enter the recipient have a high probability of being transferred before the conjugation pilus breaks. Those genes which are further away from *oriT* have a much lower probability of entering the recipient before the pilus snaps.

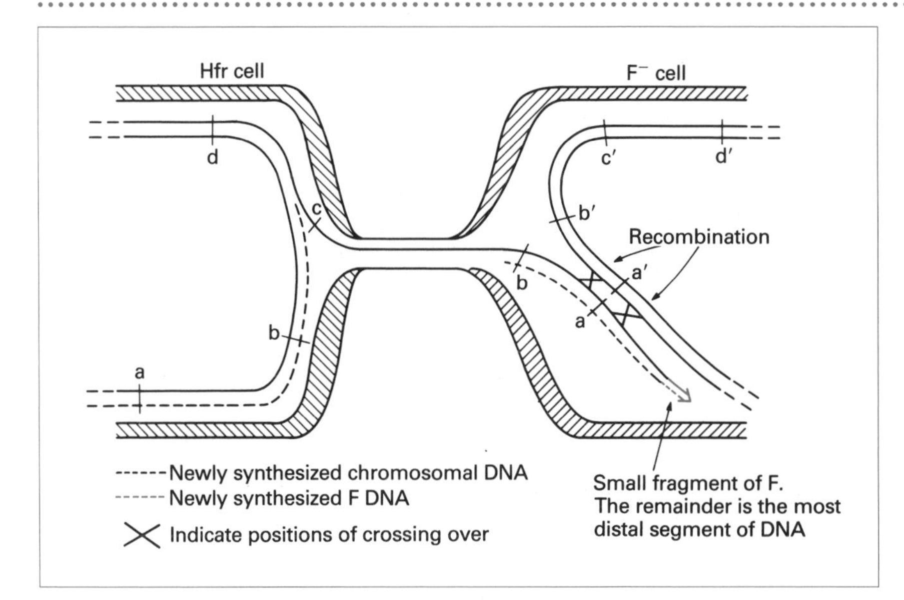


**Figure 8.6** Transfer of bacterial genes in a Hfr × $F^-$ cross. Transfer of bacterial genes occurs when the F factor breaks at its origin of transfer (*oriT*). Part of it is transferred to the recipient followed by chromosomal genes. For a gene to be assimilated into the $F^-$ genome, recombination must occur between two almost identical DNA sequences.

When the integrated F factor initiates transfer of the bacterial genome, a break occurs within it (at *oriT*) and the now linear chromosome is transferred. Consequently, part of the F factor is the first DNA to enter the recipient and the remainder is the last DNA to enter – the two fragments are separated by the entire bacterial chromosome, so the latter part of the F factor rarely enters the recipient, which remains $F^-$ (Figure 8.6).

**BOX 8.1**

**The interrupted mating experiment**

In this experiment, Hfr and $F^-$ strains of *Escherichia coli* are mixed in a liquid medium so that conjugational chromosome transfer can take place. At time intervals, samples of conjugating bacteria are mechanically separated in a blender. Samples are immediately diluted to prevent mating pairs from reforming. Diluted samples are spread on to solid selective media to establish the time required for the transfer of Hfr genes to $F^-$ cells.

The results obtained show that Hfr genes are always transferred after characteristic time intervals. Furthermore, genes which are transferred later are transferred less frequently than those transferred earlier. The first of these observations supports the idea that the circular Hfr chromosome breaks to form a linear structure which is then transferred over a period (90–100 minutes) to the $F^-$ recipient. The fall in the frequency of transfer of different Hfr genes is a reflection of the spontaneous separation of mating bacteria (because the conjugation tubes break) over the course of the experiment.

In addition to providing experimental evidence which supports the proposed mechanism of conjugational gene transfer, the interrupted mating technique enables the location of genes on the *E. coli* chromosome to be mapped in terms of time units.

A final point to be made is that if the experiment is repeated using a number of different Hfr strains, the genes transferred and their order of transfer varies from one mating to another. The only way in which gene maps from different crosses can be reconciled is if they are located on a circular map. The observation that the chromosome map of *E. coli* is circular provided one of the earliest pieces of experimental evidence that the chromosome itself is also a circular structure.

**BOX 8.1** *continued*

In the conjugation mixture of donor Hfr and recipient $F^-$ cells:

Hfr $azi^r$ $lac^+$ $gal^+$ $tet^r$ $str^s$
$F^-$ $azi^s$ $lac^-$ $gal^-$ $tet^s$ $str^r$

The Hfr strain is resistant to the respiratory inhibitor azide ($azi^r$) and to the antibiotic tetracycline ($tet^r$) but is sensitive to and is killed by streptomycin ($str^s$). It can use either lactose ($lac^+$) or galactose ($gal^+$) as carbon sources.

The $F^-$ strain is sensitive to azide ($azi^s$) and tetracycline ($tet^s$) but resistant to streptomycin ($str^r$). It is unable to utilize either lactose ($lac^-$) or galactose ($gal^-$) as carbon sources.

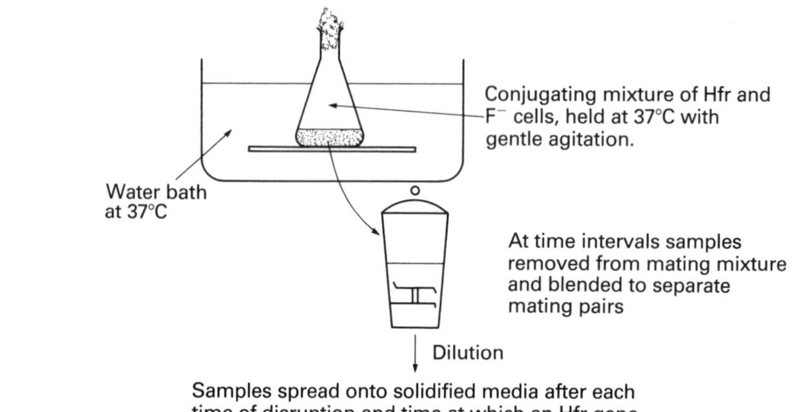


Dilution

Samples spread onto solidified media after each time of disruption and time at which an Hfr gene first appears in recombinant cells is noted

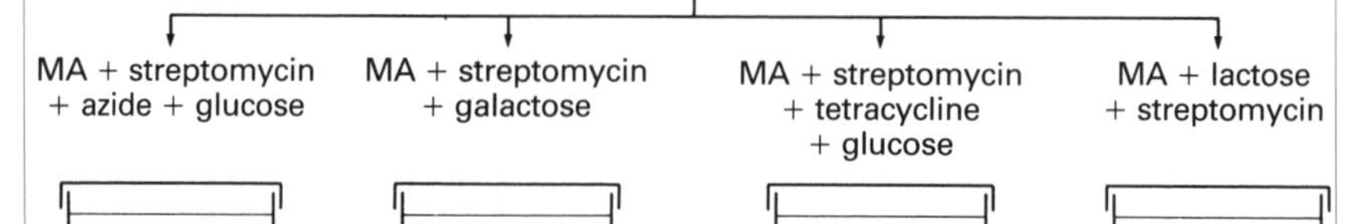


All plates contain minimal agar and a number of other components. All contain streptomycin to kill Hfr donor cells. What will be the genotypes of the recombinant colonies which grow on these four types of medium?

The results obtained are shown below.

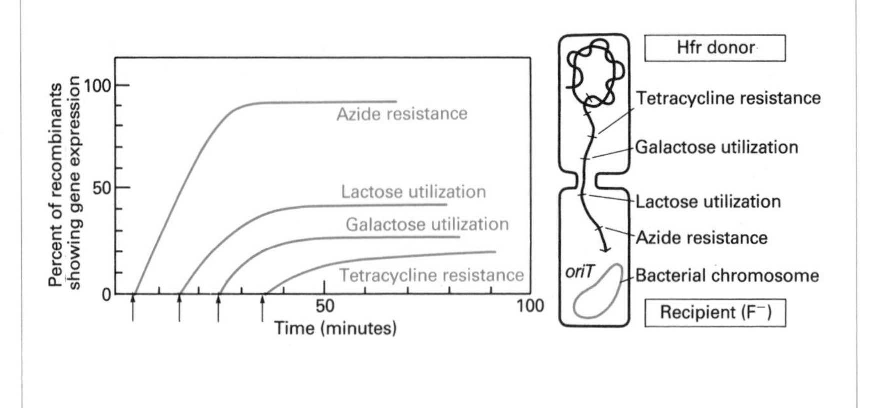

**BOX 8.1** *continued*

| Approximate time of appearance of genes in recombinants (see arrows on graph): | Azide resistance: 7 minutes<br>Lactose utilization: 13 minutes<br>Galactose utilization: 22 minutes<br>Tetracycline resistance: 32 minutes |
|---|---|

A gene map for this part of the *E. coli* chromosome could be drawn as:

*azi* *lac* *gal* *tet*

0 min 10 min 20 min 30 min

When the experiment is repeated with different pairs of Hfr and F strains, different linear maps are produced which can all be reconciled by a single circular map:

| *Hfr strain* | *Linear order of genes* |
|---|---|
| Hfr Hayes | – *thr* – *pro* – *mal* – *ile* – |
| Hfr2 | – *pro* – *thr* – *ile* – *mal* – *trp* – |
| AB311 | – *trp* – *pro* – *thr* – *ile* – *mal* – |
| AB312 | – *mal* – *ile* – *thr* – *pro* – *trp* – |

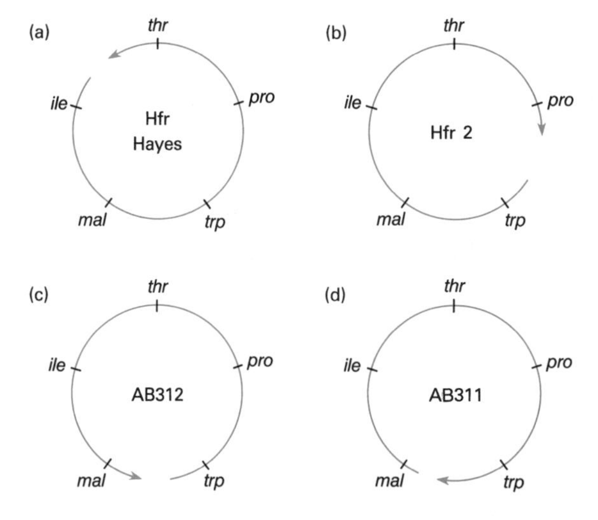


The linear order of genes was determined by interrupted mating. Although the linear permutation is different, the gene order can be seen to be identical if the genes are placed on a circular map. The linear order is the consequence of the F factor becoming inserted at different sites in the genome. Additionally, the F factor can be inserted in two orientations resulting in either anticlockwise (maps a and b above) or clockwise (maps b and d above) transfer of the genome.

**Note**: derived from Jacob and Wollman (1961).

In order that transferred donor (Hfr) genes can be transmitted to future generations of recombinant $F^-$ cells, transferred genes must be incorporated into the chromosome of the $F^-$ cell. The reason for this is that only when genes constitute part of the intact bacterial chromosome can they be replicated. Unincorporated linear fragments of transferred donor genes cannot be replicated because they do not contain the requisite replication initiation site and eventually will be enzymatically degraded.

## Transduction: bacteriophage-mediated gene transfer in bacteria

In order to exist, viruses are obliged to replicate inside other cells; in other words, they are obligate intracellular parasites. One group of viruses, the bacteriophages (or more simply phages), has achieved an elevated status in microbial genetics for two reasons. First, phages are easy to handle because their hosts (bacteria) are easily and rapidly culturable. Other viruses require tissue culture techniques, whole plant hosts, growth in hen's eggs and similar techniques which are less convenient and more expensive. Second, some bacteriophages can transfer bacterial genes. This process, like the other forms of bacterial gene transfer (transformation and conjugation), not only stimulated research into the mechanism of transfer itself but also provided a further valuable technique for mapping the positions of genes along bacterial chromosomes and plasmids.

Before examining transduction we need first to introduce and review some terms associated with bacteriophage genetics. A **virulent phage** (for example T2 and T4 of *E. coli*) injects its DNA into the bacterial cell and remains in the **replicative** form. That is, it follows its reproductive cycle and produces many copies of itself before breaking open (**lysing**) the bacterial host to release infective phage particles. This sequence of events is known as the **lytic cycle**. The biology of viruses is described in greater detail in Chapter 10.

When a **temperate phage** infects a bacterial host it becomes 'dormant' and neither bacteriophage replication nor lysis immediately occur. This condition is known as the **lysogenic cycle**. **Lysogens** are bacteria which contain a lysogenic phage. Some lysogenic phages have the ability to integrate into the bacterial chromosome, in which state they are called **prophages** (Figure 8.7). The presence of a lysogenic phage in a bacterium prevents other closely related phages from establishing themselves successfully in the host. This condition is known as **superinfection immunity**. A lysogenic

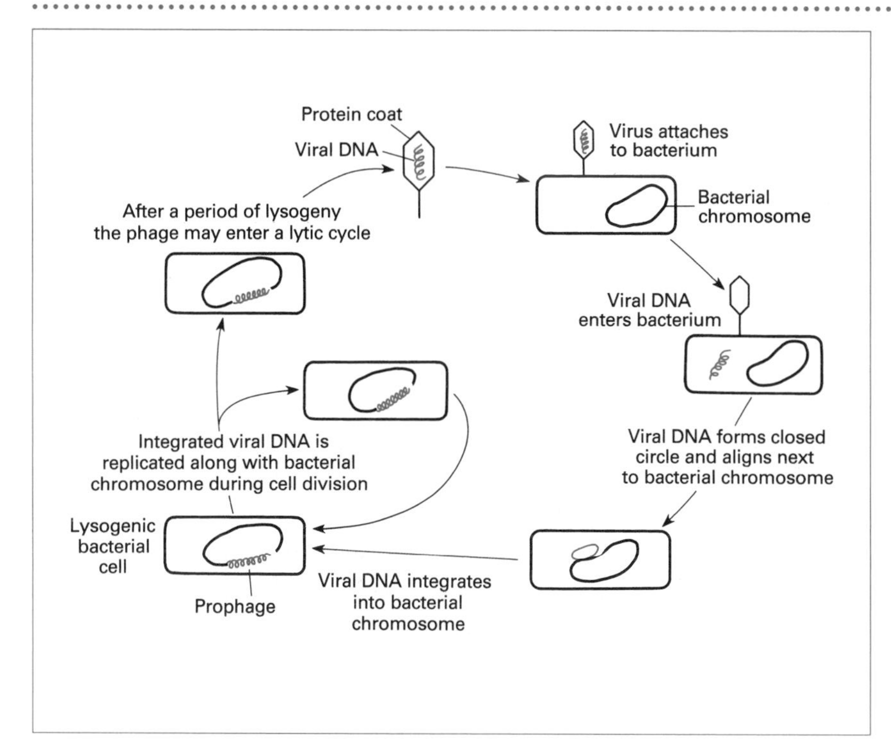


**Figure 8.7** The lysogenic cycle of a bacteriophage. Temperate bacteriophages, such as λ of *E. coli*, P22 of *Salmonella typhimurium* and PBS1 of *Bacillus subtilis*, can exist as a dormant form in the host cell—often as a prophage. Under certain environmental conditions, the dormant phage can be induced to enter a lytic cycle of replication: experimentally, this can be achieved by exposure to low doses of ultraviolet light.

bacterium may be *induced* to enter the lytic cycle. Induction is often triggered by contact with agents such as UV light, which damage DNA.

The process by which certain phages are able to transfer bacterial genes between hosts is called **transduction**. Some **transducing phages** transfer a small number of genes restricted to particular regions of the chromosome. This is known as **specialized** or **restricted transduction**. Other transducing phages transfer a small number of genes but these can originate from any region of the bacterial genome. This is referred to as **generalized transduction**.

## Generalized transduction

The first transducing phage to be identified was a generalized transducing phage. Transduction by phage P22 of *Salmonella typhimurium* was discovered by Lederberg and Zinder in 1951. In an experiment using a Davis U-tube they discovered that when the *Salmonella* auxotrophs LA2 ($phe^-$ $trp^-$ $met^+$ $his^+$) and LA22 ($phe^+$ $trp^+$ $met^-$ $his^-$) were incubated on either side of the filter, prototrophs could be isolated. Some agent in the growth medium was effecting gene transfer. The possibility that this was naked DNA (and therefore that transformation was occurring) was discounted because the process continued in the presence of DNase, an enzyme which degrades DNA. The fact that the gene transfer was unidirectional, occurring only if

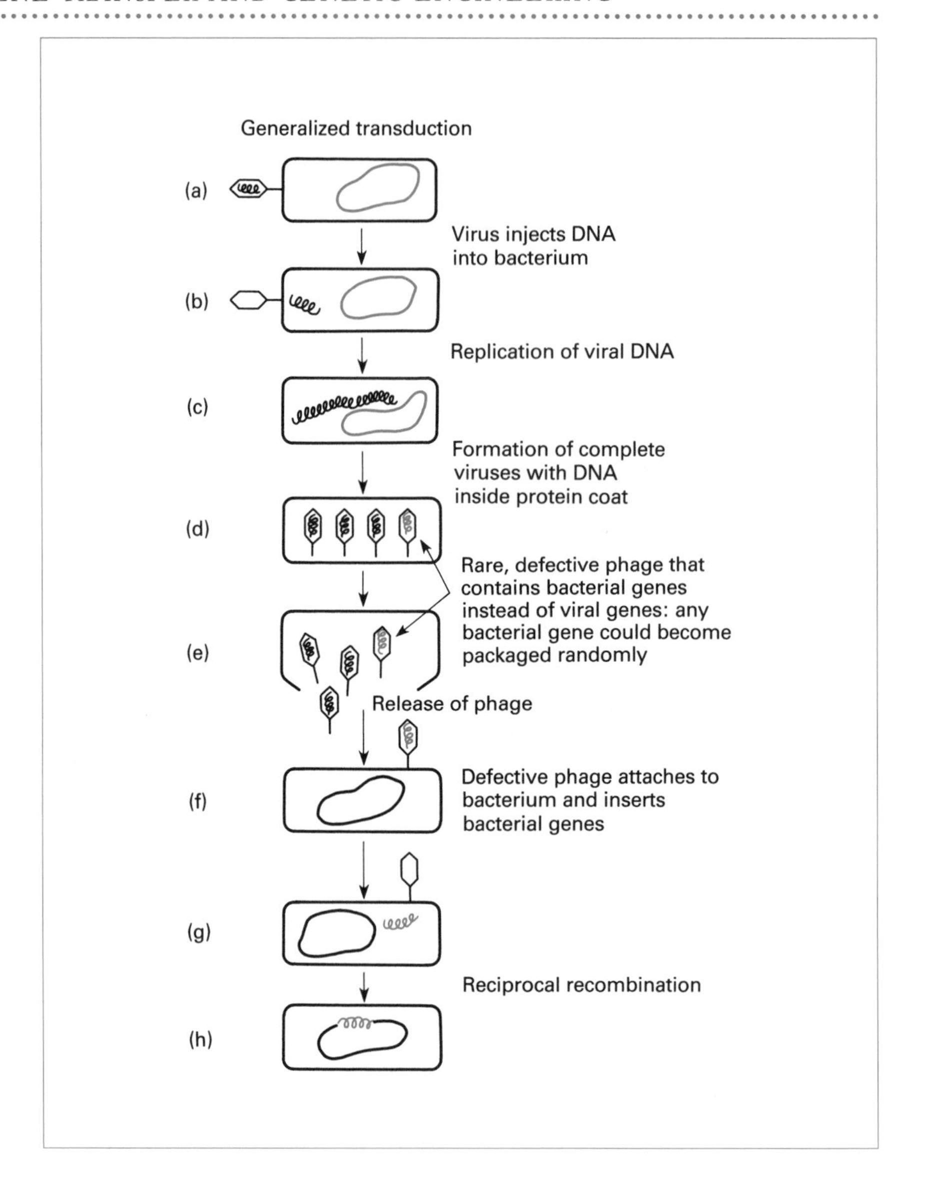


**Figure 8.8** Complete and abortive transduction in *Salmonella typhimurium* by phage P22.

a filtrate of LA22 was added to cells of LA2, and not *vice versa*, indicated that the infectious agent (or **vector**) was in LA22.

Generalized transduction can be explained by examining the reproductive cycle of P22 in *S. typhimurium* (Figure 8.8). When P22 infects the host cell it degrades the host chromosome and produces multiple copies of its own genome. These are packaged into the protein coats or **capsids** (described in Chapter 10), and the mature phages are released by lysis of the cell. The capsids identify the phage DNA by its size. Occasionally fragments of the *Salmonella* chromosome of approximately the size of the P22 genome are accidentally packaged into the capsids in place of phage DNA.

These phage capsids, which contain fragments of bacterial chromosome, are transducing phages.

Following release of the transducing phage from a donor bacterium by lysis, the capsid attaches itself to a new host (the recipient) and injects donor bacterial DNA into the recipient cytoplasm in the usual manner. Thus, a fragment of the donor *bacterial* DNA enters a recipient bacterial cell. After the chromosomal fragment has entered the recipient cell, it must integrate into the chromosome so that it can be replicated and inherited by future generations. This is referred to as **complete transduction**. Alternatively, the fragment may fail to integrate into the recipient chromosome, producing an **abortive transductant**. The unincorporated DNA fragment cannot replicate but may persist in cells which, through mutation, lack the enzymes which would normally degrade it. At cell division it is passed to only one of the daughter cells and gives rise to a single recombinant cell in the population. The transferred gene can then direct the synthesis of many molecules of its protein product. At cell division these may be distributed to both daughter cells, although only one of them inherits the DNA fragment itself. By successive cell divisions a sub-population or **clone** of cells containing the gene product will be produced, one member of which contains the gene itself. Such a clone of cells will produce a **microcolony** on a solid medium, visible only by microscopic examination.

The generalized nature of the transduction process is explained by the fact that the bacterial DNA is incorporated into the phage head purely on the basis of size: the 'headful hypothesis'. Thus, any gene on the chromosome can be transferred as long as it occurs on a fragment of the appropriate size (roughly 0.01% of the length of the *Salmonella* genome).

In mapping a bacterial chromosome, conjugation can be used to map the position of genes throughout the whole of its length but genes which are very close together (closely linked) cannot be separated. Generalized transduction can then be used to map closely linked genes within a short segment of the chromosome. In other words, used together, conjugation and transduction can pinpoint the exact position of all the known genes.

## Specialized transduction

Specialized (or restricted) transduction, first reported by Lederberg in 1956, is exemplified by phage lambda (λ) of *E. coli*. Like phage P22 of *Salmonella typhimurium*, λ is a temperate phage and can enter a lysogenic relationship with an *E. coli* cell. Again, to understand how specialized transducing phages arise we need to examine the normal infection cycle of phages (Figure 8.9). The biology of λ is described in Chapter 10.

The phage attaches itself to an *E. coli* cell and injects its DNA into the bacterial cell. Although the λ DNA has a linear structure when it forms part of a phage particle, once it is injected into an *E. coli* host cell, its ends join by complementary base-pairing of single-stranded ends to produce a circular structure. The circular λ genome can then be integrated into the circular *E. coli* chromosome by a single cross-over. Integration occurs at a specific point in the *E. coli* chromosome located between a set of genes

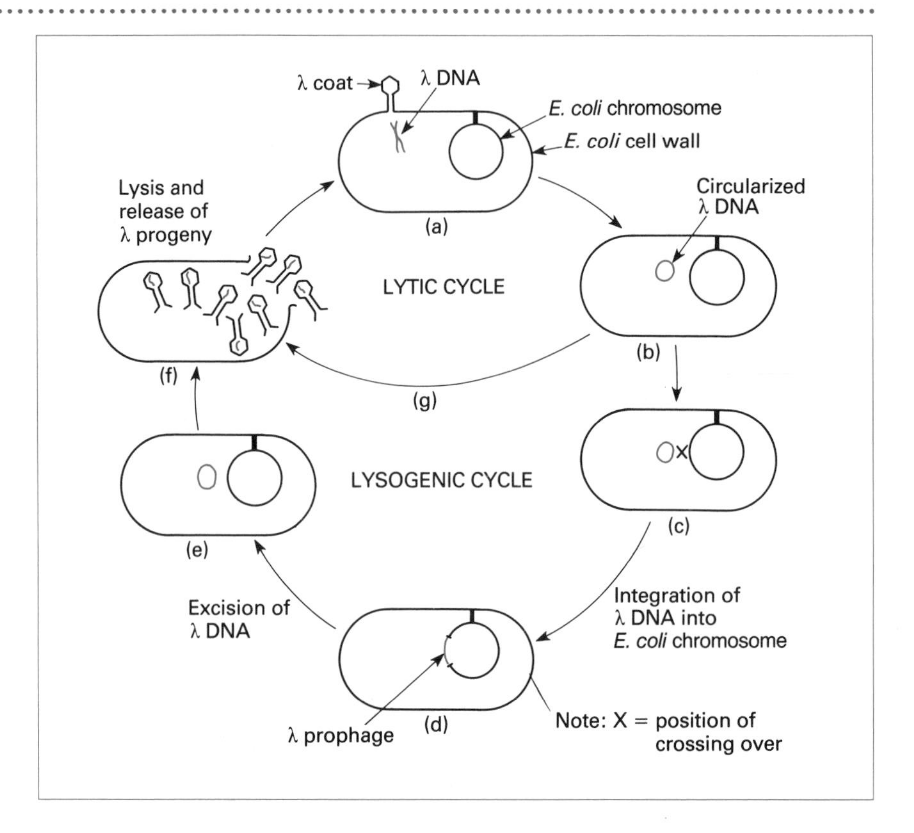


**Figure 8.9** The life-cycle of phage lambda (λ).

required for synthesis of the vitamin biotin (the *bio* genes) and a set of genes required for fermentation of the sugar galactose (the *gal* genes). Integration requires interaction between corresponding attachment sites in the genome and the *E. coli* chromosome: the site in the λ genome is designated *att*P and the one in the *E. coli* chromosome *att*B, where the P and B indicate phage and bacterium respectively. Once integrated into the bacterial chromosome, the λ genome is called **prophage**. It is replicated along with the rest of the chromosome and behaves as a set of bacterial genes.

The production of λ-transducing phages occurs following induction of a λ prophage; that is, during the lytic cycle. Only two sets of genes can be transduced, either the (*bio*) or the (*gal*) genes. This contrasts with generalized transduction where any bacterial gene can be transferred from donor to recipient cell. The reason for this difference is that the molecular events required for generalized and restricted transduction are completely different. As far as λ is concerned, transducing particles are formed by reversal of the integration process, which lead to prophage formation. If this excision process occurs incorrectly, a hybrid DNA molecule is formed which consists partly of λ DNA and partly *E. coli* DNA. The bacterial DNA may comprise either the *bio* or the *gal* genes.

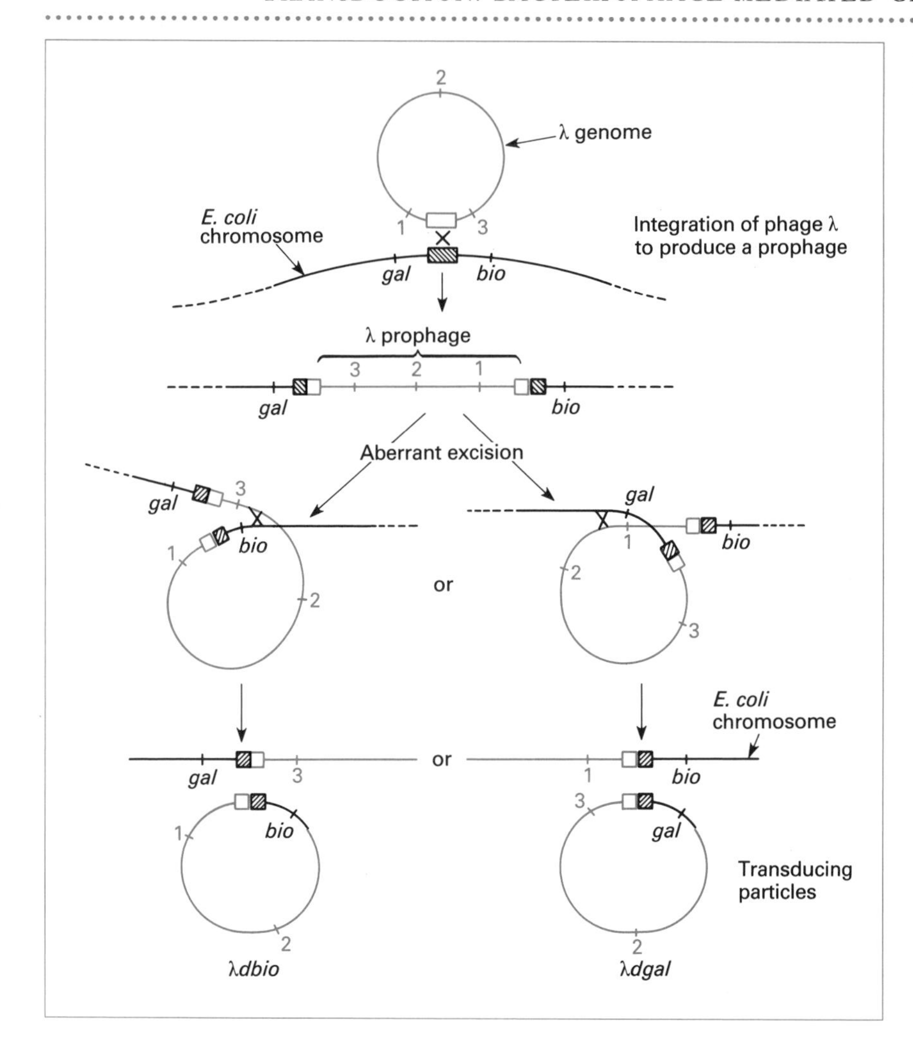


**Figure 8.10** Production of λ transducing particles.

The transducing particles thus formed are limited to a 'headful' of DNA: if they contain bacterial genes they must be deficient in part of the λ genome of corresponding size. Transducing particles of λ are, therefore, designated either λ*dgal* or λ*dbio*: the '*d*' emphasizes that transducing particles are defective for a number of λ genes; in other words they are absent (Figure 8.10).

Following infection of a recipient *E. coli* cell by either a λ*dgal* or λ*dbio* transducing particle, integration and replication can once again lead to the production of recombinant populations of *E. coli* cells.

## Sexduction or F-mediated transduction

In our description of bacterial conjugation, we saw that an F factor, located in the cytoplasm of $F^+$ cells, could become part of the bacterial chromosome to produce an Hfr cell. This integration process is reversible, and

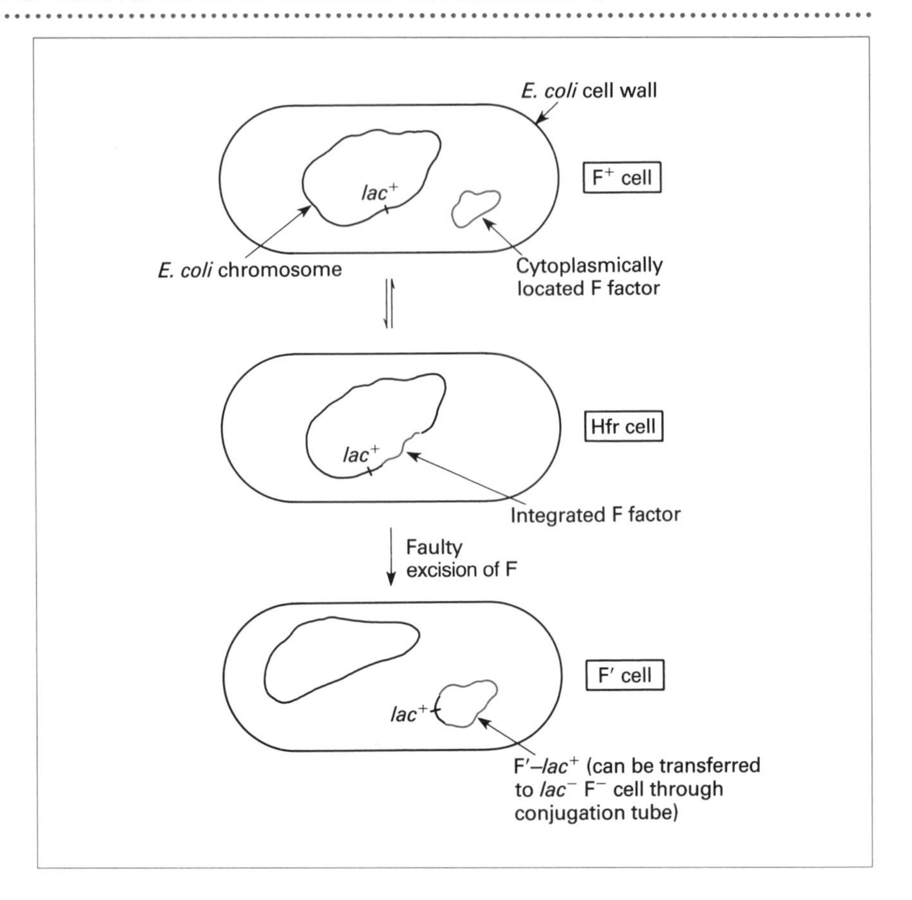


**Figure 8.11** Generation of a modified sex factor (F′).

excision of the F factor from its chromosomal location leads to the formation of an $F^-$ cell. Occasionally, the excision process occurs incorrectly (in a fashion very similar to that which we have described for excision of the λ prophage) and a hybrid of F factor DNA and chromosomal DNA is produced. Such a structure is called a modified F factor or F-prime (F′). Unlike λ DNA, the F factor can be integrated at a number of sites in the *E. coli* chromosome so that, following excision, modified F factors carrying a variety of chromosomal genes can be formed. A modified F factor carrying genes for lactose fermentation would be designated F′-*lac* (Figure 8.11).

Chromosomal genes can be transferred as modified F factors from a donor to a recipient cell through a conjugation tube.

## Bacterial transformation

This final method of gene transfer, although historically the first to be observed, has been discussed in Chapter 7 where we looked at transformation as providing evidence that DNA was the genetic material.

## Gene cloning and genetic engineering

In Chapter 7 we saw the importance of recombination as a genetic process which leads to variation in populations of organisms. In nature, recombination requires the participation of intact living individuals of the same species. Genetic engineering techniques also lead to the formation of recombinant individuals but they differ from the natural process in two important respects. First, the DNA to be recombined can be from completely unrelated species, such as *Escherichia coli* and *Homo sapiens*. Second, the two DNA molecules are recombined *in vitro* (in the test-tube, so to speak).

In practice, one of the DNA molecules concerned is able to replicate itself and is often a bacterial plasmid. If this is recombined with a sample of foreign DNA, the plasmid and its *insert* of foreign DNA can be introduced into a suitable host cell where replication can produce many copies of the plasmid and its insert. So one of the uses of this type of **recombinant DNA technology** is to produce large numbers of copies of DNA fragments for further analysis or manipulation.

Alternatively, under appropriate conditions, a foreign gene may direct the synthesis of its polypeptide product within a bacterial host cell. The value of this approach is that human proteins required for the treatment of diseases can be inexpensively synthesized by 'engineered' bacteria. Hormones are proteins, and the synthesis, by bacteria, of human insulin for the treatment of diabetes is a good example (see Chapter 11).

In the following sections we describe the techniques used for the construction of recombinant DNA molecules in the laboratory.

*In vivo* and *in vitro*
These terms are used to distinguish a biological process which is studied in an intact living organism (*in vivo*) from one which is studied by using one or more components of a living organism which have been separated from the whole. An example of this second (*in vitro*) approach includes the use of various enzymes following extraction from intact cells. The expression *in vitro* is derived from the common practice of storing such preparations in glass containers.

### Gene cloning

Gene cloning is the technique of producing a hybrid DNA (or recombinant) molecule *in vitro* which, following its introduction into a living cell, can replicate itself and be passed from generation to generation. Because the molecule into which the gene (or genes) is introduced is able to replicate independently, it is often referred to as a **replicon**. Replicons employed in gene cloning are called **cloning vectors** and usually are derived from plasmids, bacteriophages, or other viruses.

### Requirements for *in vitro* genetic manipulation

To successfully clone a gene of interest the following are required: restriction endonucleases, a cloning vector and a suitable host organism.

**Table 8.2** Examples of restriction endonucleases

| Enzyme | Bacterium of origin | Target sequence |
|---|---|---|
| *Alu*I | *Arthrobacter luteus* | A G ↓ C T<br>T C ↑ G A |
| *Bam*HI | *Bacillus amyloliquefaciens* H | G ↓ G A T C C<br>C   C T A G G ↑ |
| *Bgl*II | *Bacillus globigii* | A ↓ G A T C T<br>T   C T A G ↑ A |
| *Eco*RI | *Escherichia coli* RY13 | G ↓ A A T T   C<br>C   T T A A ↑ G |
| *Eco*RII | *Escherichia coli* R245 | ↓ G C $\mathrm{T}^{\mathrm{A}}$ G G<br>G G $\mathrm{A}^{\mathrm{T}}$ C C ↑ |
| *Hae*III | *Haemophilus aegyptius* | G G ↓ C C<br>C C ↑ G G |
| *Hind*II | *Haemophilus influenzae* $R_d$ | G T ↓ Py Pu A C<br>C A ↑ Pu Py T G |
| *Hind*III | *Haemophilus influenzae* $R_d$ | ↓ A A G C T T<br>T T C G A A ↑ |
| *Pst*I | *Providencia stuartii* | C T G C A ↓ G<br>G ↑ A C G T C |

Py, any pyrimidine; Pu, any purine;
↑ indicates points at which target sequences are cleaved (cut).

**(a) Restriction endonucleases.** These are enzymes which cut DNA at specific places called recognition sequences. Endonucleases are produced by all bacterial cells as a defence against invasion by 'foreign' DNA, such as phage DNA. The DNA of the host cell is protected from the action of its own restriction enzyme by another enzyme which methylates host DNA to distinguish it from non-methylated foreign DNA. Several examples of restriction enzymes are shown in Table 8.2.

The most common numbers of base-pairs to constitute recognition sequences are either four or six. When cut by the restriction endonuclease, a staggered break can occur giving rise to a short single-stranded overhang. For example *Bam*H1 gives rise to the single-stranded terminal sequence

GATC shown below:

```
G–3'    5'GATCC
CCTAG–5'    3'–G
```

The single-stranded overhangs produced by restriction endonucleases can form hydrogen bonds by complementary base-pairing and are called **cohesive** or **sticky ends**. Some restriction endonucleases, for example *AluI*, cut the recognition sequence cleanly, producing **flush or blunt ends** instead of staggered breaks.

```
AG–5'    3'CT
TC–3'    5'GA
```

**(b) Cloning vectors.** These are required in order to replicate and produce large amounts of a cloned gene in a population of host cells. Cloning vectors are examples of replicons; that is, they are DNA molecules with an origin of replication. This nucleotide sequence is a recognition site for the DNA-synthesizing machinery to initiate the replication process. A DNA molecule without such a site cannot be copied and is not suitable for use as a vector. Those which have been employed as cloning vectors ( plasmids and viruses) are small and easy to manipulate.

Plasmids are the most common vectors for cloning genes. Although in early experiments naturally occurring plasmids were employed, they were soon superseded by cloning vectors constructed in the laboratory.

The most commonly used plasmid-derived cloning vector is pBR322. The small size of pBR322 (4363 bp) means it can be purified easily. The

**BOX 8.2**

**Plasmid cloning vectors**

The ideal cloning vector should:

- have a **low molecular weight**. A small plasmid is easier to isolate intact than a larger one and its maintenance in the cell requires less energy and fewer metabolic intermediates.
- be present in **moderately high numbers**, perhaps around 15 copies per cell*. If there are 15 vectors bearing the gene there should also be 15 times more product. Too high a copy number can be detrimental to the cell.
- confer a readily **selectable phenotype** on the host to make it easier to distinguish between cells which contain the vector and those which do not. The most commonly employed selective traits are antibiotic resistance and the ability to ferment lactose.
- possess **unique cleavage sites** for several restriction endonucleases. It is particularly useful if, in a vector possessing genes for more than one selectable trait, some of these sites occur in one of the genes. This allows distinction of cells containing the vector only from those which possess a hybrid vector (Figure 8.12).

*The number of copies of a vector present in a cell is called the **copy number**.

copy number, about 15 when grown on a nutrient-rich medium, ensures a high yield of recombinant DNA molecules.

Viruses can also be used to clone genes. Bacteriophages have been manipulated for cloning genes in bacteria. In view of the detailed knowledge available about the genome organization of phage λ, a number of lambda-derived cloning vectors have also been developed for use in *E. coli*.

In the presence of an inhibitor of protein synthesis such as chloramphenicol, the copy number of pBR322 can be amplified to more than 1000 copies per cell, thus providing a very high yield of recombinant molecules.

## The *in vitro* stages of genetic manipulation

The manipulation of genes from other bacteria or from the lower eukaryotes into *E. coli* host–vector systems is relatively straightforward. We shall illustrate the principles by cloning a bacterial gene (*genX*) into plasmid pBR322 using the restriction enzyme *Bam*HI, which cuts the molecule in the tetracycline-resistance gene producing linear pBR322 (Figure 8.12). Purified bacterial chromosomal DNA is also cut with *Bam*HI, producing a

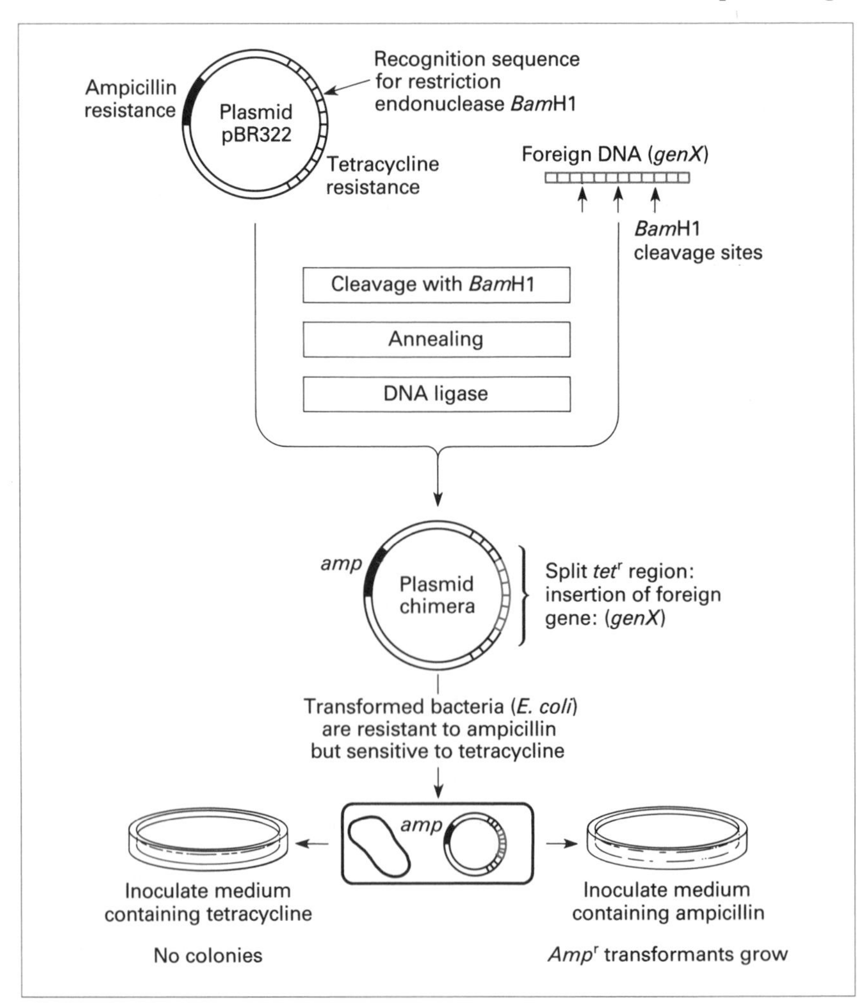


**Figure 8.12** Cloning a foreign gene into plasmid pBR322.

random collection of fragments, some of which contain *genX*. Both types of molecule (plasmid and chromosome) have similar sticky ends, CTAG-5′. When the two DNA preparations are mixed they can associate by hydrogen bonding between the complementary bases. The single-stranded breaks (or nicks) which occur in the sugar–phosphate backbones of the hybrid DNA molecule are sealed by the enzyme DNA ligase. The recombinant molecules are introduced into an *E. coli* host by transformation.

This transformation involves incubating *E. coli* cells and the hybrid DNA in the presence of $CaCl_2$ at 0°C for 1 hour, followed by a heat shock at 42°C for about 3 minutes.

To select transformed cells (those which have taken up extracellular DNA), the bacteria are plated onto a selective medium. There are likely to be four types of cell in the mixture:

1. Some cells will not have been transformed and will be sensitive to both the antibiotics for which pBR322 carries resistance genes (ampicillin and tetracycline).
2. Bacteria containing recombinant plasmids will be resistant to ampicillin but sensitive to tetracycline as a result of **insertional inactivation** of the *tet* gene. In other words, *genX* (often called an insert) has been incorporated into a restriction site situated within the *tet* gene. This disrupts the *tet* base sequence – essentially a mutation-like process occurs – and the protein product responsible for tetracycline resistance is no longer produced in a functional form.
3. In some cases the vector will have ligated without receiving a piece of chromosomal DNA.
4. Cells containing intact plasmids will be resistant to both ampicillin and tetracycline. Recombinant *E. coli*, resistant to ampicillin and sensitive to tetracycline, can be isolated by replica plating.

## The use of cDNA for cloning eukaryotic genes

Eukaryotic genes contain non-coding sequences of DNA (introns) which separate the coding sequences (exons). In a eukaryotic cell a primary mRNA transcript of both the introns and exons is produced. The regions corresponding to the introns are subsequently removed enzymatically, and the remaining exons joined to produce a secondary mRNA transcript. This transcript is translated by the ribosomes and produces the corresponding polypeptide. Because bacteria lack the enzymes for transcription of eukaryotic DNA, eukaryotic genes cannot function in hosts such as *E. coli.* This problem is circumvented by using mRNA from which the introns have been removed and employing an enzyme called **reverse transcriptase** to re-synthesize the DNA *without the introns.* The DNA produced by this process is called **complementary DNA (cDNA)**. The major steps in this process are shown in Figure 8.13.

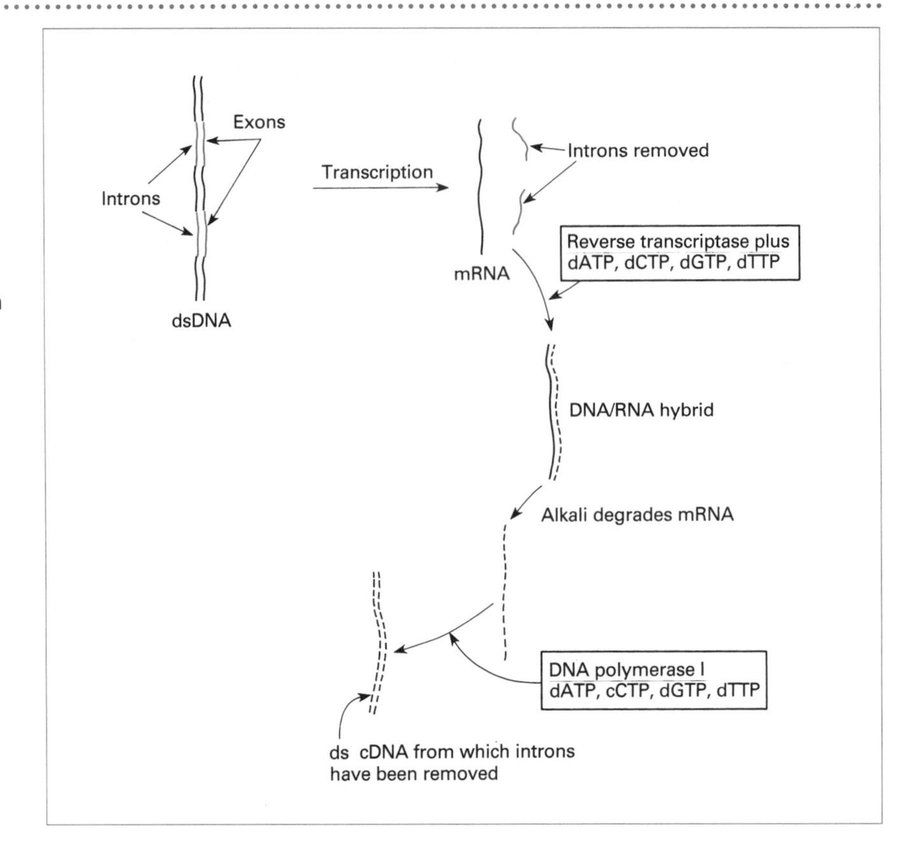


**Figure 8.13** The synthesis of cDNA from mRNA. Eukaryotic genes are split genes. That is to say they contain intervening sequences or introns. Because bacteria are unable to express this type of gene, introns are removed and a cDNA molecule is produced by the action of the enzyme reverse transcriptase. Because cDNA does not contain introns it can be expressed in a bacterial host cell.

## Identification of specific genes

We described earlier how bacteria containing recombinant pBR322 can be identified by an alteration in their antibiotic sensitivity. Although this identifies the bacterium as recombinant, it does not indicate whether it is the *desired* gene or an unwanted one which has actually been inserted into the vector. Three approaches can be adopted to screen recombinant bacteria to confirm whether the *desired* gene has been inserted.

1. **Complementation** relies on the ability of a gene cloned into a bacterium to compensate for (complement) a genetic deficiency in that host (Figure 8.14). This technique has so far been limited to bacteria and the lower eukaryotes such as *Saccharomyces cerevisiae* and *Neurospora crassa.*
2. **Immunological methods** are of value if antibodies to the polypeptide product of the recombinant gene are available. The **immunoprecipitation test** involves adding antibody to the agar on which the recombinant colonies are growing. The excreted polypeptide product of the cloned gene forms a precipitate of the antibody–antigen complex, which can be seen as a ring round colonies (Figure 8.15). An extension of this technique uses a radioactive antibody (Figure 8.16).

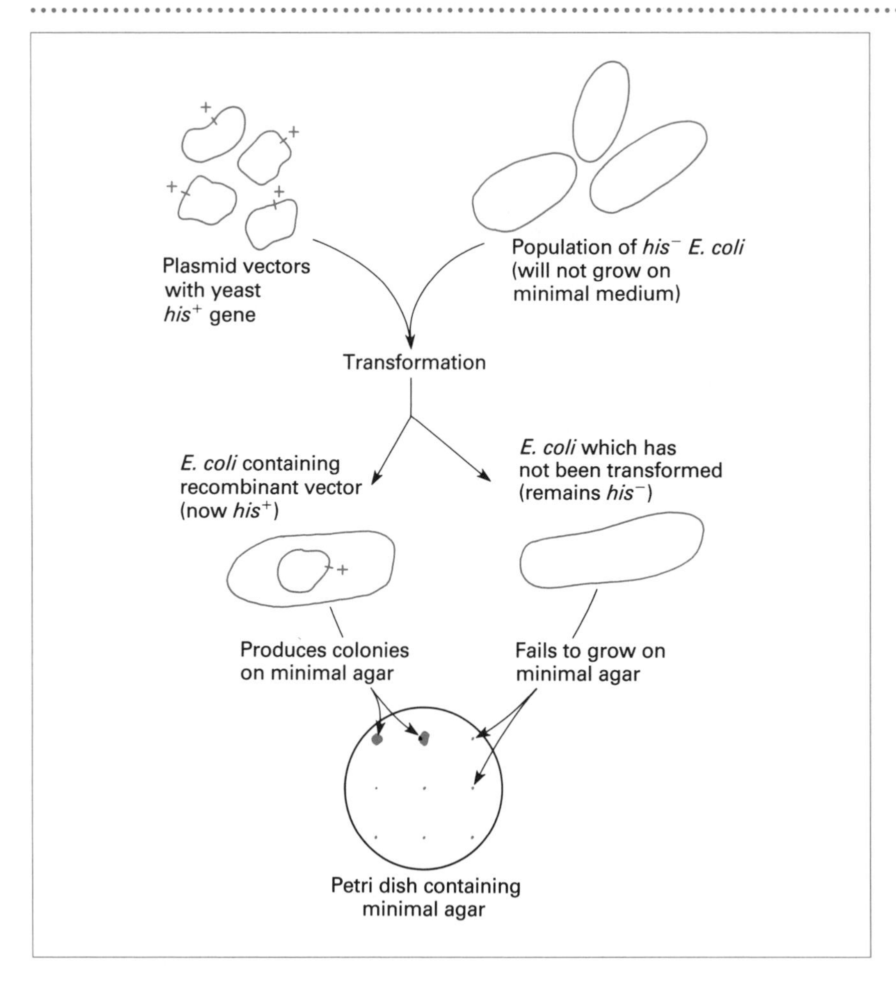


**Figure 8.14** Complementation to demonstrate the presence of a recombinant vector.

3. **Hybridization methods** using a **nucleic acid probe** can be employed with recombinant bacteria in which the gene of interest is not expressed (does not function). That is to say, the protein-synthesizing apparatus of a bacterium is unable to take instructions from a eukaryotic gene which, therefore, remains unexpressed. Complementary, single-stranded nucleic acid molecules can base-pair with each other

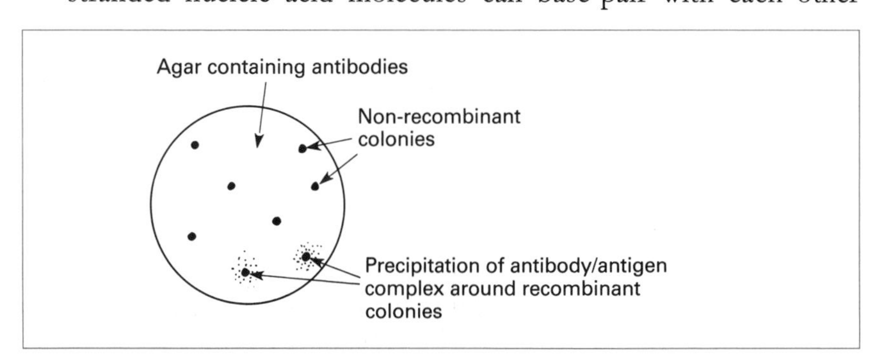


**Figure 8.15** The immunoprecipitation test for the identification of recombinant colonies.

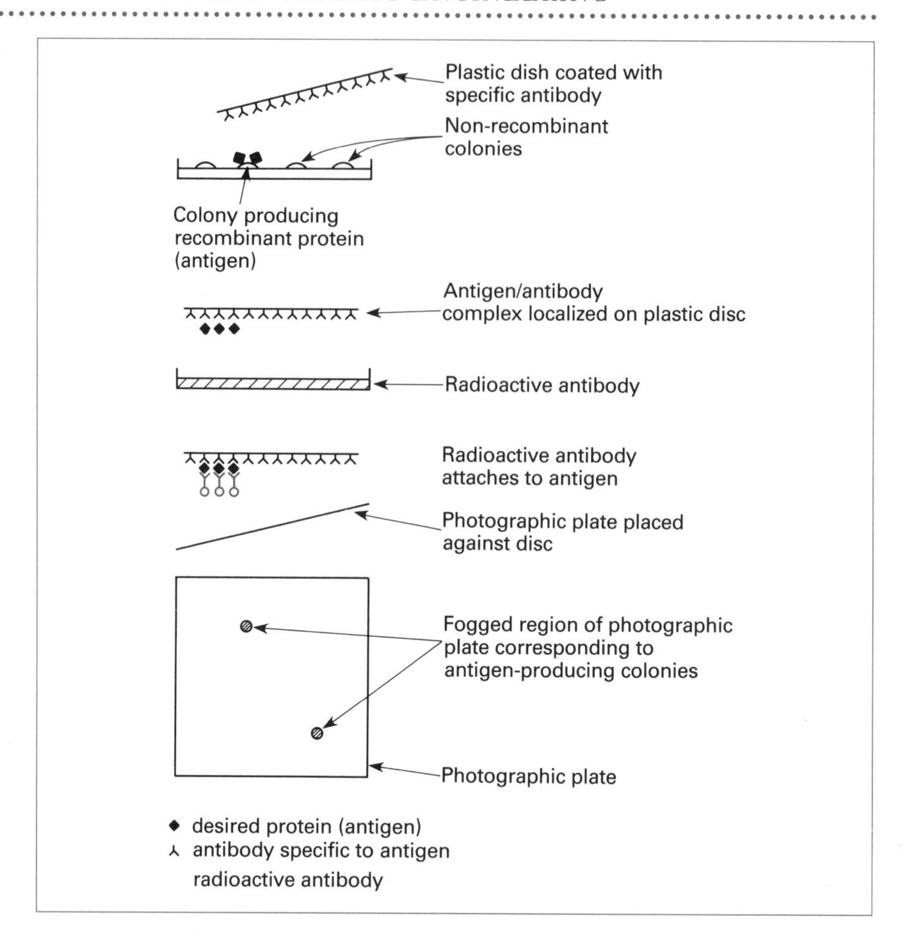


**Figure 8.16** The radioactive antibody test.

(hybridize) to produce a double-stranded molecule. Hybridization can occur between DNA and RNA or between like nucleic acids. A **gene probe** or **hybridization probe** is a nucleic acid sequence which contains all or part of the gene of interest. It is labelled, for example, radioactively, and used to identify recombinant bacteria which contain a homologous gene.

## Eukaryotic microorganisms and genetic engineering

Although our discussions have concentrated on bacteria, eukaryotic microorganisms are also extremely useful as genetic tools and as hosts for cloning foreign genes. From the point of view of genetic engineering, eukaryotic microorganisms have a number of potential advantages compared with bacteria. First, because animals and plants consist of eukaryotic cells, the way in which they direct the synthesis of polypeptide chains is the same as that of eukaryotic microbes. It is, therefore, potentially much simpler for an animal or plant gene to express itself in a eukaryotic host than in a bacterium.

Second, some microbial eukaryotes are unicellular, as bacteria are, and can be manipulated using similar bacteriological techniques. The brewer's yeast, *Saccharomyces cerevisiae*, has been much studied from the genetic and molecular biological points of view. In the laboratory it can be cultured on a solid medium to produce separate, compact colonies in much the same way as *Escherichia coli* does.

## Chapter summary

The focus in this chapter has been plasmids and their importance in microbiology. The first point you learned was that the vast majority of plasmids are small molecules of dsDNA which carry genes often described as non-essential. In fact, genes located on plasmids are responsible for a number of functions. For example, the F factor of *Escherichia coli* directs the formation of sex pili required for conjugation, and R plasmids carry genes responsible for resistance to a number of antibiotics. Plasmids can be viewed, therefore, as optional extras insofar as additional packs of genes may be acquired by bacteria which may act as survival kits during adverse environmental conditions. Possession of genes for antibiotic resistance will increase the chances of survival in an environment containing antibiotics.

Through our description of F factor function, you learned that a number of methods are available for the transfer of genes from one bacterium to another. In addition to conjugation, transformation and transduction were also described.

After describing the importance of plasmids from the bacterium's viewpoint, we moved on to examine the use of plasmids as vectors in genetic engineering. You learned that the essential tools of the genetic engineer are restriction endonuclease enzymes and a vector. The endonucleases permit an experimenter to cut very large DNA molecules into more manageable gene-sized chunks. These can then be inserted into a plasmid or a bacteriophage genome and re-introduced into a suitable host cell – commonly an *E. coli* cell. Once it is inserted into the vector, replication of the hybrid vector/insert can be carried out by the host cell. In this way large amounts of a desired gene can be produced.

Once large quantities of a gene become available they can be used either as the raw material for further analysis or they can be used to direct the synthesis of a useful protein. Potentially the most valuable use of genetic engineering is to produce bacterial 'factories' which can synthesize a whole range of therapeutic agents, including human hormones and vaccines.

Finally, we pointed out the potential importance of using eukaryotic host cells and vectors in genetic engineering.

## Further reading

T.A. Brown (1990) *Gene Cloning: An Introduction.* 2nd Edition. Chapman & Hall, London – an enjoyable account of genetic engineering techniques.

T.A. Brown (1992) *Genetics: a Molecular Approach.* 2nd Edition. Chapman & Hall, London – a lucid treatment of the molecular bases of genetics.

K. Hardy (1981) (new edition) *Bacterial Plasmids.* Chapman & Hall, London – rather old now, but still a comprehensive account of the biology of plasmids.

B. Lewin (1990) *Genes IV.* Oxford University Press, Oxford – a more advanced treatment of genetic phenomena.

J.C. Murrell and L.M. Roberts (1989) *Understanding Genetic Engineering.* Ellis-Horwood, Chichester – a gentle introduction to the techniques of genetic engineering.

J.D. Watson, N.H. Hopkins, J.W. Roberts, J.A. Steitz and A.M. Weiner (1987) *Molecular Biology of the Gene.* 4th Edition. Benjamin-Cummings, Inc., Menlo Park, California, USA – a comprehensive treatise which includes J.D. Watson (of double helix fame) among its authors. One for the enthusiast.

## Questions

**1.** Fill in the gaps:
(i) The first plasmid to be discovered was the __________ of *Escherichia coli*. Nearly all plasmids are composed of __________ stranded DNA which forms a __________ __________. In general, plasmids carry __________ genes which may nevertheless endow a selective advantage under certain conditions. For this reason, plasmids are sometimes called __________ chromosomes. Resistance plasmids carry genes which lead to

__________ resistance, enabling certain bacteria to survive in the presence of __________ -producing microorganisms.
(ii) *E. coli* cells which carry an F __________ in their __________ are designated $F^+$ and those which carry no F factor are designated __________. $F^+$ cells possess hair-like structures called __________ __________ which initially serve to establish __________ __________ __________ contact between an $F^+$ __________ cell and an __________ recipient. Once contact has been established, this structure may act as a __________ __________ for the transfer of F factor DNA from __________ to __________ cell.
(iii) In some *E. coli* cells the F factor may be integrated into the chromosome itself. If this happens, the cell is designated __________. If such a cell is mated with an $F^-$ cell, the F factor can be transferred through the __________ __________. However because the F factor is now part of the bacterial chromosome, this latter structure is also transferred. Before transfer, a break occurs within the __________ F factor so that the chromosome/F factor becomes a __________ structure. This conjugative chromosome transfer occurs in a __________ fashion and starts at a site within the F factor designated __________. This part of the F factor is the first section of the hybrid chromosome-plasmid to be transferred. Chromosomal genes close to the __________ __________ __________ will be transferred first and those furthest away will not usually be transferred. This is because the conjugation tube is a __________ structure and is likely to __________ before transfer of the whole chromosome has been accomplished.
Choose from: F factor (twice), accessory, cytoplasm, sex pili, conjugation tube (twice), double, non-essential, antibiotic (twice), donor (twice), closed loop, factor, $F^-$ (twice), cell to cell, recipient, Hfr, single, linear (twice), origin of transfer, fragile, *oriT*, break, integrated.

**2** Multiple-choice questions:
(i) Which of the following *Escherichia coli* genes can be transferred by phage λ? (a) *trp*, (b) *met*, (c) *lac*, (d) *gal*, (e) *bio*
(ii) The first example of generalized transduction was: (a) P1 in *Escherichia coli*, (b) PBS1 in *Bacillus subtilis*, (c) λ in *Escherichia coli*,

(d) P1 in *Salmonella typhimurium*, (e) P22 in *Salmonella typhimurium*

(iii) When a chromosomally located F factor is incorrectly excised it may contain chromosomal genes. Such a modified F factor is designated: (a) $F^+$, (b) $F^-$, (c) F′, (d) Hfr, (e) F

(iv) Transducing particles of phage λ may be designated: (a) λdg, (b) λb, (c) λbd, (d) λdb, (e) λg

**3** Short-answer questions:

The following questions all relate to recombinant DNA technology (genetic engineering).

(i) What is genetic engineering?

(ii) Give two examples of vectors.

(iii) What is the function of a vector?

(iv) Explain the importance of restriction endonucleases.

(v) Using pBR322 as an example, how would you identify a clone which carries an insert-containing plasmid?

(vi) How can a recombinant DNA molecule be incorporated into an *E. coli* cell?

(vii) Explain why genetic engineering is such an important technique.

# Eukaryotic microorganisms

# 9

In this chapter we will consider the structure, growth and reproduction of eukaryotic microorganisms. The features that they share will be emphasized, and we will discover the ways in which they have solved the environmental problems that affect them. In spite of their small size and relative simplicity, eukaryotic microorganisms are very important to man. They include the **algae**, the organisms responsible for the greatest proportion of primary productivity in the oceans, the **protozoa**, organisms responsible for diseases such as malaria and schistosomiasis, and the **fungi**, the organisms responsible for the breakdown and recycling of carbon on a global scale.

The fungi, algae and protozoa occupy a curious territory between the **prokaryotes** (primitive organisms such as the bacteria and actinomycetes with no membrane-bound nucleus) and the more highly evolved **eukaryotes** (organisms such as the brown algae, ferns and land plants with a membrane-bound nucleus). They can be grouped with the microorganisms on the basis of their small size, simple cell structure and inability to form true tissues. However, they also have affinity with the more highly evolved eukaryotes because they possess nuclear membranes, nucleoli, mitochondria and many other eukaryotic features. Because of this dual affinity, many authorities place eukaryotic microorganisms in a kingdom of their own called the **Protista**, and put the fungi in a separate kingdom. Figure 9.1 illustrates the five-kingdom system of classification, first suggested by Whitaker in 1969.

There are problems with the five-kingdom classification system because, in some circumstances, it appears to separate groups of organisms that have similar characteristics. Some of the primitive aquatic fungi bear a very close resemblance in both structure and lifestyle to the algae and protozoa. During their reproductive stages they have zoospores that have no cell wall and move by means of flagella (page 255). In the vegetative stages of their lives these same fungi have cell walls that contain cellulose, like the algae. Some authorities already have placed these fungi together with some species of algae in a group of their own.

The study of each of the different groups of organisms in the protists and the fungi is given a name. The study of algae is called **phycology**; the study of fungi is called **mycology**; and the study of protozoa is called **protozoology**.

Even within the Protista there are similar problems in the division between the algae and the protozoa. *Euglena* is a unicellular, motile,

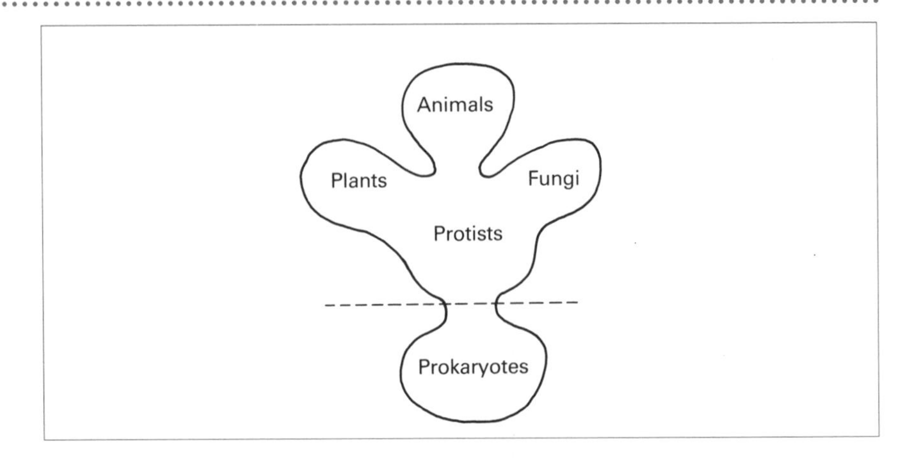


**Figure 9.1** The five-kingdom system of classification.

photosynthetic protist claimed by both phycologists and protozoologists. When it possesses a chloroplast it has clear affinities with the algae, but when it is 'cured' of its chloroplasts by chemical or heat treatment it is indistinguishable from *Hyalophacus* and *Astasia*, two flagellate protozoa.

One particular technique that is giving us more information about likely affinities between eukaryotic microorganisms is that of **RNA sequence analysis.** This technique allows us to compare sections of RNA messages for the same or similar genes in different organisms. These RNA messages (as rRNA or mRNA) tend to be conserved (kept intact) through evolution, so we can see which organisms have affinities with others. It seems likely that RNA sequence analysis of the Protista and the fungi will give us a better idea of their true relationships, and that the five-kingdom system will have to be modified in the light of this information.

**BOX 9.1**

**The 'curing' of *Euglena***

The chloroplasts of *Euglena* are surrounded by **three** membranes, and it is believed that these membranes represent the vestiges of a symbiosis that occurred between the non-photosynthetic protozoal ancestor of *Euglena* and chloroplasts from a green ancestral alga. The three membranes around the chloroplast of *Euglena* are the membranes around the chloroplast, plus the plasma membrane which formed a vesicle around it.

Evidence for this is provided by 'curing' experiments, where the chloroplasts of *Euglena* can be removed either by growing the cells at high temperatures of between 32 and 35°C, or by treatment with ultraviolet light or antibiotics.

The bleached cells are able to grow heterotrophically, but never become green again. They strongly resemble protozoal species such as *Astasia*, shown here.

There are many other examples of algae with three-layered chloroplast membranes, and these are all thought to be derived from similar symbiotic events. The species that we see today probably represent the best, most integrated, associations.

(a)

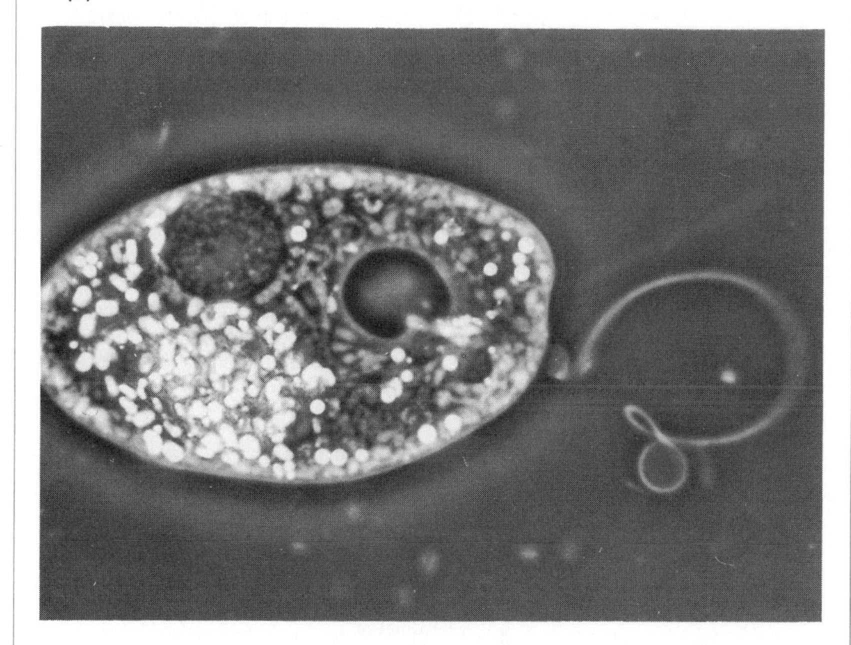

(b)

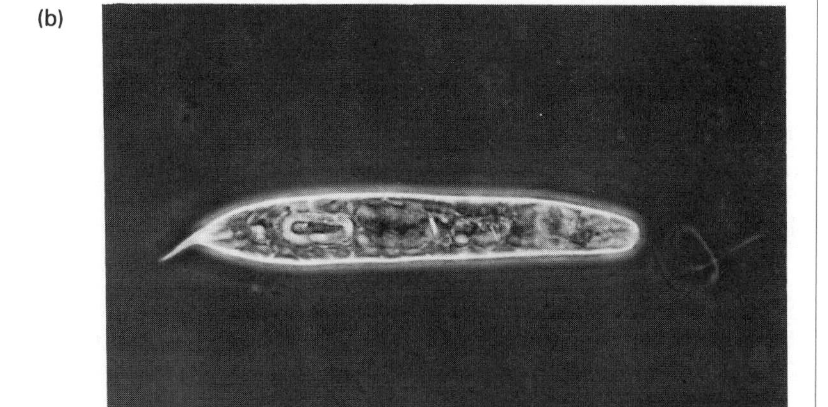

**BOX 9.1** *continued.*
(a) *Astasia*, a closely related but non-photosynthetic flagellate, and (b) *Euglena*, a photosynthetic flagellate.

**BOX 9.2**

**Ribosomal RNA analysis**

Ribosomal RNA (rRNA) is found in nuclei, mitochondria and chloroplasts. Its function, in combination with messenger RNA (mRNA) and transfer RNA (tRNA), is to synthesize the proteins that a cell needs. As very small changes in the structure of rRNA will prevent the functioning of the ribosome, alterations in its nucleotide sequences occur only very slowly through the evolution of an organism. This means that we can find out about the evolutionary history of an organism by looking at sections of rRNA (called sequences) and comparing them with sequences from other organisms.

To do this, rRNA has to be extracted from a cell and broken into fragments using an enzyme called ribonuclease. Sequences that we know to be kept constant (conserved) because they are essential for function can be picked out with a primer, a 'sticky label' of complementary nucleic acids. We can then use another enzyme, reverse transcriptase, to build up a complementary DNA version of the rRNA sequences close to the conserved regions.

The sequences of DNA can be determined using electrophoresis, and we can then compare the patterns of sequences within and between species using a computer, which can then calculate how far they are related. This information is often displayed as a dendrogram or phylogenetic tree.

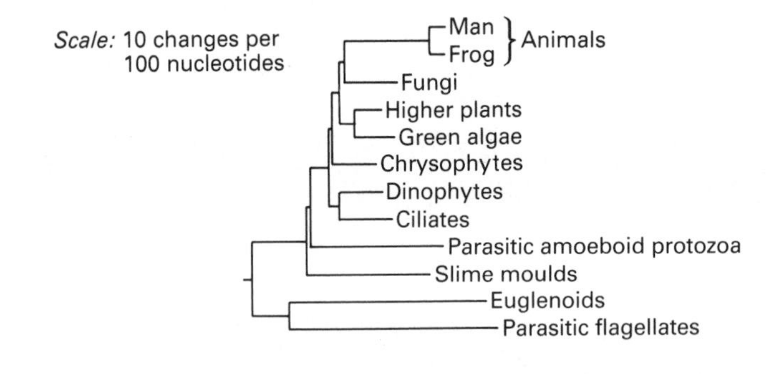


## Structure and growth of protistan microorganisms

Most kinds of algae and protozoans live in moist or aquatic habitats, either in water masses such as lakes, rivers and seas, in moist pore-spaces in soils, or in the nooks and crannies of animal bodies. In contrast, the fungi, with a few exceptions, live in comparatively dry, terrestrial habitats whether they are parasites on plants or free-living in soil. In such environments the fungi encounter desiccation, osmotic stress and extremes of ultraviolet light.

## Structure of the fungi

There are five major groups in the fungi, two in the so-called lower fungi (**Zygomycetes** and **chytrids**) with a simple morphology, and three in the

**Table 9.1** Features of the main groups of fungi

| *Group* | *Septation* | *Cell wall fibres* | *Asexual spores* | *Sexual spores* |
|---|---|---|---|---|
| Lower fungi | | | | |
| Pin moulds (Zygomycetes) | Coenocytic | Chitin | Non-motile sporangiospores | Zygospore |
| Water moulds (Oomycetes and chytrids) | Coenocytic | Cellulose | Motile zoospores | Oospore |
| Higher fungi | | | | |
| Sac fungi (Ascomycetes) | Septate | Chitin | Conidia | Ascospore |
| Club fungi (Basidiomycetes) | Septate | Chitin | Rare | Basidiospore |
| Imperfect fungi (Deuteromycetes) | Septate | Chitin | Conidia | None |

higher fungi with a much more complex morphology. These include the sac fungi (**Ascomycetes**), the club fungi (**Basidiomycetes**), and a group of fungi which are unable to reproduce sexually and which are therefore not classifiable into the other groups, the imperfect fungi (**Deuteromycetes**). Details of their taxonomy and distinguishing features are shown in Table 9.1.

In all these groups, morphology reflects the ways in which fungi minimize the effects of adverse environmental conditions on their biology. They tend to be filamentous organisms in the vegetative stage of their life-cycle. Individual filaments are called **hyphae**, and a network of hyphae is called a **mycelium**. Hyphae have a thick, multi-layered cell wall composed of **chitin**, a fibrillar carbohydrate-based polymer. The wall protects them from the outside environment. Species of fungi that are exposed to extremes of light have black pigments, called melanins, in the cell wall, which help to minimize damage caused by ultraviolet light and also increase resistance to desiccation. Many of the fungi that live on the surfaces of plants are deeply pigmented or black.

Hyphae contain the organelles that are characteristic of non-photosynthetic eukaryotes, including nuclei, mitochondria, ribosomes, endoplasmic reticulum and many types of membrane-bound vesicles (Figure 9.2). In some kinds of fungi the mycelium is divided into cells by cell walls called **septae**, and these may be perforated with pores that allow certain organelles, but not nuclei, to pass. Not all fungi have this filamentous morphology. The yeasts are unicellular fungi in the Ascomycete group. Their cell walls contain fewer microfibrillar materials and more amorphous polysaccharides such as mannans (Figure 9.2). The chytrids and their

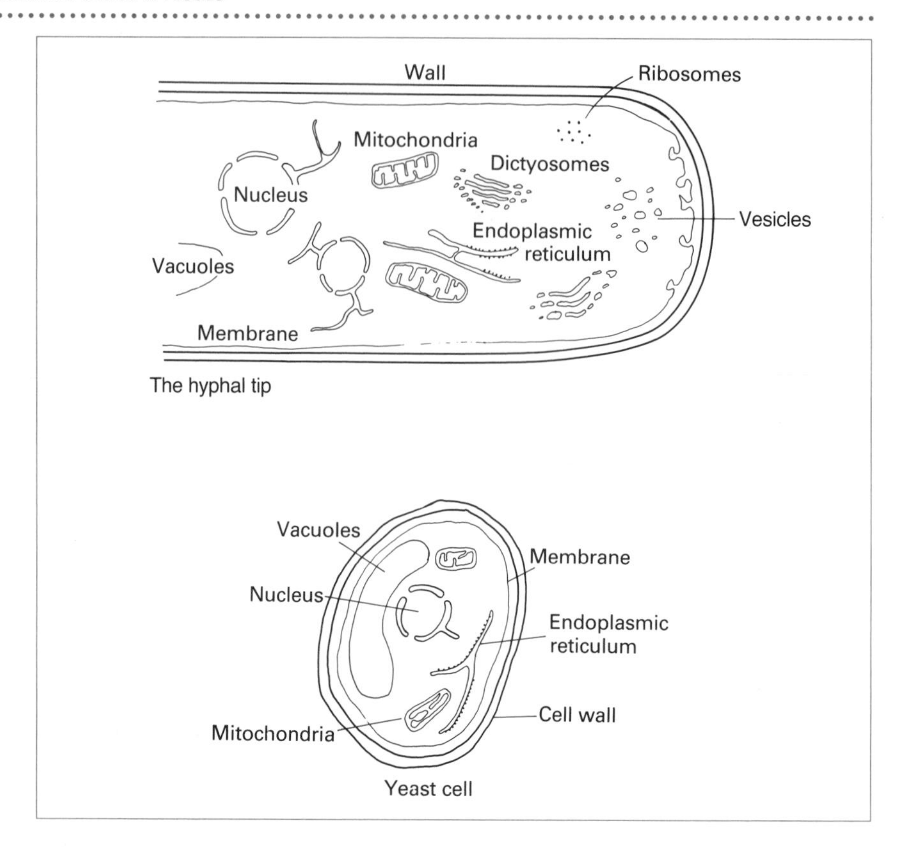


**Figure 9.2** Schematic diagrams of fungal hypha and yeast cells. After Grove, S.M., Bracker, C.E. and Morré, B.J. (1970) *Am. J. of Botany*, **57**, 245–66.

relatives are fungi that closely resemble aquatic algae, and more details about them are covered on page 267.

Fungi are heterotrophic for carbon. In other words, they cannot synthesize the carbohydrates they need using light or chemical energy as plants do. They are obliged to obtain their carbohydrate needs from outside sources. They can be classified into **saprophytes** (those fungi that live on *dead* plant and animal remains) or **parasites** (the plant and animal pathogens, that live on *living* tissue). Some plant pathogens attack the living plant and eventually cause its death. They then grow on the dead plant remains. Such a habit is termed a **facultative** type of nutrition. Dead plant and animal materials contain predominantly insoluble forms of carbohydrate which must be broken down before any absorption into a fungus can take place. Saprophytic and facultatively parasitic fungi secrete extracellular hydrolytic enzymes that break down polymers which are then absorbed by diffusion through the cell wall. Fungi are capable of producing many types of these enzymes, some of which are used commercially.

Other parasites are unable to live on dead plant remains. Instead of attacking the host with enzymes and degrading and killing it, these fungi enter the host and develop specialized feeding structures called **haustoria**

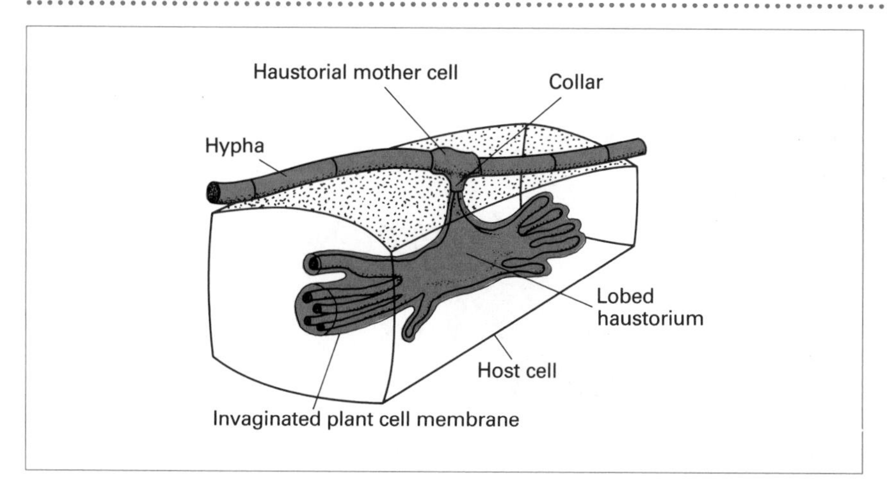


**Figure 9.3** Haustorium of a powdery mildew fungus, inside a leaf epidermal cell. Redrawn from Bracker, C. E. (1968) *Phytopathology*, **58**, 12–30.

(Figure 9.3). Haustoria are able to divert the host's metabolism without alerting its resistance mechanisms. The fungus is able to obtain carbohydrates as soluble sugars made by photosynthesis, and to absorb all other nutrients it may require from the plant. This relationship continues for as long as the pathogen remains unrecognized in the host. However, the disadvantage for the fungus is that it can live only on a living host. The reason for the dependency on living hosts seems to be that, in the evolution of this **biotrophic** type of nutrition, these fungi lost the ability to produce extracellular enzymes; without such enzymes they were unable to live saprophytically, and so became dependent on living hosts. These fungi are termed **obligate** pathogens, because they are *obliged* to grow on living plants. Some obligate plant pathogens, such as powdery mildews, are particularly good at living within plant tissues whereas others, such as rusts, tend to debilitate the plant seriously by the end of a growing season.

Fungi may live with living plants in a beneficial association called **symbiosis**. The association between the two organisms may be loose and not physically intimate (when it is termed **commensal**), or the two organisms may be so integrated they are impossible to separate, as in the **mycorrhizae**. These are associations between the roots of plants and fungi (Figure 9.4). Here, as with biotrophic relationships, there is a physical intimacy between plant and fungus, but instead of the nutritional relationship being one-way, there is a two-way exchange of nutrients. The fungi gain carbohydrates from the plants, and the plants gain phosphates and nitrogen sources from the fungi. Further details of the different sorts of nutritional relationships that fungi have with other organisms are shown in Table 9.2. There are advantages for both organisms in these associations, and plants that possess mycorrhizae grow better in nutrient-poor environments than plants without them.

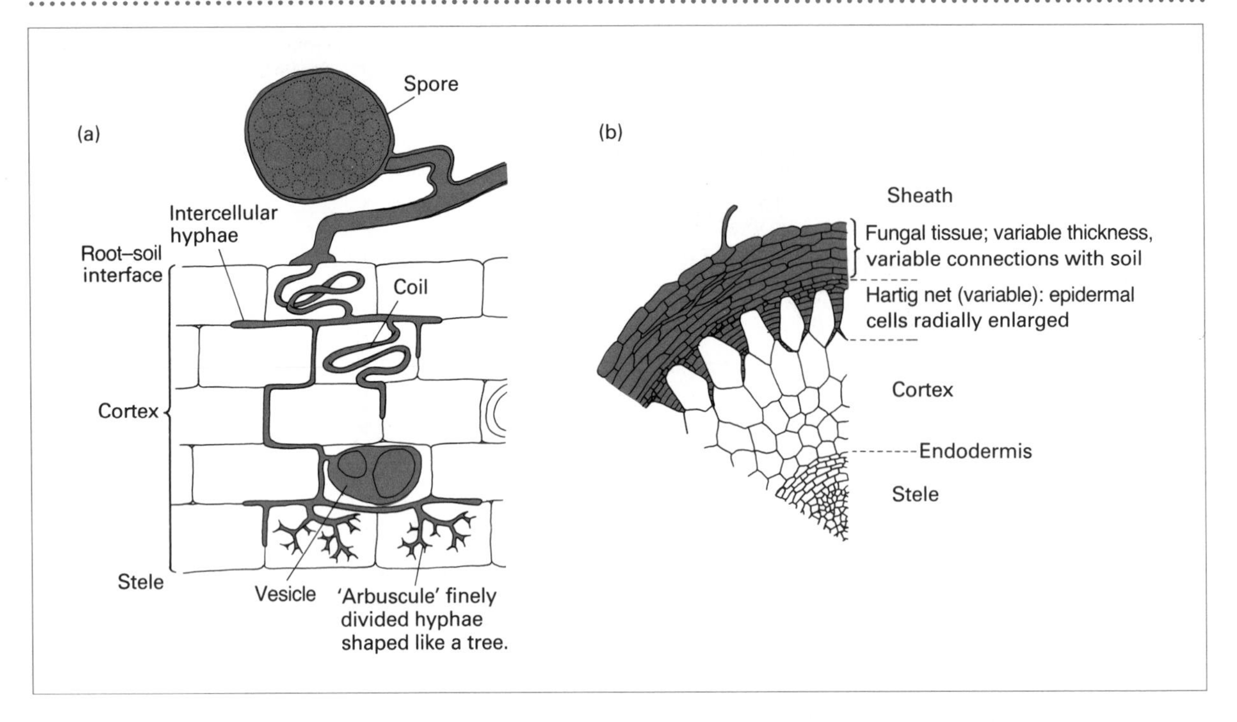


**Figure 9.4** Mycorrhizal associations of fungi with plants. (a) Endotrophic mycorrhizal fungus *within* plant root cells. (b) Ectotrophic mycorrhizal fungus around and *between* plant root cells. Redrawn from Isaac, S. (1991) *Fungal–plant interactions*, Chapman & Hall, London.

**Table 9.2** Nutritional relationships between fungi and other organisms

| *Type of symbiosis* | *Definition* |
|---|---|
| Commensalism | One population of organisms benefits from the association while another is unaffected.<br>For example, some fungi can convert complex substrates into simple sugars. Excesses of these can be used by other organisms which otherwise could not use the substrate. |
| Synergism | Both populations of organisms benefit from an association, but the association is not obligatory.<br>For example, the biological degradation of toxic pesticides is very complex, and only rarely can one organism degrade a compound completely. **Propanil**, a herbicide, can be partially degraded by *Penicillium piscarum* to propionic acid, which it uses as a carbon source, and 3,4-dichloroaniline, which is toxic. Another fungus, *Geotrichum candidum*, can detoxify this compound, so the two organisms can grow together on Propanil. |
| Symbiosis | An association between two organisms which is obligatory and highly specific.<br>For example, the **mycorrhizae** and the **lichens** are both associations between fungi and plants where there is great physical intimacy between the partners. Carbon compounds flow from the photosynthetic partner into the fungus, and nitrogen- or phosphorus-containing compounds move from the fungus into the plants. |

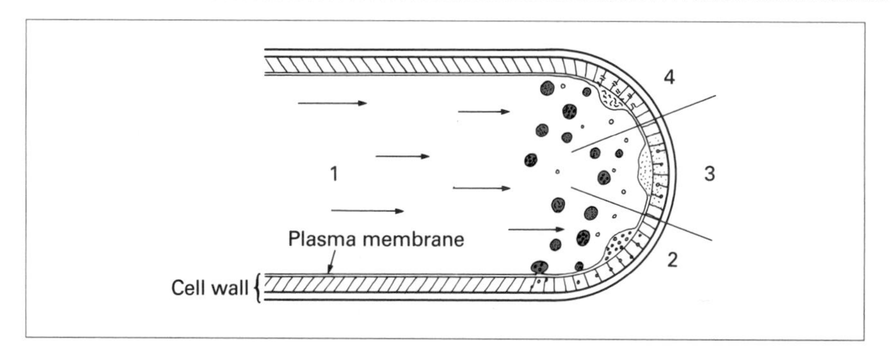


**Figure 9.5** Cell wall synthesis in the fungi. Step 1. Vesicles migrate to the apical regions of the hyphae. Step 2. Wall-lysing enzymes break fibrils in the existing wall, and turgor pressure causes the wall to expand. Step 3. Amorphous wall polymers and precursors pass through the fibrillar layer. Step 4. Wall-synthesizing enzymes rebuild the wall fibrils. Redrawn from Isaac, S. (1991) *Fungal–plant interactions*, Chapman & Hall.

Fungal metabolism is like that of aerobic prokaryotes, in that glycolysis takes place chiefly *via* the Embden–Meyerhof pathway, the hexose monophosphate pathway generating the nucleic acids and amino acids. The Entner–Doudoroff pathway has only been occasionally reported in the fungi. These pathways are shown in detail in Chapter 4.

## Growth in the fungi

In order to maximize the uptake of materials from the environment, fungi have evolved a spreading growth habit. Their hyphae constantly exploit new substrates so as not to become stranded in an area depleted of nutrients. The hyphae are full of sugars and salts and are, therefore, at a high osmotic potential relative to the environment. Water moves into the hyphae from outside by osmosis, which keeps the fungal cytoplasm turgid against the cell wall. The tip of the hyphal wall is softened by enzymes. The cytoplasm exerts pressure on the softened region and the hyphal tip pushes forwards, much like an inflating balloon. The wall is regenerated behind the growing tip (Figure 9.5). Yeasts grow in a very similar way, but instead of the cell continuing to elongate as it does in the filamentous fungi, growth is by cell division. Some yeasts divide by budding, others by binary fission; we look at these processes in greater detail on page 261.

Hyphae constantly grow into new habitats. The mycelium that is left behind the growing tip is largely emptied of cell contents as the protoplasm is pulled forward into the new growing areas. The growth pattern of a fungus on an agar plate illustrates this phenomenon on a scale visible to the naked eye (Figure 9.6a), and a fairy ring is an even larger manifestation (Figure 9.6b). Fairy rings are caused by the presence of toadstool mycelia growing in the soil among plant roots. The fungus grows from an initial point outwards as a ring and, where the fungus is in active growth, plant roots die, causing the characteristic circle of dead vegetation. The ring always grows outwards leaving an expanding centre of recovering grass, which is dark green. Ringworm infections of animals and man, also caused by a filamentous fungus, show a very similar pattern of growth (Figure 9.6c).

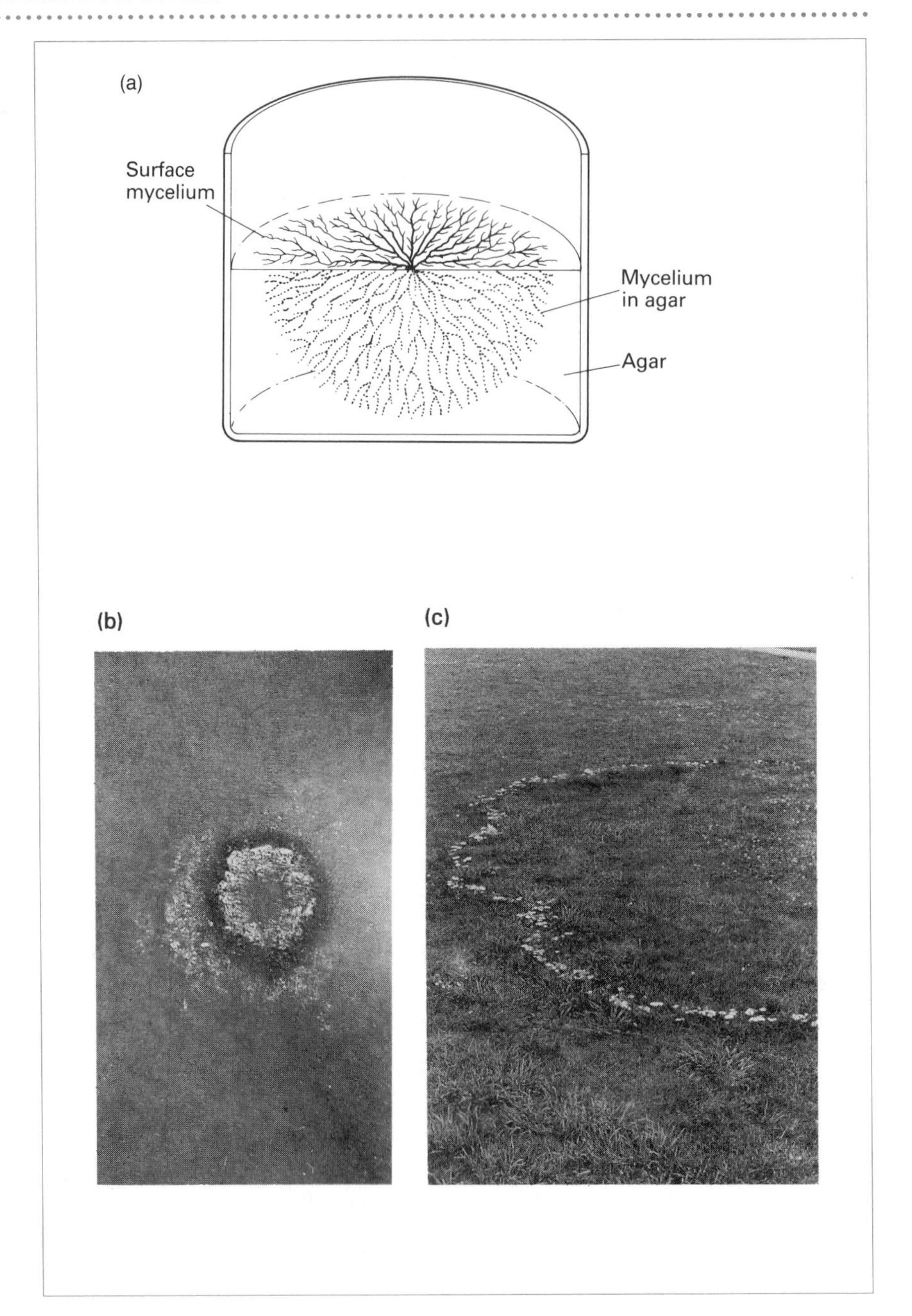


**Figure 9.6** Fungal growth patterns. (a) In agar; (b) in skin; (c) in turf. (a) Redrawn from Ingold, C.T. and Hudson, H.J. (1993) *The Biology of Fungi*, Chapman & Hall, London. © Reproduced courtesy of Biophoto Associates.

## Structure of the protistan algae

There are four major groups of algae in the Protista: the green algae, the golden brown algae, the euglenids, and the dinoflagellates. Table 9.3 lists the four major groups and summarizes their distinctive characteristics.

**Table 9.3** Features of the main groups of algae

| *Group* | *Number of species* | *Pigments* | *Storage products* | *Flagella* | *Cell wall constituents* |
|---|---|---|---|---|---|
| Green algae (Chlorophyta) | 7500 | Chlorophylls a and b, carotenes, xanthophylls | Sugars, starch, fructosan | 1, 2–8; equal, apical or subapical | Cellulose, mannan, protein, $CaCO_3$ |
| Golden brown algae (Chrysophyta) | 6000 | Chlorophylls a, c1/c2 | Chrysolaminarin, oils | 1–2; equal or unequal, subapical | Cellulose, silica, $CaCO_3$, chitin |
| Euglenids (Euglenophyta) | 700 | Chlorophyll a and b, beta carotene, xanthophylls | Paramylon, oils, sugars | 1–3; slightly apical | Proteinaceous pellicle |
| Dinoflagellates (Pyrrophyta) | 1100 | Chlorophyll a, c1/c2, beta carotene, fucoxanthin, peridinin, dinoxanthin | Starch, glucan, oils | 2; one trailing, one girdling | Cellulose or absent |

Most types of protistan algae exist as single cells, but in some species, cells are aggregated to form **colonial** types. They can also be **filamentous** (strips of cells) in structure, and some of the green algae can be almost leaf-like. Figure 9.7 illustrates the range of forms seen in the protistan algae.

Like the fungi, most kinds of algae have cell walls. They are usually composed of fibrils of cellulose, together with various other types of polysaccharide, and some are strengthened with materials such as calcium carbonate or silica. A few kinds of algae do not have cell walls at all but possess what is called a **pellicle**, a series of proteinaceous strips under the cell's outer membrane, which keeps the walls rigid but flexible.

Unlike the fungi, the algae are photosynthetic, requiring only light and carbon dioxide as their source of energy and carbon. This type of nutrition is said to be **phototrophic**. For this reason, the algae do not need to be protected from high light intensities and have auxiliary pigments as well as chlorophyll that help them trap energy from light of various wavelengths (Table 9.3).

While sharing most of the respiratory pathways of the fungi, the algae differ from them in that they are photosynthetic. This makes them primary producers rather than decomposer organisms. However, algal photosynthesis differs from that of lower prokaryotes because it is located on **thylakoid** membranes confined to specific organelles called **chloroplasts**. Chloroplasts are membrane-bound and contain their own type of ribo-

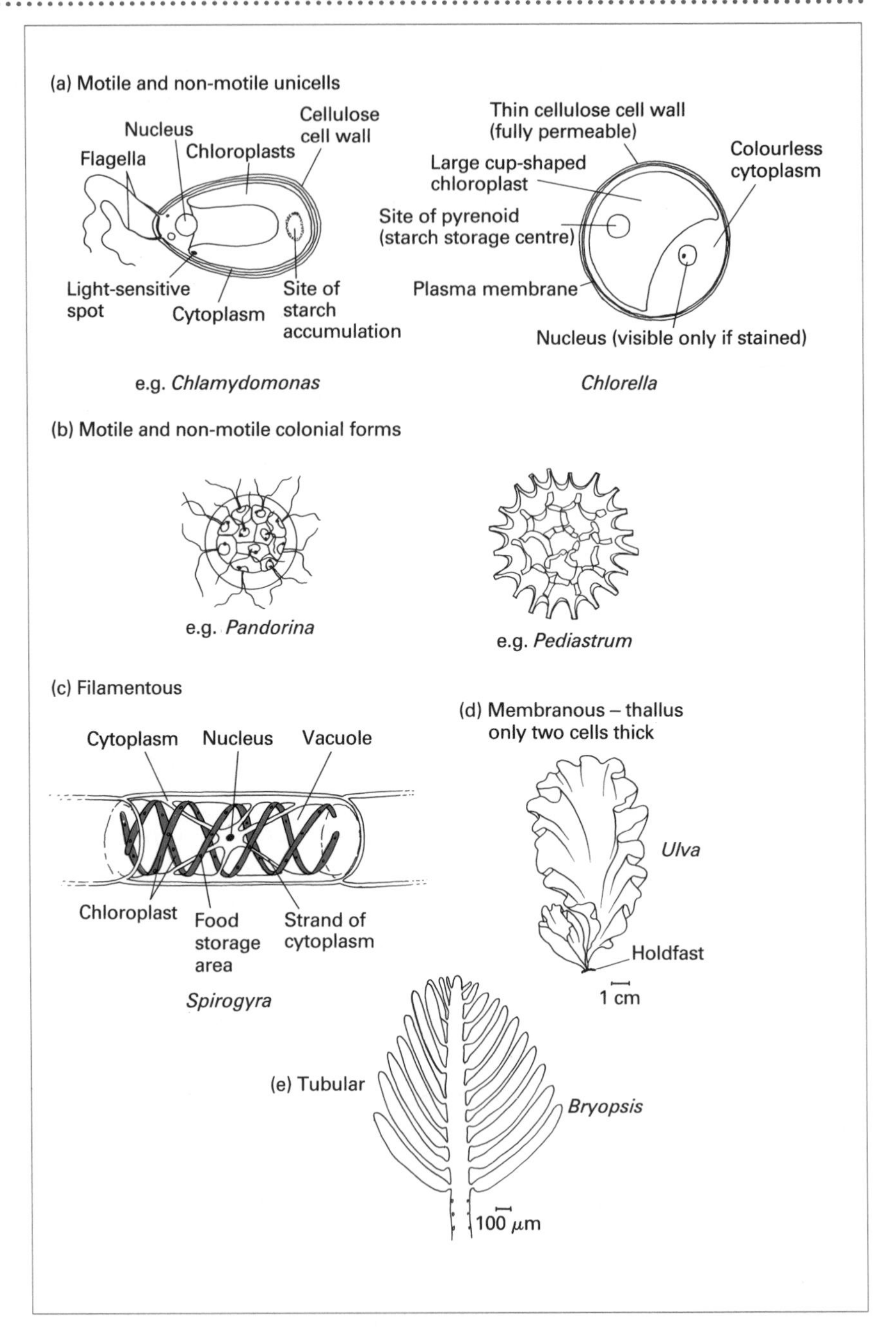


**Figure 9.7** The range of forms of protistan algae. Redrawn from Clegg, C.J. (1984) *Lower Plants,* John Murray, London.

somes and a little DNA. Algal photosynthesis uses two linked photosystems: Photosystem I, in which electrons are taken from reduced cytochromes and used to reduce NADP; and Photosystem II, in which cytochromes are reduced by electrons removed from water. Coupled to Photosystems I and II is the establishment of a proton gradient across the thylakoid membrane in the chloroplasts which generates ATP. The

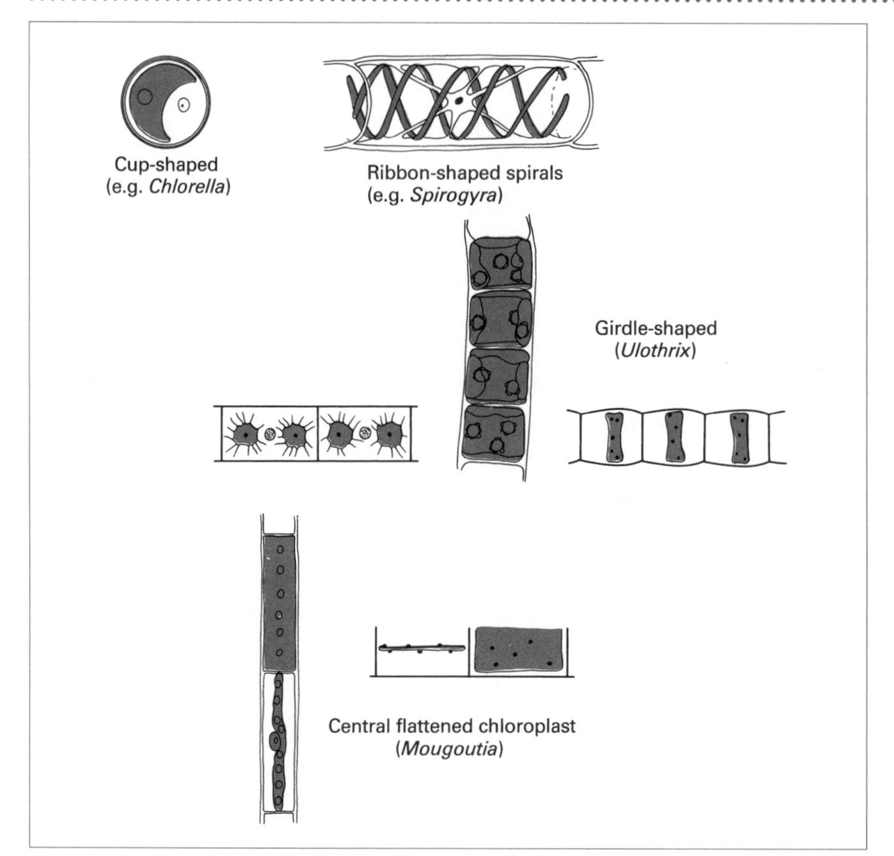


**Figure 9.8** Shapes of algal chloroplasts. Redrawn from Clegg, C.J. (1984) *Lower Plants*, John Murray, London.

NADPH and ATP made in the light-dependent cycle are then used to incorporate carbon into carbohydrate in the so-called dark reaction of photosynthesis, the Calvin–Benson cycle. These reactions are summarized in Chapter 4, pages 101–3.

Both marine and freshwater protistan algae are restricted to the top half-metre of water, below which light of the required wavelengths does not penetrate in sufficient amount to support photosynthesis. Various strategies are employed to ensure that the algae do not slip below this **photic zone**. Some are motile, using a brightly pigmented eyespot to 'sense' both the direction and strength of the light. Others are non-motile, but possess buoyancy aids such as gas vacuoles or elaborate spines and projections that utilize the constant turbulent eddies of water to maintain their position.

Biologists have found that the *shape* of algal chloroplasts is a very useful character in classifying algae into groups. For instance, some, like those of *Chlorella vulgaris*, are large and cup-shaped. Others, in the filamentous forms, can be ribbon-shaped, like those in *Spirogyra*, star-shaped as those in *Zygnema*, girdle-shaped as those in *Ulothrix*, or flat and plate-like as those in *Mougoutia* (Figure 9.8).

Some of the algae are not solely phototrophic. Their photosynthesis is supplemented by a heterotrophic nutrition in which alternative or additional sources of carbon are exploited. Some algae are **phagocytotic**, engulfing food particles and digesting them. Others obtain their extra carbon as fungi do, by absorbing it parasitically from living hosts or as saprophytes from dead ones.

In fresh water, algae are under constant threat of taking up too much water and bursting. This is because their cell sap is like that of the fungi, full of salts and sugars which exert an osmotic pull. The filamentous algae, like the fungi, have evolved a thick, rigid cell wall that prevents the cell from taking up excess water. Motile types of algae have a less rigid cell wall, and these species have a contractile vacuole that collects excess water and removes it from the cell. Controlling the uptake of water from the environment, be it by adjusting the osmotic potential of the cytoplasm as some species with rigid cell walls do, or by the activity of contractile vacuoles, is termed **osmoregulation**, and we will come across this subject again in the section on protozoa.

Many of the filamentous, colonial and leafy algae are non-motile, some being anchored to substrates and others free-floating. Many of the unicellular forms, and some of the colonial species, are motile and have evolved a variety of ways of moving, often using flagella or cilia. Like the flagella of all eukaryotic organisms, algal flagella have an internal structure of nine peripheral **microtubules** with two central microtubules. These microtubules are held in place by proteinaceous links, and are surrounded by a plasma membrane. Movement is generated by either a lateral beating thrust, as in *Chlamydomonas* (Figure 9.9a), or by a spiral wave motion which starts from the base of the flagellum and moves to the tip (Figure 9.9b). Both of these movements push the alga forwards. Cilia have a similar internal structure, but they are shorter and occur in great profusion (Figure 9.9c).

### Growth in the protistan algae

As most of the protistan algae are unicellular, growth takes place by binary fission. We will look at this in the section on Reproduction on page 261.

## The protozoa

The protozoa are divided into four major groups on the basis of the way they move, the way they reproduce, and the types of nuclei they possess (Table 9.4). The amoebas and flagellates form two groups, some of which

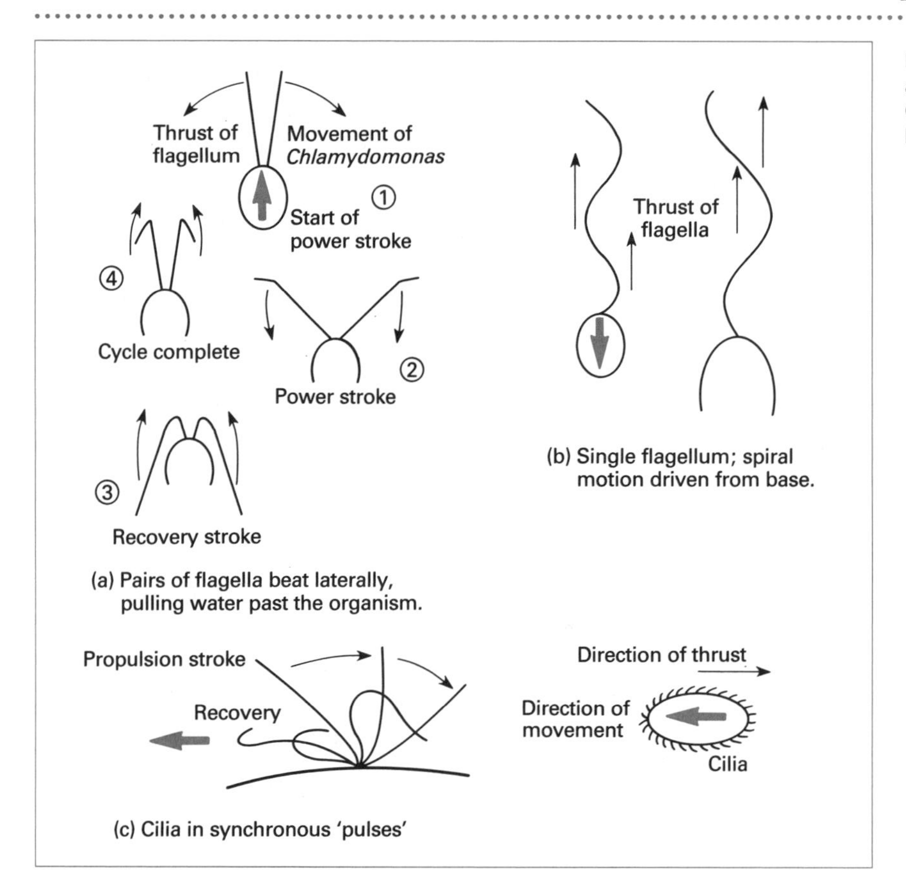


**Figure 9.9** Movement in the algae. Redrawn from Clegg, C.J. (1984) *Lower Plants*, John Murray, London.

are free-living, while others are parasites of man. The sporozoans are a group of protozoa that are predominantly parasitic on man and animals, and one group contains members that are algal parasites or have a saprophytic type of nutrition. The fourth group, the ciliates, are largely

**Table 9.4** Features of the main groups of protozoa

| *Group* | *Locomotion* | *Cell covering* | *Nuclei* | *Asexual reproduction* |
|---|---|---|---|---|
| Flagellates (Mastigophora) | One or more flagella | Naked or proteinaceous pellicle | Single | Binary fission |
| Amoebas (Sarcodina) | Pseudopodia | Naked, shells often present | Single | Binary fission |
| Ciliates (Ciliophora) | Cilia | Proteinaceous pellicle | Two, one large one small | Binary fission |
| Sporozoans (Sporozoa) | Rarely motile | Proteinaceous spore wall | Single | Multiple fission |

free-living. Many kinds live in the intestines of animals and are important in breaking down food substances such as cellulose for the host. This is a symbiotic habit, similar to those seen in the associations between fungi and other organisms.

Like the fungi and the algae, the structure of protozoans reflects their lifestyle. Many of their body forms are very like those of the unicellular algae. Free-floating types are spherical, or radially or bilaterally symmetrical. Motile species are streamlined into bullet or kidney shapes. Figure 9.10 illustrates the range of forms seen in the protozoa.

Unlike the algae and the fungi, however, protozoans have no cell wall. In many species of both parasitic and free-living forms, it appears that a naked plasma membrane is all that protects the protozoan from its environment. This is an advantage in that uptake of nutrients, be they soluble ones diffusing into the cell or particulate ones that are captured by **phagocytosis**, is unimpeded. In parasitic and most marine species the environment is roughly isotonic with the internal cytoplasm, so that osmoregulation is not a problem. In contrast, most species of freshwater protozoans have an internal osmotic pressure well above that of their environment. This causes an influx of water by osmosis. More water is taken up in the ingestion of food particles by phagocytosis. Like the algae,

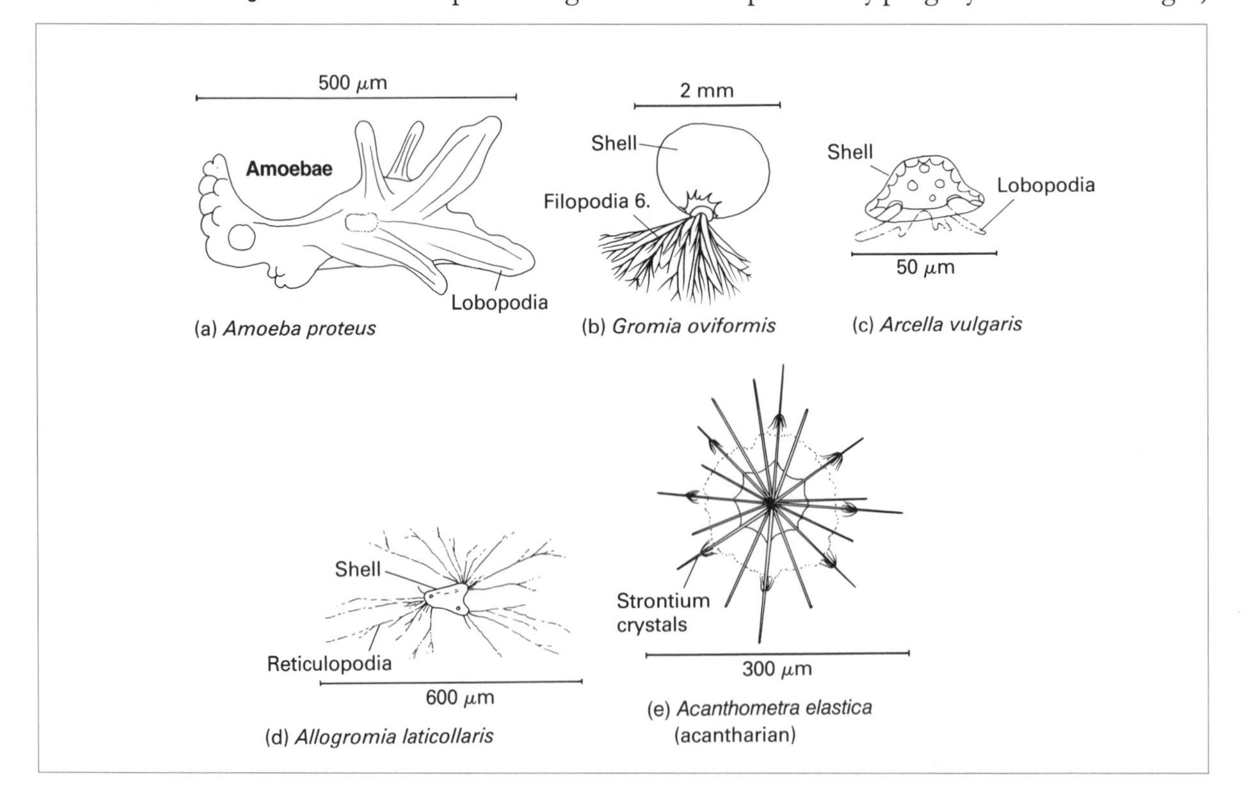


**Figure 9.10** Variations in form of the protozoa. Redrawn from Sleigh, M. (1991) *Protozoa and Other Protists*, Cambridge.

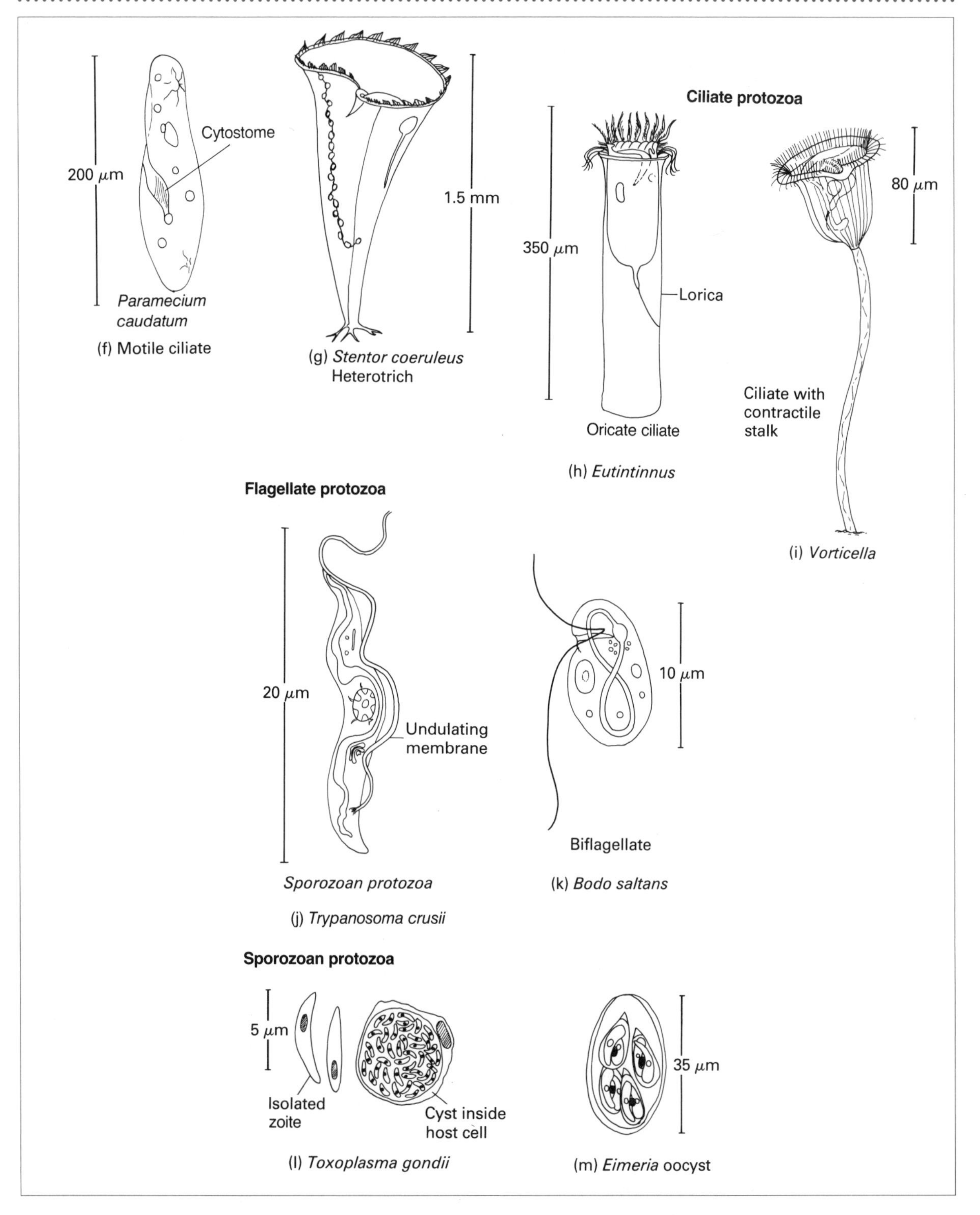


**Figure 9.10** *continued*

**BOX 9.3**

**Contractile vacuoles and osmoregulation**

In the absence of a cell wall, protists reduce the amount of water in a cell by expelling it through contractile vacuoles. Contractile vacuoles are surrounded by a membrane, and there may be one or more in each cell.

Simple contractile vacuoles appear close to the cell membrane, at any point in the cell, and they are surrounded by smaller vesicles, microtubules and mitochondria. They swell slowly as the smaller vesicles fuse with them, emptying their contents into the lumen of the contractile vacuole. This swelling is known as **diastole**. When the vacuole reaches a critical size it contracts rapidly (**systole**), expelling its contents outside the cell as its membrane fuses with the plasma membrane. This type of contractile vacuole is seen in the Amoebae.

A more complex contractile vacuole is found in species of protist with pellicles. Contractile vacuoles in these species are in fixed positions, and have a more complex anatomy with permanent pores in the pellicle. Several canals, supported by microtubules, radiate from the vacuole into the cytoplasm.

Contractile vacuoles are more active in freshwater protists than in marine or symbiotic species because the osmotic pressure of the cytoplasm is much higher than the fresh water round it.

Protists and fungi with cell walls adjust to changes in the osmolarity of their surrounding media in a different way. The cell wall forms a rigid support for the plasma membrane which prevents it from taking up too much water, rather like trying to inflate a balloon in a cardboard box.

If these organisms are placed in a strong sugar or salt solution the initial response of the cell is to *plasmolyse*, the plasma membrane pulling away from the cell wall as water is extracted from the cell by osmosis. The cell may recover from this state by accumulating salts or sugars in its cytoplasm until equilibrium is reached again.

freshwater protozoans have problems with osmotic regulation, and often have contractile vacuoles.

Close inspection of the apparently naked forms reveals that they are not quite as vulnerable as first appears. Some protozoa have a layer of cytoplasm immediately below the cell membrane that has a more rigid nature than the rest. In the amoebae it is a diffuse, ill-defined layer from which lobes of protoplasm called **pseudopodia** are extended. These can be simple, lobed structures, as in the Amoebae, or they can have an internal skeleton, as in *Acanthamoeba.* Variations in the structure of pseudopodia are illustrated in Figure 9.11.

In other species, the layer of cytoplasm beneath the cell membrane may be well-defined, forming a pellicle like those in the alga *Euglena* (page 243). This gives the cell considerable strength, and the pliability necessary to move through sand-grains, sediments and plant and animal tissues. The structures of the flagella and cilia are much like those of the motile algae and fungi. In some species they are combined to form undulating membranes, a common feature in the parasitic forms (Figure 9.10j).

Other species of protozoans have rods or plates of calcium, silica or strontium crystals beneath their cell membranes (Figure 9.10e). Some species have incomplete extracellular structures made from a mucilage

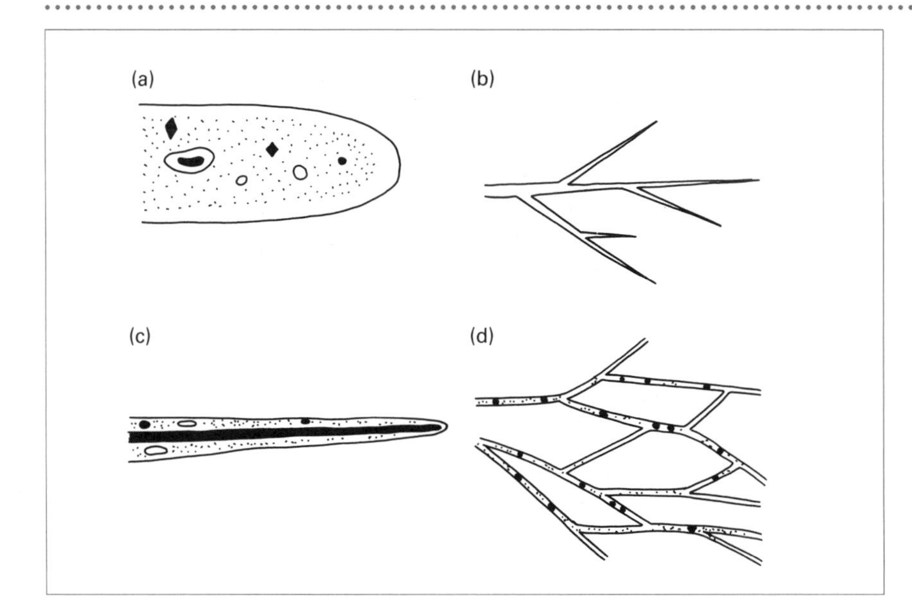


**Figure 9.11** Different types of pseudopodia. Redrawn from Sleigh, M. (1991) *Protozoa and Other Protists*, Cambridge.

excreted from the cell. This material is described as **pseudochitin** or **tectin**, and the structures are termed **loricas** (Figure 9.10h). **Keratin** stalks may be secreted by either loricate or naked forms to support themselves (Figure 9.10i). When the external layer is not complete but has a large opening it is termed a **shell** or **test** (Figure 9.10c).

Many species of protozoa form cysts or spores, the walls of which are commonly composed of phosphorylated proteins with an inner layer of cellulose or chitin. In the rabbit parasite *Eimeria* (a coccidian), the cyst is covered in a tannin–protein complex, with an inner layer of lipoprotein (Figure 9.10m). This extremely resistant cell wall accounts for *Eimeria*'s extraordinary tolerance of adverse conditions. The spore will not become active until eaten, when the spore wall is dissolved by the gastric juices of the rabbit small intestine.

Some kinds of the protozoa house internal symbiotic algae and have become partly phototrophic. For example, the radiolarians are protozoa found in the tropical plankton. These depend upon symbiotic dinoflagellate algae for their carbon compounds. However, the majority of the protozoa are entirely heterotrophic and, like the fungi, depend on external sources of carbohydrates for nutrition.

Feeding in the protozoans may be by the absorption of soluble nutrients such as carbohydrates and amino acids by diffusion, or by the binding of these compounds to the plasma membrane followed by their intake as membrane-bound vacuoles (pinocytotic vesicles). This is termed **saprozoic nutrition**. Larger particles of food can be ingested in phagocytotic vesicles – so-called **holozoic nutrition**. In those species with a pellicle, vesicles are generated at specific points on the plasma membrane called **cytostomes** (Figure 9.10f) or all over the cell membrane from **food cups**, as in the amoebae and some flagellates. Like the fungi and the algae, protozoans oxidize glucose by glycolysis and the TCA cycle, with electron transport producing the necessary ATP. They are also able in reduced oxygen environments to oxidize glucose using fermentative pathways, like the bacteria and the fungi.

Protozoa may have one nucleus per cell, two identical nuclei per cell, or even two distinctly different nuclei per cell – one large **macronucleus** and one small **micronucleus**. The macronucleus appears to be in control of non-reproductive functions. It is polyploid and contains large quantities of histone and non-histone proteins. Protein synthesis is dependent on macronuclear RNA, and this is contained in large nucleoli. The micronucleus controls reproduction. It is diploid, small and compact and contains no RNA and no nucleolus.

### Growth in the protozoa

Like the algae and the yeasts, protozoa are unicellular, and growth can take place only by cell division. As with the algae, we will look at protozoan growth in the section on Reproduction.

## Reproduction in the fungi and the Protista

All living things have a limited life-span and will die eventually, be it from accidents, disease or old age. Life continues because organisms have the ability to reproduce, giving rise to new organisms with the same basic characteristics as their parents. Central to this idea is the conservation of information at cell division, by **mitosis** and **meiosis**, during which DNA is accurately copied so that offspring are like their parents.

Cells can have one set of chromosomes, two sets or many sets. When cells have only one copy of each chromosome they are said to be **haploid**. If they have two copies of each chromosome they are said to be **diploid**. Cells that have *multiple* copies of chromosomes are termed **polyploid**.

Unlike the prokaryotes, eukaryotic microorganisms contain so much nuclear material that it cannot exist as a simple, circular chromosome, as it does in the prokaryotes. In eukaryotes, the DNA is packaged into a number of separate bodies called chromosomes, together with specialized nuclear proteins called **histones**, and surrounded by a nuclear membrane. Because of the increased complexity of organization of the nuclear material in eukaryotes, nuclear division is necessarily more complex. Mitosis is the process by which the nuclear material of a cell divides to give rise to new cells with the same number of chromosomes as the parent cell. This is the type of cell division seen in vegetative reproduction, which gives rise to growth.

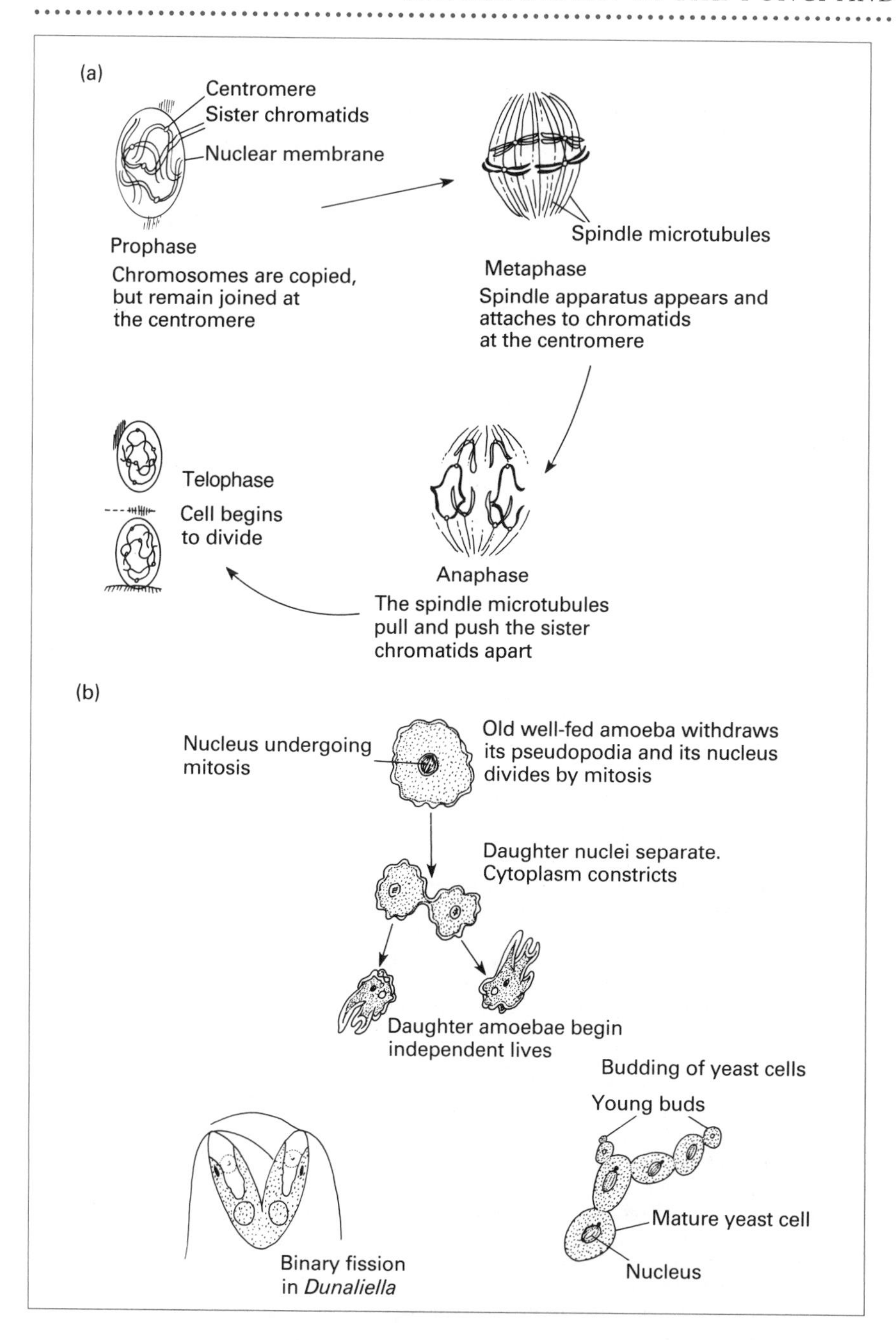


**Figure 9.12** (a) Nuclear events of mitosis. (b) Cell division after mitosis. Redrawn from Beckett, B.S. (1978) *Biology*, Oxford University Press, Oxford.

**(a) Mitosis.** Mitosis is a continuous process but can, for convenience, be divided into four phases: **prophase**, **anaphase**, **metaphase** and **telophase**. Chromosomal events during these phases are illustrated in Figure 9.12a. During prophase, the chromosomes are copied so that each chromosome becomes a pair of **chromatids**, joined at one point called the **centromere**. The centromeres are also the points where the microtubules of the **spindle apparatus** attach themselves to the chromatids. These microtubules appear during the next phase of mitosis, metaphase. During the third phase of mitosis, anaphase, they pull and push the chromatids apart, starting at the centromere. At the end of telophase, one pair of each sister chromatid finishes up at the end of the cell and, during telophase, plant cells begin to form a dividing cell wall between the daughter cells. Animal cells constrict across the mid-line to form an hour-glass shape and the plasma membranes fuse eventually to separate the cells.

Mitotic division of the nuclei followed by division of the cell into two is called **binary fission. Budding**, in which the nucleus divides mitotically but the daughter cells are considerably smaller than the parent, occurs in both the fungi and the protozoa. **Multiple fission** occurs in the parasitic protozoa as well as binary fission. Here, the nucleus divides mitotically many times giving rise to multiple daughters rather than just two. These events are illustrated in Figure 9.12b.

**(b) Meiosis**. Meiosis, followed by fertilization, gives an opportunity to increase **variation** in a population. Moreover, it offers cells the ability to

**BOX 9.4**

**The importance of meiosis in the protists**

Meiosis is the process by which cells with two sets of chromosomes (diploids) reduce their number of chromosomes to one set (haploid). Daughter cells of meiosis have half the number of chromosomes in the parent cells. The original number of chromosomes is restored only when two of these haploid cells fuse. Sexual reproduction must be preceded by meiosis, otherwise cells would have an ever-increasing number of chromosomes. At fertilization, two unique sets of genes come together, maximizing opportunities for genetic variation. In some protists there are complex devices that encourage cross-fertilization, such as multiple mating types, so that the greatest possible difference will exist between parental genes.

During meiosis there are opportunities for genes to be exchanged between homologous chromosomes, allowing for variation in the offspring. Mutation can also occur during these events, either through mis-copying of the DNA during duplication, or by the activities of mutagenic agents like ultraviolet light, chemical agents or X-ray radiation. Mutation is the source of genetic change.

In haploid organisms mutant genes are expressed as soon as they arise because there is no other copy of the gene to mask the mutant's presence. If that gene is lethal then the daughter organism dies. However, in diploid organisms mutant genes can remain hidden by a dominant gene on the duplicate chromosome (termed an **allele**). Thus a diploid organism can have copies of genes that are not identical within its genome, and these might be useful when environmental conditions change. In unfavourable environmental circumstances, protists and the fungi tend to undergo sexual reproduction to 'escape'. During sexual reproduction the different copies of the genes go to different daughter cells. If one set of genes has a mutation that allows for, say, enhanced temperature-tolerance when the environment has become too warm for the parent cells to compete successfully in the environment, that particular daughter cell may fuse with another gamete, and as long as that gene is dominant, the offspring will be temperature-tolerant and the species will compete more successfully.

rectify mistakes. It also allows for the resetting of developmental genes that have been turned off in the adult vegetative cell. Cells which have altered genes called **variants** (if the change is caused by the exchange of chromosomal material in cell division) or **mutants** (if the change is caused by an error) can be tested in the environment. If the variant or mutant has a selective value, that is, if it confers on its cell an advantage over other organisms, it can lead to an evolutionary change in the population. To understand this fully we need to look at the events of meiosis in detail.

Change in a haploid, asexual organism is slow, because the occurrence of a favourable mutation is very rare. Thus, it is a great advantage to an asexual organism if it is diploid or polyploid, because it can conserve genetic variability by having two or more copies of the same gene with slight differences.

There are eight identifiable stages of meiosis but, like mitosis, it is actually a continuous process. In the initial stages, duplicate copies of chromosomes (remember that meiotic cells are diploid) come together as **homologous pairs** and copies of the individual chromosomes are made, termed chromatids. These chromosome pairs contain almost exactly the same genes. The homologous pairs of chromosomes assemble at the **equator** (centre) of the cell, and a spindle pulls the pairs of homologous chromosomes apart. In the next phase of division, the chromatids are pulled apart on the spindle. These events give rise to a four-way split of the chromosomes. However, because there has only been one duplication, the number of copies of chromosomes in each daughter cell is reduced to one. These events are illustrated in Figure 9.13.

Not all of the fungi and Protista have the advantages of the full form of meiosis, with a two-phase meiosis and cross-fertilization (fertilization by a different individual). Some practise self-fertilization (**autogamy**), a process in which gametes from the same parent fuse to form a diploid. Autogamy allows for a *reassortment* of genes but, as this process brings together two identical sets of genes, there is no opportunity for acquiring the genetic variation that fertilization would bring. In some species of the Protista, crossing-over opportunities are limited because meiosis is a one-division process (the second phase only) which is probably the primitive type of meiotic division.

## Reproduction in the fungi

Each of the fungal groups is characterized by differences in the way they reproduce. In almost all cases this is by the production of structures called **spores**, and these spores contain products of either mitotic or meiotic nuclear division. With the nucleus in the spore are cytoplasmic organelles, including mitochondria, and energy reserves, such as glycogen and oils, that support respiration during dormancy and germination.

Asexual spores, produced by mitotic divisions, are probably the most significant way in which the lower and ascomycete fungi spread. Spores are

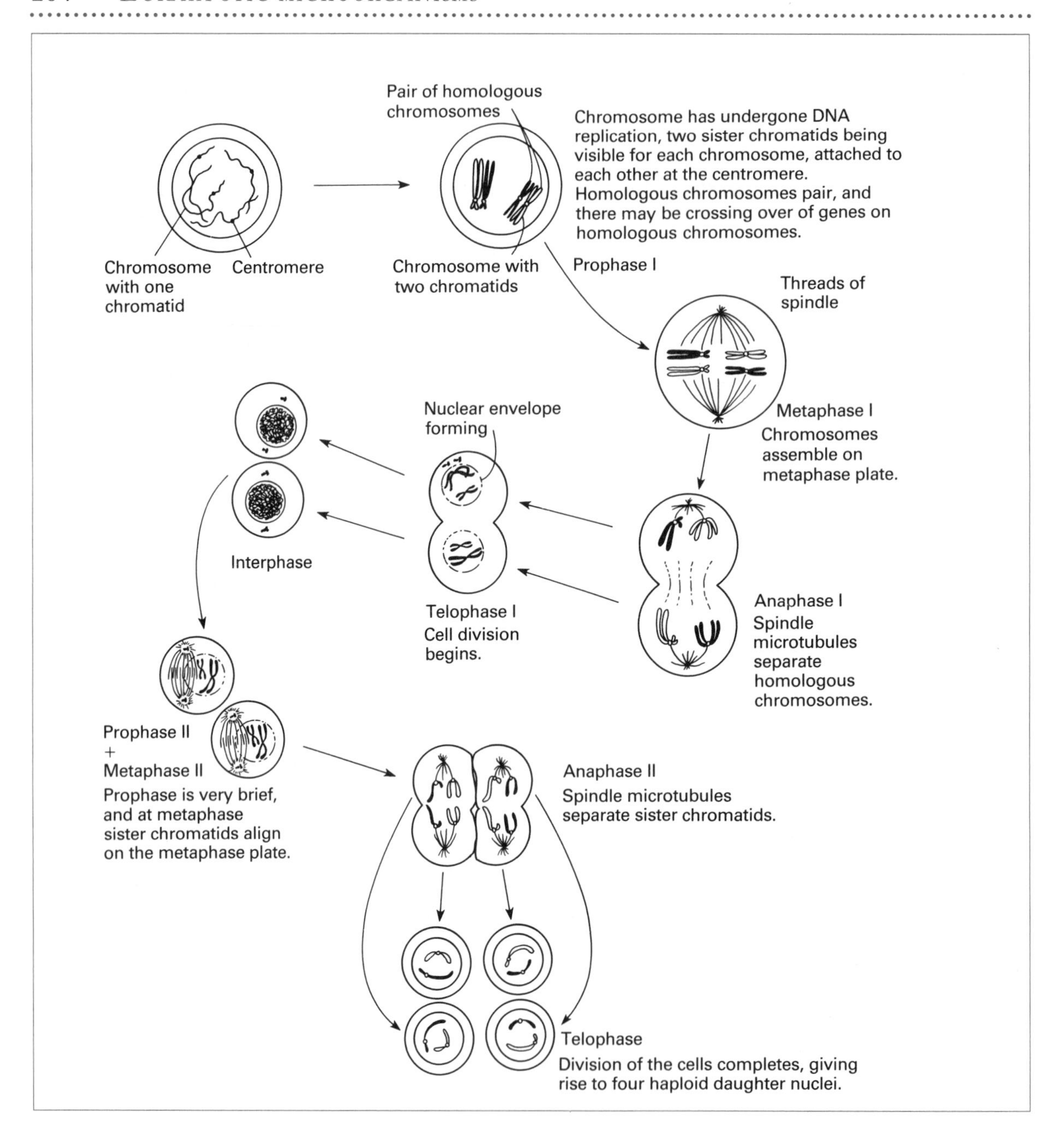


**Figure 9.13** Nuclear events of meiosis.

produced in great profusion during periods when growth conditions are good. Spore formation in the two groups probably had a common origin, but now differs in detail. In the lower fungi, asexual spores are formed in a body called a **sporangium**, a specialized hyphal tip in which multiple mitoses have occurred, giving rise to many haploid nuclei. This structure is supported by a hypha termed a **sporangiophore** (Figure 9.14). Each nucleus is packaged with a little cytoplasm and all of the necessary organelles, around which a spore wall is secreted. In some cases thousands of spores can be crammed tightly into the sporangium. The cell wall between the sporangium and the sporangiophore is termed a **columella**. As the sporangium ripens, the wall around it thins, and simultaneously the spores take up water. The pressure becomes so great that the wall ruptures and the spores are released. Many specialized mechanisms have evolved to ensure dissemination of the spores. The Oomycetes and chytrids, collectively termed the water moulds, have uniquely complicated life-cycles.

In the Ascomycetes and the closely related Fungi imperfecti, asexual reproduction is achieved by the production of spores called **conidia**. These are produced from the tips of aerial hyphae, either as a result of a **blastic** process where the tip softens locally and a spore is blebbed out, or as a result of a **thallic** process, where hyphae divide into a number of 'cells' with dividing walls, called septa, and then fragment (Figure 9.15). Once again, these spores contain mitotically produced nuclei and, like the sporangiospores, contain cellular organelles and respiration reserves. The conidiospores may be produced from single specialized hyphae called **conidiophores**, or the hyphae that bear spores may be aggregated into structures that are visible to the naked eye. Aggregated condiophores are

Asexual reproduction is the only method of division confirmed in the euglenoids, some flagellates and amoebae, and in the Fungi imperfecti. In spite of the limitations to variation and the dangers of lethal mutation, mitosis seems to provide these simple organisms with a perfectly satisfactory and successful system of reproduction. In the long run, however, asexual reproduction represents an evolutionary dead-end.

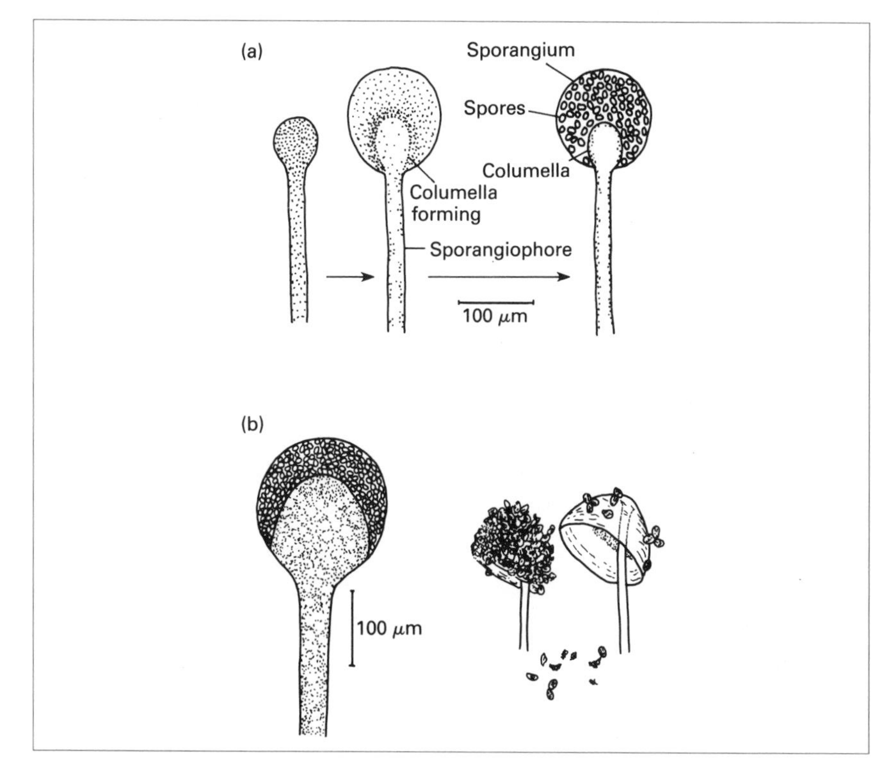


**Figure 9.14** Asexual reproduction in the Zygomycetes. The formation and dehiscence of sporangia. (a) Stages in the formation of a sporangium of *Phycomyces blakesleeanus*. (b) Sporangium of *Rhizopus stolonifer* before (left) and after (right) dehiscence. Redrawn from Ingold, C.T. and Hudson, H.J. (1993) *The Biology of Fungi*, Chapman & Hall, London.

**The dung fungi**

The dung of herbivores represents an excellent substrate for fungi to grow on, being rich in both carbon and nitrogen. Because it is such a nutrient-rich environment, there is great competition between species of fungi to colonize it, and many specialist mechanisms have evolved to give one fungus the edge over another in the race to get to it.

One problem that these fungi have is that they need to get into the alimentary tract of the herbivore, so that the dung emerges ready-inoculated with fungal spores. However, around a dung deposit there is a 'zone of distaste' where a herbivore will not graze. In order to ensure the ingestion of spores they must be placed on herbage outside this zone.

Several mechanisms have evolved to achieve this aim. The most spectacular are seen in the lower fungi, and are either based on a fungus 'gun' (*Pilobolus*) or on a stalked spore drop (*Pilaria*). In both cases the fungus is sensitive to light, and the sporangium is either shot towards or grown towards it. The sporangia of these fungi have two distinguishing features: they have a thick outer wall which is heavily pigmented, which protects them from ultraviolet light damage and damage in the animal's gut, and they have a sticky part which helps them adhere to plant leaves, ready to be eaten by the herbivore and start the cycle again.

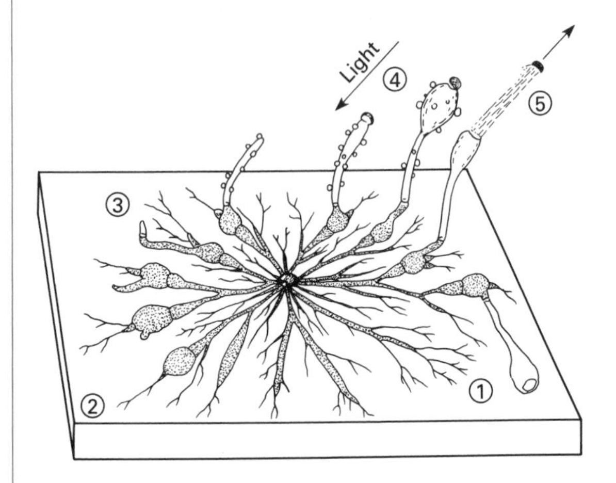


*Pilobolus:* Development of the fungus 'gun'. Stage 1. Formation of carotene-rich feeding mycelium. 2. Formation of trophocyst. 3. Sporangiophore emerges from trophocyst. 4. Development of sporangium and orientation towards light. 5. Explosive dehiscence caused by great internal pressure in the sporangiophore. (After Ingold (and Page))

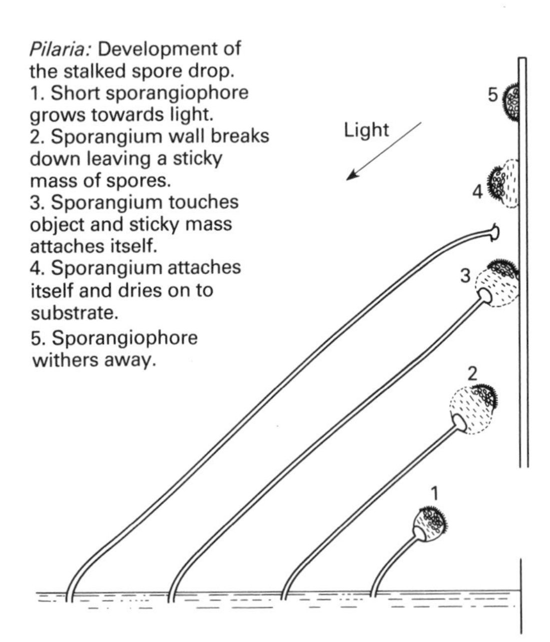


*Pilaria:* Development of the stalked spore drop.
1. Short sporangiophore grows towards light.
2. Sporangium wall breaks down leaving a sticky mass of spores.
3. Sporangium touches object and sticky mass attaches itself.
4. Sporangium attaches itself and dries on to substrate.
5. Sporangiophore withers away.

**BOX 9.5**

often found on plant tissues and substrates such as wood, and serve to protect the spores.

Although conidiophores and their aggregations are many and varied in their structure, they are relatively simple in design. They are also of a rather inconsistent character, in that a change in environmental conditions will change the sort of spore produced. Some of these fungi even produce two types of spores under identical conditions. This means that these asexual spores are of limited value in trying to classify this group of fungi. However, fungi in the Deuteromycetes have only asexual spores, so we have to use these structures to classify them.

**BOX 9.6** Redrawn from Deacon, J. (1984) *Introduction to Modern Mycology*, Blackwells, Oxford.

### The water moulds

The oomycetes and the chytrids are unlike any of the other members of the fungi. They have cell walls that contain cellulose rather than chitin and their mitochondria have tube-like cristae rather than the plate-like cristae found in the other groups.

They have many affinities with the algae; they reproduce like them, having a motile male gamete or a non-motile antheridium, which fertilizes a large egg or oogonium. They also usually inhabit aquatic environments.

They are important decomposers in water, but their greatest claim to fame is as pathogenic organisms. Some species are parasitic on fish gills, while others are pathogens of plants. The oomycete *Peronospora* species cause millions of pounds worth of damage to tobacco crops in the USA, and a close relative, *Phytophthora infestans*, was largely responsible for the Irish potato famines in the mid-1800s.

(a)

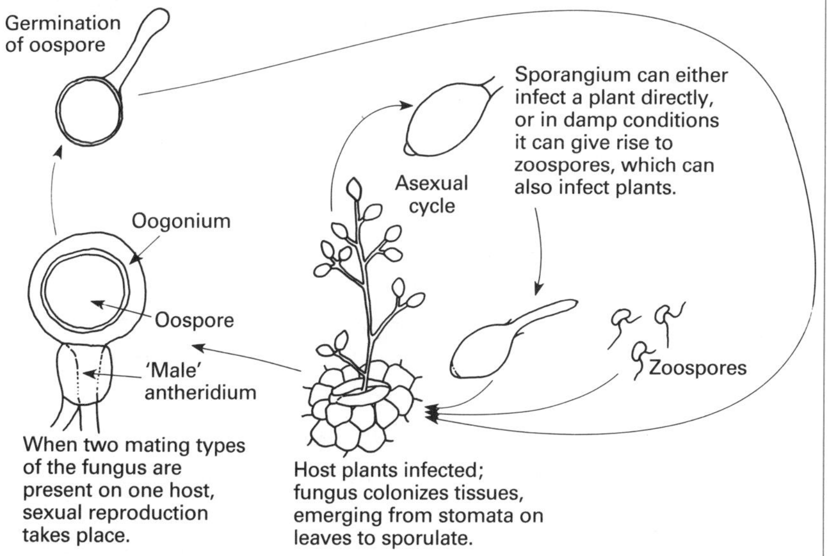

Germination of oospore
Sporangium can either infect a plant directly, or in damp conditions it can give rise to zoospores, which can also infect plants.
Asexual cycle
Oogonium
Oospore
'Male' antheridium
Zoospores
When two mating types of the fungus are present on one host, sexual reproduction takes place.
Host plants infected; fungus colonizes tissues, emerging from stomata on leaves to sporulate.

(b)

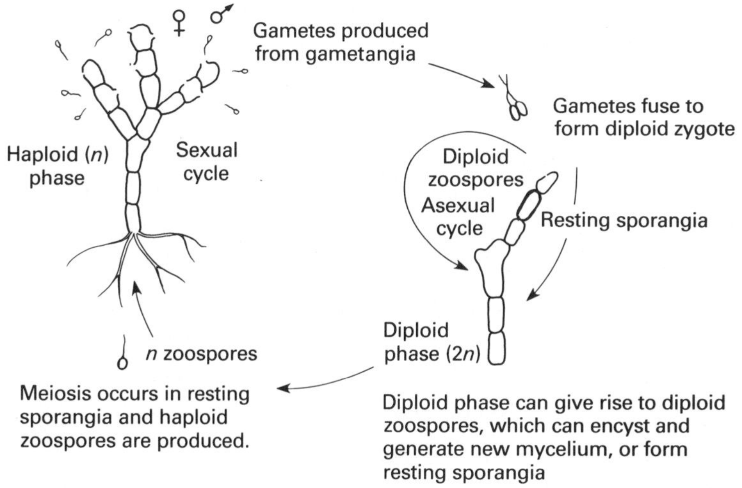

♀ ♂
Gametes produced from gametangia
Gametes fuse to form diploid zygote
Haploid (*n*) phase
Sexual cycle
Diploid zoospores
Asexual cycle
Resting sporangia
Diploid phase (2*n*)
*n* zoospores
Meiosis occurs in resting sporangia and haploid zoospores are produced.
Diploid phase can give rise to diploid zoospores, which can encyst and generate new mycelium, or form resting sporangia

Life cycle of the chytrid *Allomyces*

**Figure 9.15** Asexual reproduction in the Ascomycetes and Deuteromycetes. Redrawn from Ingold, C.T. and Hudson, H.J. (1993) *The Biology of Fungi,* Chapman & Hall, London.

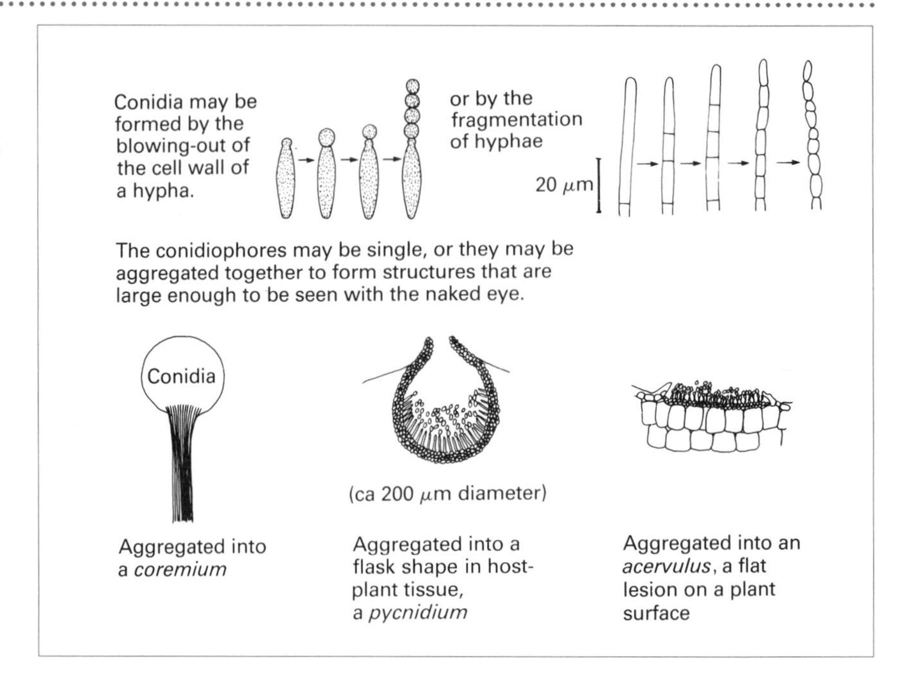


A fungal mycelium typically contains just one type of nucleus. To produce sexual spores it will need to fuse with a mycelium containing another, compatible, type of nucleus. Such fungi are termed **heterothallic**. In other species of fungi the mycelium contains at least two types of nuclei, and can produce sexual spores without having to seek out a mycelium of a different but compatible nucleus. These fungi are said to be **homothallic**. Sexual reproduction is broadly similar in both, but the sex organs are produced either on separate mycelia, or both from the same mycelium.

In contrast to the variability of the pattern of development of asexual spores, the production of sexual spores follows a much more constant pattern. A pre-requisite for sexual reproduction is the union of **compatible nuclei**, which, unless the mycelium already contains two types of nuclei, will require the fusion of the mycelia of two different mating types of mycelium.

Figure 9.16 illustrates the typical events that occur during sexual reproduction in the bread mould, *Mucor hiemalis.* This particular fungus is heterothallic, so mating between strains has to occur. The strains are designated + and − because there are no clear morphological differences between them. When the two strains are grown close to each other a series of molecular signals are exchanged, culminating in the formation of sexual organs called **gametangia** (singular, gametangium). The cell walls between the gametangia dissolve, the cytoplasms mingle, and the nuclei fuse. On fusion, the structure is called a **zygospore**, because it contains a **zygote**. A warty spore-coat is secreted around the zygospore, which helps it to survive adverse conditions. Meiosis of the diploid zygospore nucleus usually does not occur until just before germination. In *Mucor,* a tiny outgrowth called a **germ tube** emerges from the zygospore and immediately produces a sporangiophore with sporangia. The sporangia contain haploid spores that are disseminated away from the remains of the resting spore.

Reproduction in the Ascomycetes follows a slightly different pattern. These fungi also require mycelia of two different mating types to fuse, but

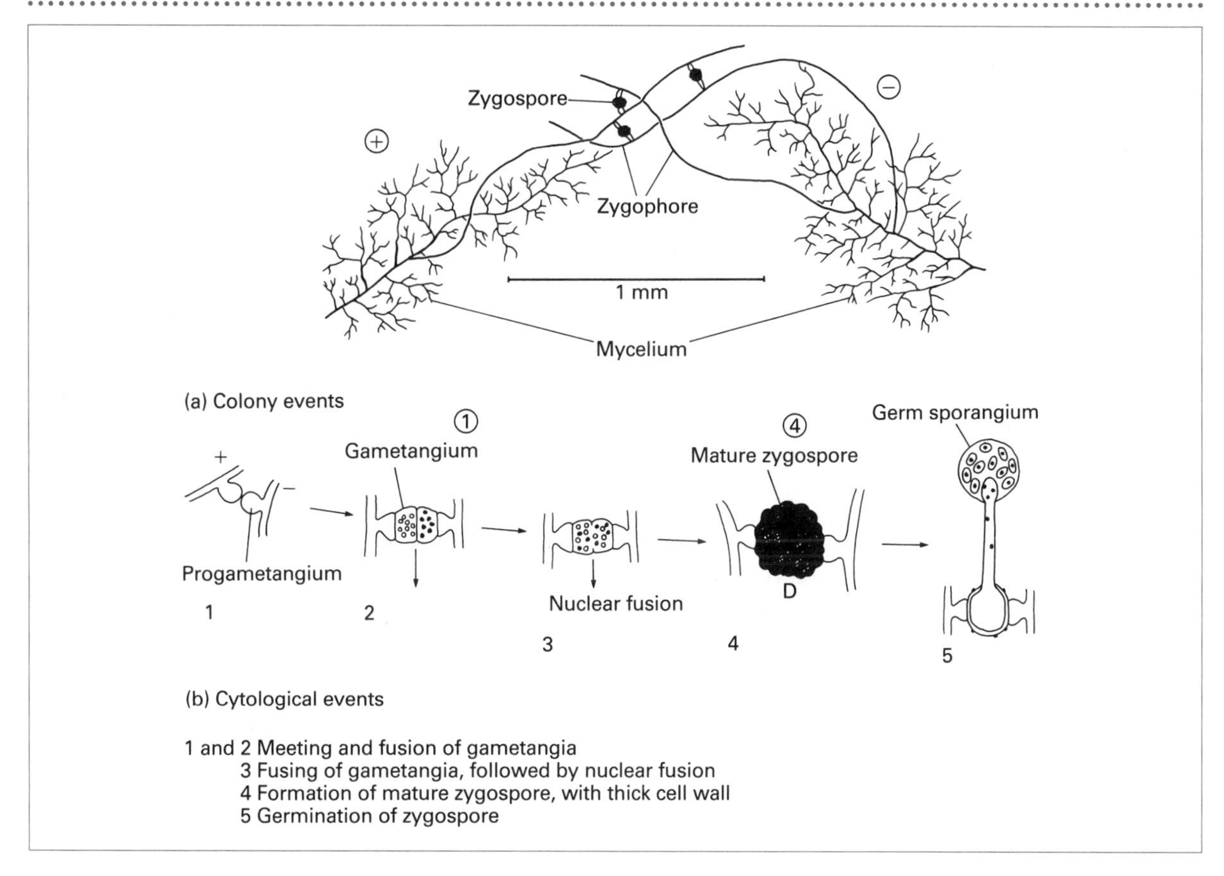


**Figure 9.16** Events in sexual reproduction in the mucorales. Redrawn from Ingold, C.T. and Hudson, H.J. (1993) *The Biology of Fungi*, Chapman & Hall, London.

only rarely do we see the sex organs. This is because they are buried within a mass of sterile tissue. Central to the onset of sexual reproduction is the formation of a body called an **ascogonium**, an enlarged cell containing both types of nuclei (Figure 9.17). From the ascogonium a specially modified

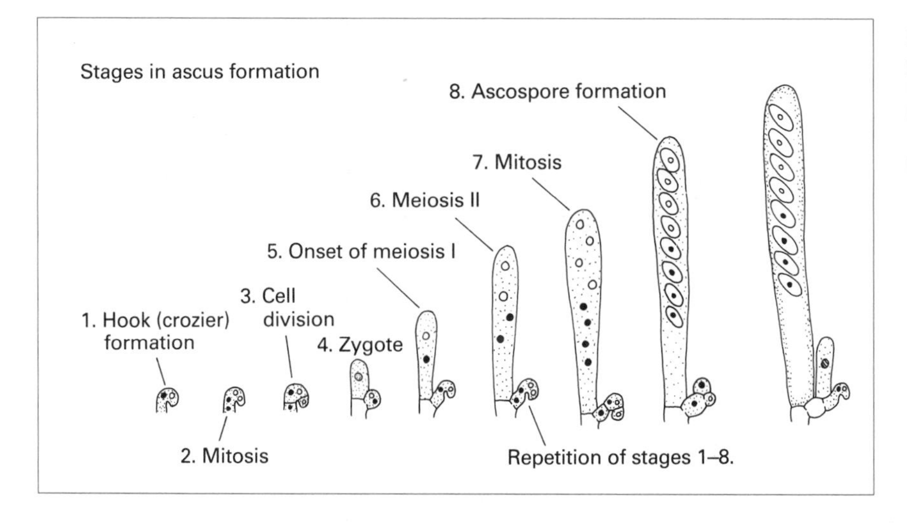


**Figure 9.17** Sexual reproductive processes of the Ascomycetes. Redrawn from Ingold, C.T. and Hudson, H.J. (1993) *The Biology of Fungi*, Chapman & Hall, London.

cell grows out with a bent tip (called a **crozier** because it looks like a shepherd's crook). A series of coordinated nuclear and cellular divisions takes place in the crozier, which leaves two nuclei of different mating types isolated in both the terminal and the sub-terminal cell of the crozier. These nuclei fuse and meiosis occurs. Meiosis is followed by mitosis, giving eight nuclei in all.

The cell in which these events take place is called an **ascus**, and the spores that form around the nuclei are called **ascospores**. The two nuclei that are left in the cell at the base of the ascus then go on to repeat the process many times. These large aggregations of asci are called a **hymenium**, or fertile layer. Protecting this fertile layer are varying amounts of sterile tissue. In the yeasts the ascus is naked and unprotected (yeasts live in comparatively 'friendly' environments such as fruit pulp, vegetable fluids and beer worts). In other groups of fungi which live in more hostile environments, elaborate structures have been evolved that both protect the asci and help them to escape the parent fruit-body once conditions have changed (Figure 9.18).

In the Ascomycetes, water is often harnessed to help the spores discharge. The actual mechanism of spore discharge uses water pressure to burst the ascus. The cytoplasm left in the ascus sac round the ascospores is

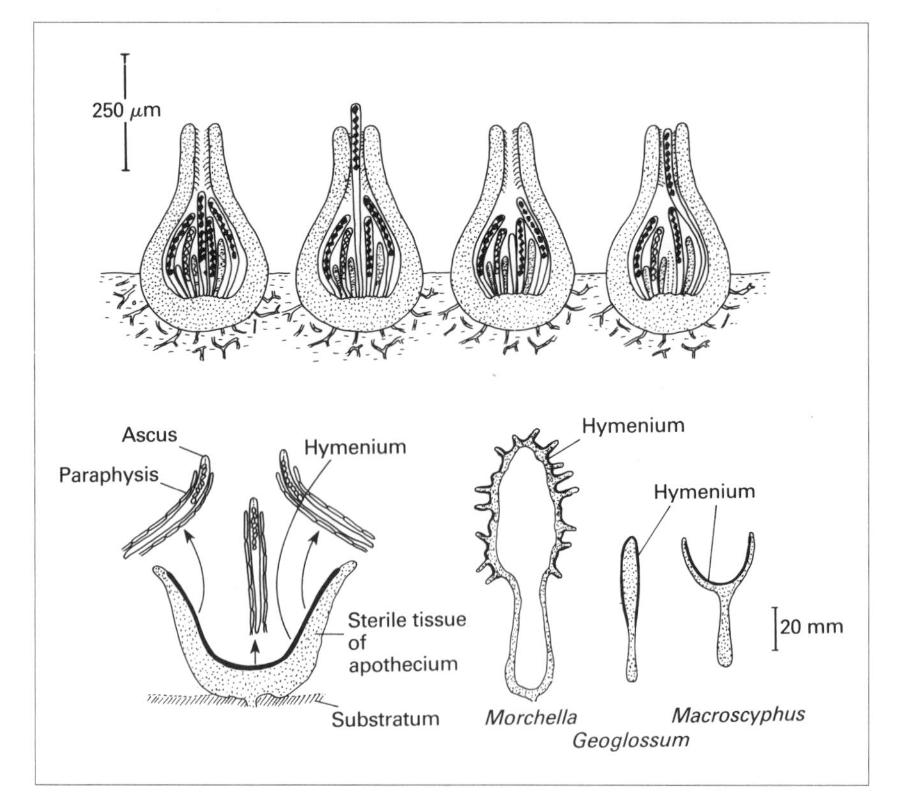


**Figure 9.18** Discharge of spores in the Ascomycetes. Discharge in the flask fungi (*Sordaria*) and the cup fungi (*Peziza*). The fertile layer (hymenium) can be arranged over sterile tissues of many shapes and sizes (for example, *Morchella*). Redrawn from Ingold, C.T. and Hudson, H.J. (1993) *The Biology of Fungi*, Chapman & Hall, London.

### The yeasts

The life-cycle of the brewer's yeast, *Saccharomyces cerevisiae*, is shown below. In sugary solutions such as beer wort or grape juice, yeast multiplies rapidly by budding. Because the supply of oxygen in these solutions is limited, yeast respires by glycolysis and anaerobic fermentation, producing carbon dioxide and ethanol (see pages 73–6).

Yeasts can be induced to form ascospores by starvation. In these conditions meiosis takes place, leading to the formation of four ascospores within the parent cell wall. The ascospores emerge from the wall once conditions improve and begin budding as haploid yeasts.

Haploid strains can fuse with each other to form a diploid cell, which continues its life-cycle by budding. It is the diploid phase which is used in the brewing and baking industries.

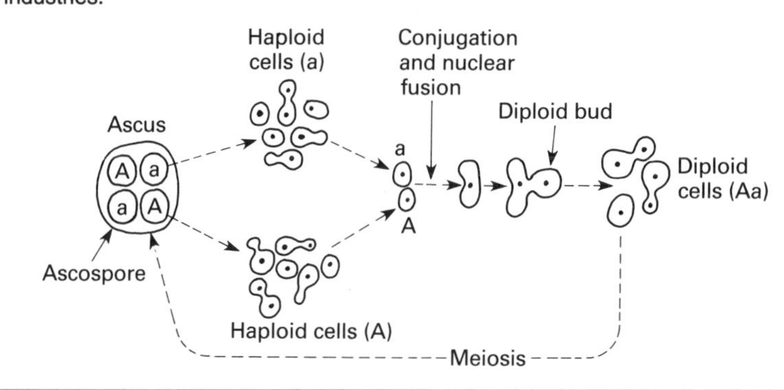


**BOX 9.7** Redrawn from Ingold, C.T. and Hudson, H.J. (1993) *The Biology of Fungi*, Chapman & Hall, London.

more or less devoid of organelles, but contains carbohydrate reserves. These are converted to simpler sugars when the hymenium is ripe, and these sugars cause the uptake of water into the ascus sac. The sac becomes turgid, and eventually bursts, in some species *via* a slit or pore, in others by a lid called an **operculum**.

Sterile tissues help to direct these discharges. In *Sordaria*, for example, the neck of the flask-shaped fruit-body (called a **perithecium**) grips each ascus in turn as it ripens, causing the maximum build-up of pressure. This energy propels the spores for a considerable distance on discharge (Figure 9.18). In the cup fungi, where there is a large flat expanse of hymenium, only the very tips of the asci extrude above the level of the sterile tissues. The tips of these hyphae are light-sensitive, and they point towards the light source and away from debris and overhanging materials. In some species the tips may bend towards the light; in others the operculum or a pore may form on the illuminated side of the fruit-body. All of these mechanisms are attempts to maximize spore discharge and dissemination.

In the other group of higher fungi, the Basidiomycetes, the fungal mycelium contains two mating types of nuclei for most of its life-cycle. Such a mycelium is said to be **dikaryotic**. In some fungi the spores themselves contain two nuclei of different mating types, so the germinating mycelium is immediately dikaryotic. In other Basidiomycetes the dikaryotic

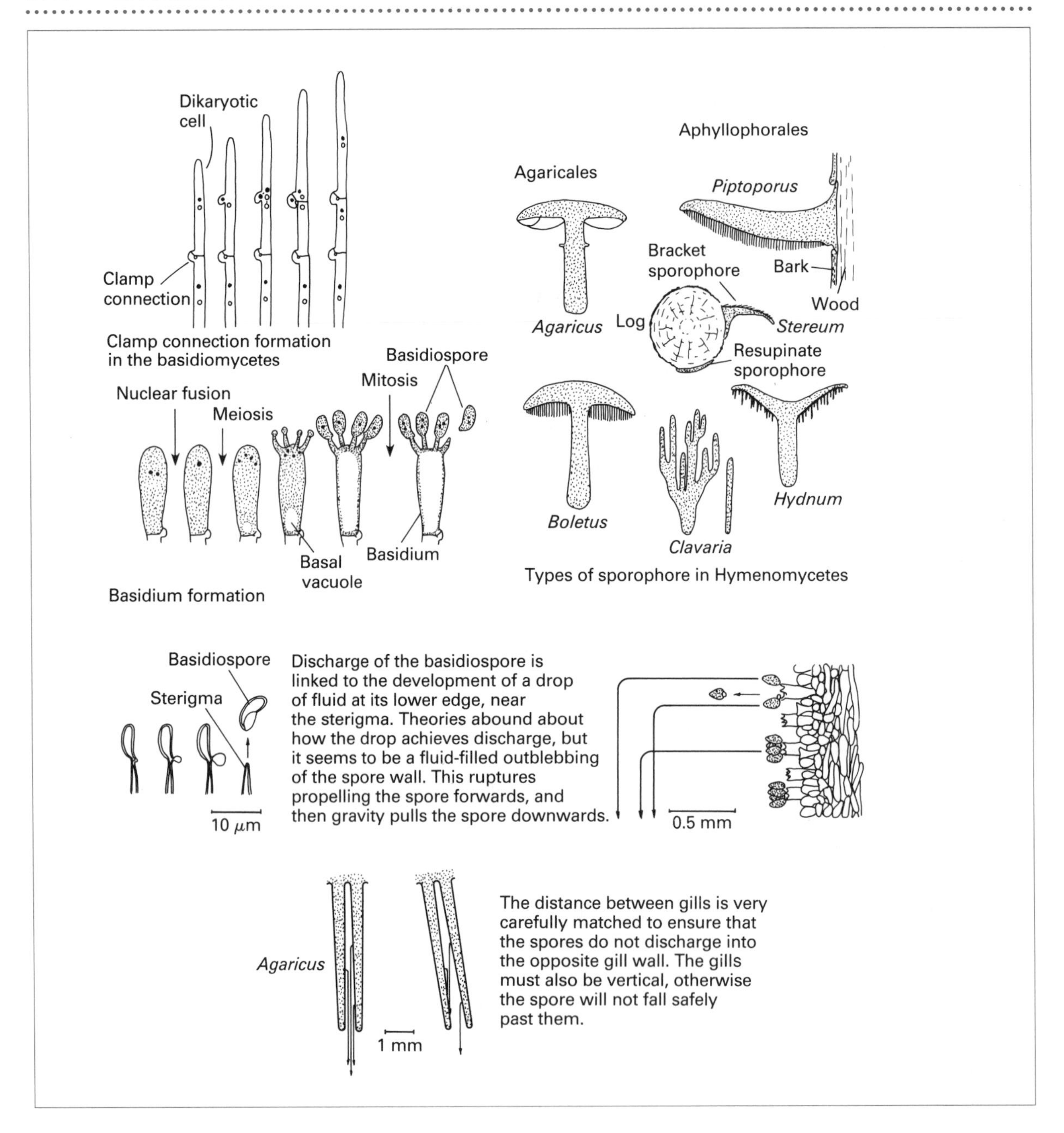


**Figure 9.19** Sexual reproduction in the Basidiomycetes. Redrawn from Ingold, C.T. and Hudson, H.J. (1993) *The Biology of Fungi*, Chapman & Hall, London.

state is achieved by the fusion of two different mating type mycelia at some point early in the fungus life-cycle. In general, Basidiomycetes do not produce asexual spores characteristic of the other groups. In some species the dikaryotic mycelium has a highly distinctive structure, and elaborate walls are formed to ensure that two different mating types of nuclei stay in one compartment at each cellular division (Figure 9.19). These structures are called **clamp connections** and they are somewhat reminiscent of the crozier in the Ascomycetes.

The stimulus to undergo sexual reproduction and fruit-body formation in Basidiomycetes is linked to a change in environmental conditions, although there may also be some sort of internal biological clock in some of the fungi because even when they are isolated from any outside influences they still undergo sexual reproduction at their usual time of the year. Whatever the stimulus, sexual reproduction appears to be initiated by the formation, in the dikaryotic mycelium, of a ball of sterile tissue called a **primordium**. Within the primordium, fusion of the two different nuclei takes place in specialized cells called **basidia**. Meiosis then occurs, creating four haploid nuclei. These nuclei migrate to four points at the tip of the basidium, where they are ballooned out through narrow necks called **sterigma**, together with cytoplasm, organelles and food reserves. These spores are termed **basidiospores**, and they secrete a thick, protective cell wall.

As in the Ascomycetes, fertile hymenia develop, surrounded by sterile dikaryotic tissues to protect them. Unlike in the Ascomycetes, however, the hymenium is very susceptible to water damage. For this reason, the basidiomycete fruit-body (**sporophore**) is constructed to protect the hymenium from water as well as to maximize the discharge and dispersal of the spores (Figure 9.19). Such requirements have led to the evolution of many wonderful forms of fruit-body in this group, including the familiar mushrooms and toadstools, the stinkhorns and puffballs, and the catapults and splash cups.

Despite the large size of some of the fungi, they have a very simple

**BOX 9.8**

**Mushrooms and toadstools**

Mushrooms and toadstools are the fruit-bodies of Basidiomycetes that we see every year under forest trees and on lawns. Many of them are in mycorrhizal association with the trees they are growing beneath.

The way in which these fungi discharge their spores is very sensitive to water damage, and their fruit-bodies have evolved to protect the spore-bearing hymenium from water damage: hence the familiar umbrella shape of the cap.

However, the design of the gill has one major fault. If the cap is anything but vertical, the discharged spores will fall not out into the air below, but on to the opposite gill. To overcome this problem both the cap and its stalk are able to adjust their positions to grow precisely vertically.

Many of these large fruit-bodies are edible, and some are delicious. Others are poisonous, and a few are deadly.

Unlike the short-lived fruit-bodies of the mushrooms and toadstools, bracket fungi produce fruit-bodies that are tough and long-lived. Their spore-producing hymenium covers masses of tiny tubes or spines, and has to respond to changes in orientation of the fruit-body by regrowing at the appropriate angle every year.

You will find bracket fungi on forest and amenity trees, and some of them cause serious damage, eventually killing them. For example, the horse's hoof fungus (*Fomes*) attacks and kills oak trees, and a much smaller bracket (*Chondrostereum purpureum*) causes silver leaf of pear trees and can devastate orchards.

**BOX 9.8** *continued*
Different fruit body shapes of the Basidiomycetes. (a) Toadstools (b) a morel and (c) fungi.

biology, which reveals them to be protists. They never develop true tissues. Even the largest of them still contains aggregations of mycelium cemented together by cell wall materials. Their life-cycles are very simple and only the water moulds (which probably should not really be in the Fungi anyway) show the complexities of life-cycle that we see in the higher plants.

## Growth and reproduction in the algae

Growth in the four major groups of algae is by binary fission, and reproduction either by conjugation or by fusion of gametes (sex cells), followed by meiosis. Details of these events are slightly different in each group, and we will look at each in turn.

Unicellular green algae grow by increasing their size until a limit is reached. The cell then divides by binary fission. In the case of

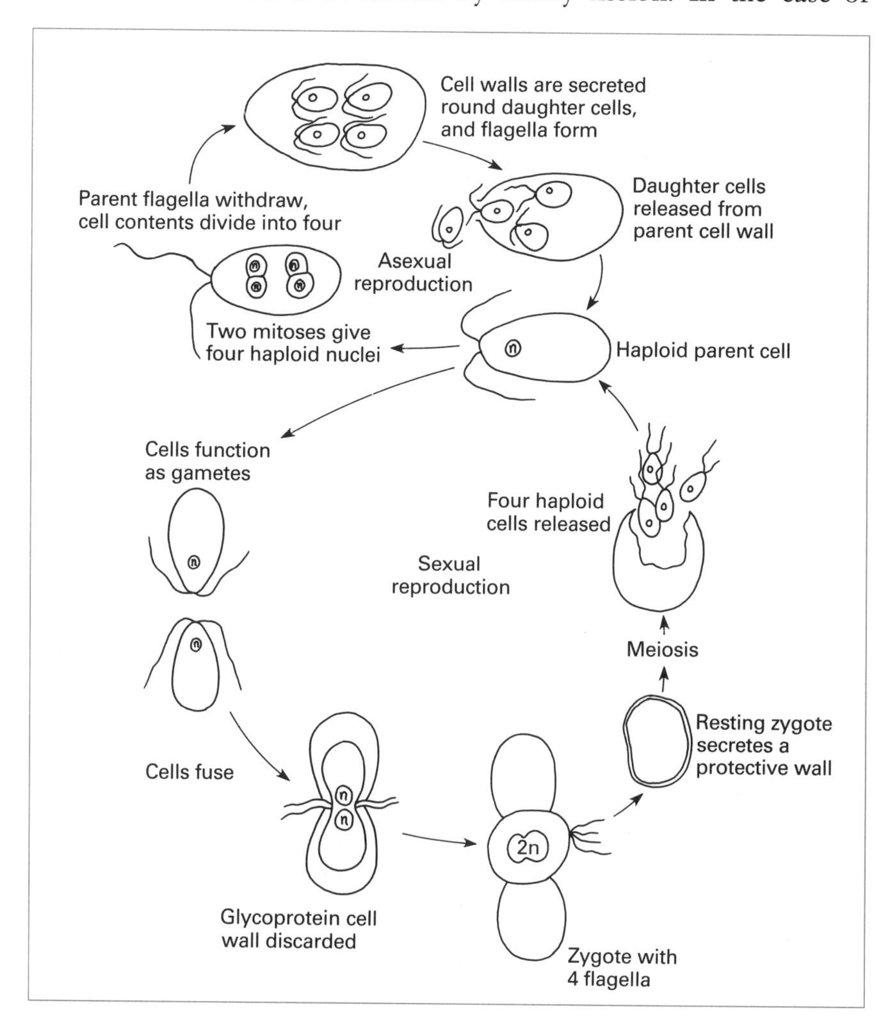


**Figure 9.20** The life-cycle of *Chlamydomonas*.

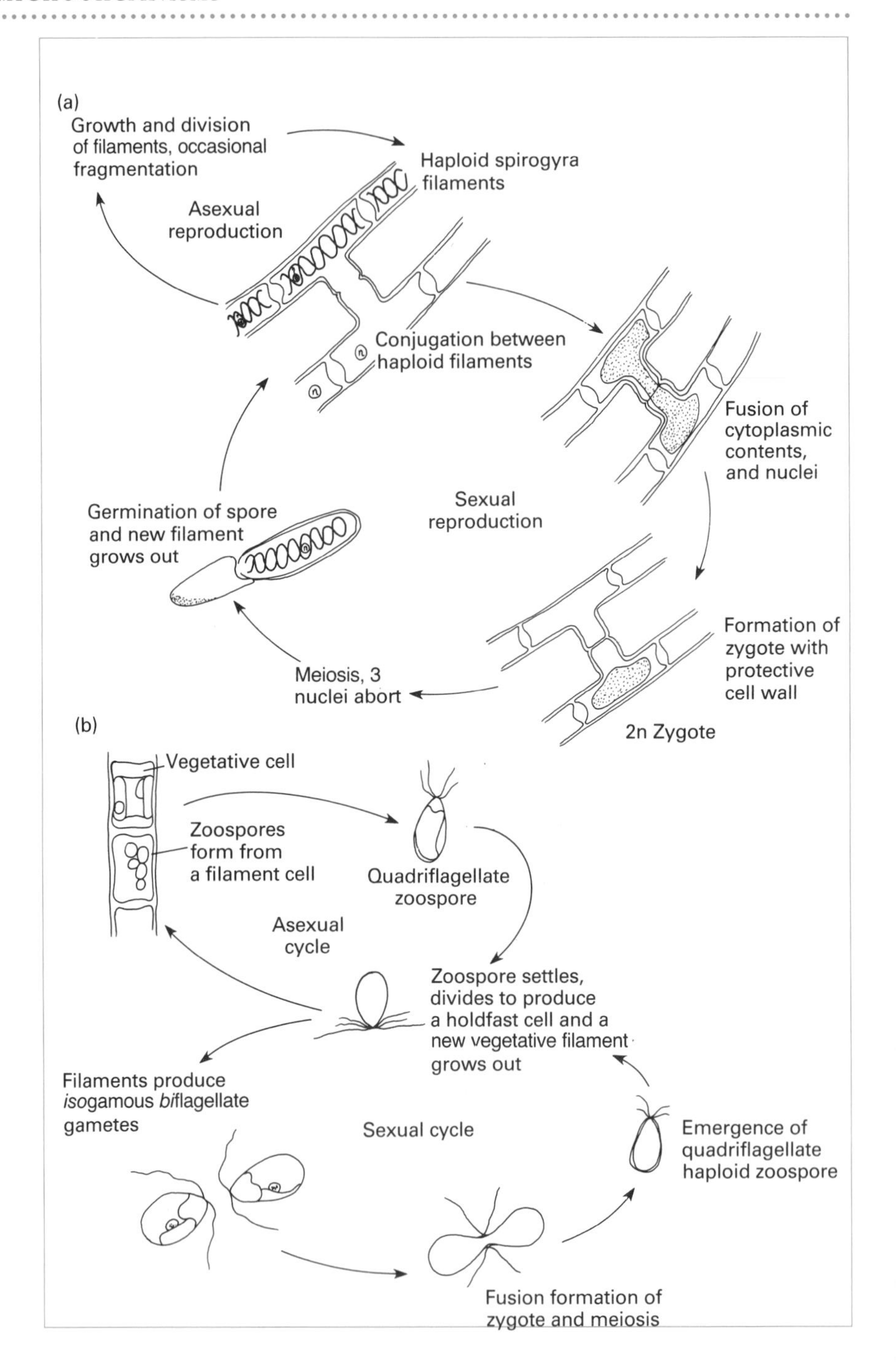


**Figure 9.21** The life-cycle of filamentous green algae. (a) *Spirogyra*; (b) *Ulothrix*

*Chlamydomonas*, the nucleus undergoes two haploid divisions to produce four daughter nuclei round which four equal-sized identical cells are formed (Figure 9.20). In some species of green algae only this type of asexual reproduction has been reported. *Chlamydomonas* is haploid. Sexual

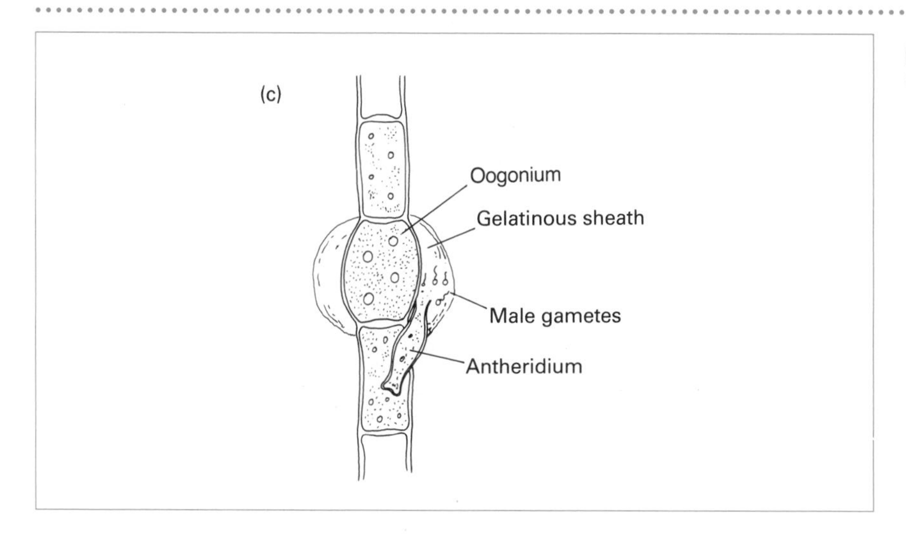


**Figure 9.21** (c) Oogamy in *Oedogonium*.

reproduction takes place when two motile cells act as gametes and fuse. The two nuclei fuse and divide by meiosis to produce four motile daughter cells.

Filamentous green alga species such as *Spirogyra* can grow from any cell by binary fission to produce two non-motile vegetative cells, extending the filament. These filaments can be broken by wave action and turbulence to help spread the species. Sexual reproduction takes place by means of the **conjugation** (fusion) of two vegetative cells (Figure 9.21a). The zygote which is formed encysts to form a spore which will eventually germinate to produce more *Spirogyra* filaments.

In some filamentous species two motile gametes may be produced from vegetative cells and these may fuse, as happens in *Ulothrix* (Figure 9.21b). In other species a motile gamete will fuse with a receptive non-motile cell in the filament, as in *Oedogonium* (Figure 9.21c). Such a process is termed **oogamy**. These structures and events bear a strong resemblance to those seen in the water moulds (page 267).

Cell division in other motile green algae is by longitudinal division, typically starting near the rear end of the cell and moving forwards, towards the flagellum. In species with plates on the outside of the cell, plates may either be shed before cell division, or shared between daughter cells and new ones synthesized to complete the cells' surfaces.

The golden brown algae, including the diatoms, are responsible for about 20% of the earth's primary production of carbon in the oceans. Reproduction in this group is predominantly asexual, by binary fission (Figure 9.22), but this is complicated by their possession of two, interlocking, shells of silica called **frustules**. At cell division, one side of the parent frustule goes to each daughter cell, and the second side of the frustule is

**Figure 9.22** The life-cycle of the diatoms.

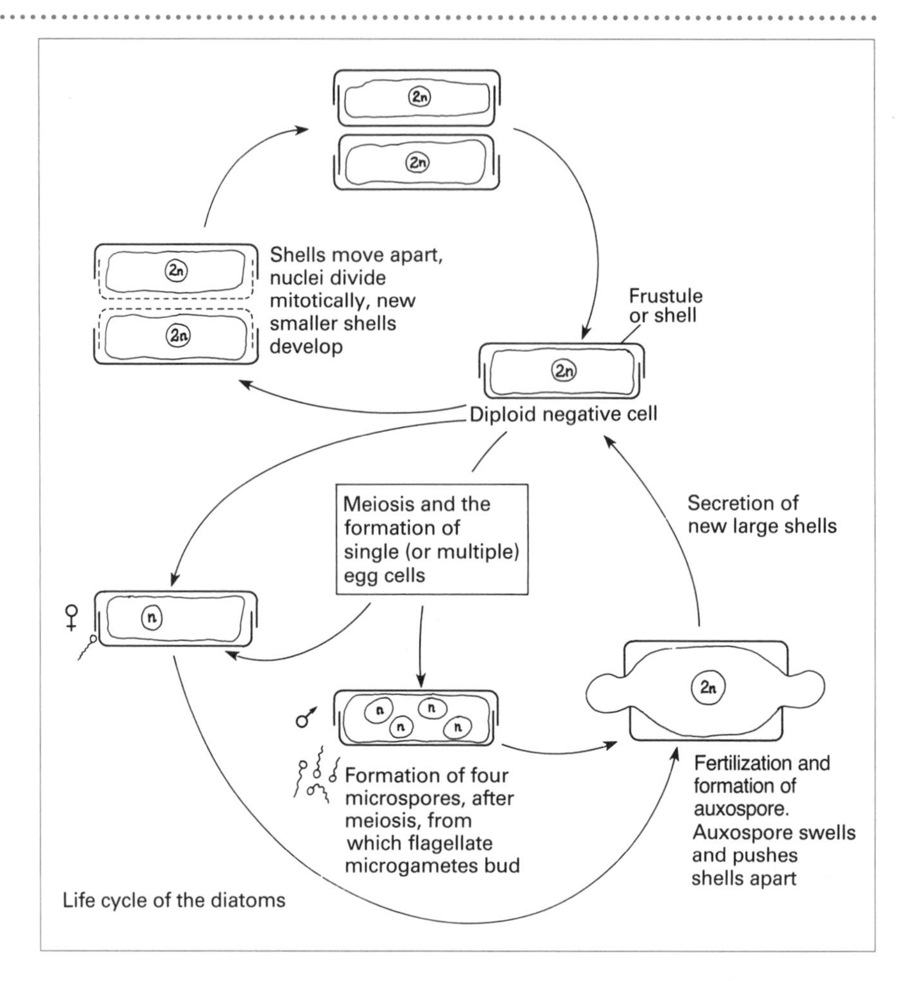


synthesized by the daughter cell to fit inside the one from the parent. So while one of the daughter cells will be the same size as the mother cell, the other one will be smaller because it inherited the smaller of the two frustules. During one growing season, cells reduce in size with each cell division from 400 μm in diameter down to 50 μm. Incredible shrinking diatoms overcome this problem by means of sexual reproduction.

Diatoms are diploid, and meiosis occurs within the small cells to produce gametes. In some species of **centric** (round) diatoms, whole cells are converted into either a single gamete (**oogonium**) or multiple, non-motile gametes (**oogonia**), or into motile microgametes. Once fusion of the gametes has occurred, the zygote forms a body called an **auxospore**, which enlarges and, on germination, secretes a new, large pair of frustules. In other species of diatoms, particularly the long, thin **pennate** diatoms, pairs of diatoms conjugate and then the products of fusion form an auxospore.

The dinoflagellates and euglenoids also reproduce asexually by binary fission. Some of the dinoflagellates have a pellicle like *Euglena* and its

BOX 9.9

**The secret life of the dinoflagellates**

Dinoflagellates have two unusual properties. One is dangerous to higher animals and man, and the other useful.

Dinoflagellates can grow very quickly, and in conditions where nutrients are plentiful their numbers can become so high in coastal waters that they give colour to the sea. The so-called 'red tides' are caused by the carotenoid and xanthophyll pigments in their cells.

Some of the species causing red tides contain toxic materials. Toxins fall into three groups: those that kill primarily fish, those that kill primarily invertebrates, and those that kill few marine animals but concentrate in the digestive glands or siphons of filter-feeding bivalves, such as clams and mussels. When such foods are eaten they can cause paralytic shellfish poisoning.

In their favour is the fact that some of the dinoflagellates are bioluminescent. They have the capability to store energy in a chemical from and release it later as light. Some dinoflagellates contain the luminescent compound *Luciferin*, which emits light when split by the enzyme luciferase. There are two types of luminescence: flash emission, which occurs when cells are shaken, and glow emission which occurs without mechanical stimulation and seems to be related to a diurnal cycle, peaking in the middle of the dark part of the cycle. We can use these bioluminescent compounds in disease detection.

relatives, but additionally many species are covered in multiple plates which may either be shed or inherited by the daughter cells, as in the green algae. They can grow so quickly in certain circumstances that so-called 'blooms' can occur which give rise to coloured tides.

Dinoflagellates can also reproduce sexually. Most, but not all, are haploid, and produce gametes by means of mitosis of vegetative cells. Their life-cycle

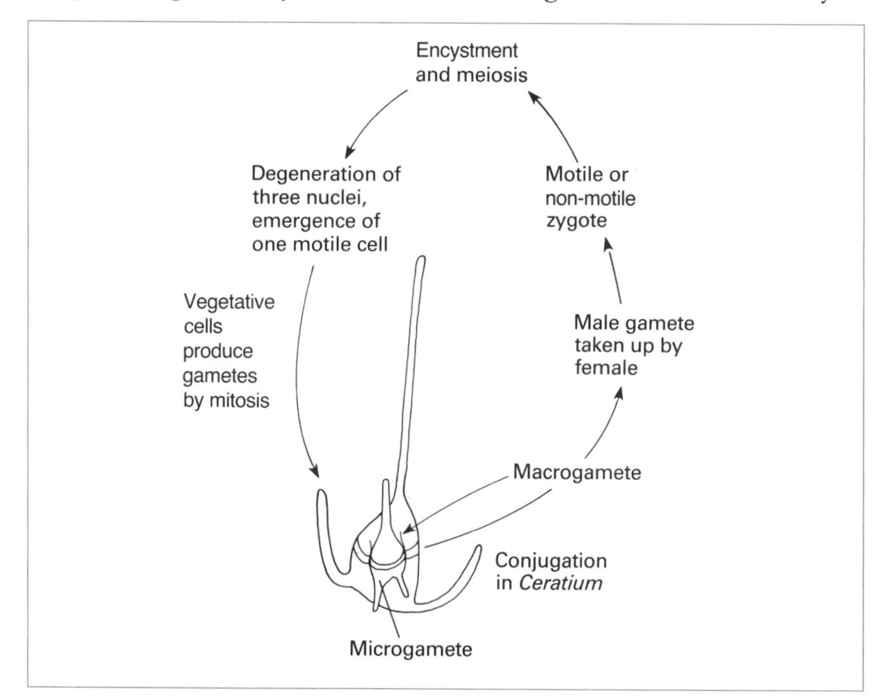


**Figure 9.23** The life-cycle of a dinoflagellate.

is illustrated in Figure 9.23. In those species that are diploid, meiosis occurs in the formation of the gametes, and the diploid state is regained in the zygote.

## Growth and reproduction in the protozoa

As all protozoa are unicellular, they must grow by cell division. Binary fission is the most common method of asexual reproduction, although budding is also found in some species. Details of growth and reproduction differ in each group so we will look at each of the four groups in turn.

Amoeboid protozoans have a world-wide distribution, living free in water and soil or as parasites of animals and man. Only asexual reproduction has been seen in some of these species, and this is by binary fission. Amoebae have the ability to encyst to avoid adverse conditions, and in parasitic species the encysted stage may be the means of transfer between one host and another. **Excystation** (escape from the cyst) occurs when the external environment has improved. In parasitic species the degradation of the cyst wall depends on the action of host gut enzymes.

Amoeboid protozoa include the **slime moulds**, often placed in the fungi because of their habit of producing spores. There are two very different groups within the slime moulds, the cellular slime moulds and the acellular slime moulds. Both kinds of slime moulds have a vegetative stage during which they feed and grow on soil bacteria. However, the cellular slime moulds form a colony of aggregated amoebae called a **pseudoplasmodium**, though the cells remain as individuals. In contrast, the acellular slime moulds form a multinucleate plasmodium as a result of mitotic division of a diploid nucleus from one cell. Details of the life-history of the cellular and acellular slime moulds are shown in Figure 9.24.

Protozoa which exhibit sexual reproduction form haploid gametes. If the gametes are of unequal size the gametes are called **anisogametes**. The larger is called a **macrogamete** and the smaller is called a **microgamete**. If the two mating type gametes are identical they are termed **isogametes**. The process of gamete fusion is termed **syngamy** when it is in the external environment, and **autogamy** when it occurs within the cell of one parent. If there is an exchange of gametes while two parent protozoa are fused together it is called **conjugation**.

In other free-living amoebic forms, such as the Foraminifera, the life-cycle is very simple, with mature haploid amoebae producing zygotes which grow into diploid cells. Meiosis occurs in these cells, giving rise to the haploid generation again.

Asexual reproduction in flagellate protozoans is by mitosis followed by binary fission, which occurs longitudinally along the longest body axis, very like the euglenoids. Sexual reproduction is reported in some groups. There are both free-living and parasitic forms of flagellate protozoa. The parasitic species include the trypanosomes, leishmanias, giardias and crithidias, which are all very important pathogens of man and animal. Figure 9.25 illustrates a typical life-cycle of one of these parasites.

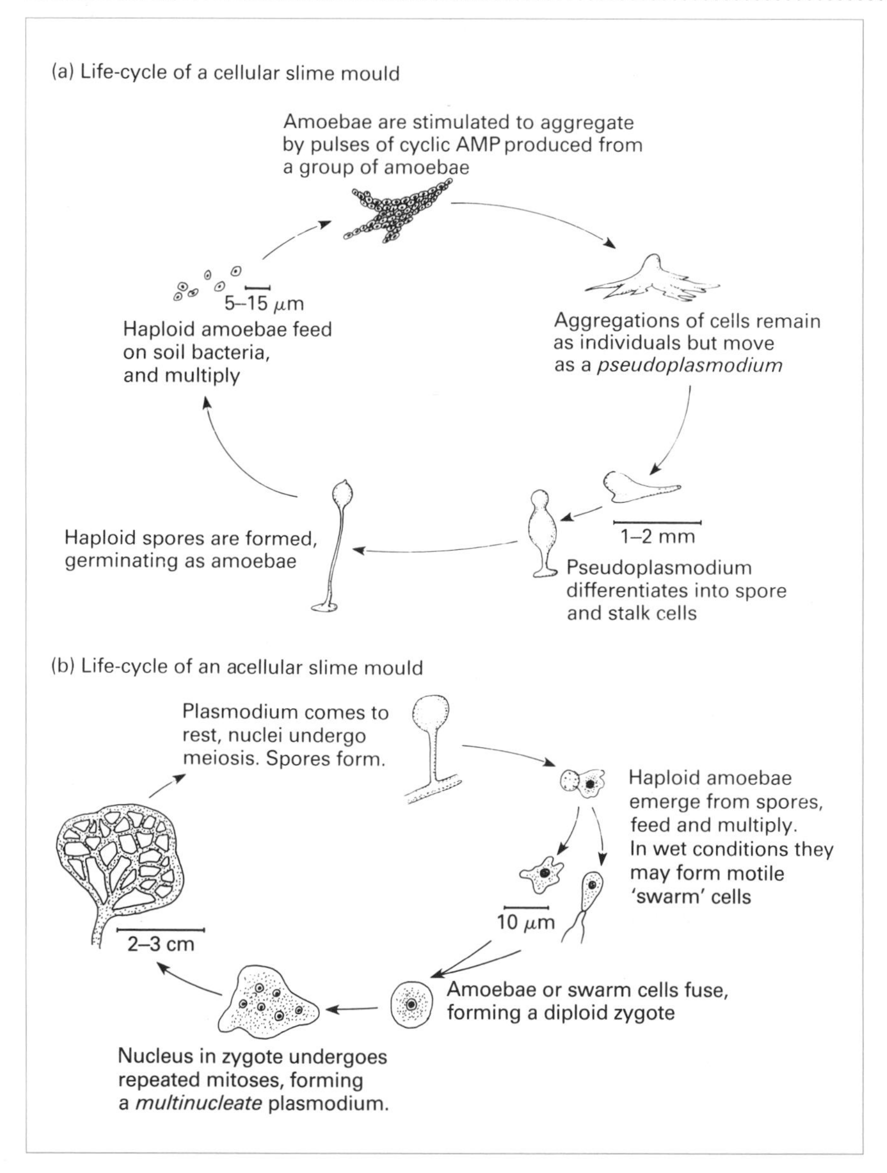


**Figure 9.24** The life-cycle of the slime moulds. (a) Cellular slime mould; (b) Acellular slime mould. Redrawn from Sleigh, M. (1991) *Protozoa and Other Protists*, Cambridge.

Asexual reproduction in the ciliate protozoa is by binary fission but it is important to remember that ciliate protozoans have two nuclei, which makes nuclear division just that bit more complicated than in other groups. The diploid micronuclei and the macronuclei divide mitotically, and one nucleus of each type goes to each daughter cell. The events of conjugation and sexual reproduction in the ciliate protozoan *Paramecium* are shown in Figure 9.26. During sexual reproduction only the micronucleus divides, giving rise to gametes, while the macronucleus degenerates. After syngamy, nuclear fusion and meiosis, the micronucleus divides mitotically to form a new macronucleus.

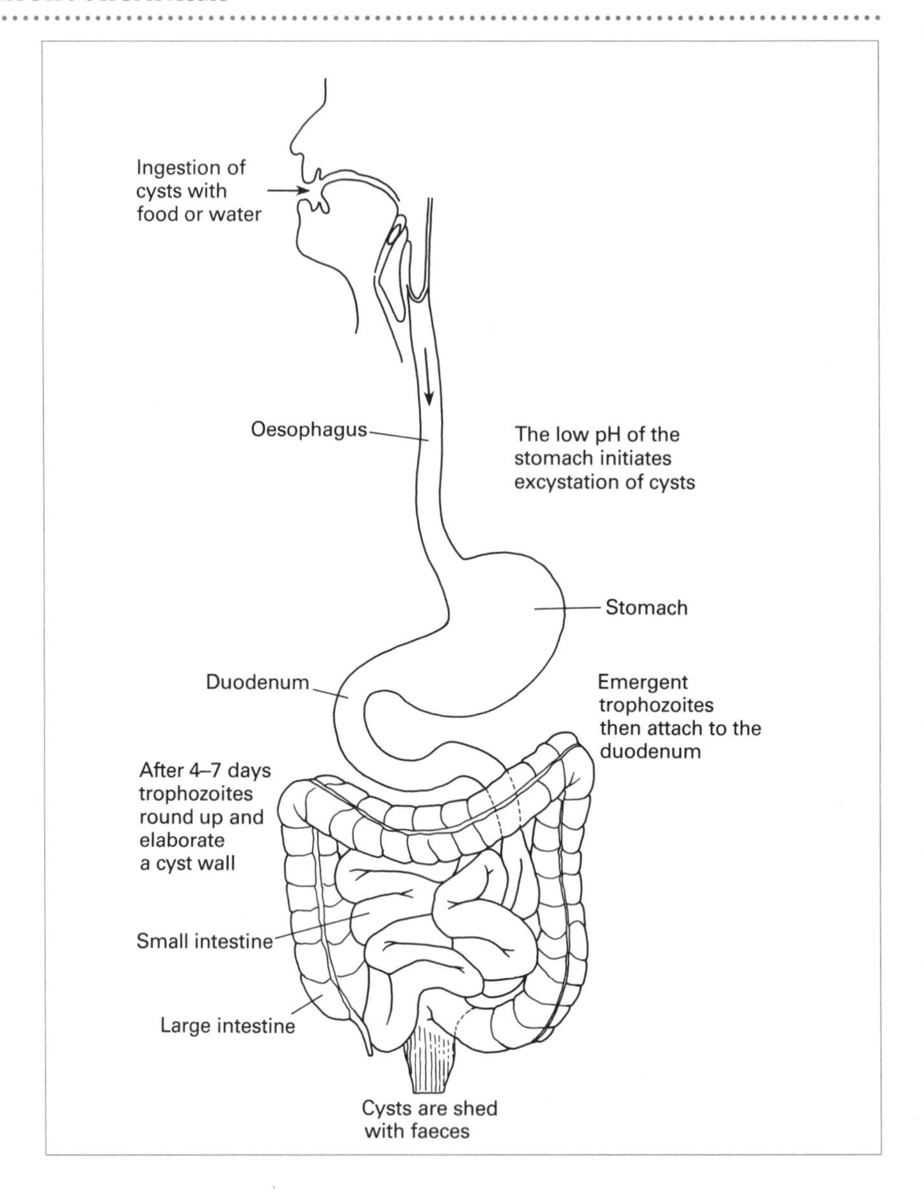


**Figure 9.25** The life-cycle of *Giardia intestinalis.* Redrawn from Dr T. Paget, Biology Department, Birkbeck College, University of London.

If you think carefully about this, it is obvious why this happens. The old macronucleus contained genetic material of only one of the parent gametes. If it had been inherited by one or all of the daughter cells after meiosis of the micronucleus, the macronucleus would have contained genetic material that the micronucleus did not. The degeneration of the macronucleus ensures that any genetic changes that have occurred during mating and meiosis are present in the new macronucleus derived from the new micronucleus. Quite why such a small, unicellular organism should need two nuclei is obscure. In the derivation of the new macronucleus, many copies of micronucleus chromosomes are made, the macronucleus becoming polyploid.

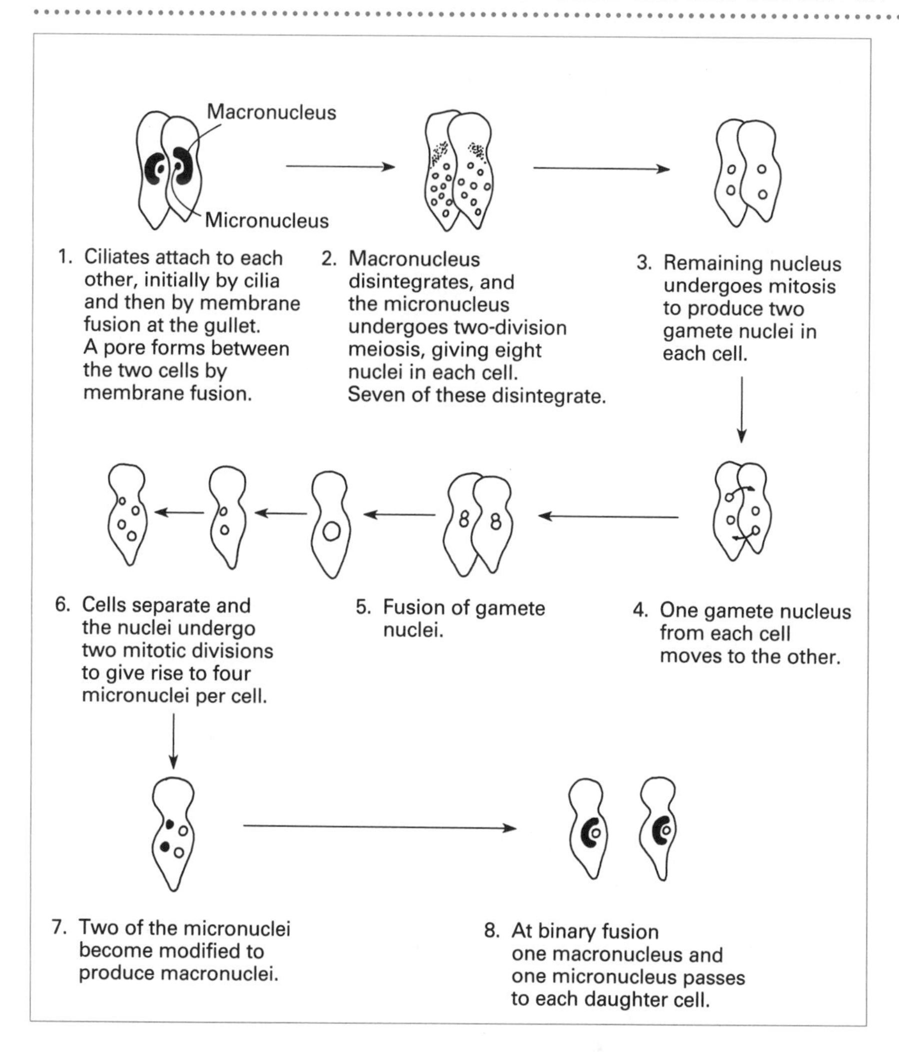


**Figure 9.26** Sexual reproduction in the ciliates.

Sporozoan protozoa have complex life-cycles, often in two or more hosts. The life-cycle is characterized by both haploid and diploid phases, including an asexual division called **schizogony**, in which multiple mitotic divisions of the nucleus produce many infective organisms. Sexual reproduction occurs by the production of a large female macrogamete, which is fertilized by a small, motile male gamete producing, in some species, a thick-walled oocyst. Meiosis occurs within the oocyst, producing infective haploid spores, in an event called **sporogony**. The resulting sporozoites can be naked or have a spore coat. There are four very important animal and human parasites in this group that cause malaria, toxoplasmosis, coccidiosis and pneumocytic pneumonia. The life-cycle of malaria is shown in Figure 9.27.

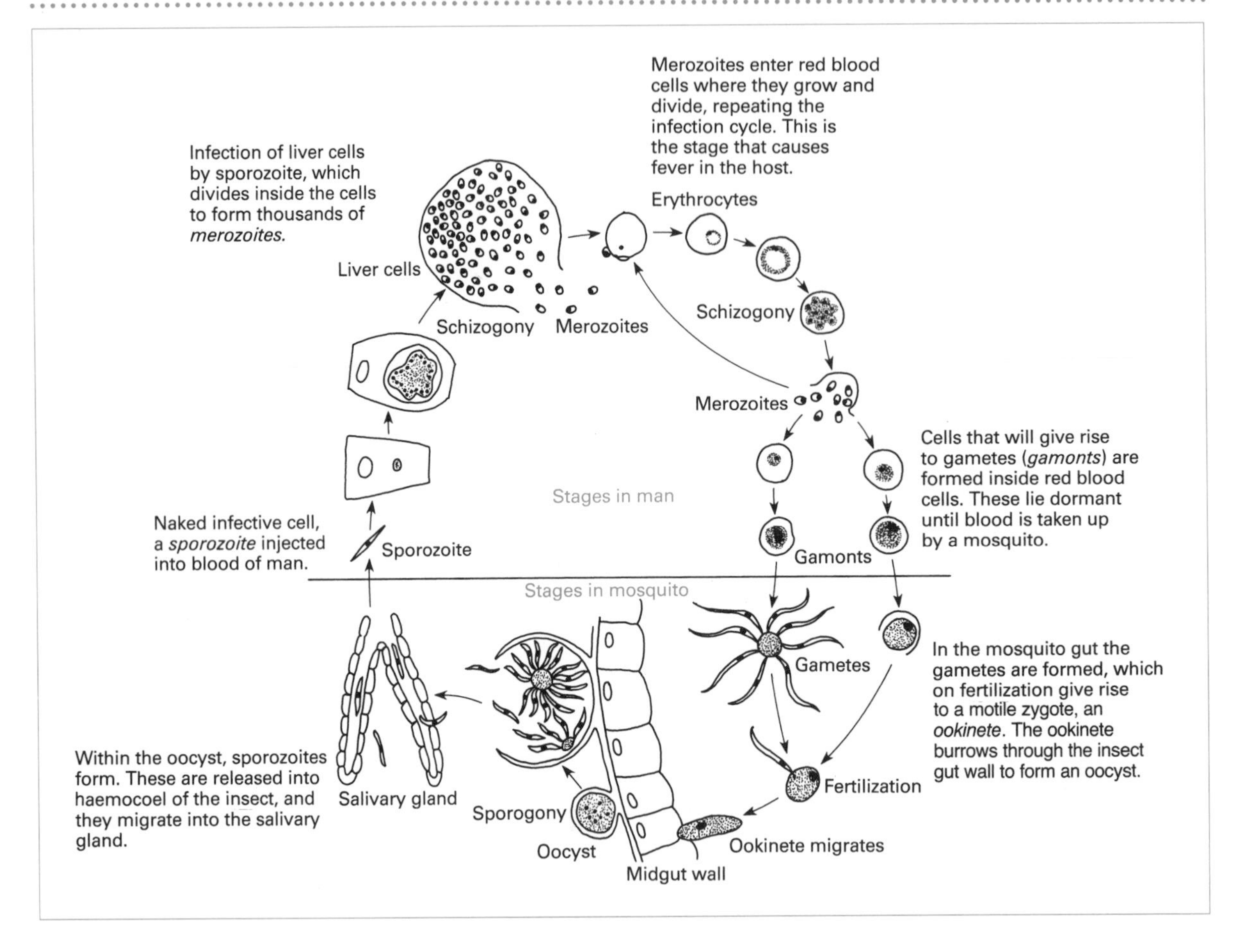


**Figure 9.27** The life-cycle of the malarial parasite, *Plasmodium falciparum*. Redrawn from Sleigh, M. (1991) *Protozoa and Other Protists*, Cambridge.

## Chapter summary

In this chapter we have looked at the structure and growth of the fungi, algae and protozoa. We have seen that all three groups have to withstand similar environmental stresses.

Water, in both deficit and surfeit, poses a problem to all three groups. Contractile vacuoles cope with problems of osmoregulation in the protozoa and some algae. In other kinds of algae and in the fungi, a rigid cell wall prevents the uptake of excess water. At the other end of the scale, desiccation has to be avoided, and cell walls also help to minimize water-loss in terrestrial environments.

To the fungi and Protista, high light-intensity can be a cause of genetic damage and death. Organisms that occupy niches where light intensity is high are pigmented, fungi with dark pigments that absorb and dissipate the light, algae with pigments that utilize light energy to power photosynthesis.

Nutrition in the fungi and protozoa is predominantly heterotrophic, so they must occupy niches where there are plenty of dead or living substrates for them to use. In contrast, the algae, and some of the protozoa that have symbiotic algae within them, need light environments where photosynthesis operates at its greatest efficiency.

In all three groups we have seen that reproduction serves two functions. It produces offspring that ensure a wide distribution of the organism in the environment, but it also provides an opportunity for change and variation in the organism itself. Mechanisms of asexual reproduction are remarkably similar throughout all groups, binary fission being common to all three.

It is apparent that sexual reproduction is rather more idiosyncratic, each group having its own special characteristics. The higher fungi have the greatest size of reproductive body, but we see the greatest complexity of sexual reproduction in the algae.

The significance of these organisms to man and the ecosystem of the planet cannot be over-estimated. They are responsible for primary and secondary productivity in the oceans, and for the recycling of carbon and nitrogen in terrestrial ecosystems. They are also the cause of many important diseases of plants, animals and man.

## Further reading

C.T. Ingold and H.J. Hudson (1993) *The Biology of Fungi.* 6th Edition. Chapman & Hall, London – an excellent introductory text for anyone wishing to learn more about the fungi. It covers their morphology and reproduction in detail, and gives detailed consideration of their importance in the ecosystems they inhabit.

M. Sleigh (1991) *Protozoa and Other Protists.* Cambridge University Press, Cambridge – This book covers both the protozoa and the protistan algae to some depth. As well as details of structure and function, the author gives some thought to the similarities between members of the two groups.

S. Isaacs (1992) *Fungal–Plant Interactions.* Chapman & Hall, London – the ecology of the fungi on plants is considered in this book, from parasites to symbionts. An excellent book for environmental scientists.

R.E. Lee (1989) *Phycology.* Cambridge University Press, Cambridge – this book covers the algae extensively, including morphology, physiology and ecology of all the major groups. It provides interesting details of famous phycologists too.

## Questions

1. Define the following terms: obligate pathogen, phototrophy, osmoregulation, holozoic nutrition, binary fission, autogamy, dikaryon, oogamy, schizogony.
2. What characters can be used to distinguish different members of the fungi?
3. Why should *Euglena* be placed in the algae and not the protozoa?
4. How do protozoa reproduce?
5. Describe in a few short sentences the following techniques or processes: (i) RNA sequence analysis; (ii) hyphal growth; (iii) feeding in the protozoa; (iv) spore liberation in the dung fungi; (v) sexual reproduction in the diatoms; (vi) the life-cycle of cellular slime moulds.

# Viruses 10

The viruses form a unique group in the microbial world. What makes viruses special is that they are **non-cellular** entities. This means that in terms of design and function, viruses bear no similarity at all to the other microorganisms which have either prokaryotic or eukaryotic cellular organization.

## Distinctive properties of viruses

### Size of virus particles

Virus particles, or virions as many American authors call them, are often described as 'ultramicroscopic' structures. The use of this term is to draw attention to the fact that they are smaller than most structures that might be regarded as just 'microscopic'. It was this property of ultramicroscopic size that led to the initial discovery of viruses as agents of disease that were different from other pathogenic bacteria or cellular microorganisms. Almost all virus particles are so small that they can be examined only in an electron microscope. A light microscope is not powerful enough! It is useful to remember that the limit of resolution of the light microscope is about 0.2 μm (200 nm).

All virus particles are smaller than 200 nm in their longest dimension, with one notable exception – the poxviruses, which are usually described as brick-shaped and measure 400 × 240 × 200 nm. The poxviruses infect only animals and include the smallpox virus and cowpox virus. At the other extreme, the smallest complete viruses particles are those formed by the parvoviruses which are roughly spherical with a diameter of about 24 nm. A good example of this group is canine parvovirus, which was not known before 1978. Since then it has become important because it causes enteritis or myocarditis in dogs. Dog-owners will have heard of this virus because it is usual to vaccinate puppies against it at an early age and then to give adult dogs annual booster inoculations.

**BOX 10.1**

**Naming viruses**

Traditionally the names of **animal viruses** have been based on the disease caused in the principal host followed by the word virus. For example: influenza virus, poliovirus, human immunodeficiency virus (HIV).

Similarly, the **plant viruses** are named according to the major host plant with which a virus is associated and the main symptoms of disease caused. For example: tobacco mosaic virus (TMV), tomato bushy stunt virus (TBSV), raspberry ringspot virus.

Here some help is needed with the descriptive terms: *mosaics* are irregular blotches of yellowing tissue on the normal green leaves of plants; *stunts* are retarded growth forms which give rise to low crop yields; *ringspots* are small concentric circles (a few millimetres in diameter) of infected cells on the leaf which usually die and turn brown, but with zones of normal green tissue in between them.

This system for naming plant viruses is far from ideal because a virus can infect different varieties or species of plant and produce different sets of symptoms. This makes it difficult to decide on the major host plant. For example, cucumber mosaic virus (CMV) can give rise to 80 sets of symptoms in its various host plants!

**Bacterial viruses** are named by code letters or by a system of letters and numbers. For example: phage λ, Qβ, T2, φX174.

There is no apparent consistency in this system. It has arisen from practice in the laboratories in which the viruses were first isolated. For instance, MS2 was so named because it was a second isolate of a virus found to infect *m*ale *s*pecific strains of *E. coli*.

## Some viruses can be crystallized

Many, but not all, virus particles can be obtained in a **crystalline** or **paracrystalline** state (in which the particles do not arrange themselves in a perfect lattice but in an orderly manner in one direction to give needle-shaped crystals). The ability to form these types of crystals is confined to the smallest types of virus particles which have the simplest construction.

**BOX 10.2**

**Discovery of viruses**

The first scientists to discover that agents of infection existed that were different from bacteria, protozoa and fungi were the Russian botanist Ivanowski (in 1892) and the Dutch worker Beijerinck (in 1898).

What Ivanowski did, and this was repeated by Beijerinck, was to take sap from tobacco plants infected with the mosaic disease (tobacco mosaic virus) and pass it through a fine-pored ceramic filter known to retain cellular microorganisms. The clarified extract was free from any cellular material, yet still highly infectious if rubbed onto the leaves of healthy tobacco plants. These plants subsequently developed symptoms of the mosaic disease. This simple experiment demonstrated the existence of a class of agents smaller than anything known at the time.

This type of experiment was repeated to show that other diseases were caused by similar agents. Loeffler and Frosch, for example, demonstrated that the foot-and-mouth disease of cattle was caused by a similar agent that was so small it passed though a filter known to trap cellular microorganisms.

It was this property of 'filterability' that proved that these pathogens were entirely novel. Filterability was simply an indirect way of showing that the viruses concerned were of very small size – too small to be seen in any microscope of the time. These agents were at first known as 'filterable viruses' and later just 'viruses'.

The fact that some viruses can be crystallized was first discovered by Wendell Stanley in 1935 who was developing a method to purify tobacco mosaic virus (TMV) from tobacco plant sap using ammonium sulphate precipitation (a standard method for protein purification). Other viruses were soon crystallized, including poliovirus and some of the bacterial viruses or phages, for example MS2. This simple observation implied that virus particles are very different from other microorganisms in that they behave as if they were simple chemical substances. Cellular microorganisms, of course, do not have this property.

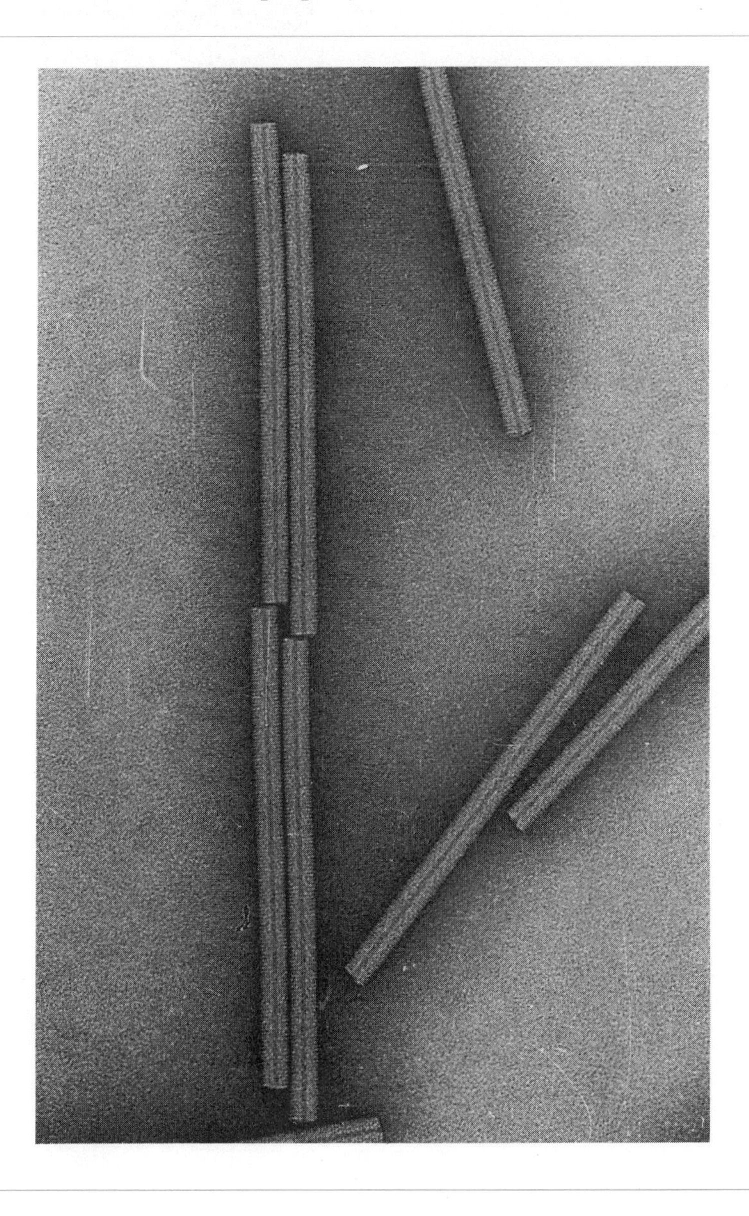

**Figure 10.1** An electron micrograph of five complete particles of tobacco mosaic virus. The particles are about 300 nm long. The virus particles have been negatively stained to reveal the subunits that make up the protein coat. Notice the stain has penetrated into the hollow central channel of the particles. This electron micrograph is reprinted with permission by John Finch, Laboratory of Molecular Biology, Medical Research Council, Cambridge.

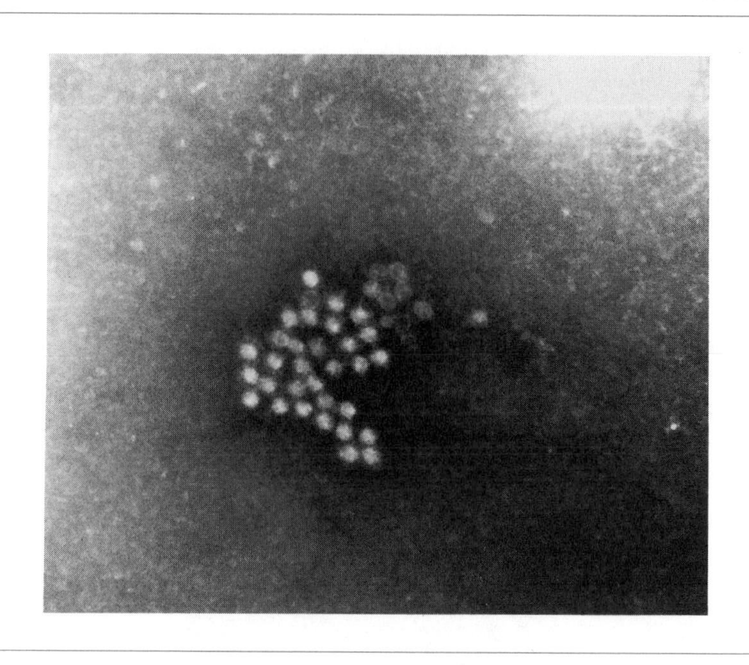

**Figure 10.2** An electron micrograph of particles of poliovirus. These particles have been negatively stained. The isometric particles have a diameter of about 27 nm which is roughly the same as a ribosome in the cell. Reproduced with permission from Alan Curry, Manchester Public Health Laboratory.

## Shapes of virus particles

Virus particles can be divided into three main types according to their morphology and architecture. The three basic shapes are shown in Figures 10.1, 10.2 and 10.3. Figure 10.1 is an electron micrograph of particles of the plant virus TMV. These particles are described as **filamentous**. The particles of poliovirus shown in Figure 10.2 are roughly spherical, but not quite perfect spheres and for this reason are described as **isometric**. Lastly there is an electron micrograph of the phage T4 in Figure 10.3. This type

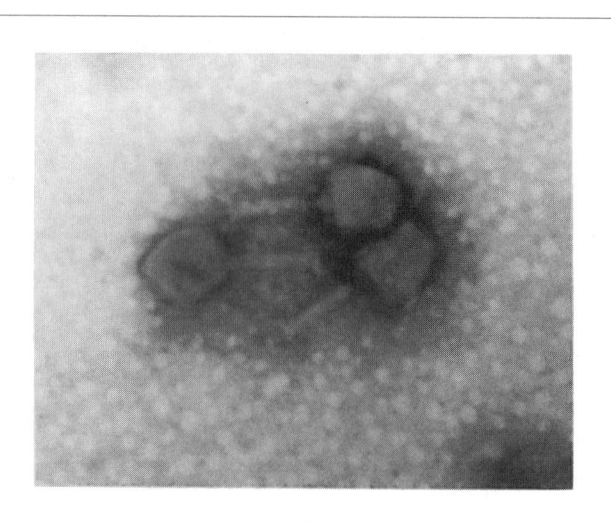

**Figure 10.3** An electron micrograph of particles of phage T4. The particles have complex morphology with heads and tails. The tail fibres attached to the base plate (at the tip of the tail) are clearly visible. The length of the particles from the top of the head to the base plate is about 200 nm. Reproduced courtesy of Dr V. Virrankoski-Castrodenza.

of particle construction is far more complicated than that found in the majority of viruses. Each virus particle has a large hexagonal head with a tail of quite complex structure and six tail fibres. Particles like these and those of the poxviruses are described as having **complex morphology**.

The shapes of these virus particles seen in the electron microscope are determined by the structural protein shell, most usually referred to as the **capsid**. In TMV, the capsid is constructed from many identical molecules of a single protein called the **coat protein**. In poliovirus, the isometric shell is built from multiple copies of four different proteins called the **virion proteins**. T4 is exceptionally complicated in structure with more than 30 different structural proteins making up the capsid.

### Viral genetic information

The virus particle normally carries the viral genetic information inside it and so the capsid might be regarded as a protective structure. The **viral genome**, as it is called, carries all the viral genes necessary for construction of the capsid proteins and replication of the virus. These viral genes products are described as virus-specific proteins. A special feature of the viruses is that this information is not always encoded in DNA, as it is in every other kind of organism, but can be of RNA (but never both together). The genome can code for as few as three or four proteins, as in certain of the bacterial viruses (phage MS2), or for more than 200 products, as in the large complex poxviruses. The combined structure of capsid and the viral genome (DNA or RNA) is described as the **nucleocapsid**.

### Virus particles are stable structures

The majority of virus particles are relatively stable structures. This means that they can persist in the environment without breaking down and retain the important property of being able to infect new host cells; in other words they retain their infectivity. The 'environment' can be the circulating body fluids of an animal, the sap of a plant, aerosol droplets, water, or in some type of vector that transmits the virus from an infected organism to a healthy one. For example, particles of TMV have been known to survive for at least 24 years in dried leaf material in herbarium specimens. Canine parvovirus is regarded to be a particularly stable animal virus. It retains its infectivity in faecal material on the street for 2 to 6 months, and this is one of the reasons why this virus caused world-wide infections of epidemic proportions in dog populations in the first few years after its emergence.

### Function of virus particles

One important point to note about virus particles is that their function is simply to introduce the viral genome into suitable host cells in which the genome may be replicated and new virus particles assembled. It is the viral genome that is the essential part of the virus, and the structural components around it may be regarded as simply a means to promote survival between cycles of infection and to facilitate the introduction of the genome into the host cell.

### Viruses are parasites

An important feature that sets the viruses apart from almost all the other bacteria, fungi and protozoa is the fact that they are *all* **obligate intracellular parasites** and, as such, can multiply only inside living host cells. In this respect it might be argued that they are not unique because there are some minor groups of bacteria that multiply only inside living cells. These include the chlamydiae that invade human host cells and the bacterial genus, *Bdellovibrio*, that parasitically attacks other bacteria. However, what is special about the viruses is that no matter how complex a cocktail of nutrients is provided, the virus particles will not multiply in any cell-free medium.

The intracellular mode of multiplication is illustrated in Figure 10.4 which shows a generalized cycle of infection of the bacterial virus T2 in cells of *E.*

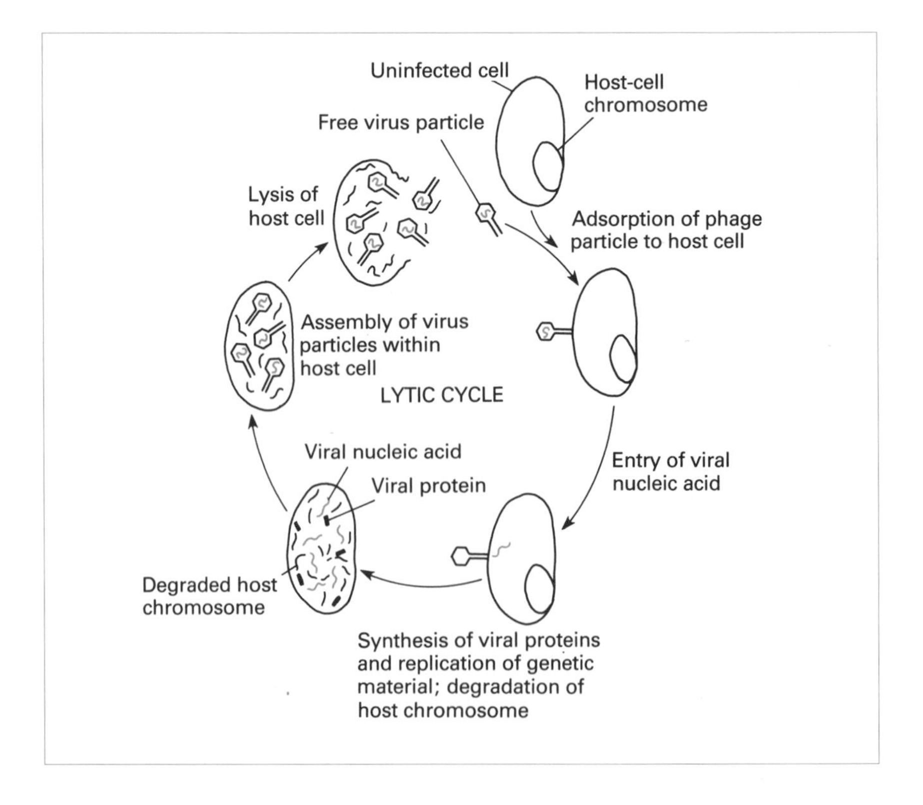


**Figure 10.4** Multiplication of phage T2 in host cells of *E. coli.* Phage T2 has tadpole-shaped virus particles with a hexagonal head and tail. The viral DNA is packaged into the head.

**Earliest records of virus diseases**

Probably the first reports of virus disease were made by Chinese writers about 2000 years ago. They described the symptoms of what we now recognize as smallpox, a virus disease declared by the World Health Organization as eradicated in 1980.

The first records of plants infected by viruses appeared in Dutch paintings and descriptions of 17th century tulip flowers. Tulips were highly prized flowers at that time and the ones most fashionable were those with bold, multicoloured striped petals. These show symptoms of 'tulip break', a disease that is now known to be caused by a plant virus (tulip break virus) transmitted by an aphid vector.

The photograph shows Tulip Zomerschoon with typical symptoms of tulip break. This is probably the oldest tulip in cultivation and its registration dates back to 1620. This photograph is from the 1992 Van Tubergen catalogue, a specialist supplier of flower-bulbs.

J Tulip Zomerschoon

**BOX 10.3** Reprinted with permission of Van Tubergen UK Ltd., Diss, Norfolk.

*coli*. This cycle is typical of most viruses in that the process gives rise to the production of a large number of progeny virus particles from probably just a single infecting particle. The way in which viruses multiply is distinctive and quite unlike the binary fission characteristic of many cellular microorganisms. In the host cell the viral genome is replicated, and structural capsid proteins and virus-specific enzymes are synthesized separately, using host-cell ribosomes. Finally, progeny particles are assembled from the component parts. The entire process relies upon the host-cell biosynthetic machinery and, in particular, uses host-cell nucleotides, host-cell amino acids, and host-cell ATP to fuel these reactions. It is this draining of cellular resources for viral biosynthesis that leads to cell damage and ultimately to symptoms of infection. In multicellular organisms it is these effects that collectively lead to the symptoms of disease.

## Major groups of viruses

There are now descriptions of thousands of viruses and generally they are divided into animal viruses, plant viruses and bacterial viruses. The study of animal virology includes all those viruses that infect vertebrates, including those human viruses of medical importance and others of veterinary importance. Also there is a large body of information on invertebrate viruses, especially insect viruses. Of the plant viruses, most is known about the virology of flowering plants.

All the bacterial viruses are usually referred to as **phages**, an abbreviated form of 'le bactériophage', the name given to them by one of the scientists who first discovered these viruses at the Institut Pasteur in Paris (d'Hérelle, 1919). Most is known about the phages that infect *E. coli* but, of course, there are many viruses known that infect other prokaryotes. There are viruses that infect all the major groups of organisms, including protozoa, algae, filamentous fungi and yeasts. However, these viruses are not so well studied and far less is known about their biology.

This division of viruses into three major groups is generally satisfactory because most viruses are specific to just certain host species or strains. There are just a few viruses that cross these major boundaries. An example is the rhabdovirus group which has members with very wide host ranges. One representative of the rhabdoviruses is lettuce necrotic yellows virus (LNYV). This virus multiplies in a plant host (as its name suggests) and is remarkable because it also replicates in aphids which act as vectors to transmit the virus from infected to healthy plants. The aphid vectors acquire virus when they feed by sucking sap from infected plants and can go on transmitting for the rest of their lives.

### Definition of a virus

From the introductory survey of the properties of viruses, it will have become apparent that the life-histories of viruses can be divided into two phases. The first phase, the **dormant** or **extracellular phase**, is the one represented by the **virus particles**. The other phase is the part of the life-history when the virus is multiplying inside its host cell and this is the **vegetative** or **intracellular phase**.

It is helpful to try and encapsulate these distinctive properties of viruses in a simple and concise definition. A number of eminent virologists have attempted to devise a satisfactory definition that conveys the essential properties of the group. Probably the best of these is the definition made by the distinguished virologists S.E. Luria (a Nobel prize winner for his work on viral and bacterial genetics) and James Darnell (1967):

**Viruses are entities whose genomes are elements of nucleic acid that replicate inside living host cells using the cellular synthetic machinery and causing the synthesis of specialized elements (virus particles) that can transfer the viral genome to other cells.**

The value of this definition is that it places the viral genome in a central role which is crucial to the infection process and the production of progeny virus particles in the host cell. Another valuable feature is that it recognizes the function of the virus particles, in the extracellular phase, as vehicles which carry the viral genome to new host cells. What this definition does *not* do is draw attention to the very small size of virus particles, which is a characteristic that distinguishes them from other types of microorganisms.

## Types of virus infections

Some virus infections are undetectable (and are described as **asymptomatic** or **silent infections**) while others give rise to symptoms of disease, which can be of different degrees of severity, possibly even culminating in death. In general, animals and plants respond to virus infection in one of two ways. The virus can spread throughout the whole organism. These infections are described as **generalized** or **systemic** infections. Or the virus can remain highly **localized** and confined to a system within the organism, or alternatively close to the point of entry into the host, surrounded by healthy tissues.

### Animal infections

Some animal viruses produce highly localized infections. The most obvious example is the human wart (papilloma) virus which causes keratinized **benign** (non-malignant) tumours in the epidermal layer of skin on hands and feet. This virus is spread by contact, hand to hand, or in the case of warts on feet, in swimming pools. Similar human papilloma viruses cause genital warts and are implicated in cervical cancer. In common warts on the skin, virus particles can be seen only in the surface layers and these are the main source of virus that spreads to cause new infections. The virus particles are shown in Figure 10.5 and can be seen to be isometric particles with a diameter of about 55 nm. The genomic material is DNA.

Other animal viruses cause infections which are localized to a system in the body. For example, influenza virus causes a disease that is localized to the respiratory tract. The virus gets inhaled into the body through the nose and mouth from droplets in the air. Once they have managed to pass through the respiratory mucous secretions, they infect and destroy the

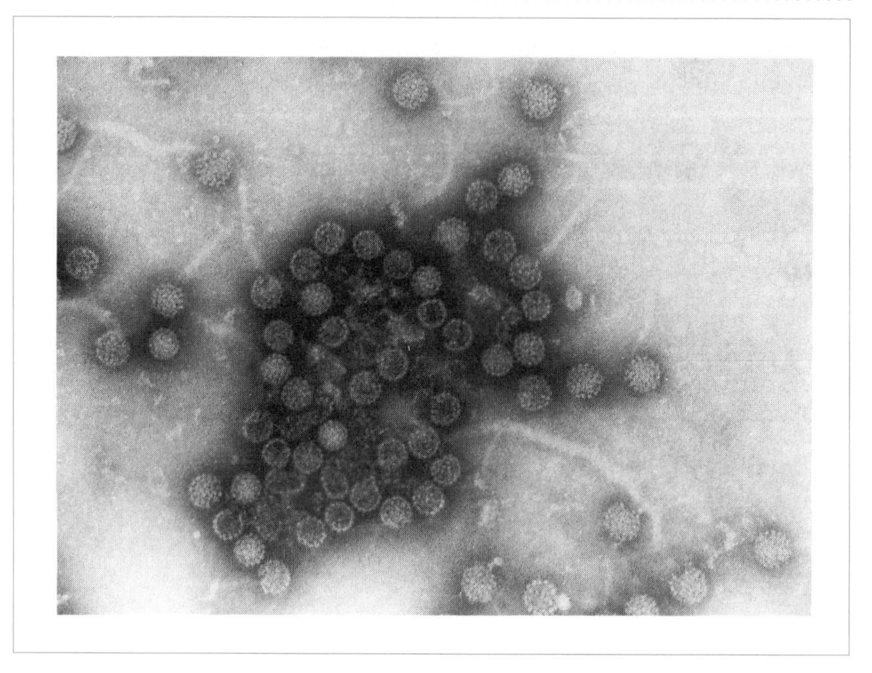

**Figure 10.5** Electron micrograph of human wart (papilloma) virus particles that have been negatively stained. The particles are 55 nm in diameter. The stain has penetrated some particles and these probably lack DNA. Reproduced with permission from Alan Curry, Manchester Public Health Laboratory.

ciliated epithelial cells lining the upper respiratory tract, trachea and bronchi. The infection process results in the production of virus particles in infected epithelial cells and, in general, the severity of the disease can be correlated with the amount of virus particles produced. These are shed in the nasopharyngeal secretions.

In cases of generalized or systemic infections, virus particles spread to the different parts of the body and they are carried to all the different sites in the blood and lymphatic system.

Smallpox (variola) virus produces a **generalized** infection which leads to the formation of characteristic skin eruptions called 'pocks'. This was a most disfiguring disease that had a high mortality rate, commonly up to 25%. Thankfully smallpox has now has been eradicated by a programme of vaccination administered by the World Health Organization and the last naturally occurring case was recorded in Ethiopia in 1977.

During the course of smallpox infection, the virus multiplies first in the upper respiratory tract and then finds its way into the regional lymph nodes. Virus particles spread to the liver, spleen and lungs where they again multiply. Multiplication is followed by a period of invasion through the bloodstream, usually associated with the onset of fever, headache and malaise. The virus finally spreads to the skin where it attacks epidermal cells and, after 3 or 4 days, characteristic pocks erupt.

Smallpox is spread by person-to-person contact. At first, virus is shed from infected individuals in the form of tiny aerosol droplets that come from lesions in the upper part of the respiratory tract. Later when the skin lesions erupt, these may be another source of virus that can contaminate water or household utensils.

Many other viruses infect pets, livestock and wild animals and can cause diseases that result in severe economic loss. Foot-and-mouth disease virus is a highly contagious disease of cattle. Infected herds have to be destroyed to prevent the spread of the disease. Other animal viruses are known to infect humans, yet the source of the infection is wild animal stocks which may or may not show signs of disease. The rabies virus multiplies in tissue of the central nervous system where it causes extensive damage and, usually, death. Infection usually follows a bite from a rabid dog. A variety of animals act as hosts of rabies virus, forming **reservoirs of infection**, and these include bats, foxes, squirrels and badgers. In Europe the fox is the most important reservoir of infection. Infected foxes show signs of rabies which is usually fatal, and these pass the virus on to dog populations.

New kinds of virus infections appear from time to time. One that has caused considerable concern is **phocine distemper virus** (a morbillivirus) which infects grey seals living in coastal waters. This virus was responsible for thousands of seal deaths in North European waters (the Baltic Sea, North Sea and Irish Sea) around the UK in 1988. There has been much debate on the possible link between this virus infection and pollutants such as polychlorinated biphenyls (PCBs) which accumulate in the fatty tissues of marine animals. It is claimed that PCBs may depress the immune systems of affected seals and so render them more susceptible to virus attack. However, the evidence for this is not clear cut.

### Plant infections

In some plants the response to virus infection is highly localized and the virus does not spread far from its original point of entry. The resulting patch of infected cells is described as a **local lesion**.

Local lesions are most commonly seen on the leaves of plants. The lesions are typically only a few millimetres in diameter and can take a number of different forms. Commonly the infected cells die and turn brown giving rise to **necrotic** lesions. Figure 10.6 shows necrotic lesions formed by TMV on the leaves of a commercial variety of tobacco (*Nicotiana tabacum*) 'Xanthi'. In other varieties of tobacco, TMV would normally give the systemic mosaic disease rather than local lesions.

With other virus–plant combinations the infected cells in the lesion may lose their chlorophyll and other pigments and so turn pale green or white. These are **chlorotic** lesions. Ringspot lesions have already been described on page 288. Where symptoms are limited to local lesions, there is no

**Figure 10.6** Necrotic local lesions formed by tobacco mosaic virus (TMV) on the leaves of *Nicotiana tabacum* var. *xanthii*. The plant has been cut down to just three leaves and the different halves of each leaf have been inoculated with TMV at different concentrations. Notice the different number of lesions on each half-leaf.

spread of virus to the rest of the plant and the host is said to show a **hypersensitive response**.

With many virus–host plant combinations the virus invades the whole plant and it becomes systemically infected. In these cases virus spreads from cell to cell via the fine cytoplasmic connections known as **plasmodesmata**. Long-distance transport throughout a plant takes place in the plant's conducting system, especially in a tissue called **phloem**. Systemically infected plants usually display symptoms of disease and these can include reduction in plant size (**stunts**), yellowing diseases of whole plants, wilting, and abnormalities in the development of leaves, fruits and stems. On leaves, signs of infection include the formation of mosaic patterns in broad-leaved plants and yellowish stripes or streaks on narrow-leaved plants.

Systemic plant virus infections cause economic losses to both field and greenhouse crops. For example, barley yellow dwarf virus (BYDV) infects cereal plants (Figure 10.7). This virus is important throughout the world. Aphids (greenflies and blackflies) transmit virus to plants when they feed. In the UK, 60 to 70% of cereals may be infected with BYDV (ADAS, 1978) and this may cause a 5 to 10% loss in yield.

**Figure 10.7** Barley plants infected with barley yellow dwarf virus.

Sometimes the symptoms of virus infection are exploited horticulturally in decorative or specimen plants. Some species of *Abutilon* are commonly grown as ornamental shrubs (Figure 10.8) and in the variegated form have attractive green and white leaves caused by **abutilon mosaic virus**. Other plants of economic importance, such as *Cassava* (tapioca) are also affected (Figure 10.8).

## Bacterial infections

Infections of bacterial cells by phages are broadly classified into two types: **lytic infections** and **lysogenic infections**.

**(a) Lytic infections.** These are caused by phages such as T2 or T4 that infect *E. coli* (specific strains, such as *E. coli* B). Here the virus multiplies in the host cells and causes it to burst open (lyse) to release progeny virus particles (Figure 10.4). These types of phages are commonly found wherever there are dense populations of suitable host bacteria. For example, they can be readily isolated from sewage where they attack *E. coli* of faecal origin.

Phages that attack **cyanobacteria** (blue–green bacteria) are proving to be of great importance. In the sea one of the principal components of the

**Figure 10.8** Mosaic symptoms on the leaves of *Cassava* caused by African Cassava mosaic virus. The mosaic pattern on the leaves makes an attractive feature. Reproduced by courtesy of Rothamsted Experimental Station, AFRC.

photosynthetic phytoplankton is the cyanobacterium *Synecoccus*. Lytic phages which attack *Synecoccus* can be isolated from seawater and may be present in numbers of $10^8$ or $10^9$/ml. *Synecoccus* and other phytoplankton in the sea grow and develop into 'blooms' and then disappear in cycles. It seems likely that these phages play a part in regulating natural populations of *Synecoccus*.

**(b) Lysogenic infections.** These are caused by the so-called **temperate phages**, such as phage λ which infects *E. coli* K12 (see page 225). Infected cells show no obvious signs of infection but retain the ability to produce infective virus particles (Figure 10.9). Such infected cells are described as being **lysogenized** and they carry the viral genome in a repressed form. Certain treatments, such as non-lethal doses of ultraviolet light, will induce a lytic infection in which virus particles are produced and the host cell is killed and breaks open.

These temperate phages are important because they can cause changes to the phenotype of the lysogenized cells. Two sorts of changes are well known.

A group of phages, known as the **epsilon phages**, cause changes to the surface antigens formed by lipopolysaccharide (LPS) on the host-cell surface. In *Salmonella*, different strains can be distinguished by tests with antisera specific for this LPS in the outer membrane of the cell wall. Epsilon phages interfere with the synthesis of LPS and so the antigenicity of the cell

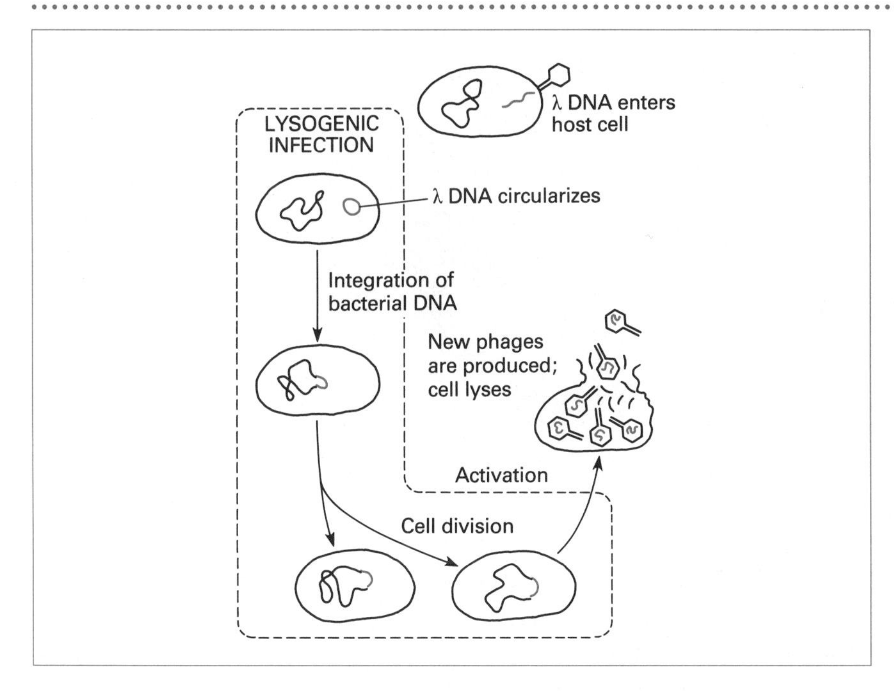


**Figure 10.9** Infection of *E. coli* by a temperate phage, such as phage λ. The virus commonly establishes a lysogenic infection in which the viral DNA is integrated into host-cell DNA. It is maintained there and replicates with the host-cell DNA so that when the cell divides, each daughter cell gets a copy of the integrated viral DNA. Under certain conditions, the viral DNA is activated. It comes out of the host-cell DNA and starts to replicate in the cells, leading to eventual lysis and release of progeny phage particles.

surface is changed. What is interesting is that the LPS molecules also form receptors by which these phages attach themselves to the cells of *Salmonella* and so, on infection, they modify their own receptors, eliminating further attack from the same phages.

The second type of changes brought about by lysogenic phages are **pathogenic conversions**. *Corynebacterium diphtheriae* is the bacterium that causes diphtheria, and the way it induces this severe disease of the pharynx is by the production of potent protein toxins. The bacterial strains which cause diphtheria are described as **toxigenic** and all of them have been infected by a group of **β phages** which have brought about the conversion from non-toxigenic to toxigenic strains. Strains not infected by these phages do not cause diphtheria. The diphtheria toxin is, in fact, coded for by the *tox* gene on the β phage genome. This type of conversion of pathogenic bacteria may be quite common and there are other examples of pathogenic conversions, notably involving toxigenic strains of bacteria that cause scarlet fever and botulism.

## Structure of virus particles

Two experimental techniques have proved especially useful in determining the structure of virus particles: electron microscopy and X-ray crystallography.

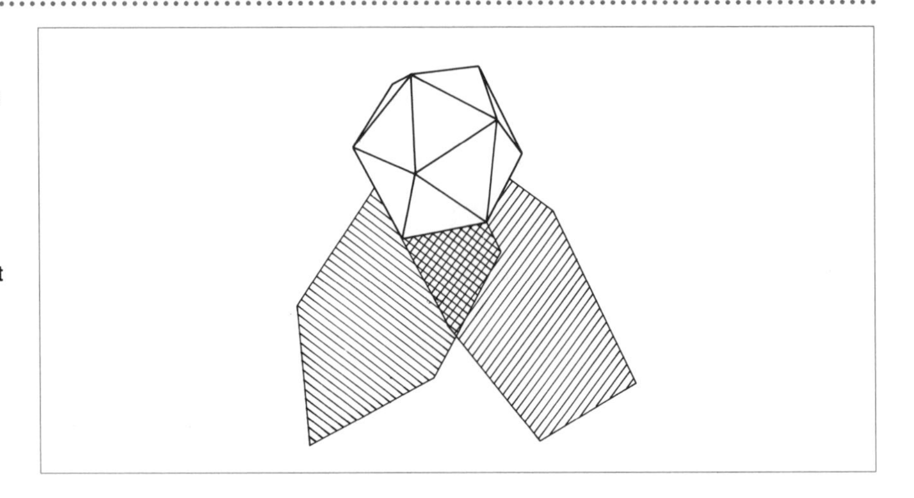

**Figure 10.10** How shadow-casting works. In this case light is being directed at the body from two different angles and the shadows cast can be seen. The icosahedral shape of the body can be deduced from the shapes of the shadows. The same principle applies when heavy metal vapour is directed at virus particles.

The first virus particles to be detected in the electron microscope were those of TMV in 1939. Virus particles are transparent in the electron beam in the microscope and so special staining methods are needed to 'see' their shape and detailed structure.

In **shadow casting**, an electron-dense metal, usually gold, is vaporized and sprayed over the specimen at a known angle. The metal thereby forms a thin film over the virus particle, but 'shadows' are created behind it where no metal is deposited. From the image of the shadows resolved in the microscope, the three-dimensional shape of the virus particles can be deduced. The principle of shadow casting is illustrated in Figure 10.10.

A simpler method of sample preparation is **negative staining**. This simply involves mixing a negative stain with the virus suspension and then allowing it to dry on a special support film for the electron microscope. Negative stains are all heavy metal salts and those most frequently used include potassium phosphotungstate and uranyl acetate. Under appropriate conditions, these stains do not combine chemically with the particles, but instead penetrate into all the gaps and crevices in the virus particles. Hence the details of particle structure stand out against a dark background of stain which is electron dense. As shown in Figure 10.1, the negative stain can also penetrate into hollow virus particles, such as the rods of TMV.

X-ray crystallography is a powerful means of investigating the structure of those virus particles that can be obtained in a pure crystalline or paracrystalline form. This has been very successfully applied to particles such as poliovirus and TMV. The essence of X-ray crystallography is that deductions can be made about virus structure by interpreting the patterns of X-ray diffraction obtained when an X-ray beam is directed at a virus crystal.

On the basis of morphology and structure, virus particles can be divided into four main types.

## Virus particles with icosahedral symmetry

These include the small isometric viruses that infect animals (poliovirus), plants (turnip yellow mosaic virus) and bacteria (ϕX 174). In these virus particles the protein subunits that make up the capsid (protein shell) are arranged in a regular way with a symmetry that resembles an **icosahedron**. They are said to have **icosahedral symmetry**.

Evidence for this type of symmetry in virus particles first came from X-ray crystallographic studies on viruses such as turnip yellow mosaic virus (TYMV). When these virus particles are carefully examined in the electron microscope, it can be seen that their capsids are built from structural subunits that are regularly arranged with each one touching its immediate neighbours. These structural subunits are described as **capsomeres**.

The capsomeres are, in fact, groups of protein molecules. In some virus particles the capsomeres consist of aggregates of just one type of protein. This is the case in TYMV and, in this example, the protein is called TYMV coat protein. In other icosahedral particles there may be several types of protein in the capsomeres. For example, in the particles of poliovirus, each capsomere contains four different structural proteins designated VP (virion protein) 1, VP2, VP3 and VP4 in equimolar amounts.

The bonds which hold the protein molecules together in the capsomere are weak non-covalent bonds, including electrostatic attractions, hydrophobic bonding and hydrogen bonding. The bonds which hold the capsomeres together to form the capsid are similar non-covalent bonds which can be disrupted easily under appropriate conditions.

**Icosahedral symmetry**

An icosahedron is a regular three-dimensional body with 20 triangular faces, 12 pointed corners (called vertices) and 30 edges. The important characteristic of an icosahedron is that its symmetry is rotational. This means that it can be viewed from three different directions and three different types of symmetry will be seen. When viewed directly at one of its corners, the icosahedron can be rotated about a **fivefold axis** of symmetry. This means that the same symmetry appears every time the body is rotated by one-fifth of a complete revolution (72°). The same body viewed at the centre of one of its triangular faces can be rotated about a **threefold axis** and, if viewed along a central position, vertical to one of the edges, shows a **twofold axis** of symmetry. The icosahedron is therefore described as having 5:3:2 rotational symmetry.

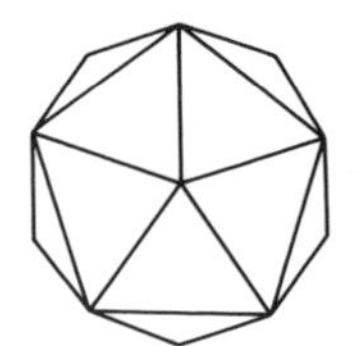

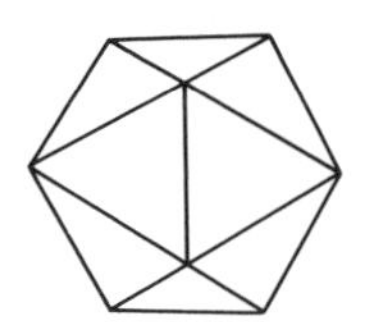

**BOX 10.4** An icosahedron viewed from three different positions to show rotational 5:3:2 symmetry.

**BOX 10.5 (A)** The structure of turnip yellow mosaic virus. Notice that the particle is built from clusters of protein subunits that form the morphological units or capsomeres. Look carefully: some of the morphological units contain five protein subunits (pentamers), and others contain six protein subunits (hexamers). The RNA is packaged inside this protein shell and cannot be seen.

**(B)** The structure of adenovirus. Particles of adenovirus have icosahedral symmetry. They are built from 252 capsomeres (morphological units). Of these, 240 are hexamers that make up the faces and edges of the triangles of the icosahedron. There are 12 pentamers at the pointed corners. The pentamers have a fibre which acts as the attachment organ to the host cell.

**A closer look at turnip yellow mosaic virus**

TYMV is a good example of a virus particle with icosahedral symmetry.

Particles of TYMV are 28–29 nm in diameter. In the electron microscope these particles show 32 morphological units or capsomeres. The coat protein of TYMV has a molecular weight of close to 20 000 and this is folded to give a protein subunit that is somewhat banana-shaped.

It is easiest to envisage the capsid being built from capsomeres positioned on the icosahedron. Twelve of these morphological units are located at the pointed corners of the icosahedron and these are pentamers (aggregates of five banana-shaped subunits). The remaining 20 morphological units are hexamers (aggregates of six of the banana-shaped subunits) and these are positioned on the triangular faces of the icosahedron.

Thus, each virus particle is built from 12 pentamers ($12 \times 5 = 60$ coat protein molecules) and 20 hexamers ($20 \times 6 = 120$ coat protein molecules). This accounts for the 32 capsomeres and gives a total of 180 protein subunits.

The capsid contains the viral genome. In this case it is a strand of RNA and the polynucleotide is folded up within the capsid, although the exact details of this arrangement are not clear.

Many other viruses have capsids of icosahedral symmetry that are larger and far more complex. Generally speaking the number of capsomeres increases in larger virus particles, although the basic pattern of symmetry is maintained. One of the largest virus particles of this type is adenovirus which has 250 capsomeres. A model of this particle is shown. The icosahedral particle has a diameter of 75 nm.

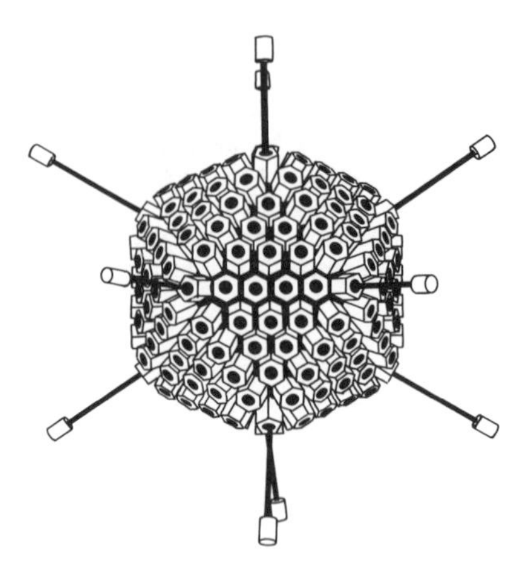

### Virus particles with helical symmetry

Some viruses form particles which are cylindrical in shape and these are commonly described as **rod-shaped**. These viruses are found infecting plants, bacteria and animals. The rod-shaped virus particles only ever contain just one type of structural protein and this forms subunits that can be clearly seen in the electron microscope in a helical arrangement. These particles have **helical symmetry**.

Rod-shaped virus particles can be rigid, as in the case of TMV, or flexible, as found with potato virus X. Another type of virus which has a flexible nucleocapsid is mumps virus but, in this example, the nucleocapsid is coiled up inside a special lipoprotein envelope.

This type of virus structure was first deduced from X-ray crystallographic work on viruses such as TMV.

### Enveloped viruses

Enveloped viruses have a lipid-rich outer membranous coat surrounding the nucleocapsid in the form of a regular phospholipid bilayer, similar in organization and chemistry to the membranes of eukaryotic cells. The envelope is not a rigid structure and so these virus particles can often take different forms and are described as **pleomorphic** particles.

This membrane usually has glycoprotein structures inserted into it which project from the surface of the envelope. These structures are called **peplomers** or **spikes** and the glycoproteins from which they are made are coded for by the viral genome. The spikes may serve a number of purposes. Some may act as **receptors** to attach the virus particles to suitable host cells at the start of the infection process. Others have enzymatic activity, such as the **neuraminidase spikes** in influenza virus. Some enveloped virus particles with surface spikes show **haemagglutinating activity**. Virus particles with haemagglutinin (HA) spikes stick to the surface of red blood cells and cause them to link together in a lattice, thus causing visible clumping (haemagglutination) of the red cells.

The envelope is essential for the infection process and any treatment that damages it causes a loss of infectivity. Hence, treatment of enveloped viruses with detergents or organic solvents, such as chloroform or ether, damages the lipoprotein layer and inactivates the virus.

Envelopes can surround nucleocapsids of helical symmetry, as found in particles of mumps virus or influenza virus. In addition, enveloped viruses can have nucleocapsids of icosahedral symmetry, as is the case in **human immunodeficiency virus (HIV)**. The envelope is added to the particles during the final stage of the virus multiplication process when whole virus particles are formed from ready-assembled nucleocapsids by budding

**BOX 10.6** The structure of tobacco mosaic virus. The protein subunits are helically arranged with the RNA in the middle of the particle. There is a fine channel down the centre of the particle.

**A closer look at tobacco mosaic virus**

Particles of TMV are approximately 300 nm in length, with a diameter of 18 nm. The particles are hollow and each one has a central hole of about 4 nm diameter.

The coat protein of TMV has a molecular weight of about 17 500 and forms the protein subunits seen in the electron microscope. The protein subunits are regularly arranged in a helical fashion, such that there are 16.33 protein subunits for every turn of the helix. This number is specific to TMV, and other viruses may have their protein subunits arranged in different numbers per turn of the helix. Non-covalent bonds are formed between adjacent subunits on the same turn of the spiral, as well as between adjacent turns. It is this bonding that holds the particles together and, under special conditions, it is possible to get the TMV coat protein to assemble into rod-like structures without any of the TMV-RNA in the middle.

Within the virus particle the TMV-RNA lies in a helical groove between the protein subunits. The RNA takes the form of a compact right-handed helix and each protein subunit is associated with three nucleotides of RNA.

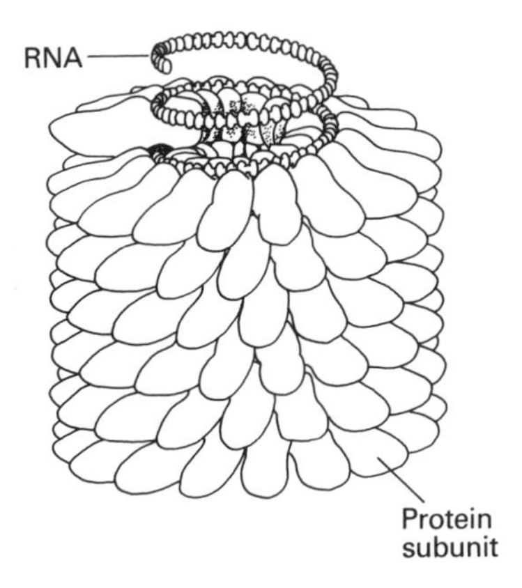


In virus particles which have this type of construction, the diameter of the nucleocapsid is determined by the characteristics of the structural protein subunits, while the length depends on the size of the viral nucleic acid. The arrangement does not rely on the viral nucleic acid holding the protein subunits together, like beads on a string.

Many of the rod-shaped particles infecting plants and bacteria are as described for TMV, and consist of just **naked nucleocapsids**. However, the viruses of helical symmetry that infect animals are somewhat more complex because these have helical nucleocapsids enclosed in a lipoprotein envelope. These are **enveloped viruses**.

through a cell membrane. This is the way in which they are released from infected cells. Most often, as with influenza virus and HIV, the viral nucleocapsids bud through the cell surface membrane (Figure 10.11). However, in a few cases, such as herpesviruses which multiply in the host cell nucleus, the virus particles mature by budding through the nuclear membrane.

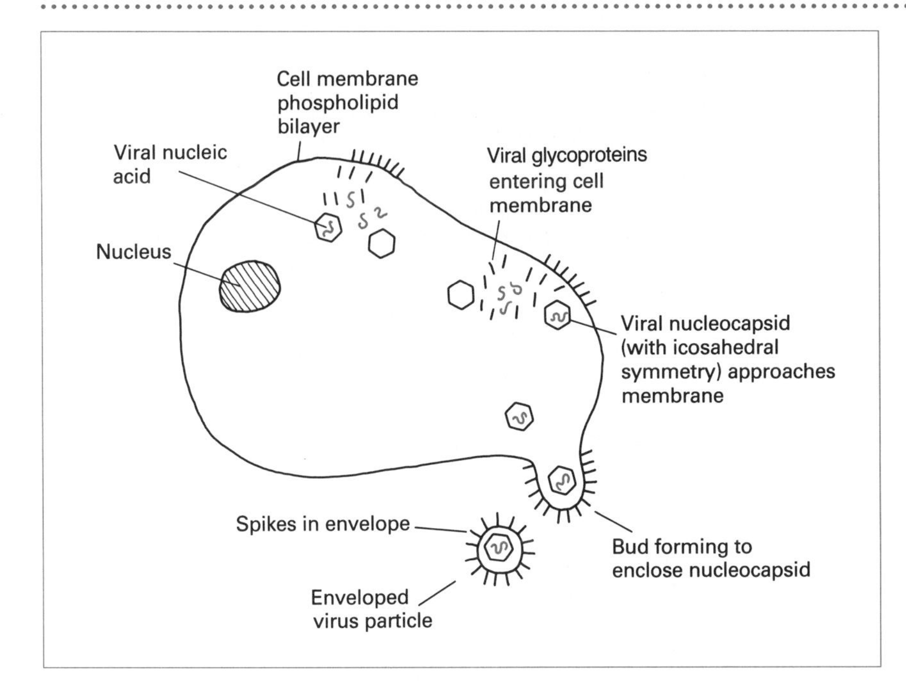


**Figure 10.11** Release of enveloped viruses by budding through the membrane of the infected host cell. Viral glycoproteins are inserted into the host-cell membrane and project through it. Viral nucleocapsids bud through the membrane and acquire a lipid envelope with spikes.

## Complex virus particles

Complex virus particles can be divided into two types: poxvirus particles, and phages with head and tail morphology.

**(a) Poxviruses.** The poxvirus particles are described as brick-shaped (smallpox) or ovoid (as in orf, a poxvirus that infects lambs or kid goats) and they do not have clearly defined protein shells (Figure 10.12).

The virus particle contains DNA that forms a central nucleoid core. This is biconcave in shape, rather like a red blood cell, and it is covered by a lipoprotein membrane. On either side of the core are two lateral bodies. A layer of coarse fibrils near the surface of the particle gives the virus a characteristic striated appearance when looked at in negatively stained preparations in the electron microscope. The particle contains very many structural proteins.

**(b)Phages with head and tail morphology**. These are phages such as T2, T4 or λ that are tadpole-shaped, with an obvious head and tail. These virus particles have very elaborate structures and are only ever found infecting bacterial cells. Presumably this design has evolved to facilitate the infection process which involves penetrating the bacterial cell wall. There are a number of phages with this type of structure and some have short tails, others have long non-contractile tails, and yet others have complicated contractile tails. Such phage particles are described as having **binal sym-**

**BOX 10.7** The structure of influenza virus. The genome of influenza virus is segmented. The particles contain eight segments of viral RNA, each one being separately encapsidated.

**A closer look at influenza virus**

The structure of influenza virus is shown. Lipids are a major component in these virus particles and make up the basic structure of the envelope. The usual types of phospholipids are present, arranged in a bilayer. The virus particles contain five structural proteins and three enzymes coded for by the viral genome. The shape of the particles is not very uniform. Mostly they are more or less spherical with a diameter of about 100 nm, but occasionally long filamentous forms may be seen.

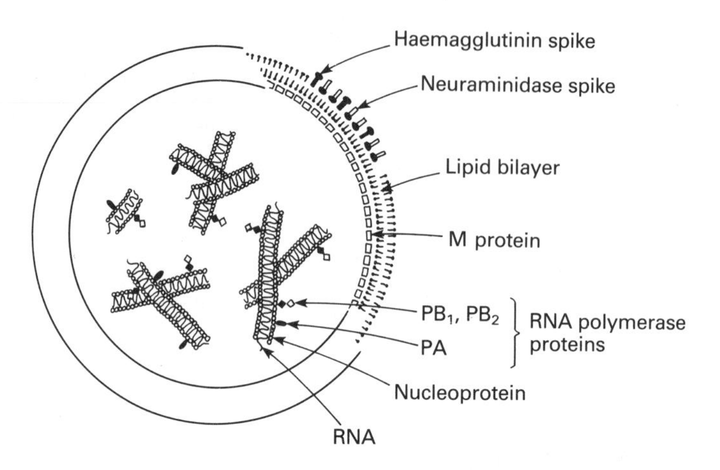


The most abundant protein in the virus particles is the M (for matrix) protein that underlies the envelope. In the centre of the particle there are segments of nucleocapsid, eight in all. These contain the viral RNA with the NP (for nucleoprotein) associated with it and they show helical symmetry. The three largest proteins are designated PA, PB1 and PB2 and these are all involved in the replication and transcription of the viral genome.

Glycoproteins HA1 and HA2 make up the haemagglutinin spikes that project from the surface of the envelope. The HA spike is rod-shaped, about 10 nm long, with a round head. The base of the rod is embedded in the phospholipid bilayer of the envelope. These spikes are responsible for the initial attachment of the virus particle to a host cell. If this structure is treated with a specific antibody against it, the virus infectivity is lost and the virus is said to be 'neutralized'. The other glycoprotein is the N (neuraminidase) protein. This forms the neuraminidase spikes. N spikes possess enzymatic activity that destroys the virus receptors on host cells. The function of this neuraminidase does not seem to be entirely clear. The neuraminidase does release excess virus particles that get stuck onto host cells in the first stages of infection and so may facilitate the spread of the virus from one host cell to another. Also, they facilitate the release of newly formed virus particles from infected host cells and so aid virus spread.

Since both the HA and N proteins are such obvious surface proteins on the virus particles, it is not surprising that they can both induce the formation of specific antibodies against them in infected animals. The HA and N proteins both form effective viral antigens that are very useful in dividing influenza viruses into different types and strains.

**metry** because they have two sorts of symmetry: a head which is of icosahedral symmetry with a tail with helical symmetry. Phages such as T2, T4 or T6 (the so-called **T even phages**) have the most complicated type

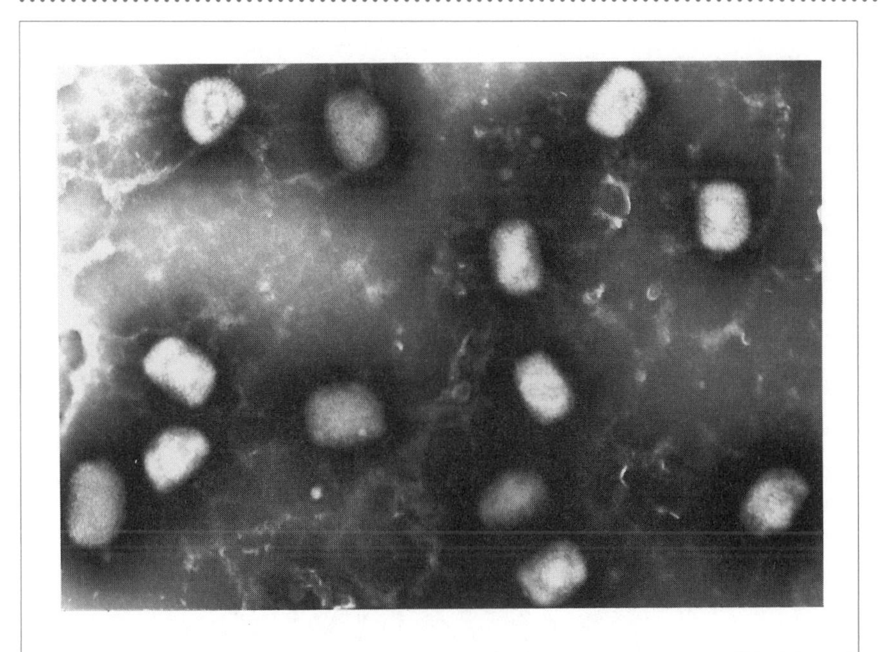

**Figure 10.12** An electron micrograph of pox virus particles. These particles have a complex structure. The surface shows a fibrillar appearance. There is a core in the centre that contains DNA and several enzymes. Reproduced with permission from Alan Curry, Manchester Public Health Laboratory.

of head and tail morphology and we will now look at an example in more detail.

## Enzymes associated with virus particles

In addition to structural proteins, some virus particles contain enzyme proteins. These are enzymes made by the virus and packaged into particles during the assembly stage in the infected host cell. All these enzymes are products of the viral genome and so are referred to as **virus-specific enzymes.** Usually these enzymes are not made by host cells and so the virus has to take them into a new host cell to initiate infection. In some other cases, these enzymes in virus particles are similar to host-cell enzymes and are needed early in the infection process. At this stage the host-cell enzyme may not be available to the virus (it might be in a different part of the cell) and so the virus particle carries its own enzyme into the cell with it.

The range of enzymes associated with virus particles is fairly restricted. Some virus particles (influenza virus, poxviruses) contain RNA polymerase enzymes needed for the transcription of the viral genome to give the first molecules of viral mRNA formed during infection. Others (hepatitis B virus, the causative agent of serum hepatitis) may contain DNA polymerases needed for the replication of the viral DNA. Particles belonging to the retroviruses (HIV) contain a very characteristic enzyme, reverse transcriptase, which makes a complementary DNA copy of the viral RNA genome. A few virus particles (influenza virus) have enzymes that affect

**BOX 10.8** The structure of phage T4.

**A closer look at phage T4**

The head of phage T4 is roughly hexagonal in shape and measures 85 × 110 nm. Its shell is built from capsomeres that are 5 nm in diameter.

The tail is 110 nm long and 25 nm in diameter. This phage belongs to the group that has **contractile tails**. It has a central hollow tube with a contractile sheath surrounding it. The proteins which make up the contractile sheath can change their shape, causing the sheath to contract. This plays a vital part in the infection process and provides the mechanism by which the virus pushes the central tube into the bacterial cell wall to puncture it. The virus then injects its DNA into the cytoplasm of the host cell.

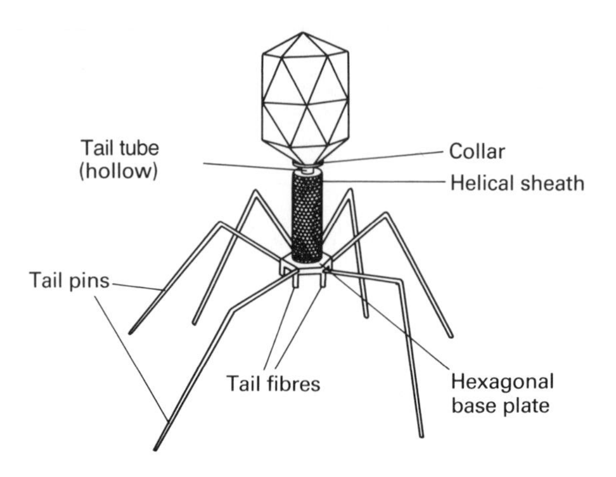


The tail is connected to the head by a thin disc or collar which has **whiskers** attached. At the tip of the tail is a hexagonal **base plate** which carries a tail pin at each corner. There are also six long thin tail fibres attached to the base plate. Each part of the virus particle is built from several kinds of structural proteins and, in all, it contains over 30 structural proteins.

the interaction of particles with the surface of the host cell; for example, spikes with neuraminidase activity.

## The viral genome

Virus particles carry inside them the viral genome. This genomic material is in the form of either DNA or RNA, but never both together. For this reason, we tend to distinguish **DNA viruses** and **RNA viruses**. However, a further sub-division is also made. The nucleic acid within the virus particles exists as either a single-stranded molecule or in a double-stranded (duplex) form. The distribution of these four types of viral genomes among different types of viruses is shown in Table 10.1.

**Table 10.1** Distribution of viral genomes

| *Type of nucleic acid* | *Bacterial viruses* | *Plant viruses* | *Animal viruses* |
|---|---|---|---|
| ssRNA | Common | Most common | Common |
| Examples: | RNA coliphages (MS2, Qβ) | Majority of plant viruses | Poliovirus<br>Influenza virus |
| dsRNA | Very rare | Very rare | Not common |
| Examples: | | Reovirus | Reovirus |
| ssDNA | Not common | Very rare | Rare |
| Examples: | Small DNA phages (ϕX174) | Maize streak virus | Parvoviruses |
| dsDNA | Common | Very rare | Common |
| Examples | Large complex phages (T2, T4, T6) | Cauliflower mosaic virus | Herpesvirus<br>Poxvirus<br>Adenovirus |

ds, double-stranded; ss, single-stranded

Some further explanation of the term **single-stranded** is needed. What this means is that the DNA or RNA molecules consist of one polynucleotide strand. However, the molecules will usually contain sequences in one part that are complementary to those in another. Hence, the two sets of complementary sequences will base-pair to give short sections of double-stranded nucleic acid within the one molecule. Thus, the genome is described as single-stranded, but it will contain sections of double-stranded nucleic acid where the molecule is folded upon itself. These structures usually have some functional significance. For example, they may play a part in the initiation of the replication process, or they may be involved in the initiation of the synthesis of virus-specific proteins.

Canine parvovirus contains single-stranded DNA in its particles and, characteristically, the sequences at its 5′ and 3′ termini fold back to form double-stranded hairpin structures at each end (Figure 10.13). In this case, these structures pay a vital part in replication of the viral DNA. This is a relatively simple type of secondary structure. In the small RNA phages (MS2, Qβ), much of the RNA molecule is folded upon itself and base-paired together, and this secondary structure is involved in regulating the expression of the viral genes.

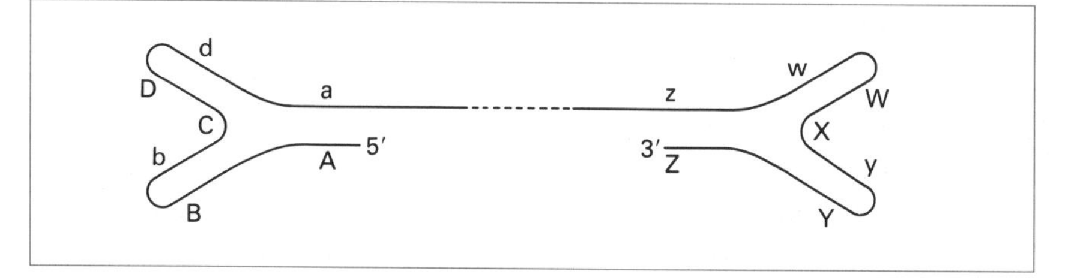


**Figure 10.13** Diagram to show the genome structure in canine parvovirus. The ends of the DNA molecule have complementary sequences that allow them to fold back and base-pair to give hairpin structures that are Y-shaped.

## Size of viral genomes

Viral genomes range in size from a minimum of 3.5 kilobases (kb) to a maximum of about 250 kb. These are useful units in which to measure viral genomes because 1 kb (a run of 1000 bases) of genome will code for a polypeptide of about 33 000 molecular weight, and this is roughly the size of an average protein.

The viruses with the smallest complete genomes are the RNA phages, such as MS2 or Qβ, that infect *E. coli.* These have genomes which code for just four virus-specific proteins. At the other end of the scale are the poxviruses which have genomes that code for 200 or 300 viral proteins.

Some care is needed with this type of approach because in certain small viruses the coding capacity on the viral nucleic acid is increased by arranging the genes so that they overlap with one another. This is a very distinctive feature of phage ϕX174 which has nine or ten genes carried on just 5.2 kb of viral DNA.

Virus particles almost always contain only one copy of the viral genome and so can be thought of as **haploid.** However, there is one notable exception to this rule and this is the retrovirus group, which includes HIV. These virus particles are **diploid** containing two almost identical copies of the viral genome held together near one end by a short region of complementary base-pairing.

**Overlapping genes**
Different genes can be read off a single DNA sequence by starting transcription of the mRNA at different points. In this way, a given sequence of DNA can code for more than one gene product.

5′ end
START gene 1 here
↓
G G C T A A T C G A . .
↑ .
START gene 2 here .
. . . . . . . . . . . . .
.
. 3′ end
. END gene 1 here
. ↓
. . . . T A A G G C G C
↑
END gene 2 here

## Viral DNA

The single-stranded DNA inside virus particles can be in the form of a simple linear molecule with two free ends (canine parvovirus) or it can be a circular molecule formed by covalently joining the 5′ and 3′ ends together (phage ϕX174).

Many viruses have circular double-stranded DNA in their particles in which both strands of the duplex are covalently joined together. These include the tumour-forming polyoma viruses of animals. Such viruses are unusual in that they have histone-like proteins associated with their DNA, making it resemble the chromatin in eukaryotic cells.

The double-stranded DNA of phage λ should also be mentioned. This DNA is really linear double-stranded DNA, but has the characteristic property of being able to circularize under suitable conditions (Figure 10.15). The linear form of the molecule has short single-stranded extensions to the strands at each 5′ end. These contain inverted complementary sequences of bases. Thus, the two ends of the molecule can combine together and base-pair to give the circular form, although obviously the circular molecule formed has a gap in each strand, not too far apart, and so it is not covalently closed. The DNA from this phage is said to have

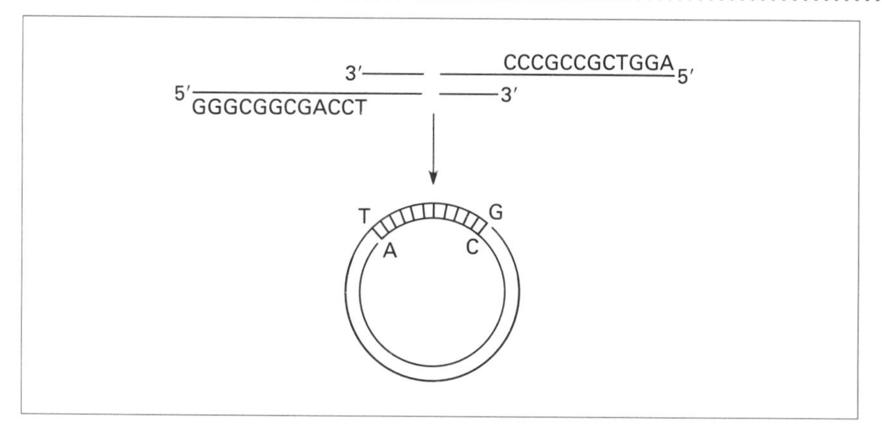


**Figure 10.14** Diagram of phage λ DNA. The linear molecule has single-stranded 'sticky' ends. The molecule can circularize and the complementary base sequences pair.

cohesive or 'sticky' ends. The ability to circularize in this way is an important requirement if the phage DNA is to establish a lysogenic relationship with its host cell (see page 225). The circular form is a prerequisite for integration into the host cell DNA.

The DNA of all viruses contains the usual four bases: guanine, cytosine, adenine and thymine. However, the DNA of some of the large complex phages is different because it can contain unusual bases. In phages T2 and T4, the usual cytosine is replaced by 5-hydroxymethylcytosine (HMC) (Figure 10.15). The presence of this modified base in the viral DNA protects it from attack by host cell nuclease enzymes. Similar types of modification are found in the DNA of other phages. For example, the methylated base 5-hydroxyuracil replaces thymine in *Bacillus subtilis* phage SP01, and 2-amino adenine replaces adenine in phage S-2L (which attacks the marine cyanobacterium *Synecoccus*). These modifications presumably all serve the same function as that of HMC. The DNA of animal and plant viruses is free from any modifications of this sort.

## Viral RNA

The RNA viruses can carry either single-stranded or double-stranded RNA in their particles, but the RNA molecule is always linear. The RNA genome contains all the genetic information needed for the infection

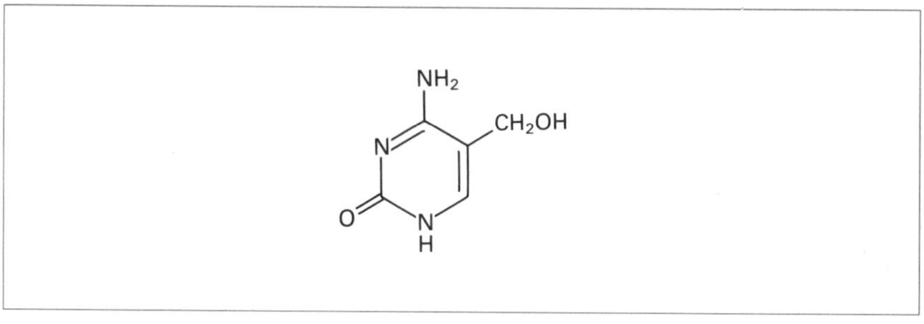

**Figure 10.15** 5-Hydroxymethylcytosine (HMC). HMC does not occur in uninfected bacterial cells and so its synthesis must be directed by phage genes. The enzymes to make HMC are synthesized early in infection. HMC is made by methylating the nucleotide before it is incorporated into the phage DNA.

process and the production of new virus particles. This important concept was first demonstrated by Gierer and Schramm in 1956 at Tübingen. This German group adapted a method to use phenol to strip the coat protein from TMV particles and then showed that the pure RNA, when inoculated into healthy tobacco plants, caused not only mosaic disease but also the formation of progeny particles. This simple type of experiment has since formed the basis for a fundamental division of the single-stranded RNA viruses into two major groups.

The first group comprises all those viruses whose RNA, when purified from virus particles and introduced into susceptible host cells, cause infection. This group includes the majority of plant viruses, some animal viruses – notably the picornaviruses (poliovirus, foot-and-mouth disease virus) – and the simple RNA phages that infect *E. coli* (MS2, Qβ). In all of these, infectivity of the naked viral RNA is taken to mean that inside the cells, the RNA acts directly as mRNA and so is translated on the host cell ribosomes. These viruses are called **plus-strand RNA viruses** because their genomes consist of RNA that has the same sequence as the viral mRNA (which is defined as having positive-strand RNA because it can be translated).

The other major group of single-stranded RNA viruses is formed by those in which the RNA, when extracted from virus particles, is *not* infectious if introduced into suitable host cells. This group includes the orthomyxoviruses (influenza viruses), the paramyxoviruses (mumps and measles virus) and the rhabdoviruses (rabies virus, lettuce necrotic yellows virus). The RNA in these virus particles is complementary to the viral mRNA that appears in infected cells. These are **minus-strand RNA viruses** because their genomes consist of RNA that is of negative polarity with respect to their mRNA. The RNA inside the virus particles cannot be translated directly into virus proteins by the host-cell ribosomes.

In the plus-strand RNA viruses, the RNA in the particles is often modified in structure to allow it to function in the host cell directly as mRNA. It will be recalled that eukaryotic mRNA molecules typically have a 5′ cap structure and a 3′ poly(A) tail. The RNA of viruses that infect eukaryotic host cells commonly carries these structures.

Caps are present on the 5′ ends of all the animal plus-strand viruses with one exception, the picornaviruses. The members of this group have a small protein called VPg (virion protein g) instead of the cap. In these cases VPg is not related to the mRNA function, but is implicated in initiating the RNA replication process.

In plus-strand plant viruses, some viral RNA molecules have caps (TMV, TYMV) but others do not. Likewise, a few of the plant viruses have small protein molecules attached to their RNA at the 5′ ends which are believed

**Generalized cap structure in eukaryotic mRNA**

5′ end
$m^7G^{5'}\text{-ppp-}^{5'}X^{(m)}\ p\ Y^{(m)}$
pApBpCpD . . . .

The terminal 7-methylguanosine residue (the guanine base is methylated) is linked to the nucleotide X by an unusual 5′ to 5′ linkage, spanned by a triphosphate bridge.

The last-but-one nucleotide X may also be methylated at its 2′ position (on the ribose sugar component). Sometimes the 2′ position of neighbouring nucleotide Y is also methylated.

(Plant viruses which have caps do not methylate either nucleotides X or Y in their RNA molecules.)

to serve a similar function to the VPg in picornaviruses. The RNA molecules from phage particles, like bacterial mRNA, always lack any cap structures.

Most species of mRNA from eukaryotes normally have poly(A) tails which are, typically, runs of between 20 and 100 adenylic acid residues. Such poly(A) tails are present on the RNA of most of the animal plus-strand viruses. However, poly(A) tails are rarely present on the RNA of the plant plus-strand viruses. The viral RNA from TMV and TYMV lacks poly(A) tails.

## Segmentation of the viral genome

Not all the RNA viruses have their genomes carried on a single unbroken molecule. Some have their viral genes distributed in a specific way on a number of different segments of RNA which collectively form the viral genome. This type of organization is described as a **segmented genome**.

A common feature of all the double-stranded RNA viruses is that they have segmented genomes. These viruses collectively form the **reovirus group**, which has members that infect vertebrate animals, insects and plants. One of the reoviruses that infect humans, rotavirus, is becoming

BOX 10.9

**Multipartite plant viruses**

Multipartite plant viruses are all single-stranded RNA viruses which have segmented genomes. But in these cases, the segments are not packaged together in the virus particles. Instead, the segments are separately encapsidated with coat protein to form different types of virus particles.

There are two types of multipartite viruses. Some have their viral genome divided into **two** pieces (bipartite); others have their genome divided into **three** pieces (tripartite).

We will consider just one example of a multipartite virus. **Tobacco rattle virus** (TRV) has the bipartite arrangement which is quite common among plant viruses. TRV forms rod-shaped particles with a diameter of 23 nm. From infected plants, it is possible to separate two types of particles:

| Long particles | Short particles |
|---|---|
| 180–210 nm long | 45–110 nm long |
| These contain RNA-1 | These contain RNA-2 |

Both types have the same coat protein.

The long particles contain RNA-1 which carries all the genetic information for virus replication but does not contain the gene for the virus coat protein. The short particles carry RNA-2 which has the gene for coat protein but none of the information needed to replicate it.

Plant infections occur only when both types of particles are present together. Only then can properly coated long and short particles of TRV be formed in infected tissues.

increasingly important as an agent of diarrhoeal disease. The various reoviruses have their double-stranded RNA genomes divided into 10, 11 or 12 segments. The precise number is characteristic for a particular virus and each segment is unique and carries one or two specific viral genes. The reovirus particles carry one copy of each segment.

Besides the double-stranded RNA viruses, single-stranded RNA viruses can also have segmented genomes. Of these, the most important is the orthomyxovirus group (influenza virus). The genome of influenza virus is divided into eight different segments, each one forming a segment of nucleocapsid inside the enveloped virus particles. In the influenza virus, each segment codes for just one specific viral protein, with the exception of the two smallest segments which each specify two proteins using overlapping genes. In the influenza virus the different viral genes have thus been ascribed to individual segments of viral RNA.

**Rotaviruses**
Rotaviruses are so-called because in the electron microscope their particles (70 nm diameter) resemble wheels with a central axle and radiating spokes. They belong to the reovirus group and the particles contain 11 segments of double-stranded RNA.

Rotaviruses are being increasingly recognized as the major causative agent of non-bacterial diarrhoeal disease. Infants and young children are at greatest risk. They are commonly transmitted through water. In routine water analysis they may be separately enumerated from **enteroviruses** which are the most important group of water-borne viruses. In developing countries rotaviruses may be responsible for nearly two million deaths per year.

## Viroids

Viroids were first discovered by T.O. Diener in 1962. He was working on the spindle tuber disease of potatoes, which caused abnormal spindle-shaped potatoes and reduction in crop yields. The causative agent seemed to resemble a plant virus in many respects. It could be easily transmitted to healthy plants in extracts of diseased tissue. However, if these extracts were centrifuged at high speed, which would normally spin virus particles down to a pellet in the bottom of the tube, the infectious material remained in the supernatant fraction. The reason for this is that the potato spindle tuber disease is caused by an agent that is very much smaller than any plant virus. It is unique in that it consists of just RNA (with no coat protein).

**BOX 10.10**

**How influenza viruses vary**

Influenza viruses can be divided into three **types**: A, B and C. Type A is the most common type and is responsible for the most serious outbreaks of influenza in humans.

Each type, but especially type A, undergoes genetic variation naturally and new **strains** of influenza virus arise from time to time that can be distinguished by tests with specific antibodies (these strains are called **serotypes**). The strains differ in the antigenic structure of two of the surface proteins on the virus particles. These are the glycoproteins that make up the surface HA and N spikes (see page 308). Both of these surface antigens can undergo two types of variation:

1. **Antigenic drifts** are minor changes in antigenic structure caused by point mutations (see page 184) and are caused by simple amino acid substitutions in the HA and N proteins. So the structures of HA and N change slightly over the years because of the slow accumulation of point mutations. As these proteins change, so their antigenicity alters and immunity built up against them over the years becomes less effective or useless. These changes explain local, relatively mild influenza epidemics that appear every few years.

**BOX 10.10** *continued*

2. **Antigenic shifts** are major and dramatic changes in the antigen structure of virus particles and can give rise to the appearance of new strains of virus. These are responsible for the world-wide epidemics of influenza that appear less frequently.

There are several different versions of the HA and N proteins. There are 13 known serotypes of HA and nine known serotypes of N. Before 1968 the predominant strain of influenza virus was strain H2N2 called **Asian 'flu** and most people had developed immunity to it. In 1968, a new strain appeared, called **Hong Kong 'flu**. In these virus particles, the HA protein had changed dramatically to the H3 variant. Many of the population were not immune to it and a new epidemic appeared.

How did this new strain of 'flu virus appear? Because of segmentation of the viral genome, it is reasonable to believe that H3N2 (the Hong Kong strain of influenza virus) arose by re-assortment of the viral genes caused by a double-infection of a host cell with H2N2 and a strain with H3. In this way different segments of the influenza genome from different strains could be combined to give a new type.

Strain H3N2 has been circulating since 1968, but since then has undergone considerable antigenic drift.

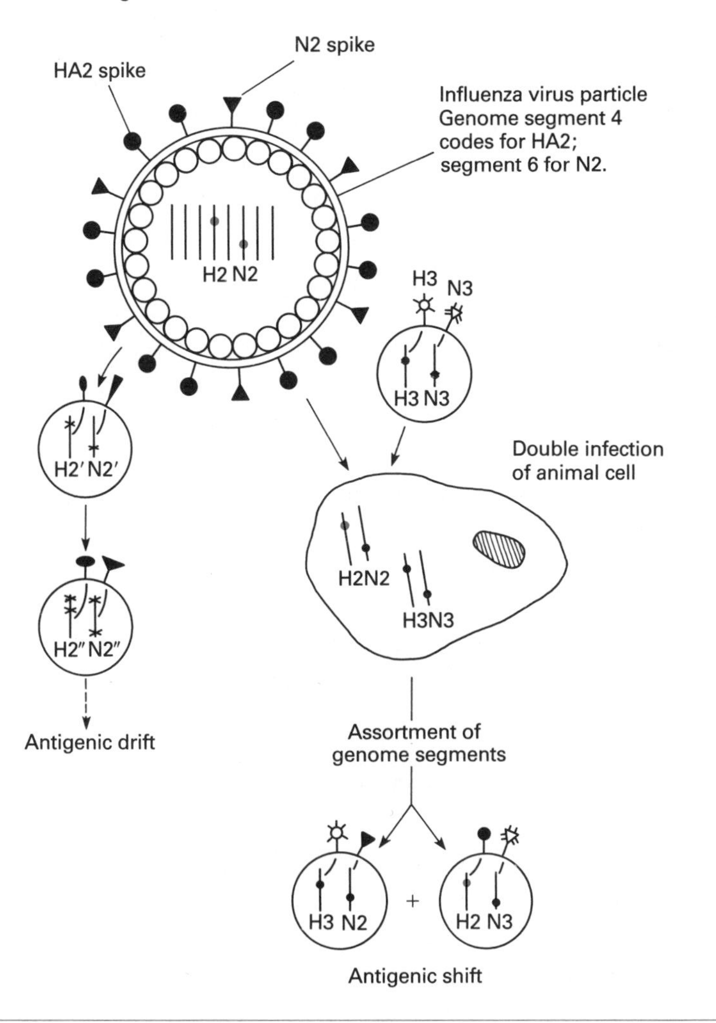

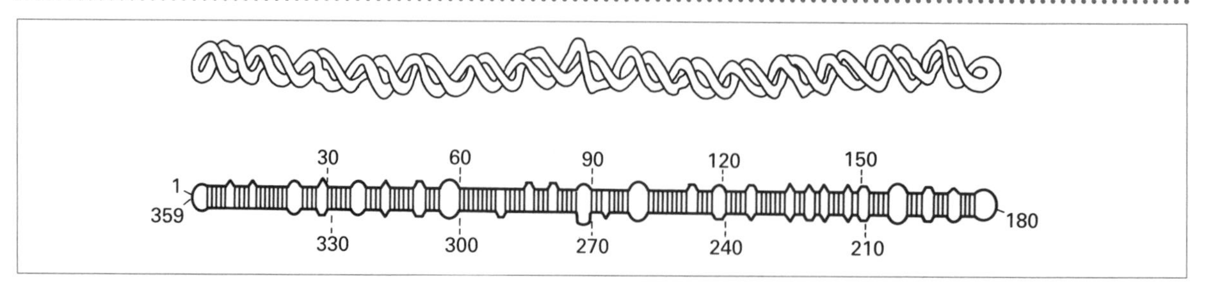


**Figure 10.16** Structure of potato spindle tuber viroid. The upper diagram shows how the molecule of viroid RNA twists naturally to form a rigid rod (about 30 nm long). The lower diagram illustrates complementary base-pairing in the molecule.

This RNA has the ability to replicate in host tissues *on its own.* It does not code for any coat protein and it never encapsidates its RNA to form structures that resemble virus particles. Diener called this type of agent a **viroid** to distinguish it from a conventional virus. This first such viroid to be characterized was **potato spindle tuber viroid** (PSTV).

Other viroids have since been isolated and identified. There are at least 11 others that cause plant diseases. They cause symptoms such as discoloration of leaves and fruits, retardation of growth and plant malformations. They all cause economic loss in crop plants (tomatoes, hops, avocado, coconut) and horticultural plants (chrysanthemums).

What is unique about viroids is that their RNA molecules are so very small. All plant viroid RNA molecules range in size from 240–380 bases; PSTV has 359 bases. For comparison, TMV-RNA has about 6340 bases. Viroids all have single-stranded RNA in the form of a covalently closed circular molecule. Within this RNA are complementary sequences that allow most of the molecule to base-pair with itself. Overall, the viroid RNA takes the form of a short, rigid, rod-shaped molecule. The structure of PSTV is shown in Figure 10.16. All viroids show this characteristic arrangement with short stretches of double-stranded RNA and alternate unpaired regions, giving rise to internal 'bubbles'. At each end of the molecule is a small unpaired loop.

Other viroids have different RNA sequences, but they all show this type of secondary structure and form rod-shaped molecules about 30 nm long. These simple RNA molecules are infectious if inoculated into suitable host plants. During infection these viroid molecules are replicated and, somehow, although the details are far from clear, give rise to the symptoms of disease.

If the RNA sequence in PSTV is carefully examined, we find that the RNA does not contain the codon AUG (the start codon for translation of mRNA). Even though the viroid RNA is infectious, it appears that it cannot be translated in the host cell and so does not directly have a mRNA function. The same applies to a complementary copy of the viroid RNA. No viroid-specific proteins have ever been detected in infected tissue.

The possible origins of the plant viroids have attracted a great deal of speculation. The view at present seems to be that they represent introns that have 'escaped' from their host plant genomes during evolution. The evidence for this comes from the fact that both viroids and introns in certain types of plant genes contain some common characteristic RNA sequences.

There is very little evidence to show that viroid-like agents infect animal cells. To date, the closest example seems to be the **human hepatitis delta virus** (HDV), formerly referred to as human hepatitis delta agent. This virus is associated with a particularly virulent form of hepatitis. HDV has single-stranded RNA which resembles a viroid in structure. It is in a closed circular form, but contains only 1678 bases, making it far larger than any plant viroid. HDV-RNA also contains complementary sequences in its RNA that allows it to base-pair and form rigid, partially double-stranded, rod-shaped molecules. However, HDV is different because it does encapsidate its RNA forming virus particles with a diameter of 30–40 nm.

HDV is reliant upon a 'helper' virus to function in its host cells. The helper virus is hepatitis B virus (HBV, a DNA virus) and HDV can replicate only with the help of this virus. HDV multiplies in liver cells with its helper and increases the severity of the symptoms that would otherwise be caused by HBV alone. The particles so formed by HDV seem to contain the same structural protein as HBV. HDV is certainly a novel agent, unlike anything else that infects animals. It may not closely resemble a plant viroid, but it can be described as a **sub-viral agent**, in just the same way as a viroid, since they both lack the full attributes of a complete virus.

## Assaying viruses

Various methods can be used to determine the number of virus particles in a sample. The problem with all of them is that in a given suspension there will always be a small proportion of virus particles that are **non-infectious**. There are several possible reasons for this: some particles may be damaged or imperfectly formed; others may be mutant virus particles; yet others may

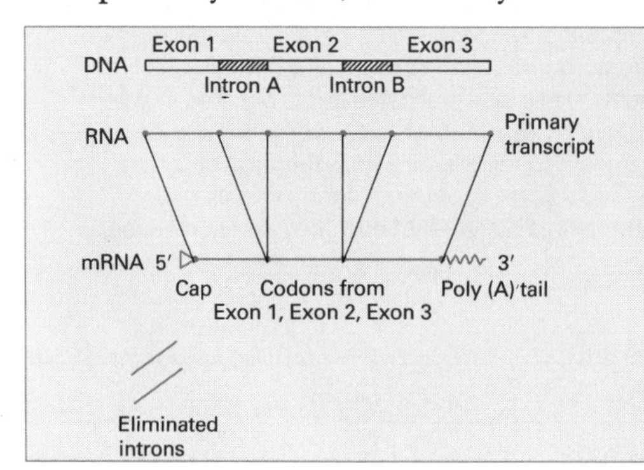


**Introns and exons**

In eukaryotes, the coding sequences in genes are split into pieces. The expressed sequences in the DNA are called **exons**. In between these are non-functional sequences or intervening sequences called **introns**. The size of introns range from 10 to 600 base-pairs of DNA.

These introns have to be removed from the final mRNA molecules for the gene. They are cut out of the primary RNA transcript and the coding sequences (exons) are joined together (spliced) to give the final functional mRNA molecule.

**BOX 10.11**

**Prions**

Prions are a unique class of infectious agent that are quite different to either viruses or viroids. The first prion to be characterized was the agent associated with **scrapie**.

Scrapie is a disease of sheep and goats that is widespread throughout Europe, Asia and North America. The disease is called scrapie because infected animals rub and scrape their skin against walls and trees.

Diseases of this type are technically described as **subacute transmissible spongiform encephalopathies**. They are characterized by a loss of motor control in affected animals, causing loss of coordination. Animals stagger, generally become unsteady and cannot walk. This type of disease also leads to dementia. The symptoms are caused by a gradual degeneration of the tissues of the central nervous system. Characteristically, such diseases have very long incubation periods. In the case of scrapie, it is at least 6 months before symptoms start to show.

There are two human disease of this type that should be mentioned. **Creutzfeldt-Jacob disease** affects people in their late middle age and causes dementia and death within 6 months. **Kuru** is a disease limited to the Foré tribe in Papua, New Guinea. This disease shows an interesting pattern of infection, with women and children mostly affected. Its prevalence seemed to associated with the unusual burial rites of this tribe, since banned. Formerly, it was the custom for women to scoop out the brains of deceased kinsmen and then to rub them into the body (to instil wisdom). Sometimes they were possibly even eaten!

All diseases of this type are caused by an infectious agent that is filterable and will cause disease if infected into a healthy animal. But this agent is unlike any conventional virus or viroid. No virus particles have ever been detected in diseased central nervous system tissue and no RNA or DNA has ever been associated with infectious material.

Agents of this type are also far more resistant than viruses to inactivation by chemical agents (formaldehyde), heat and ionizing radiation. Another peculiarity is that, unlike viruses, these agents do not elicit any immune response in infected animals.

The only macromolecule associated with infectivity is a protein of 27 000–30 000 molecular weight isolated from amyloid fibrils in infected brain tissue. This protein co-purifies with scrapie infectivity from extracts of diseased brain. It is, in fact, a glycoprotein.

The infectious agent is destroyed by digestion with a proteolytic enzyme, but not by treatment with RNase or DNase enzymes. This reinforces the view that no nucleic acid is associated with infectivity.

This 27 000–30 000 protein seems to represent a novel type of infectious agent and Prusiner has called it a **prion** (for *pro*teinaceous *in*fectious agent).

In 1986, the scrapie agent attracted renewed interest when it became apparent that '**mad cow disease**' in Britain and Ireland was caused by a very similar type of pathogen. The evidence for this came from histological examination of brain tissue from affected animals which revealed the tell-tale sign of spongiform encephalopathy never seen before in cattle. The disease was named **bovine spongiform encephalopathy (BSE)**. The original source of infection of BSE is thought to be commercial protein concentrates fed to cattle, particularly dairy cattle. The protein concentrate was manufactured from sheep meat and bone meal contaminated with scrapie. It is probably in this way that cattle herds became infected and so the scrapie agent seems to have crossed the species barrier from sheep to cattle. This source of infection has since been eliminated from the food chain.

Prions seem to be a novel type of infectious agent and, if they really do just contain protein, it will be of great interest to discover just how they cause disease.

be assembled as just empty shells with no viral genome. Some assay methods will include all of these particles in the total. However, there are other methods that aim to measure just the number of infectious virus particles,

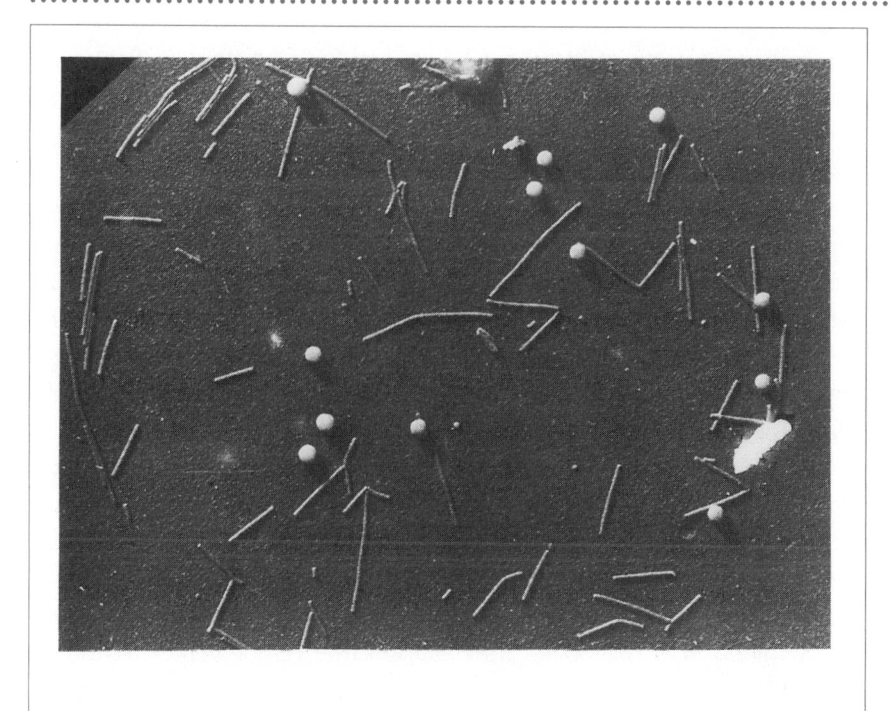

**Figure 10.17** A direct count of tobacco mosaic virus in the electron microscope. The virus suspension has been mixed with a suspension of latex beads (264 μm in diameter). The concentration of the latex beads is known and from the ratio of the beads and rod-shaped particles it is possible to make a direct count of the TMV particles. Reproduced with permission of Rothamsted Experimental Station, AFRC.

and these methods are called **infectivity assays**. We shall look at three of the most useful types of assay for viruses.

### Direct counts in the electron microscope

Virus particles can be counted directly in the electron microscope. However, the difficulty with this technique is to relate the number of virus particles to the volume of the original sample applied to the microscope grid. Remember that, in the microscope, only a small area of the specimen grid can be examined at once. The way round this problem is to mix the unknown virus suspension with an equal volume of a suspension of tiny latex beads (at a known concentration). In the microscope, counts can be made of the number of latex beads seen in a particular field of vision, and of the number of virus particles in that same field (Figure 10.17). By making these counts, it is possible to determine the number of virus particles from the original bead concentration. Thus, if there were $10^8$ latex beads per ml of original suspension, and the microscope count showed 15 virus particles for each bead, then the virus particle concentration would be:

$$15 \times 10^8 = 1.5 \times 10^9 \text{ virus particles/ml}$$

Obviously, this type of assay will not distinguish between infectious and non-infectious particles, but it can be used with virus particles of any type, provided they can be recognized and counted in the electron microscope.

### Haemagglutination assay

Many different animal viruses, both with and without envelopes, have the ability to agglutinate red blood cells of various animals. This feature forms the basis of a quick and simple method called the **haemagglutination (HA) assay**. Haemagglutination is caused by surface proteins on intact virus particles, but can also be brought about by proteins released from damaged virus particles. For example, haemagglutination can be caused by the HA spikes released from influenza virus particles on disruption with ether.

Because of the huge differences in size between a virus particle and a red blood cell, a given virus particle can bind only two red blood cells together. (It is rather like trying to join balloons together with tiny bits of sticky chewing gum; one piece of gum can join only two balloons together!) In a mixture of the two, if the number of virus particles (pieces of gum) exceeds the number of red blood cells (balloons), they will link the cells together to form a three-dimensional lattice which will settle out of suspension and cause haemagglutination. But, if the number of cells (balloons) exceeds the number of virus particles (pieces of gum), there will not be not enough virus particles to link up the cells (other than in pairs) and agglutination will not occur.

The HA assay is performed by *diluting the virus to the end point.* What this means is that the virus suspension is diluted to the point that *just* gives haemagglutination. This is called the **HA titre**. In its simplest form serial two-fold dilutions of the virus (1/2, 1/4, 1/8, and so on) are made in wells in a special plastic plate. These are mixed with a standard amount of red blood cells and incubated for 1 hour. After that time, non-agglutinated cells will have settled to the bottom of the wells to form a compact 'button' of cells. Agglutinated cells form a thin even 'shield' over the bottom of the well. In this way, the titre is determined from the greatest dilution which *just*

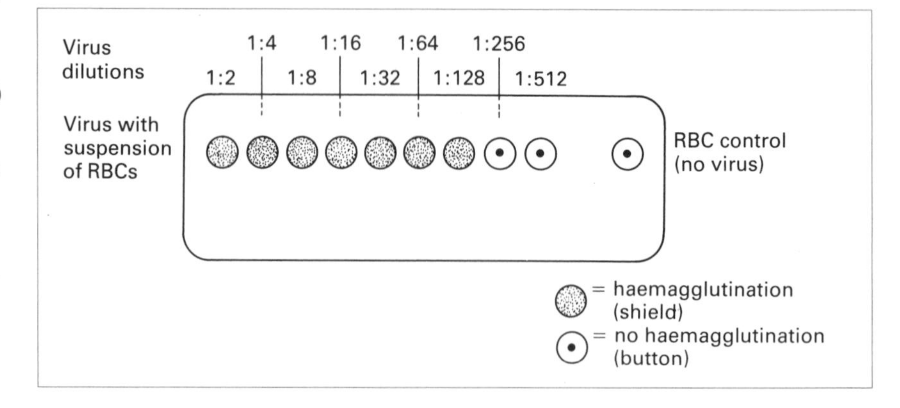


**Figure 10.18** Haemagglutination assay. Serial 1/2 dilutions of the virus have been prepared in wells on a plate. Standard numbers of red blood cells (RBC) were added. The plate was then incubated for 1 hour. Shields of agglutinated cells have formed in wells down to the 1/128 dilution. There are buttons in the 1/256 and 1/512 dilutions. The titre of the virus sample is 128 HA units. The RBC control contains no virus, just red blood cells. It is to show that cells settle and form a button without adding the virus.

gives haemagglutination. Usually this is expressed as the reciprocal of the dilution; that is, 128 HA units rather than 1/128 (Figure 10.18).

The HA assay was first devised by Hirst in 1941 for influenza virus and provides a quick method to determine the amount of virus particles in a sample. In practice, the HA assay is a relatively insensitive way of measuring the numbers of virus particles. With influenza virus, for example, as many as $10^7$ virus particles are needed to cause visible agglutination of chick red blood cells.

## Infectivity assays

Infectivity assays are all biological assays and depend upon virus infection of some sort of living cellular system. Infectivity assays give results in numbers of **infectious units**. An infectious unit is the amount of virus particles needed to initiate an infection and cause a biological response that can be observed in some way, and scored. An infectious unit may be just one virus particle, but more usually it is a larger number of particles, and in extreme cases may represent as many as one or two million particles. There are several types of infectivity assay and we will describe a few examples.

**(a) Plaque assay.** The plaque assay was developed in the 1930s for the assay of bacteriophage particles.

About $10^7$ cells of host bacteria are seeded into 2–3 ml of molten agar gel. The bacteria should greatly outnumber the phage particles that will be added next. A standard volume (usually 0.1 ml) of the unknown phage suspension is then added to the molten agar and the mixture is poured over the surface of a plate of nutrient agar. The overlay is allowed to solidify and the plate is then incubated for 24 hours.

At the start, the phage particles diffuse through the top layer of agar until they encounter and infect a bacterial cell. Such host cells are immobilized in the gel. The infection causes the bacterium to burst and the progeny virus particles spread to the neighbouring bacterial cells to infect them (Figure 10.19). At the same time, uninfected bacteria grow in the top layer of agar to form a visibly turbid 'lawn'. At the sites of infection by the phage, holes appear in this lawn. These holes are called **plaques**. They do not usually get larger than a few millimetres in diameter because the agar limits the spread of the virus. Each plaque is assumed to have arisen from one infectious unit of virus.

By counting the number of plaques, the number of infectious units can be obtained. Normally, though, a suspension of phage has to be diluted through a dilution series (typically $1/10$, $1/10^2$, $1/10^3$, and so on) before undertaking a plaque assay, otherwise the plates may have too many

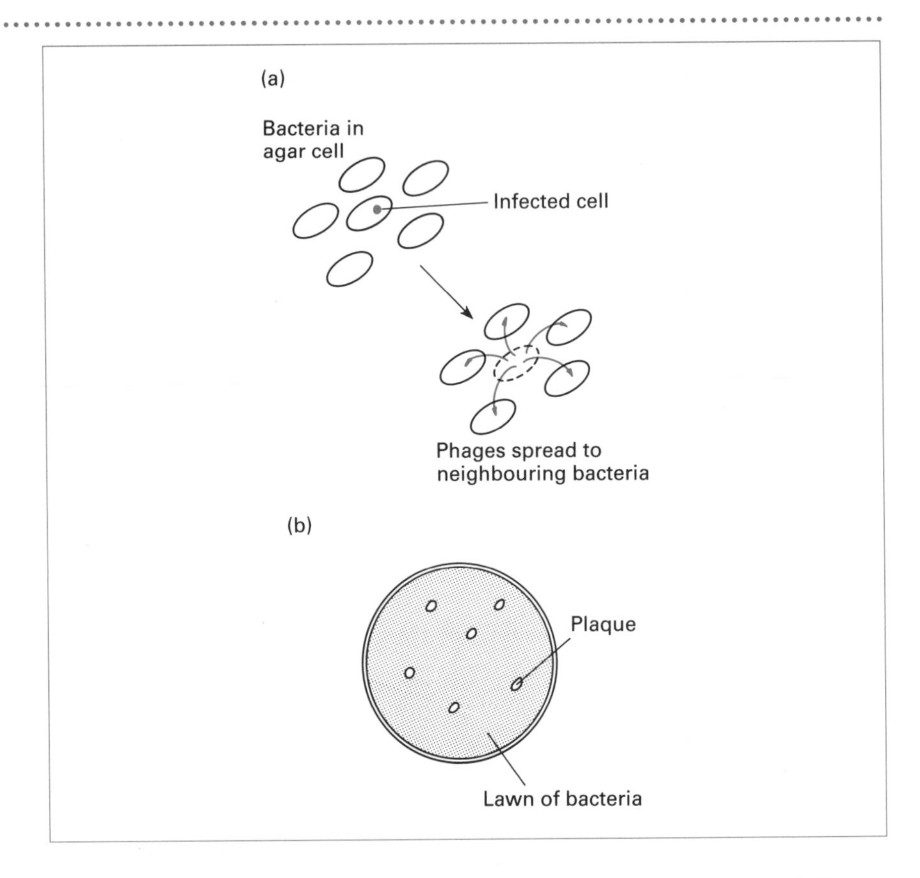


**Figure 10.19** Plaque assay of bacteriophage. (a) Particles of phage are added to excess bacterial cells in molten agar. The mixture is poured onto a nutrient agar plate and allowed to solidify. In the gel, infected bacterial cells are immobilized. They burst open and phage particles can spread only to neighbouring cells. (b) The bacteria multiply to form a 'lawn' in which are small, round, clear zones called 'plaques'. Each plaque is derived from a single infected bacterium. The number of plaques can be counted to determine the concentration of 'plaque forming units' (pfu's) in the original sample.

plaques to count. The results of this assay are expressed in terms of **plaque forming units (pfu)** to indicate that these are counts of units of infectivity rather than numbers of virus particles.

Suppose that 0.1 ml of phage suspension diluted to $10^{-5}$ of its original concentration gives 125 plaques. Then the concentration in the original suspension is:

$$125 \times 10 \times 10^5$$
$$= 1.25 \times 10^8 \text{ pfu/ml}$$

Here you have to correct for two things: only 0.1 ml of virus is added; the suspension has been diluted by $10^5$.

In 1952 the American virologist Renato Dulbecco developed the plaque assay for animal viruses and this is used routinely in animal virology. This research in virology was acknowledged by the award of a Nobel Prize in 1975. This method relies upon growing monolayers (single layers) of cells attached to the bottom of culture dishes or bottles. The technology for growing animal cells in this way is rather complicated and the cells grow much more slowly than bacterial cells in culture. A known volume of virus sample is added to the cells and the monolayers are allowed to stand for a while so that the virus particles can attach themselves to the cells. The cells are then covered with a special thin gel. This gel is to restrict the spread of virus particles over the plate as they are released from infected cells. Plaques develop on the plate at sites of infection and can be counted from 1 day to 3 weeks after inoculation, depending on the virus and the cells. The results of this assay are given in pfu/ml, in just the same way as phage plaque counts.

Plaques may not always be visible to the naked eye and sometimes they have to be revealed in special ways. For example, the plates may be flooded with special stains (called vital stains) that stain just the living cells, but not the dead ones.

Cultured animal cells may respond to virus infection in a number of ways. The changes that occur in cells during infection may make them swell (a process called 'ballooning'), cluster together, form **syncytia** (multinucleate cell masses), or even die. In all these cases cells may detach themselves from the surface of the plate. These changes are all described as **cytopathic effects** ('cpe's).

Tumour-forming viruses can also be assayed in the same way. These viruses do not necessarily destroy their host cells and so they do not produce plaques with obvious cytopathic effects. Instead they *transform* their host cells. This means that they change a cell from a normal state to one with the properties of a malignant or cancerous cell. These transformed cells tend to multiply at a faster rate than uninfected cells and so they proliferate to form little mounds of cells, rather like microtumours on the plate. These are visible to the naked eye and can be counted. The infectious units in these assays are described as **focus forming units** and so infectivity may be expressed as ffu/ml.

**(b) Local lesion assay for plant viruses**. The local lesion assay for plant viruses was first devised by F.O. Holmes in 1929. He inoculated TMV onto the leaves of *Nicotiana glutinosa* (a close relative of tobacco) and observed that the number of necrotic lesions formed on leaves was related to the concentration of the virus sample applied. This formed the basis of the local lesion assay for plant viruses.

The local lesion assay for a plant virus requires a test plant that will give a hypersensitive response to the virus and produce lesions that can be counted. Host plants like tobacco and French bean are preferred because they can be grown quickly and easily in a greenhouse. For the assay, the virus sample is inoculated onto the surface of a leaf of the test plant. Typically this is done by dipping a finger into the virus suspension and rubbing it onto the leaf. The rubbing action creates wounds to the cells of the upper epidermis and leaf hairs through which virus enters the leaf cells to initiate infection. Depending on the virus and plants, lesions will appear after about 4 days. Figure 10.6 shows lesions formed by TMV on tobacco.

The infectivity of a virus preparation can be assessed by counting local lesions formed on leaves. The result is usually given as lesion number per leaf or half-leaf (if just half of a leaf on one side of the mid-rib is inoculated). In this way the method allows comparisons to be made of infectivity in two or more samples.

This method is far more crude and far less sensitive than the plaque assays for phage and animal viruses. However, it is the only type of infectivity assay available for plant viruses.

**(c) End-point titrations.** Some viruses do not form plaques on monolayers of cultured animal cells or local lesions on leaves and, in these cases, the only type of infectivity assay available may be an **end-point titration.** For these assays, some type of biological response is needed, such as some visible signs of disease or death.

Serial dilutions of a virus sample are prepared and each dilution is inoculated into cell cultures, test plants, or even laboratory animals. For each dilution a number of replicates are used and, after incubation, the proportions of healthy and infected individuals are scored. Those cultures or organisms inoculated with virus diluted the least should all show signs of infection, but those inoculated with virus diluted the most should all remain healthy. Between these two extremes it should be possible to estimate the 50% end-point, which is the dilution that causes 50% of the host cells or organisms to become infected (Figure 10.20). The result is expressed as **Infectious Dose ($ID_{50}$)** or **Lethal Dose ($LD_{50}$)**. This is the dilution that contains an infectious dose or lethal dose large enough to infect or kill 50% of the host cells or organisms.

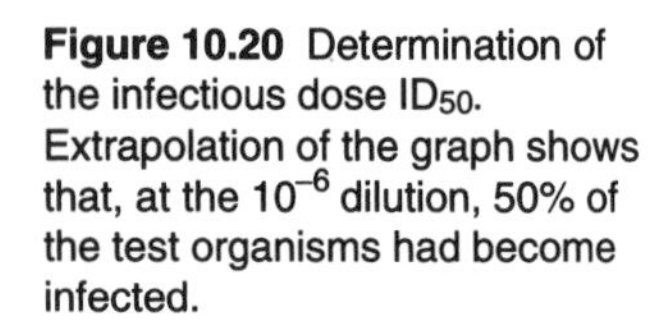

**Figure 10.20** Determination of the infectious dose $ID_{50}$. Extrapolation of the graph shows that, at the $10^{-6}$ dilution, 50% of the test organisms had become infected.

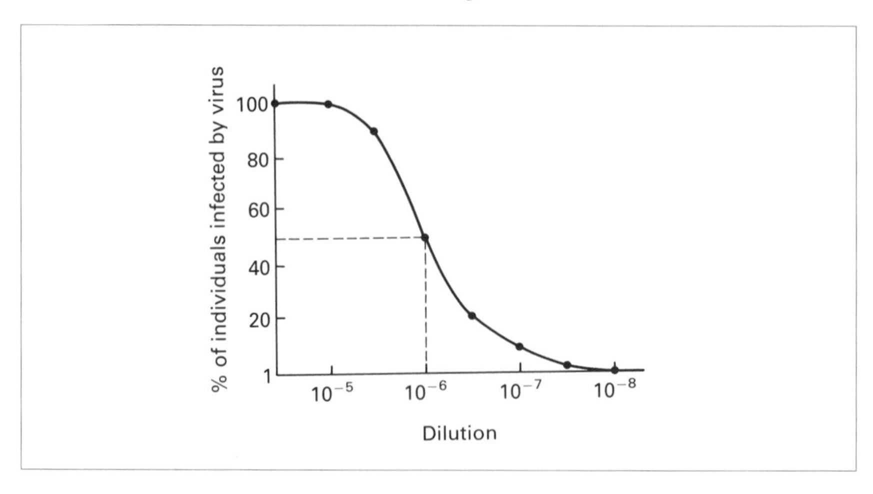

The obvious drawback with this type of assay is that it involves the use of large numbers of test cultures or organisms. It is only used if there is no alternative means of virus assay.

## Comparison of assay methods

If, for a given sample, the number of virus particles is determined by a direct count in the electron microscope and then the number of infectious units is determined in an infectivity assay, then the **efficiency of infection** can be determined.

$$\text{Efficiency of infection} = \frac{\text{number of virus particles}}{\text{number of infectious units}}$$

This ratio gives an indication of the number of virus particles needed to give a single infection. It varies widely among different viruses, and even for the same virus in different hosts (Table 10.2).

For most combinations of virus and host cell, the efficiency of infection in an assay is greater than 1.0. In other words, more than one virus particle is needed to initiate an infection. There may be several reasons for this. Suspensions of virus particles may contain a proportion of non-infectious particles. Also, on entry into a host cell, a potentially infectious particle may fail to multiply.

For bacteriophages and animal viruses, infection of the host cell is usually reasonably efficient. The virus particle attaches itself to specific receptors on the surface of the host cell at an early stage and the infection process then follows automatically. Plant viruses are clearly very different and far less efficient at infecting their host cells. The reason here seems to be that plant cells have no known receptors to which virus particles can bind. Instead, virus particles have to be driven through the cellulose cell

**Table 10.2** Efficiency of infection. The efficiency of infection is represented by the ratio of: Number of virus particles/Number of infectious units. The following values are obtained for some of the viruses that we have mentioned.

| *Virus* | *Infectivity assay* | *Efficiency of infection* |
|---|---|---|
| Phage T4 | Plaque assay (bacteria) | 1.0 |
| Phage T7 | Plaque assay (bacteria) | 1.5–4.0 |
| Poliovirus | Plaque assay (cell monolayers) | 36 |
| Influenza virus | Plaque assay (cell monolayers) | 7–10 |
| Influenza virus | Dilution to end point in eggs* | 10 |
| Tobacco mosaic virus | Local lesion assay | $5 \times 10^4$–$1 \times 10^6$ |

*The sample for assay is inoculated into embryonated hens' eggs (with developing chicks inside); if virus is present, it multiplies in the egg leading to death of the embryo.

wall of the plant cell. During inoculation, the cell walls are damaged and the cells themselves wounded. This creates sites for virus entry into plant cells, but is obviously a very inefficient process. This is why, out of about one million TMV particles on a leaf, only one, on average, gets through during inoculation to initiate local lesion formation.

## Virus multiplication

Viruses multiply only inside living host cells. To study virus multiplication the ideal approach would be to follow the process inside a single cell. Technically this would be very difficult. Another way is to follow the different phases of virus multiplication in a batch of identical cells that have all been infected at the same time and which are all going through the infection process together. It is then possible to take samples of cells from the culture at various times after inoculation with the virus to find out what is happening inside them.

This type of experiment is called a **one-step growth curve** and it relies upon achieving infection of a very high proportion of cells in the culture and ensuring that they go through the infection process simultaneously. The one-step growth curve (for phage T4 growing in *E. coli*) was first devised by Ellis and Delbruck (1939) who were both working as visiting research fellows at California Institute of Technology.

### The one-step growth curve

This technique relies on having a liquid culture of bacterial host cells. To ensure that a high proportion of bacterial cells are infected by phage and that the particles attach themselves to the cells quickly, it is necessary to inoculate at least ten phage particles for every bacterial host cell. Of course, not all virus particles attach themselves to the cells and it is, therefore, necessary to eliminate unattached viruses soon after inoculation. This is usually done by adding a specific antiserum against the virus which will bind to unattached virus particles and render them non-infectious. The culture is then highly diluted (by 1000-fold) so that any virus particles released at the end of the first cycle of infection cannot attach themselves to remaining uninfected cells. By diluting the culture, the chance of a phage diffusing through the medium and finding an uninfected host cell is very small and so, essentially, only one round of virus multiplication is possible.

The procedure for the one-step growth curve is quite simple. Cells of *E. coli* (strain B) are inoculated with particles of phage T4 at time zero. The culture is incubated for a sufficient time to allow virus particles to attach themselves to the bacterial cells. During this time, much of the input phage becomes attached to the cells. Unadsorbed phage is removed with anti-serum. The culture is diluted 1000-fold. At various times, samples from the culture are removed to determine the infectivity by a plaque assay. The results are then plotted on a graph to give the one-step growth curve as shown in Figure 10.21.

The graph shows several distinct phases. Looking at the curve marked 'extracellular virus', we can see that, during the first 10 to 15 minutes, there is no release of virus particles from the cells into the medium and the number of plaques remains constant. This phase is called the **latent period**. It is followed by the **rise period**, during which newly made virus particles are released from infected cells into the medium. The plaque count increases until a plateau is reached when all the infected host cells have burst open and released their progeny phage.

From this growth curve it is possible to determine the average number of virus particles produced per infected cell. With phages this is called the **burst size**. It is calculated from the plaque count at the plateau, divided by the initial count in the latent period. With phage T4, this value is between 100 and 150 phages per host cell, but with the small RNA phages (MS2, Qβ) it is far greater and more likely to be 10 000 to 20 000 per cell!

The latent period represents the time before cells start to release progeny phage into the medium. However, events inside the cells can be monitored during this time by taking samples and breaking the cells open (with chloroform). In this way the **intracellular virus** can be measured. This shows that, immediately after infection, there is a short period during which no infectious phage can be detected at all. This is the **eclipse period**

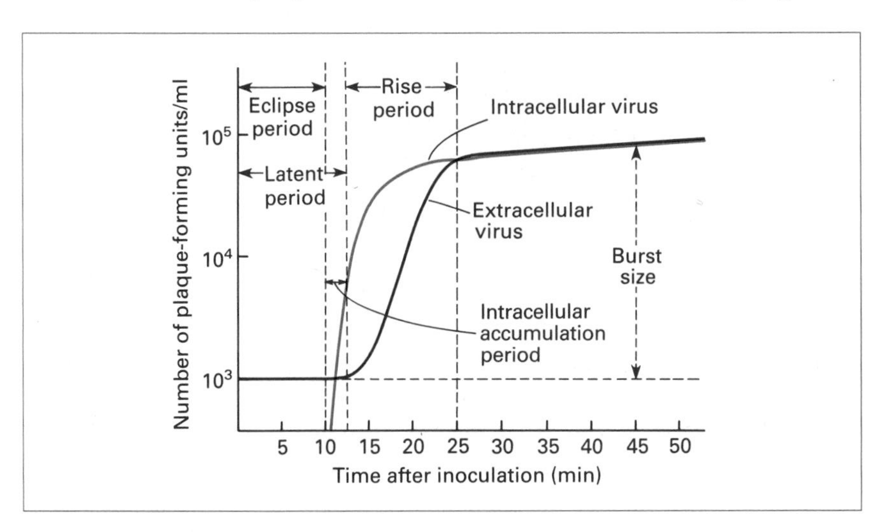


**Figure 10.21** One-step growth curve for phage T4 in *E. coli* B. The curve marked *Extracellular virus* represents phage released from infected cells as they burst open at the end of one cycle of infection. The label *Intracellular virus* represents phage production inside cells which have been artificially broken open with chloroform to release the virus particles.

and it lasts for a shorter period than the latent period: with T4 it is about 11 minutes. This represents the period after adsorption, during which the naked viral DNA is injected into the cell. Viral DNA starts to replicate and RNA and proteins are being made, but no whole virus particles have yet been assembled. The eclipse period ends as soon as the first whole virus particles form.

After the eclipse period, phage particles start to accumulate inside the host cell and come to a peak at about the same time as the first infected bacteria start to burst open. This is the period of **intracellular virus accumulation**.

From this curve the average time needed for one complete cycle of virus multiplication can be determined. It is the time from the start of the experiment to the end of the rise period. With T4, this is about 25 minutes. It really is quite remarkable to think that perhaps 150 particles of phage T4 can be made in each cell during this time.

A point to note about the kinetics of the appearance of virus particles in the host cells is that the formation of virus particles is not exponential, but is linear with time. This is different from the exponential growth characteristic of bacteria in which cells divide by binary fission. The reason is that virus particles are formed in a different way, by assembly from component parts, and this gives the process a different type of kinetics.

One-step growth curves can be produced by the synchronous infection of cultured animal cells by many animal viruses in a similar type of experiment. However, in these cases the time needed for one cycle of infection is longer. With poliovirus, for example, the complete cycle of infection lasts between 5 and 10 hours and this is one of the fastest-growing animal viruses. Others may take much longer. It is not really possible to achieve one-step growth conditions with plant viruses.

## Stages in the virus multiplication cycle

Staying with phage T4, we will take a more detailed look at the events in the virus multiplication cycle. These can be divided into five stages. The multiplication cycles of other viruses can also be divided into similar stages:

### Stage 1. Adsorption

Phages do not randomly attach themselves to the surface of bacterial cells, but stick to specific surface receptors. With T4 the receptors are lipoproteins of the outer envelope of the cell wall and there are more than $10^5$ receptors per host cell. Other phages bind to different types of receptors on the cell surface. For example, the small RNA phages (MS2, $\phi\beta$) attach themselves to the filamentous sex pili on male strains of *E. coli*. With T4 it

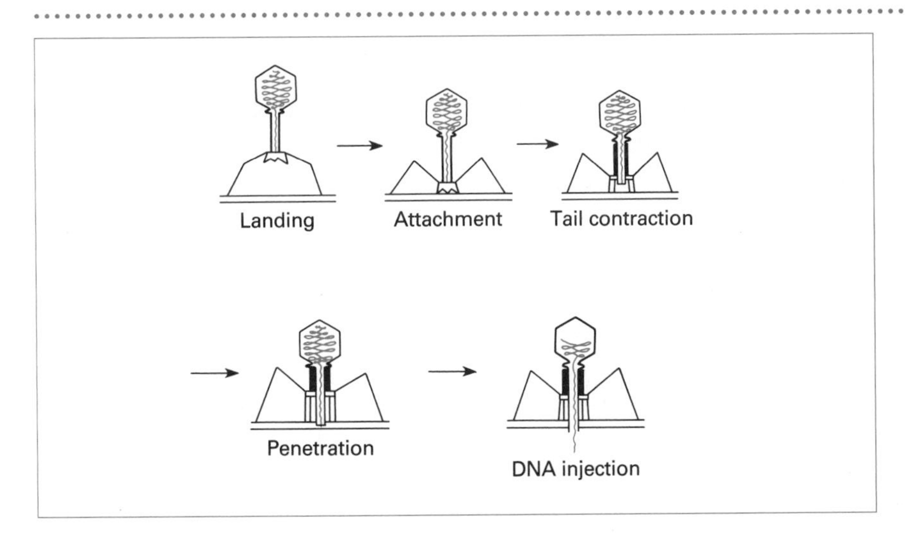


**Figure 10.22** Adsorption and penetration by phage T4.

is the tips of the tail fibres that first make contact with the host-cell receptor. The tail fibres bend at joints at their mid-point so that the base plate can become firmly attached to the surface of the cell wall (Figure 10.22).

**Stage 2. Penetration**

The penetration of the cell wall by phage T4 is exceptionally complex. A conformation change in the proteins forming the tail sheath causes the central tube of the tail to be pushed through the cell wall. The sheath contains ATP and this provides the energy for this mechanism. The tail is also thought to contain some lysozyme enzyme to help dissolve peptidoglycan in the cell wall and soften it so as to allow the tube to penetrate the wall. Phage DNA then moves from the head of the particle to the cytoplasm of the host cell (Figure 10.22).

**Stage 3. Synthesis of viral nucleic acid and proteins**

Soon after phage DNA has entered the cell, the synthesis of host-cell DNA, RNA and protein ceases. Phage mRNA is made early in the cycle, using the *E. coli* RNA polymerase enzyme for transcription. This mRNA is called **early mRNA** and codes for virus-specific enzymes and protein factors needed to redirect nucleic acid synthesis in the host cell. Some of this early mRNA specifies enzymes to degrade the host-cell DNA to generate nucleotides needed for viral DNA synthesis. This explains why phage T4 DNA contains 5-hydroxymethylcytosine (HMC) instead of the usual base cytosine, to render it resistant to this type of attack. Another early protein is a special $\sigma$ factor (page 119) needed to modify the *E. coli* RNA polymerase enzyme so that it will transcribe viral genes faster than host-cell genes. Viral DNA synthesis begins about 5 minutes after infection.

### Stage 4. Assembly

The assembly of particles of phage T4 is very complex. After about 10 minutes, the production of **late mRNA** begins. This mRNA codes for three types of virus-specific proteins: (1) phage structural proteins; (2) special scaffolding proteins which help with the assembly of the virus particles, but which do not become part of the particles; and (3) proteins involved in cell lysis and the release of virus particles.

The phage particles are assembled on three main production lines (Figure 10.23). The base plate, with its spikes, is built from about 15 gene products. The tail tube is built onto it and the sheath is assembled around the tube. Phage heads are assembled on a second assembly line. The head requires more than ten proteins. The assembled heads spontaneously attach themselves to the assembled tail. The tail fibres are built on a third line from four proteins and they attach themselves to the base plate after the head and tail have joined together.

DNA is packaged into the phage head before it joins onto the tail. From estimates of the volume of the phage head and the amount of space that the 5.6-μm-long molecule of phage DNA must occupy, it is clear that the DNA must be a tight fit in the head. Exactly how this happens is not at all clear at present.

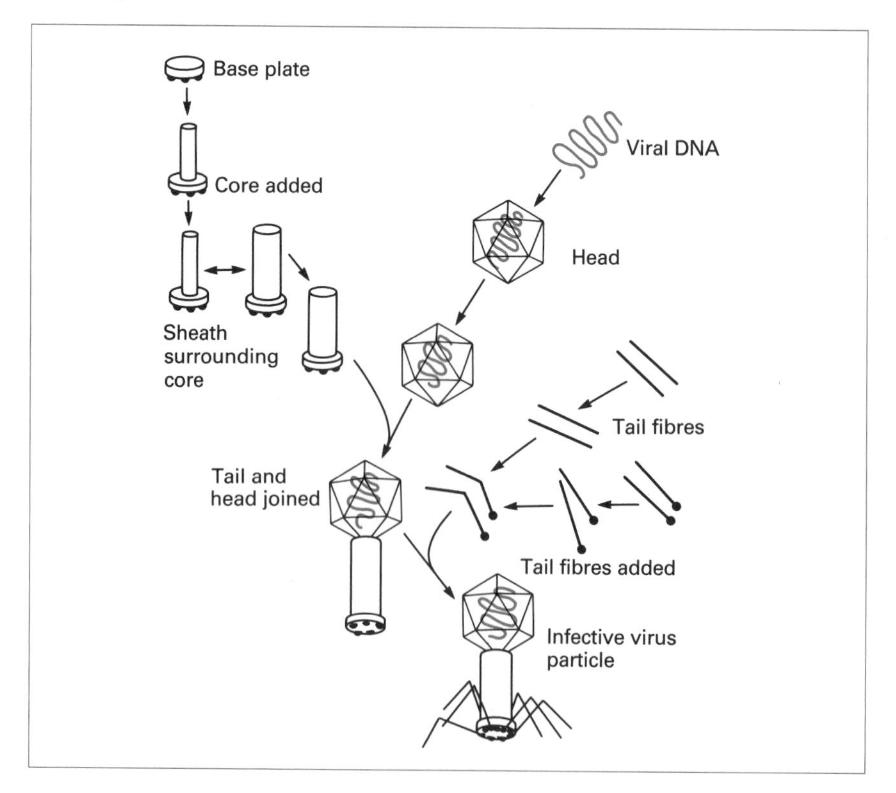


**Figure 10.23** Simplified diagram to show the assembly of phage T4. Particles are assembled on three main production lines: one for the head, one for the tail, and one for the tail fibres. Note that DNA is packaged into the head before it joins onto the tail. The tail fibres join onto the tail only after it has combined with the head.

**Stage 5. Release**

Infection by T4 leads to lysis of the host cell. Several phage genes are involved in this process. One of the gene products is a lysozyme enzyme that weakens the structure of the cell wall peptidoglycan. Another damages the cell membrane to allow the lysozyme through from the cytoplasm to get to the wall. With T4 the infected cell bursts open quite suddenly to release the progeny phage particles.

## Pathways for mRNA formation: the Baltimore classification

So far, most attention has been given to the events that take place in the multiplication of phage T4 in bacterial cells. This virus has double-stranded DNA as its genome. During infection, viral mRNA is synthesized from the viral DNA template. However, as pointed out earlier, other viruses have genomes which are not of DNA, but of RNA. In these viruses the pathway of mRNA formation is quite different.

The nature of the viral genome (DNA or RNA, single-stranded or double-stranded) and the pathway for viral mRNA synthesis in infected cells form the basis of an elegant system of classification for viruses. This can be applied to all viruses, irrespective of whether they infect bacteria, animals or plants. This scheme was first put forward by the eminent American virologist David Baltimore in 1971. Among Baltimore's many achievements was the discovery of the reverse transcriptase enzyme in the retroviruses (along with Howard Temin) for which he was awarded a Nobel prize in Physiology and Medicine.

In this scheme viral mRNA is assigned a central role, since the subsequent formation of viral proteins always takes place by the same basic mechanism. The relationship between the viral mRNA and the nucleic acid inside the virus particles forms the basis of the Baltimore classification into six major groupings. In this scheme the convention is to designate the viral mRNA as a positive strand and the complementary sequence that cannot be translated as a negative strand. A strand of DNA complementary to viral mRNA is a negative strand. Obviously, the formation of mRNA can occur only if a negative strand of DNA or RNA is used as the template (Figure 10.24). Using this approach the six classes can distinguished as follows:

**Class I viruses**

These are the viruses with double-stranded DNA genomes. The designation of this DNA as positive or negative is not appropriate as both strands are present in the viral genome and so mRNA can be transcribed from one or other strand. This is a broad grouping with viruses such as human

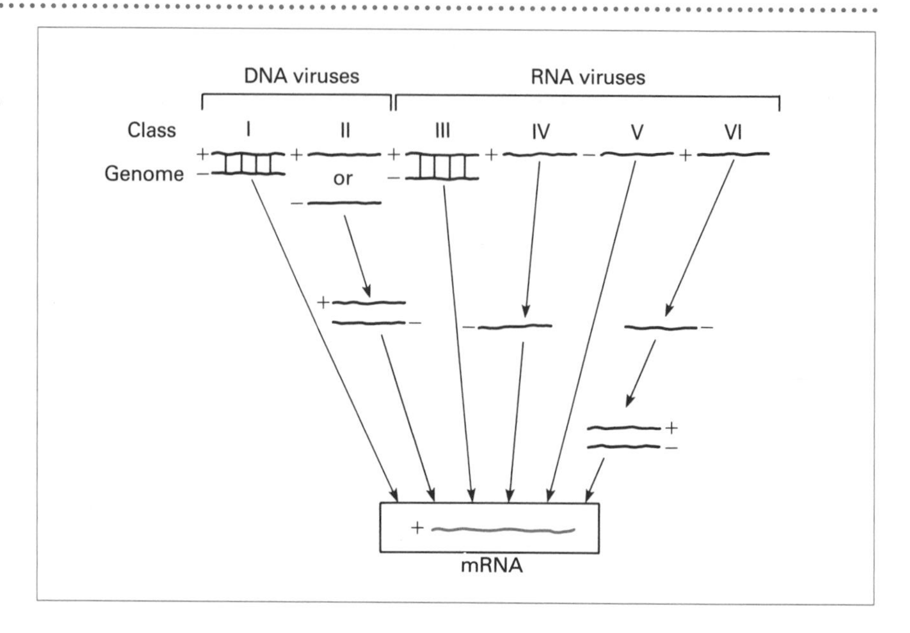


**Figure 10.24** The Baltimore classification of viruses is based on the pathways by which mRNA is made from the DNA or RNA inside virus particles.

papilloma (wart) virus, adenovirus and phages T2 and T4. In all these cases, the viral DNA enters the host cell and can be transcribed by host-cell enzymes to produce viral mRNA. In these examples, the purified DNA from the virus particles is always infectious.

The other major group of Class I viruses is represented by the poxviruses and, in these, the purified DNA alone is not infectious. Here, the virus has to take its own transcriptase enzyme (RNA polymerase) into the host cell, packaged in the virus particles, to initiate the synthesis of viral mRNA.

### Class II viruses

These are the viruses which have single-stranded DNA in their virus particles. The single strand of DNA in the virus particles can be a positive strand (as in phage ϕX174) or a negative strand (as in canine parvovirus). Some of the other parvoviruses can form a mixture of particles containing either positive or negative strands.

In the Class II viruses, the initial strand of DNA is copied to form a double-stranded molecule as a template, the negative strand of which can be used for the synthesis of viral mRNA. In these viruses, the purified DNA from the virus particles is usually infectious and so existing host-cell enzymes must be able to copy the single strand of viral DNA to make the double-stranded DNA template molecule.

### Class III viruses

These viruses contain double-stranded RNA in their virus particles. The negative strand of this molecule acts as a template for the synthesis of mRNA. This type of transcription of RNA from an RNA template requires a special type of RNA polymerase enzyme that is not normally present in cells. This explains why purified RNA from these virus particles is not infectious. These viruses have to take the virus-specific RNA polymerase, present in particles, into the host cell with their genomes to initiate infection and allow synthesis of mRNA. These viruses include the reoviruses that infect animals and plants.

**Class IV viruses**

Class IV viruses are single-stranded RNA viruses in which the RNA in the virus particles can act directly in the cell as mRNA. These are the **RNA plus-strand viruses** which include most plant viruses, poliovirus and the small RNA phages (MS2, Qβ). Viral RNA from these particles is always infectious.

**Class V viruses**

Class V viruses are those with RNA in the virus particles that is complementary in sequence to the viral mRNA. These are the **RNA negative-strand viruses**. Viral RNA from the particles is not infectious. The virus particles have to carry a special RNA polymerase enzyme into the cell in order to start transcribing mRNA from the negative-strand template. Normal host cells do not have such a polymerase enzyme to copy single-stranded RNA templates.

Examples of Class V viruses fall into two categories: influenza virus is a negative-strand virus that has a segmented genome. Each segment of RNA acts as a template for the synthesis of a single type of mRNA (except for the two smallest ones which can each specify two different types of mRNA). Rhabdoviruses (rabies and lettuce necrotic yellows virus) have genomes which consist of single undivided molecules of negative-strand RNA and the different types of mRNA can be read off this one molecule.

**Class VI viruses**

Class VI viruses are the **retroviruses**. These are viruses which have a genome of single-stranded RNA which is positive strand, the same as mRNA. However, what is interesting is that this RNA is not itself infectious. The reason is that this RNA molecule directs the formation of a double-stranded DNA intermediate and this acts as the template for the production of viral mRNA. This process requires the action of several enzymes and these are taken into the cell with the virus particle during infection.

The first stage in the infection process is to copy the RNA in the virus particle to give a complementary single strand of DNA. This is done by the

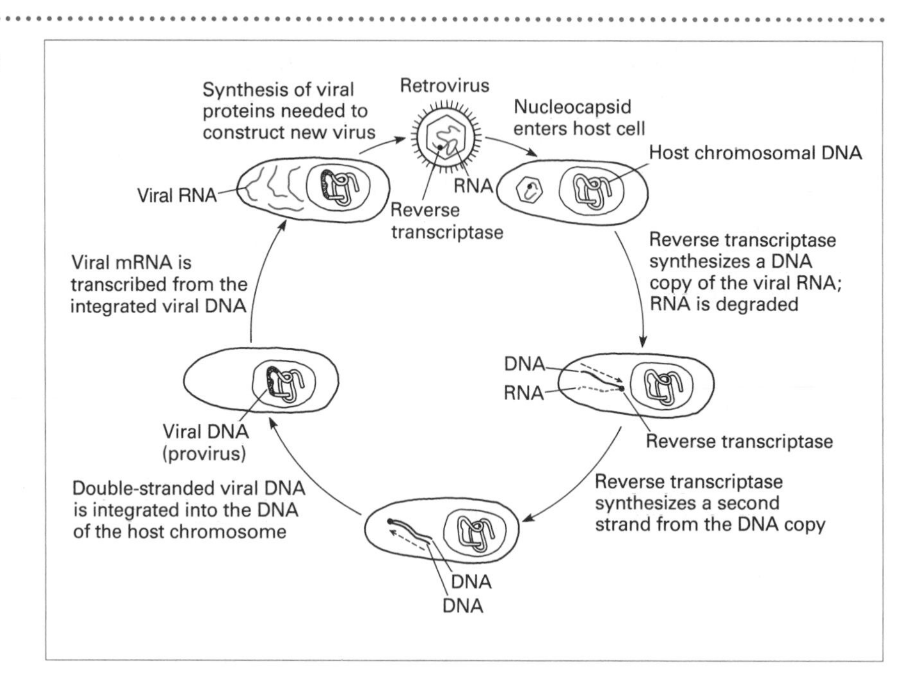


**Figure 10.25** The multiplication cycle of a typical retrovirus.

most characteristic enzyme in the virus particle: **reverse transcriptase.** The same enzyme then goes on to eliminate the RNA strand from the hybrid molecule and synthesize a second complementary strand of DNA. This double-stranded DNA form is called the **provirus** and is integrated into the chromosomal DNA in the host cell. Viral mRNA can then be transcribed from the provirus template using the usual transcription machinery of the host cell (Figure 10.25).

The retrovirus genome can sometimes carry genes that cause cancer (so-called **oncogenes**) if integrated into the genome of the host cell. The introduction of an oncogene right into the heart of the host-cell genome can transform a normal host cell to a tumour cell. For example, **Rous sarcoma virus**, which naturally infects chickens and causes solid tumours in connective tissue ('sarcomas'), carries a *src* gene that codes for an enzyme that phosphorylates cellular proteins. This is thought to affect cell growth and behaviour and so cause transformation. Other retroviruses are known which can cause certain types of leukaemia. Human T cell lymphotropic virus (HTLV) causes adult T cell leukaemia in patients, especially in south western Japan and in the Caribbean. However, by far the most is known about another retrovirus that does not transform its host cells but kills them instead. This is **human immunodeficiency virus (HIV)** which causes AIDS.

The Baltimore classification is useful because it can be applied to all viruses infecting animals, bacteria and plants. However, it is clear that some of the classes contain very wide assemblages of different sorts of viruses. Class I is a good example. It puts together smallpox and phage T4 which are totally different in their structure, function and biology.

BOX 10.12

## AIDS

AIDS (acquired immune deficiency syndrome) is primarily a disease of the immune system caused by human immunodeficiency virus (HIV). This virus must now surely be the most intensively studied virus ever.

HIV attacks particular lymphocytes called T4 helper lymphocytes. The virus binds specifically to the CD4 receptors on these cells using the glycoprotein spikes on the surface of the virus particles.

AIDS is a disease with a remarkably long incubation period. The median time for the development of AIDS after initial infection by HIV is between 8 and 10 years. However, the development of antibodies against the virus occurs between 2 weeks and 3 months after infection, but may take up to 6 months. This is called **seroconversion** and infected individuals become **HIV positive**.

HIV infection kills the T4 lymphocytes in the immune system and during the course of the disease the immune function is progressively reduced, leading to a profound immune deficiency.

The disease seems to progress though a number of stages. At an early stage, perhaps 14 days after infection, patients may show a rash, enlarged lymph nodes and high fever. This infection resembles glandular fever. At this stage there may be mouth and genital ulcers. As the symptoms subside, antibodies against the virus appear.

There follows a period which usually lasts for between 2 and 7 years, without any symptoms, although patients may show a reduction in T4 lymphocyte counts.

The third stage is described as a persistent generalized lymphadenopathy. The main clinical feature of this stage is the development of enlarged lymph nodes which persist for at least 3 months. Because of the impaired immune function, patients are susceptible to other infections called **opportunistic infections**. These include yeast infections of the mouth, and diarrhoea. Progression to AIDS correlates with a further decrease in T4 lymphocyte counts and appearance of the p24 antigen (an internal component of the HIV particles) in the blood.

The final stage is **full-blown AIDS**. This is characterized by severe immune deficiency and a whole series of opportunistic infections. These include an unusual form of pneumonia caused by *Pneumocystis carinii* (originally thought to be a protozoan, but now known to be more like a fungus), and secondary infections by other viruses. AIDS patients also commonly develop an unusual form of cancer called **Kaposi's sarcoma**. HIV can also gain access to the patient's brain and cause severe mental deterioration. Death is usually caused in the final stages by an opportunistic infection or secondary cancer.

HIV spreads between individuals through contaminated blood and by sexual contact. The disease can be passed on by intercourse between homosexuals and indeed the disease was first recognized by the fact that young homosexual men in the USA were showing an unusually high incidence of Kaposi's sarcoma and *Pneumocystis carinii* pneumonia. There is also strong evidence for heterosexual transmission from males to females and *vice versa*. Genital ulcer disease has been shown to enhance sexual transmission.

Transfusion of blood infected by HIV has provided a means of transmission of the virus. Haemophiliacs who have been given clotting factor (Factor VIII) made from contaminated blood have also been infected. These sources of virus have now been eliminated in most countries because all blood donations are screened for HIV. Spread among intravenous drug users, who share needles or use contaminated syringes, has also become a serious problem. In Scotland, where HIV was introduced to drug abusers at an early stage, around 30% of those infected are young women. Transmission from an infected mother to child is also possible because the virus can cross the placenta and cause infection of the baby before it is born; infection can also occur during the baby's delivery.

HIV is not spread by casual contact. There is no evidence for transmission by insects (biting insects, such as mosquitoes) or through saliva. Social contact (hand shaking) or sharing of eating and drinking utensils is also perfectly safe.

AIDS is one of the most devastating diseases ever to have afflicted mankind. The World Health Organization (WHO) have estimated that, unless there is some

**BOX 10.12** *continued*

effective intervention (an effective drug to treat the disease or effective vaccination), then by the year 2000, between 30 and 40 million men, women and children will have been infected by the virus. It is likely that most of these will eventually die from AIDS. In the UK alone, by the end of 1991, 16 828 cases of HIV infection had been reported. At the end of the same year, a total of 5451 cases had been reported of patients suffering from AIDS, of whom 3391 had died.

AIDS is a world-wide disease. The WHO publishes figures for the number of AIDS cases reported in various countries. However, these figures are not always reliable because numbers are under-reported in some countries for a variety of political or medical reasons. There are now serious problems in several African countries and as many as 5 to 10% of the population may be HIV positive. In Thailand there is rapid spread of AIDS among intravenous drug users and female prostitutes, with rebound waves into their male clients, wives and children. In the large cities of India up to 60% of prostitutes may be seropositive for HIV.

AIDS was first recognized as a new disease in 1981, although the causative agent, HIV, was not discovered until 1983. It is now clear that there are two strains of HIV. HIV-1 is the main cause of the world-wide AIDS epidemic. The second strain, HIV-2, was identified in West African patients in 1986 and has not spread so widely. The two strains of HIV differ slightly in their genetic structures. AIDS caused by HIV-2 seems to be milder than that associated with HIV-1.

AIDS is not a disease that is unique to humans. There are other animal immunodeficiency diseases as well. Feline immunodeficiency virus is known to infect a significant proportion of wild and domestic cats, notably in the Oxford area and in the west country in England. This virus was first isolated in California in 1986. Bovine immunodeficiency virus was first isolated in the UK in 1972 and is known to infect 24% of American cattle. Similar viruses are also known in goats and horses.

At present there is no cure for AIDS. There are two main drugs used for treating AIDS patients. These are AZT (azidothymidine) and DDC (dideoxycytidine) which are both base analogues designed to inhibit the reverse transcriptase enzyme. These do not cure the disease but slow its progress.

Education on how to prevent transmission (including 'safe' sex, single use of syringes and needles by drug users, safe handling of blood and blood products) is an important means of control, but has little success in some of the poorest regions of the world.

The major means of control will be by vaccination. Development of vaccines against HIV is difficult because it is rapidly evolving with a high degree of variability between isolates of the virus, and even between sequential isolates of virus from the same patient. A huge amount of research is being directed towards the development of vaccines that are safe and will give effective protection.

## Chapter summary

In this chapter we looked at the properties of viruses that make them a distinctive group in the microbial world. We used as examples different viruses that attack animals, plants and bacteria.

The first part of the chapter drew attention to the fact that virus particles are non-cellular entities and looked at their distinctive size, shapes and properties. We noted that all virus particles have a protein shell (the capsid) and that the genome is very simple and can consist of either DNA or RNA, enclosed in that shell. The fact that viruses can multiply only inside living cells means that they are all obligate intracellular parasites.

The effects of virus infection were surveyed. In animals and plants, we saw examples of highly localized infections, as well as generalized or systemic infections in which virus spreads throughout the whole organism. In bacteria, the consequence of virus infection is either cell lysis or conversion to the lysogenic state.

The structure of virus particles was looked at in some detail. On the basis of their architecture, virus particles were divided into (1) those with icosahedral symmetry; (2) those with helical symmetry; and (3) the complex particles which include (a) the poxviruses and (b) the phage particles with head and tail morphology.

We went on to consider the structure and organization of the viral genome. Examples were considered of different sorts of DNA viruses. The viruses which have single-stranded RNA were divided into two types: positive-strand and negative-strand viruses. Other RNA viruses were described with segmented genomes, paying particular attention to influenza virus. Then we looked at two important classes of sub-viral agents: the viroids, which cause plant diseases, and prions, associated with brain disease in humans and other animals.

The chapter includes a section on how to assay viruses. We discussed particle counts in the electron microscope, haemagglutination assays for animal viruses, and infectivity assays, which are generally the most important of all.

We considered how viruses multiply. This is usually investigated in a type of experiment called a one-step growth curve. Looking at the infection of *E. coli* by phage T4, we saw how the multiplication cycles of viruses can be divided into five stages: adsorption, penetration, synthesis of viral nucleic acid and proteins, assembly, and release.

In the final section the ways in which viruses replicate their genomes were reviewed and, on this basis, we found that we could divide all the different types of viruses into six classes: I to VI. Of these, we spent most time considering class VI which contains the retroviruses, now so important because it includes human immunodeficiency virus (HIV), the causative agent of AIDS.

## Further reading

B.D. Davis, R. Dulbecco, H.N. Eisen and H.S. Ginsberg (1990) *Microbiology*. 4th Edition. J B Lippincott Company, Philadelphia, USA – a useful textbook with good coverage of phage and viruses infecting animals. A standard work for any serious microbiology student.

N.J. Dimmock and S.B. Primrose (1987) *Introduction to Modern Virology.* 3rd Edition. Blackwell Scientific Publications, Oxford – a general virology book with emphasis on molecular aspects of the subject.

B.M. Fields and D.M. Knipe (editors) (1991) *Fundamental Virology.* 2nd Edition. Raven Press, New York, USA – a detailed treatment of animal virology.

H. Fraenkel-Conrat, P.C. Kimball and J.A. Levy (1988) *Virology.* 2nd Edition. Prentice-Hall International Inc., New Jersey, USA – an excellent overview of bacterial, animal and plant virology. Heinz Fraenkel-Conrat is a famous name in virology.

C. Vella and S.W. Ketteridge (1991) *Canine Parvovirus: A New Pathogen.* Springer-Verlag, Berlin – a general account of the biology of this virus that infects dogs.

R.E.F. Matthews (1990) *Plant Virology.* 3rd Edition. Academic Press, San Diego, California, USA – the definitive work on plant virology.

## Questions

1. List five distinctive features of the viruses that set them apart from other groups of microorganisms.
2. What is the relationship between: capsomeres, capsids and nucleocapsids in virus structure?
3. How do viroids differ from typical plant viruses?
4. List four characteristics of the causative agent of scrapie.
5. State whether the RNA purified from particles of the following viruses is **infective** or **non-infective** if inoculated into susceptible host cells: tobacco mosaic virus (TMV), influenza virus, phage MS2, poliovirus, lettuce necrotic yellows virus (LNYV).
6. How do the RNA plus-strand viruses differ from the RNA minus-strand viruses in the way they make their mRNA?
7. In an infectivity assay for phage T4 in *E. coli* B, how do the number of virus particles relate to the plaque number?
8. In an infectivity assay for phage MS2 in host cells of *E. coli*, 0.1 ml of the $10^{-9}$ dilution of the phage suspension gave 139 plaques on a plate. Calculate the infectivity of the original phage suspension in pfu/ml.

9. Where would you look for a virus that naturally infects cells of the yeast *Saccharomyces cerevisiae*?
10. Answer *true* or *false* to the following:
(i) Virus particles of complex morphology with heads and tails infect bacteria only.
(ii) During virus multiplication the viral genome is replicated by binary fission in the host cell.
(iii) Plant viruses only ever contain RNA genomes.
(iv) Some large virus particles contain viral ribosomes that they take into their host cells to make virus-specific proteins early in infection.
(v) Virus particles of influenza virus are released from their host cells by budding through the cell membrane.

# 11 Microbial biotechnology

In the previous chapters we have been concerned with the **properties** of microorganisms: their structure, biochemistry and genetics, and the variety of different types. This chapter is different in that we are concerned with the ways in which we can *make use* of microorganisms.

Microbial biotechnology is a very large subject, and we do not attempt to provide any sort of full coverage. Instead we look at a few topics chosen to give an idea of the scope and importance of the subject. The topics fall into three broad groups. The first consists of foods and drinks made using microorganisms. These have been produced by man for thousands of years and modern biotechnology has been concerned with bringing the production methods up to date, making them more efficient and the products more consistent. The second group consists of microbial products which are used in medicine or in industry. These are mostly of more recent origin and include the antibiotics, enzymes, and proteins produced for medical use by gene manipulation. The third group includes methods for the purification of organic wastes. These, too, are mostly of relatively recent origin and the main product is relatively uncontaminated water.

One point to bear in mind is that, in biotechnology, getting the science right is not enough to ensure success. Another important consideration which determines success or failure is the economic factor. Can the product be made at a low enough price so that the customer will buy it? This may be a matter of **absolute** price (can the customer *afford* to buy it?) but is more often a matter of **relative** price (can the customer buy a *cheaper* product from some other source?). In the manufacture of penicillin and citric acid, for example, there is fierce competition from other microbiological producers. Lactic acid production meets severe competition from chemical sources of supply. Some microbial products, such as chloramphenicol, acetone, butanol and glycerol, are not produced by microbial methods because chemical synthesis is cheaper.

In general, substances with simple structures tend to be cheaper to make by chemical synthesis. This is because the chemist can work with much higher concentrations of starting materials than those tolerated by living cells, so smaller reaction vessels are needed for a given output and the products are obtained in high concentration and are easier to purify.

Chemical synthesis requires many steps to make more complex molecular structures. Often a separate vessel is needed for each step, making the process more expensive, and each step is likely to produce extra unwanted by-products which reduce the yield of product. Living organisms can easily construct complex molecules in a single vessel because all the reactions go on inside the cell. We shall see that by carefully choosing the organism and the growth conditions, the yield of product can be increased and the quantities of unwanted by-products can be kept to a minimum. For these reasons products with complex molecular structures are usually cheaper to make by microbial methods.

## Alcoholic beverages

The production of alcoholic beverages is one of the oldest of man's biotechnological activities. Its origins are uncertain but it was practised by the Egyptians some 8000 years ago. It is also the most important in terms of value, accounting for more than 60% of the value of sales of all biotechnological industries.

Alcoholic beverages are produced in almost every country of the world. They are made by fermenting a carbohydrate starting material to ethanol and carbon dioxide. In almost all cases the organism used is a strain of the yeast *Saccharomyces cerevisiae*. The carbohydrate may be in the form of simple sugars from grapes or apples, or from the sap of plants such as sugar cane or cactus, or from starch-containing raw materials such as barley, wheat, maize, rice and potatoes.

Alcoholic beverages are divided into classes according to their ethanol content and method of manufacture. **Wines, beers** and **ciders** are produced directly by fermentation. High concentrations of ethanol inhibit yeast growth, so beverages containing more than about 18% volume/volume

**BOX 11.1**

**Taxonomy of the genus *Saccharomyces***

There are several hundred recognized strains of yeasts with distinct properties. The most important species is *Saccharomyces cerevisiae*, strains of which are used for the production of beer, wine, spirits and bread. The taxonomy of the genus *Saccharomyces* has been revised in recent years. Organisms formerly described as *S. carlsbergensis, S. diastaticus* and *S. uvarum* are now classified with *S. cerevisiae*. They may still be mentioned in books by their old names.

BOX 11.2

**Carbohydrate sources for alcoholic beverages**

Each type of alcoholic beverage has its own traditional raw materials. In many countries these are specified by law. Thus the Bavarian beer laws require that only yeast, malt, hops and water are used in the production of beer. Where not forbidden, other materials may be used, either because they are cheaper or because they correct some deficiency in the normal materials. Thus, in brewing, malted barley may be partly replaced by cheaper carbohydrates such as wheat, maize, rice or unmalted barley. Wine-makers in cool climates may add cane sugar to their grape juice when poor weather during the growing season has reduced the sugar content of their grapes. Conversely, citric acid may be added in very hot climates where grapes may contain too much sugar and not enough natural acids.

(v/v) ethanol cannot be produced directly by fermentation. Distillation produces much stronger beverages (usually sold at 35–50% v/v ethanol) known as **spirits**. **Fortified wines** are made by adding spirits to wines to give intermediate ethanol levels. The concentration of ethanol in spirits and fortified wines prevents bacteria growing and spoiling them. Bacteria can grow in table wines and beers, so these beverages don't keep long after the bottle is opened.

### Yeast metabolism

*S. cerevisiae* can utilize glucose and a variety of sugars ranging in size up to trisaccharides. Some strains (for example those previously known as *S. diastaticus*) secrete amylases into their surroundings to break down starch into sugars, but yeasts traditionally used for baking and wine and beer

BOX 11.3

**Ethanol concentrations**

People who work with ethanol solutions express the concentrations in several different ways. The strength of spirits is often described in terms of percentage **proof**. This terminology originated before modern methods of analysis were available. Excise officers had to assess the strength of spirits to calculate the duty to be paid. The spirit was added to a little gunpowder and a lighted match was applied. Proof spirit was defined as the concentration of ethanol *just* sufficient to allow the gunpowder to burn. Ethanol was added to weaker solutions until they ignited; stronger solutions were diluted until they would no longer ignite. In this way the ethanol concentration could be calculated.

Other ways of describing ethanol concentration are as **percentages by volume** (% v/v – the volume of pure ethanol per 100 volumes of solution) and **percentages by weight** (% w/v – the weight of ethanol per 100 volumes of solution). An ethanol solution which is exactly proof contains 57.1% v/v at 15.5°C. Ethanol is less dense than water, so 57.1% v/v is the same as 49.3% w/v. In the USA, proof is defined as 50% v/v.

Most beers contain between 3 and 5% v/v ethanol. Wines are stronger, usually between 9 and 13% v/v. In the UK, spirits are normally sold at about 70% proof (40% v/v) and many people like to drink them at this concentration. Brands imported into the UK are sometimes available with strengths up to 120% proof: this is too strong for any but the most determined drinker.

production cannot do this. Where the carbohydrate in the medium is starch, a special pretreatment is necessary to convert the starch to fermentable sugars.

The simplest sugars (such as glucose, fructose and galactose) are known as **monosaccharides**. **Disaccharides** (such as maltose, sucrose and lactose) consist of two monosaccharide units; **trisaccharides** (such as maltotriose) consist of three monosaccharide units.

Carbohydrates can be metabolized either by aerobic pathways or by fermentation (Chapter 3). In the absence of oxygen, the main metabolic activity is the fermentation of hexose sugars by the **glycolytic pathway** to ethanol and carbon dioxide, with the production of two molecules of ATP per hexose.

$$C_6H_{12}O_6 \rightarrow 2C_2H_5OH + 2CO_2$$

Aerobic metabolism takes place partly by way of the **pentose phosphate pathway** (page 79) and partly by way of glycolysis, the tricarboxylic acid cycle and the electron transport pathway (pages 80–8). Hexoses are oxidized to carbon dioxide and water by molecular oxygen, with the generation of up to 38 molecules of ATP for each hexose oxidized.

Yeast can thus use either *aerobic* or *anaerobic* pathways of metabolism but the situation is complex. If high concentrations of sugars are present, the synthesis of mitochondria and many respiratory enzymes is prevented by **catabolite repression** (page 161) so metabolism takes place by fermentation, even in the presence of oxygen.

In the absence of oxygen, yeast cannot synthesize sterols and unsaturated fatty acids, which are important components of its cell membrane. If these are not present in the medium in adequate amounts, yeast can achieve only very limited growth without oxygen. Each time cell division takes place these compounds are divided between the parent and daughter cells, and their concentrations in the membrane decrease until growth stops. Special measures are often taken in commercial fermentations to ensure that the yeast gets sufficient oxygen at the start of the fermentation to synthesize sterols and unsaturated fatty acids, and it is normal to use a very heavy inoculum of yeast to make up for the fact that each cell will only be able to divide a few times.

More yeast cells are produced when yeast grows aerobically than when growth depends on fermentation. When yeast cells themselves are the desired product rather than ethanol, conditions will be chosen so that metabolism is aerobic. Thus, yeast for use in baking or in whisky fermentations is grown aerobically, even though it is to be used in anaerobic conditions. Vitamins, mineral nutrients and nitrogen compounds (amino acids, peptides and nitrate can be utilized, but not protein) are almost always present in the fermentation media in adequate concentrations and do not need to be added.

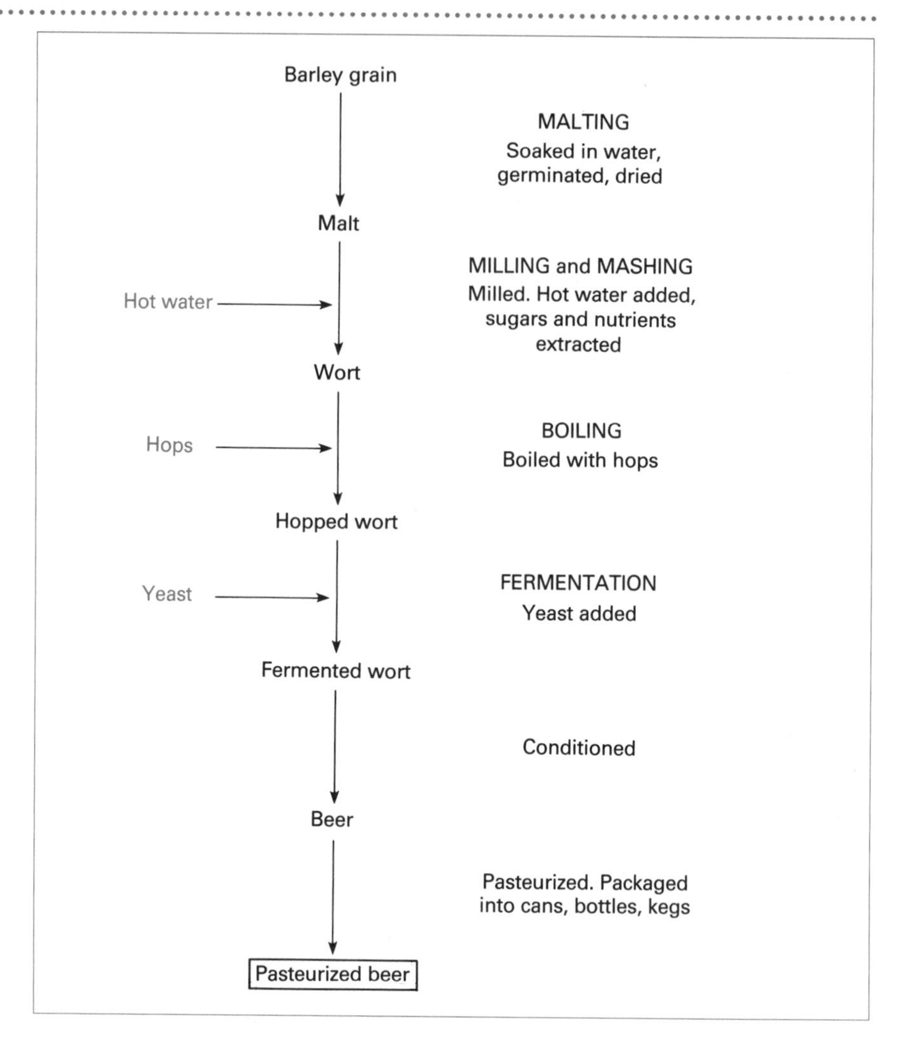


**Figure 11.1** The stages in brewing beer.

## Production of beer

The traditional starting material for beer production is the barley grain. This contains substantial reserves of starch and protein, and the **embryo**, which would give rise to a new barley plant. Two preliminary steps are necessary to convert starch to fermentable sugars, and protein to amino acids.

In the first stage, known as **malting** (Figure 11.1), barley is soaked in water to start germination. The embryo synthesizes hydrolytic enzymes (especially amylases and proteases) and these start to break down the starch and protein. After about 4 days, germination is halted and the embryo is killed by careful drying using hot gases. The product, known as **malt**, is stable and can be stored and transported. Most of the food reserves are intact, and most of the enzymes survive the drying process.

Hydrolytic enzymes are tougher than most.

The second stage is known as **mashing**. The malt is crushed between steel rollers to break open the impermeable outer husk of the grain. It is then mixed with hot water and maintained at a temperature of 65°C for about 1 hour. The malt enzymes break down starch to sugars, and proteins to amino acids and peptides. The products are extracted into the water. Carbohydrate breakdown is not complete, so some non-fermentable carbohydrate survives to contribute to the flavour of the finished beer. The

This is the method of mashing used in the UK for ales; a different regime is used for lagers.

**Fermenting vessels**

Fermenting vessels were traditionally made of wood, often lined with slate or copper, usually rectangular in shape, and open to the atmosphere. Most of these have now been replaced by enclosed stainless steel vessels which are easier to clean. For very large scale production of wine, concrete vessels lined with epoxy resin are sometimes used.

**BOX 11.4** Inside view of a beer fermentation. This is a traditional British top fermentation. The fermenting wort is covered by a foamy layer of yeast. Printed with permission from Bass Brewers.

**BOX 11.4** *continued.*
A cylindro-conical fermenter.

Cylindro-conical fermenters offer several advantages. The relatively tall, narrow shape encourages a vigorous natural circulation of fluid. Gas bubbles form at the bottom and rise at the centre, dragging fluid up with them. The gas is released at the surface and fluid flows back to the bottom in the outer part of the vessel. This circulation keeps the contents well mixed and the yeast in suspension, so fermentation is much faster than in a wider vessel.

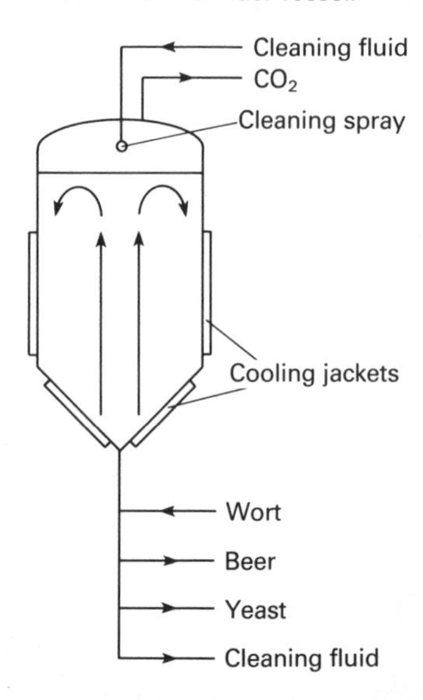


When fermentation has gone far enough, the liquid temperature is rapidly lowered by means of the cooling jackets. Gas production by the yeast stops, and the cells quickly sediment to the bottom where they collect in the cone and can easily be pumped off.

The vessel is easily cleaned by internal sprays: there are no awkward corners. Because the vessel is closed, the carbon dioxide can be collected and used or sold.

liquid, known as **wort**, is filtered off and extraction is completed by spraying the malt with more hot water.

Beer is flavoured by **boiling** the wort with the flower cones of the hop plant, *Humulus lupulus*. The complex mixture of bitter substances extracted into the wort gives beer a bitter flavour and inhibits the growth of many bacteria. The boiled wort is cooled to between 5 and 15°C and a substantial amount of yeast is added. Quite a lot of heat is generated during fermentation and, although the temperature is normally permitted to rise, it must

**BOX 11.5**

**Achieving anaerobic conditions in alcoholic fermentations**

No special steps are needed to achieve anaerobic conditions, even when open-topped fermenting vessels are used. The high concentrations of sugar in the medium repress respiratory enzymes, so sugars are metabolized by fermentation. Large amounts of carbon dioxide are produced, and a layer of the gas collects on the surface of the liquid, preventing oxygen from dissolving in the wort.

not be allowed to get so high that it affects the flavour or kills the yeast. Excess heat is removed by cooling coils. Fermentation takes between 2 and 14 days, depending on the temperature and other factors. Yeast is removed either by allowing it to settle, by filtering it off, or by sedimenting it in a centrifuge. Some of the yeast is used to start later fermentations. The surplus is used to make yeast extracts to be used as flavourings, or may be sold to distillers, who do not re-use their own yeast.

The beer is matured (**conditioned**) by storing it at a low temperature in contact with carbon dioxide under pressure. During this period the beer becomes saturated with the gas, and chemical changes occur which improve the flavour. Traditionally, beer was conditioned by carrying out a **secondary fermentation** in a cask or a bottle. **Pasteurization** (briefly heating the beer) is commonly used to kill bacteria and remaining yeast cells.

Named after the French chemist and microbiologist Louis Pasteur who first developed the technique.

**BOX 11.6** Alcoholic drinks contain many trace components. This graph shows the volatile components from cider, when separated by gas chromatography. Taken from Williams *et al.* (1978) *Journal of the Institute of Brewing*, **84**(2), 100.

**Flavours in alcoholic beverages**

Alcoholic beverages consist of complex mixtures of components, many of which contribute to the flavour of the product. For example, more than 550 volatile compounds have been identified from wines alone. Many components, such as glycerol, alcohols, organic acids, esters and carbonyl compounds, are products of yeast metabolism. Some are derived from the raw materials. Examples of these are the bitter compounds in beer derived from hops; the phenolic compounds responsible for the peaty aroma of malt whisky, which are derived from the peat smoke used to dry the malt; the sulphur compounds originating from malt and hops; terpenes derived from grapes in wines; and the flavour compounds extracted from juniper berries in gin. Wine matured in oak barrels picks up vanilla-like flavours from the oak-wood.

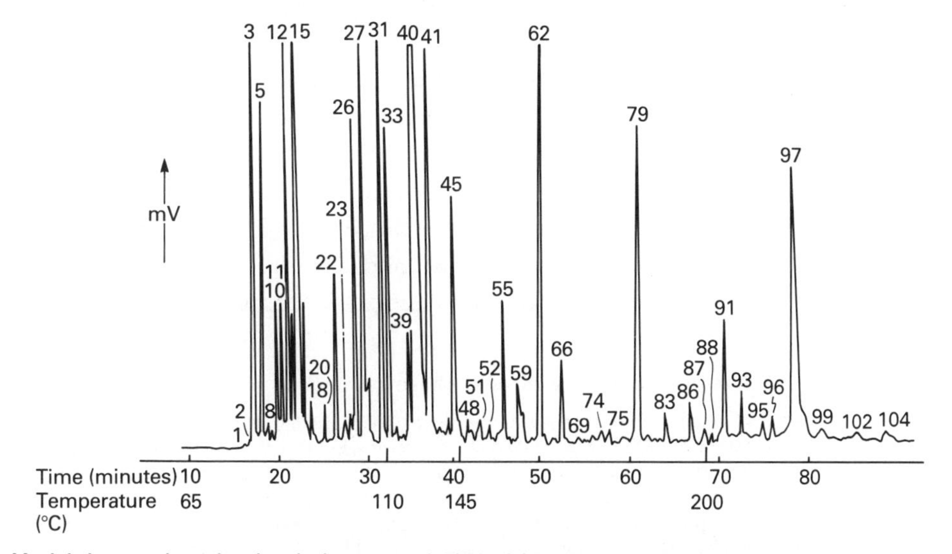


Much is known about the chemical nature and origin of the minor components in alcoholic beverages but the complex ways in which they affect flavour are only partly understood. The matter is of some economic importance because the exact balance between flavour components will determine whether a bottle of wine will fetch a high price or be fit only for distillation into industrial ethanol.

**Secondary fermentation**

A few alcoholic beverages undergo **secondary fermentation** in the container in which they are sold or dispensed for sale. For example, **champagne** and other high-quality sparkling wines are bottled when they still contain some unfermented sugar and a little yeast. The secondary fermentation takes place slowly over a period of months and produces the carbon dioxide which will make the wine fizzy when it is opened.

The difficult part of this process is to remove the yeast sediment without losing the wine. The yeast sediment in the bottle has to be got next to the cork without stirring it up. To do this, the bottle, which has been stored on its side, is moved gradually into an upside-down position. Each day, the bottle is given a gentle twist and tilted a little further. Finally, if all has gone well, the bottle will be upside down and the yeast sediment will be sitting firmly on the cork. The neck of the bottle is then chilled to freeze the sediment into a block. The cork is removed and gas pressure forces out the frozen block of sediment. More wine is added to replace the volume lost, and a new cork is put in place. The wine will now be free of yeast but will not have lost its carbon dioxide. The operation is labour-intensive, and skill is required. It is possible to carry out the secondary fermentation in a large vessel and bottle the wine without losing the carbon dioxide, but this method is considered to give an inferior flavour. Recently, some champagne producers have tried bottle fermentation with yeast **encapsulated** in agar beads. If the bottle is inverted, the beads drop to the neck, so removing them from the bottle is easy.

**Cask-conditioned** draught beer also undergoes a secondary fermentation (called conditioning) in the barrel from which it is to be sold instead of the normal **tank-conditioning** process, and a few **bottle-conditioned** beers are still produced where the carbon dioxide is derived from a secondary fermentation in the bottle. In both cases an improved flavour is the aim.

## Wine

Most wine is made from the berries of *Vitis vinifera*, the European grape. Hundreds of grape varieties are in use, each contributing individual flavours to the finished wine. Wine can be produced from grapes of just one variety or from a blend. For example, about 40 varieties are used in the production of port wine.

Ripe grapes are crushed to release the juice, known as **must**, which contains large numbers of yeasts and bacteria. Traditionally, fermentation was allowed to begin with the yeasts which happened to be present. Today,

**BOX 11.8**

**Microbial contamination of alcoholic beverages**

Alcoholic fermentations are remarkably resistant to contamination by unwanted microorganisms. Fermenting media are anaerobic, acidic (pH about 3 or 4), and become supersaturated with carbon dioxide. Most microorganisms are unable to grow under these conditions so there are no problems with pathogens or unwanted toxic products. However, many **wild yeasts** (those not inoculated deliberately) can grow under these conditions, as can some bacteria such as lactic acid bacteria. After fermentation it is important to prevent air getting to the beverage, otherwise **acetic acid bacteria** (which are strict aerobes) will be able to grow. All of these organisms cause problems by making the product turbid (cloudy) and producing unwanted flavours. Acetic acid bacteria oxidize ethanol to acetic acid, making the product taste vinegary.

BOX 11.9

**US vines saved the European wine industry from ruin**

When Europeans settled in the USA they found vines growing wild. These are different species from the European grape, *Vitis vinifera*, and give wines which most people consider vastly inferior. The European grape could not be grown in the USA because a soil aphid, *Phylloxera vastatrix*, attacked its roots. In the late 1860s *Phylloxera* appeared in France. It spread rapidly throughout Europe and many other parts of the world, devastating the vineyards. It even spread to California, which had previously been free from infection. The wine industry was saved by grafting the European grape onto the rootstock of American vines. This is how the vast majority of the world's wine is produced today. The European vine is grown on its own rootstock only in Chile, which is free from *Phylloxera*, and in parts of Portugal where vines are grown in deep sand, in which *Phylloxera* cannot survive.

except for very small-scale operations, the must is treated with sulphur dioxide to inhibit growth of the natural yeasts, and a heavy culture of *S. cerevisiae* var. *ellipsoideus* is added to start the fermentation.

Red wines are produced from highly pigmented grapes. By fermenting the must in the presence of the skins, the red **anthocyanin** pigments are extracted from the skins. White wines are produced either from non-pigmented grapes, or from pigmented grapes with the skins removed.

Fermentation temperatures between 17 and 33°C were once considered normal for red wines but recent practice has been to control temperatures more closely towards the lower end of the range to improve flavour. White wines require lower temperatures: 10 to 15°C is now common. As the fermentation nears completion and slows down, the yeast falls to the bottom of the vessel. The fermented must is separated by running it off carefully.

A period of maturation in wooden casks is usually necessary. The process appears to depend on the slow diffusion of oxygen through pores in the wood, and a variety of trace compounds including esters are formed. Sometimes flavour compounds are extracted from the wood. The wine may be given further treatment before it is bottled. For example, excess tannins may be precipitated by adding gelatine, egg albumin or isinglass, and excess proteins may be absorbed by the addition of **bentonite**, a kind of clay. Some wines are filtered or centrifuged to make them clearer. Many wines are ready for drinking soon after bottling but some, especially the better reds, require further maturation in their bottle before they are at their best.

It is said that, in the Middle Ages, lead acetate (sugar of lead) was sometimes added to wine to sweeten it!

## Scotch whisky

Whisky is produced in several countries but that made in Scotland will be taken as our example. Scotch malt whisky must be made only from malt (germinated barley), yeast and water, with caramel permitted as a colouring agent. Production begins with malting, mashing and fermentation stages

**Figure 11.2** Traditional copper stills at a malt whisky distillery, Aultmore. Printed with permission from United Distillers.

broadly similar to those used for making beer. Hops are not used, so the boiling stage is omitted. Enzymes extracted from the malt continue their activity during fermentation, and all the starch is converted to ethanol. Ethanol and other volatile components are concentrated by two stages of distillation in onion-shaped copper stills (Figure 11.2). These stills are thought to be important in producing the characteristic flavours of malt whisky. The concentrated product is matured in wooden barrels to improve the flavour. The minimum legal maturation period is 3 years but high-quality malt whiskies are usually matured for considerably longer.

Grain whiskies are cheaper to produce than malts, since the use of any convenient carbohydrate source is permitted, and highly efficient continuous distillation is used. Grain whiskies are less esteemed products than malt whiskies and are produced almost entirely for blending. Some malt whisky is sold unblended under the name of the distillery, rather like a fine wine, but most is sold as branded blends which may contain up to 40 kinds of malt and grain whiskies.

## Other yeast products

### Vinegar

Vinegar is a solution of acetic acid (about 40–50 g/l) with a large number of minor components. As with alcoholic beverages, the minor components make an important contribution to flavour. Vinegar is used as a condiment and as a preservative for foods. It is made by fermenting a carbohydrate source to ethanol, and oxidizing the ethanol to acetic acid with the bacterium *Acetobacter.* The most common carbohydrate sources are malt in the UK, apples in the USA, grapes or wine in France and rice in Japan and China. Spices and herbs are often added for flavouring. Acetic acid is an important bulk chemical, but almost all of this need is satisfied by chemical synthesis which is cheaper than fermentation. Many countries forbid the use of synthetic acetic acid for vinegar production.

To make malt vinegar, malt is extracted with hot water to make a wort. This is not boiled, so amylases extracted from the malt continue to act during fermentation, completing the breakdown of starch to fermentable sugars. Fungal enzymes can be used to aid starch breakdown but their use is not permitted in malt vinegar manufacture in the UK. Pressed brewer's yeast is used to start the fermentation. Fermentation takes 2 to 3 days at between 20 and 30°C. By the end of this period the yeast is close to dying, and fresh yeast must be used for each fermentation.

After fermentation, the beer is centrifuged to remove the yeast and precipitated proteins, and passed to the **acetification** process. Acetification is the oxidation of the ethanol to acetic acid in the presence of *Acetobacter.*

$$CH_3CH_2OH + O_2 \rightarrow CH_3COOH + H_2O$$

Since *Acetobacter* is widely distributed, it is necessary only to leave an alcoholic beverage open to the air to have it eventually converted to vinegar. The ways of doing this have been steadily improved. In the **field process**, wine or beer was stored in wooden casks in the open air and left to be oxidized by accidental inoculation by *Acetobacter.* In the **Orleans process**, the casks were kept in warm cellars and the access of air was made easier by leaving an air space above the liquid and by boring holes in the tops of the casks. The process was started off by adding a portion of vinegar to a small amount of wine. Further wine was added at weekly intervals. After 4 weeks acetification was complete and a portion of vinegar was drawn off to be replaced by more wine. A floating mat of *Acetobacter* grew on the surface. Pasteur introduced a modification in which a wooden grating was floated on the surface: it provided a support for the cells and prevented them being submerged or dispersed each time vinegar was drawn off. The Orleans process is thus an example of continuous (or at least semi-continuous) culture with immobilized cells. It dates from 1670!

The Orleans process was superseded by the **quick vinegar** process (Figure 11.3), which originated in Germany in the early 19th century. Large

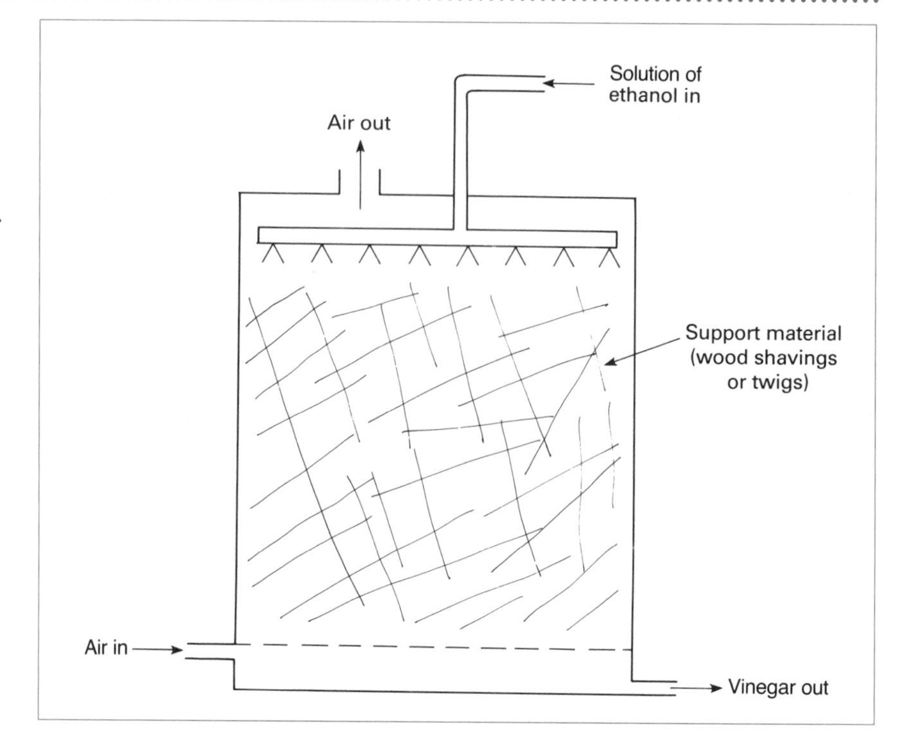


**Figure 11.3** The quick vinegar process. A solution containing ethanol, such as beer, wine or cider, is pumped in at the top and sprayed over the support material. *Acetobacter* oxidizes the ethanol to acetic acid. The vinegar trickles out at the bottom. If the ethanol is not completely oxidized, the solution is pumped round again.

wooden vats are filled with a porous support material, such as wood shavings or twigs. Wine, cider or beer is sprayed on to the top of the support, and trickles down over the surface. At the bottom it is collected and recirculated until conversion of the ethanol is complete. The temperature is maintained between 15 and 35°C. Air can penetrate freely. *Acetobacter* grows on the surface of the support and oxidizes the ethanol to acetic acid. Conversion can take as little as 5 days. During operation, the pores through the support material gradually became blocked with bacterial slime, hindering the access of fluid and air, and the conversion efficiency decreases. After a few months it becomes necessary to empty out the vat and start the whole process again – a laborious and unpleasant task.

More recently, the technique of **deep liquid culture** was introduced. In this method the bacterial cells are freely suspended in containers of well-mixed fluid with vigorous aeration. This process is more difficult to operate correctly: bacterial growth stops completely if aeration is stopped even for a short period, and temperature has to be carefully controlled. Operation can be in batch mode, or beer or wine can be fed in steadily while vinegar is removed to give continuous mode.

There are two major advantages to deep liquid culture. Oxidation is about 30 times faster than it is in the quick vinegar process, so a much smaller vessel is required for the same rate of production. Also, the conversion efficiency is much greater, so more acetic acid is obtained from a given amount of malt. Cleaning out the vessel is much easier too.

The raw vinegar is stored for several weeks, during which time bacteria, yeast and protein settle out. It is then filtered, pasteurized and bottled. Often, sodium chloride is added to prevent growth of any surviving *Acetobacter* and to improve the flavour.

**Baker's yeast**

Yeast was in use in Egypt for baking 6000 years ago. Each time bread was prepared, a portion of the dough, which contained the yeast, was kept back. This was then mixed with the next batch of dough, inoculating it with yeast. Today, most bread is produced on a large scale in bakeries. The yeast is cultivated separately and added to each batch of dough.

The main function of yeast in baking is to make the dough rise. Yeast is mixed with flour, water and sucrose (table sugar) to make the dough, and ammonium salts are added to act as the nitrogen source for the yeast. Amylases from the flour break down some of the starch to produce more fermentable sugars. The yeast ferments the sugars, and produces carbon dioxide which makes the dough rise, lightens the structure, and improves the texture. Modern baking methods rely on the dough rising in a very short time. The yeast must have a high growth rate so that it can multiply rapidly in the dough, and high enzyme activity (invertase and maltase) to break down the sugars in the dough. Yeast must keep for at least a couple of weeks with no significant changes and one batch should be very much like another.

Production of baker's yeast is a specialized operation which is carried out on a very large scale. The growth medium is based on molasses, supplemented with mineral salts and sometimes with the vitamins **biotin** and **thiamine**. Sterile air is forced through the medium to ensure that growth is aerobic to produce a high yield of yeast cells. Sugar concentrations are kept low to prevent fermentation taking place, but sugar is added at regular intervals to build up a high concentration of yeast. Starting from cultures preserved on agar, the yeast is grown up in several successive stages.

The process is not as resistant to bacterial contamination as alcoholic fermentations, so careful checks are made for purity. The volume is increased with each culture until there is enough yeast to inoculate a full-sized vessel (between 50 and 200 cubic metres). After growth, this is used to inoculate five full-sized vessels. Each of these is used to inoculate five more vessels, so finally 25 vessels are in use. The whole process takes about 10 days, and up to 500 tonnes of yeast (wet weight) are produced. Between 3500 and 7000 tonnes of sterile air will have been used. When growth is complete the yeast is sedimented in a centrifuge, and then more water is removed by filter presses until the yeast is solid enough to be cut up into blocks for sale.

## Fermented foods

### Cheese

Cheese manufacture is another process with very ancient origins; it is known to have been carried out 8000 years ago in what is now Iraq. It probably originated as a means of preserving the fat and protein of milk which would quickly spoil in a hot climate. Cheese production is now a major industry, accounting for about a quarter of the value of biotechnological products and ranking second in value to the production of alcoholic beverages. There are several hundred different kinds of cheeses, varying in properties such as water content, fat content and the flavours produced by the various types of bacteria and fungi used in their processing.

The starting material is milk: cows' milk is usual, but milk from sheep and goats can be used to make cheese with distinctive properties. The fat content can be adjusted by skimming off cream with a centrifugal separator, or by adding skimmed milk.

Normally, the milk is pasteurized as a first step to kill any harmful bacteria present, although some manufacturers prefer to use unpasteurized milk in the belief that it gives a better flavour. Fermentation is started by adding the active **starter culture** of lactic acid bacteria to the milk at a temperature between 20 and 35°C. The main milk sugar is **lactose** and the starter bacteria ferment this to lactic acid. After about 2 hours, the pH has dropped to below 4.6, at which value **casein**, the main milk protein, is insoluble. Casein **micelles** now start to clump together and precipitate. At this point, the enzyme **chymosin** is added. It hydrolyses one specific peptide bond in one of the polypeptide subunits of casein. This causes the micelles to break up. The polypeptide subunits which are released are insoluble, so precipitation of the casein is very rapidly completed, forming insoluble **curds**, and a fluid **whey**.

Can you think why milk from cows which have been treated with antibiotics (for diseases like mastitis) is not much use for making cheese?

**BOX 11.10**

**Casein**

Casein is the major protein in milk and is present at a concentration of about 3%. It consists of **micelles**, clusters of molecules 50 to 600 nanometres in diameter, made up of five types of polypeptide subunits. Milk contains $10^{13}$–$10^{14}$ micelles/ml, forming a stable suspension.

BOX 11.11

**Microorganisms used in cheesemaking**

Several species of *Lactobacillus* and *Streptococcus* are used as **starter cultures.** They can metabolize sugars only by fermentation and produce lactic acid as the major end-product. The starter bacteria are added to the milk at the rate of $10^6$–$10^7$ cells/ml. Many cheese producers buy commercial starter cultures, which are available as frozen suspensions containing about $10^{11}$ cells/ml, rather than maintain and prepare their own. The cell concentrations are so high that they can be added to the fermentations direct without any preculturing, and they can be kept frozen for several months.

The starter bacteria are vulnerable to attack by *bacteriophages.* Phages multiply much faster than their bacterial hosts, and lead to failure of the fermentation. Phage-resistant mutants usually arise, but they do not provide a fully satisfactory answer to the problem because they are typically resistant only to a limited range of the numerous strains of phage. Often, resistant mutants grow more slowly than the original strain. However, good hygiene in the factory will reduce the chances of contamination, and phage can be prevented from multiplying as the starter culture is being grown up by using media with low calcium contents. Since phage strains are highly host-specific, an infection can be countered by having several different starter cultures available, and switching to one which is resistant to the phage concerned.

With some cheeses, other organisms are used to produce the special flavours, textures and colours of each type. For example, strains of *Penicillium* are grown on the outside of **Camembert** cheese. They secrete proteases which diffuse into the cheese and change the texture and flavour. During the manufacture of **Roquefort** cheese, the curd is inoculated with the fungus *Penicillium roquefortii.* When the curd hardens, needles are forced into the cheese to permit air to penetrate. The fungus and its spores cause the blue-green veining characteristic of the cheese, and milk lipids are broken down by the fungus to form methyl ethyl ketone which is part of the characteristic flavour. In the manufacture of Swiss cheese, strains of *Propionobacterium* are used. These produce carbon dioxide which forms bubbles in the curd, making the holes in the cheese, and propionic acid which contributes to the flavour.

The curds are now subjected to various treatments (heating, stirring, cutting, pressing) to complete the removal of whey. The degree of heat treatment and pressing determines how much moisture will remain in the cheese, and whether it will be hard or soft.

BOX 11.12

**Chymosin**

Chymosin (formerly known as rennin) is a highly specific protease (an enzyme which breaks down proteins). It rapidly attacks a particular peptide linkage in one of the casein subunits but hydrolyses other peptide bonds at a very low rate, so the curds are not dissolved. The only natural source of the enzyme is the stomach of an unweaned calf. But we need fewer cows today because of the substantial increases in the yield of milk from each cow. So the number of calves slaughtered has been declining for many years, and chymosin has been in short supply. Most other proteases are unsuitable because they are less specific, so the curd is broken down, and peptides are produced which give the cheese a bitter taste. The best substitute comes from the bread mould *Mucor*, but it is inferior to chymosin. The gene for chymosin has been cloned in *Escherichia coli*, *Bacillus subtilis* and yeast, and chymosin has been produced on a small scale, but it has not been used for cheese manufacture because regulatory authorities and manufacturers are cautious about the use of a cloned product for food manufacture.

Finally, the cheese is *matured* by storing it at low temperature for weeks or sometimes years. If organisms other than the starters have been added they will grow and affect the flavour, texture and appearance. In any case, some non-specific protein hydrolysis will take place, caused by chymosin and other proteases. This will lead to a loss of elasticity and the development of flavour.

### Yoghurt

Like cheese, yoghurt is manufactured by fermenting milk with lactic acid bacteria. The final product is a viscous gel, which will keep for several days. The concentration of non-fat solids, which determine the consistency of the finished product, is increased to between 14 and 15 g/100 ml. This is done either by adding powdered milk or by concentrating the milk by vacuum evaporation. The fat content can be reduced by skimming it off the surface of the milk. A variety of materials (plant gums, seaweed extracts, starch, gelatin and cellulose derivatives) can be added to improve the texture.

The milk is vigorously mixed at a temperature between 50 and 70°C to reduce the size of the fat globules and distribute them evenly throughout the fluid. This step increases the viscosity by causing casein micelles to interact with fat globules and denatured whey proteins. Pasteurization or sterilization follows to kill unwanted bacteria, and the milk is then cooled to fermentation temperature, and inoculated with starter culture.

Bacterial fermentation produces lactate from lactose, and the resulting drop in pH causes casein to precipitate and the product becomes a gel. Yoghurt is naturally flavoured by lactic acid, acetaldehyde, diacetyl and other bacterial metabolites, but most yoghurt in western countries has fruits or flavourings added.

### Foods fermented with salt

Many foods other than milk are preserved by microbial fermentation in the presence of salt, usually involving members of the lactic acid bacteria which produce lactic acid to lower the pH. The salt inhibits the growth of toxin-producers such as *Clostridium botulinum*. The growth of fungi is often

**BOX 11.13**

**Yoghurt starter cultures**

Starter cultures for yoghurt production consist of mixtures of two organisms, *Lactobacillus bulgaricus* and *Streptococcus thermophilus*. The mixed culture grows and ferments substantially faster than either of the two organisms alone because of the symbiotic relationship between them, each producing substances that stimulate the growth of the other.

**Table 11.1** Some examples of foods fermented with salt (reproduced from Moses, V. and Cape, R.E. (1991) *Biotechnology. The Science and the Business*, Harwood. London, p. 283.)

| *Food* | *Raw material* | *Principal microorganisms* |
|---|---|---|
| Soy sauce | Soybeans, wheat | *Aspergillus oryzae* |
| | | *Pediococcus soyae* |
| | | *Saccharomyces rouxii* |
| | | *Torulopsis* spp. |
| Tempeh | Soybeans | *Rhizopus* spp. |
| Miso | Soybean paste, rice | *A. oryzae* |
| | | *A. soyae* |
| | | *P. halophilus* |
| | | *Saccharomyces rouxii* |
| | | *Torulopsis* spp. |
| | | *Streptococcus faecalis* |
| Natto | Boiled soybeans | *Bacillus subtilis* |
| Sufu (Chinese cheese) | Soybean cubes | *Actinomucor elegans* |
| Ang Khak (Chinese red rice) | Rice | *Monascus purpureus* |
| Gari (Nigeria) | Cassava | *Candida* spp. |
| | | *Leuconostoc* spp. |
| Idli (India) | Rice, black gumbeans | *L. mesenteroides* |
| | | *Streptococcus faecalis* |
| | | *P. cerevisiae* |
| Kenkey (Ghana) | Corn | *Saccharomyces* spp. |
| | | *Leuconostoc* spp. |
| Fermented fish | Cooked fish strips | *Aspergillus* spp. |
| Olives | Olives | *Pediococcus* spp. |
| | | *Streptococcus* spp. |
| Pickles | Cucumbers | *Lactobacillus* spp. |
| Sauerkraut | Cabbage | *Lactobacillus* spp. |
| | | *Leuconostoc* spp. |

encouraged. These produce amylases and proteases, which soften the material and improve flavours.

Table 11.1 shows some examples of foods preserved in this way.

## Antibiotics and their production

An **antibiotic** is a metabolic product of one microorganism which, in very small amounts, kills another microorganism or inhibits its growth. More than 5000 antibiotics are known. A very wide range of organisms produce antibiotics but most come from the **streptomycetes**, a group of filamentous bacteria. Most known antibiotics are unsuitable for medical use,

**Sir Alexander Fleming**

Sir Alexander Fleming discovered penicillin in 1929. Fleming was a bactèriologist working at St Mary's Hospital in London. He was interested in antibacterial substances and had earlier discovered the antibiotic effect of the enzyme lysozyme, which turned out to be of no clinical use. He discovered penicillin accidentally when some Petri dishes of the bacterium *Staphyllococcus aureus* became contaminated with the fungus *Penicillium*. He noticed that the bacteria were unable to grow in the region of the fungal colonies. He found that a soluble substance released into the growth medium by *Penicillium* was a powerful agent against staphylococci. Although they believed it to be of great potential importance, Fleming and his collaborators were unable to make much progress with the study of penicillin, because its activity was lost whenever they attempted to purify it.

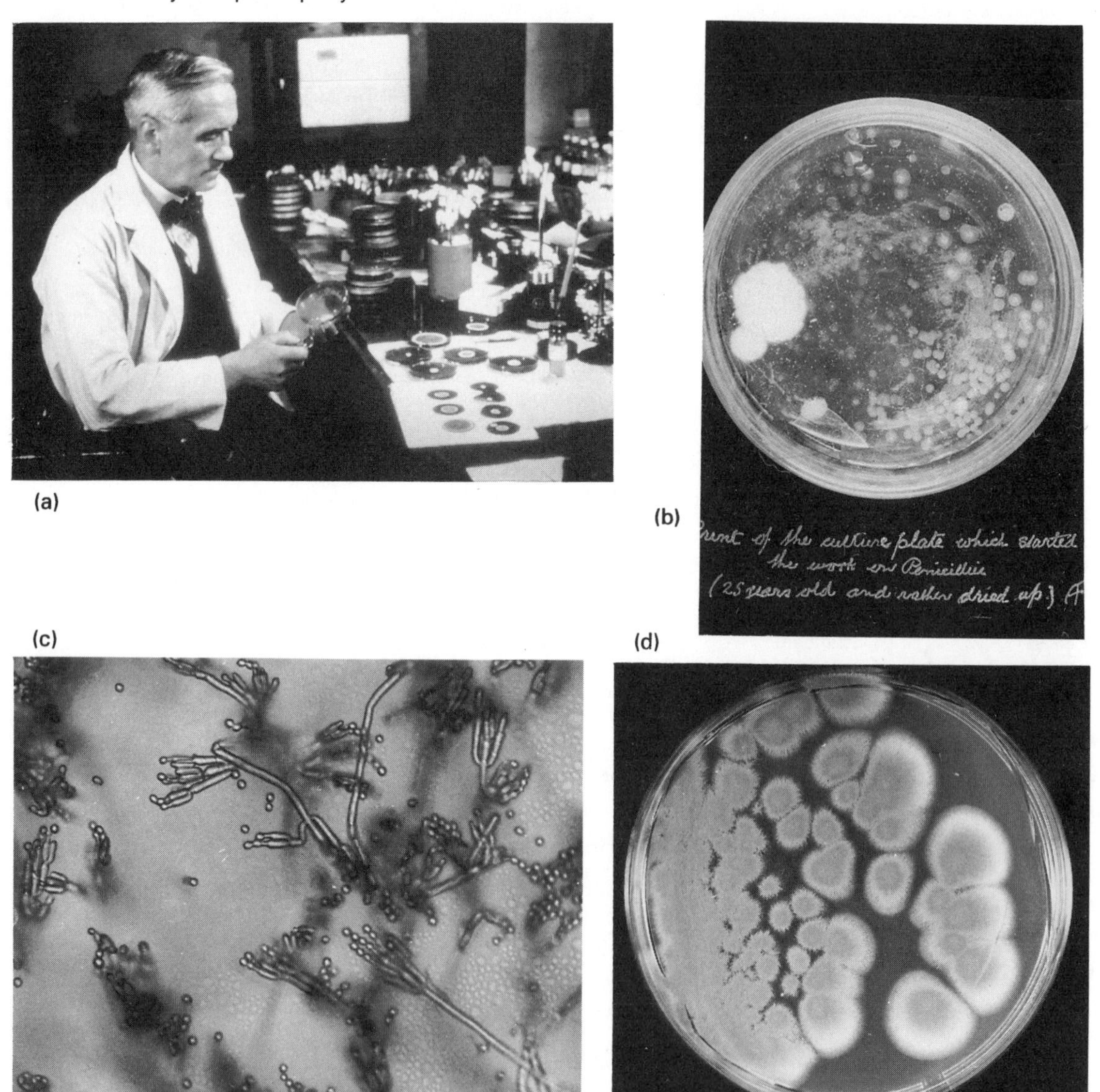


**BOX 11.14** (**a**) Sir Alexander Fleming, the discoverer of penicillin. In 1945 Fleming shared the Nobel Prize for Physiology and Medicine with Howard Florey and Ernst Chain, members of the group at Oxford University who purified the antibiotic and demonstrated its clinical effectiveness. Reprinted with permission from the Wellcome Institute Library, London.
(**b**)This is the plate on which Fleming first noticed the effect of penicillin. The *Penicillium* which contaminated his plate was killing the *Staphylococcus* bacteria which he was trying to cultivate. Reprinted with permission from the Wellcome Institute Library, London. *Penicillium notatum* (**c**) a colony growing on an agar plate (magnification x 1.5) and (**d**) mycelium and spores ( x 200). Reproduced courtesy of J. Faull and J. Smale.

BOX 11.15

**Early knowledge of antibiotics**

Some knowledge of antibiotics existed long before they were produced on an industrial scale and they appear to have played some part in folk medicine. The Mayans are reported to have used a fungus grown on corn to treat ulcers and intestinal infections. The Chinese have long used soybeans infected with fungi as a treatment for boils. Peasants in Brazil used a fungus fruiting body to treat open wounds while those in the Ukraine used a fungus grown on bread.

Numerous authors reported on the ability of one type of microorganism to kill or inhibit the growth of others. The physicist Tyndall found, in 1876, that when *Penicillium* grew on the surface of liquid bacterial cultures the bacterial cells were killed. Doehle, in 1889, appears to be the first author to have published a photograph showing antibiotic action. His illustration shows the inhibition of anthrax bacilli on a Petri dish, apparently caused by an antibiotic diffusing from an area inoculated with a bacterium which he called *Micrococcus anthracotoxicus.*

In some cases, attempts were made to use antibiotics for therapy. Starting in 1899, pyocyanase, an antibiotic from *Pseudomonas aeruginosa*, active against *Bacillus anthracis* and other bacteria, was used to treat eye infections, as a throat-spray against diphtheria and for other surface applications. It was rather toxic and this may be why it was not much used after about 1914. Actinomycetin, an antibiotic from antinomycetes, was discovered in 1924 but was not used for therapeutic purposes because it is too toxic.

usually because they are too toxic towards humans. The fairly small number which are used are of immense importance and have revolutionized the treatment of infectious diseases. Antibiotic production is one of the most valuable biotechnological industries.

Most known antibiotics are used against bacteria. They may be either **bacteriocidal** (kill bacterial cells) or **bacteriostatic** (inhibit growth so that infections can be easily dealt with by the bodies' natural defence mechanisms).

BOX 11.16

**Selective toxicity**

The ideal antibiotic would be very toxic towards bacterial cells but have no harmful effects on the human host. Penicillin, the first antibiotic put into major clinical use, comes close to this ideal. Its mode of attack is to inhibit the synthesis of the peptidoglycan component of the bacterial cell wall. Since there is no peptidoglycan in human cells, penicillin has no comparable effect in humans, and for most people it is a very safe antibiotic. However, a small proportion of the population suffers an immune reaction to penicillin. The symptoms are usually mild, but in some cases very serious immune reactions have been experienced leading to rapid death. Once a patient has experienced an adverse effect, however mild, it is not safe to use penicillins again.

In many cases antibiotics are toxic to humans but only at concentrations much higher than those used to treat infections. They can be used safely provided the dosage is carefully controlled. Some antibiotics produce severe side-effects in a small number of cases (for example, chloramphenicol which causes a disease of the bone marrow in about one patient in 40 000), and their use is restricted to instances where the infection has failed to respond to safer drugs, and is severe enough to justify the risk.

The main ways in which antibiotics attack bacteria are by inhibiting cell wall synthesis, by inhibiting synthesis of nucleic acids and proteins or by damaging the cytoplasmic membrane.

## Types of antibiotic

**(a) Penicillins.** Penicillin is not a single substance: some hundreds of related compounds have been produced, several of which are in regular clinical use. They all have the same **penicillin nucleus** containing the **β-lactam ring** (Figure 11.4), with different acid side-chains. Differences in side-chain structure can lead to useful differences in clinical properties. Thus **penicillin G** is unstable in acid solution so it is rapidly destroyed in the stomach and must be given by injection. **Penicillin V** is more stable and can be given by mouth.

Penicillins are generally most useful against Gram-positive bacteria (page 142); Gram-negative bacteria are generally much less susceptible. Penicillins such as **ampicillin** are much more effective against Gram-negative

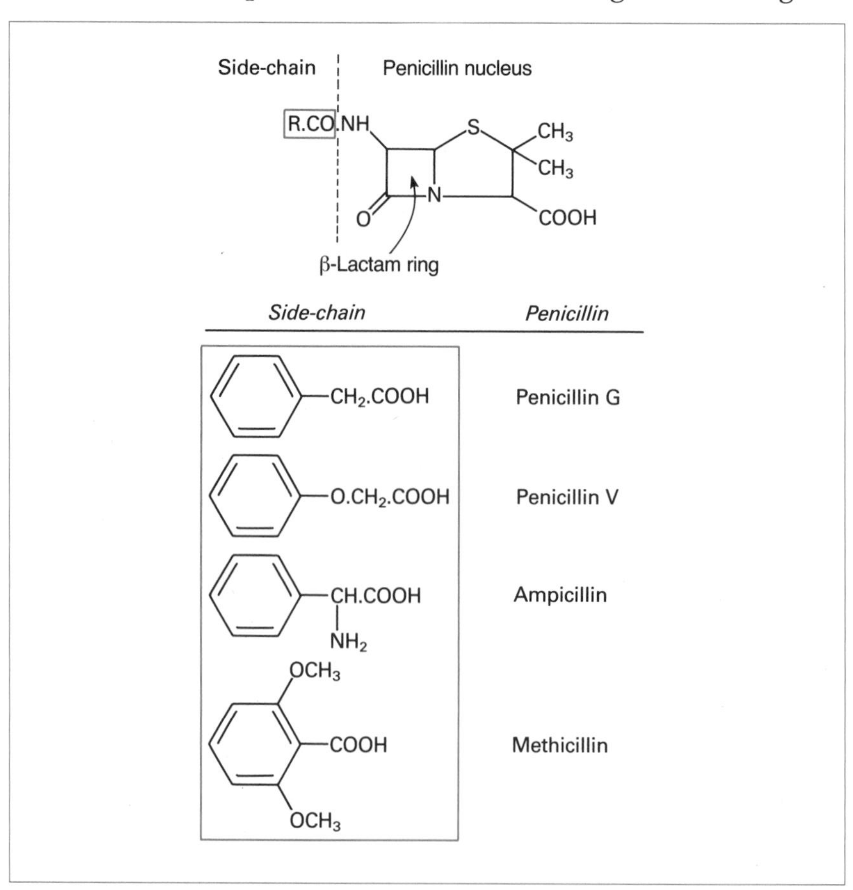


**Figure 11.4** The structures of some penicillins.

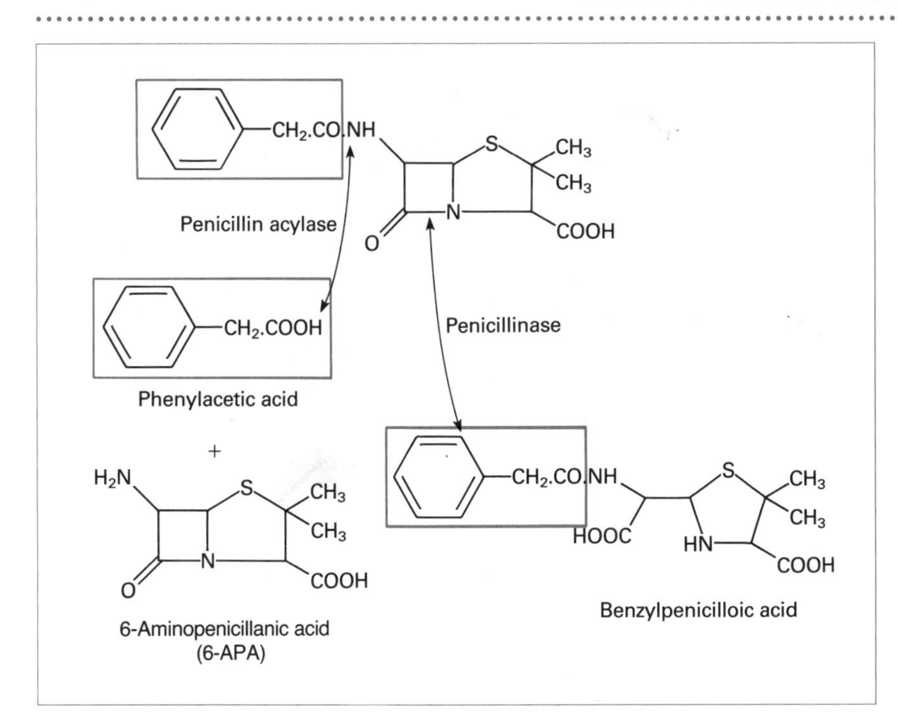


**Figure 11.5** The effects of enzymes on penicillins. Redrawn from Rivière, J. (1977) *Industrial Applications of Microbiology*, translated and edited by Moss, M.O. and Smith, J.E., Surrey University Press.

organisms. Some bacteria produce **penicillinases** (β-lactamase enzymes) (Figure 11.5) which inactivate penicillins: **methicillin** has greatly enhanced resistance to these enzymes.

Penicillins act by inhibiting the final stages of synthesis of the peptidoglycan molecules in bacterial cell walls. Cells continue to grow but the cell walls become progressively weakened. (Cells with defective cell walls are known as **spheroplasts.**) The spheroplasts swell and eventually burst (Figure 11.6) because water enters the cells by osmosis and stretches the weakened walls. This mode of attack is only effective against growing cells.

**(b) Cephalosporins.** Cephalosporins are produced by a marine fungus called *Cephalosporium acremonium.* They are chemically related to the penicillins and contain the β-lactam ring (Figure 11.7). They are effective against both Gram-negative and Gram-positive bacteria and act on cell wall synthesis in the same way as penicillins. A whole range of compounds has been made by modifying the side-chains.

**(c) Cycloserine.** Cycloserine is similar in structure to the amino acid alanine (Figure 11.8). It acts as a competitive inhibitor of cell wall synthesis. It can be used in the treatment of tuberculosis but its use is limited by undesirable side-effects. Originally discovered in streptomycetes, it is now produced by chemical synthesis.

**Figure 11.6** The effects on *E. coli.* after exposure to penicillin G. The cells have grown but penicillin has inhibited the synthesis of peptidoglycan. The cells have swollen and ruptured because they are too weak to resist pressure caused by water entering the cell by osmosis. Reproduced by permission from Schwartz, U. *et al.* (1969) *J. Mol. Biol.*, **41**, 419–29.

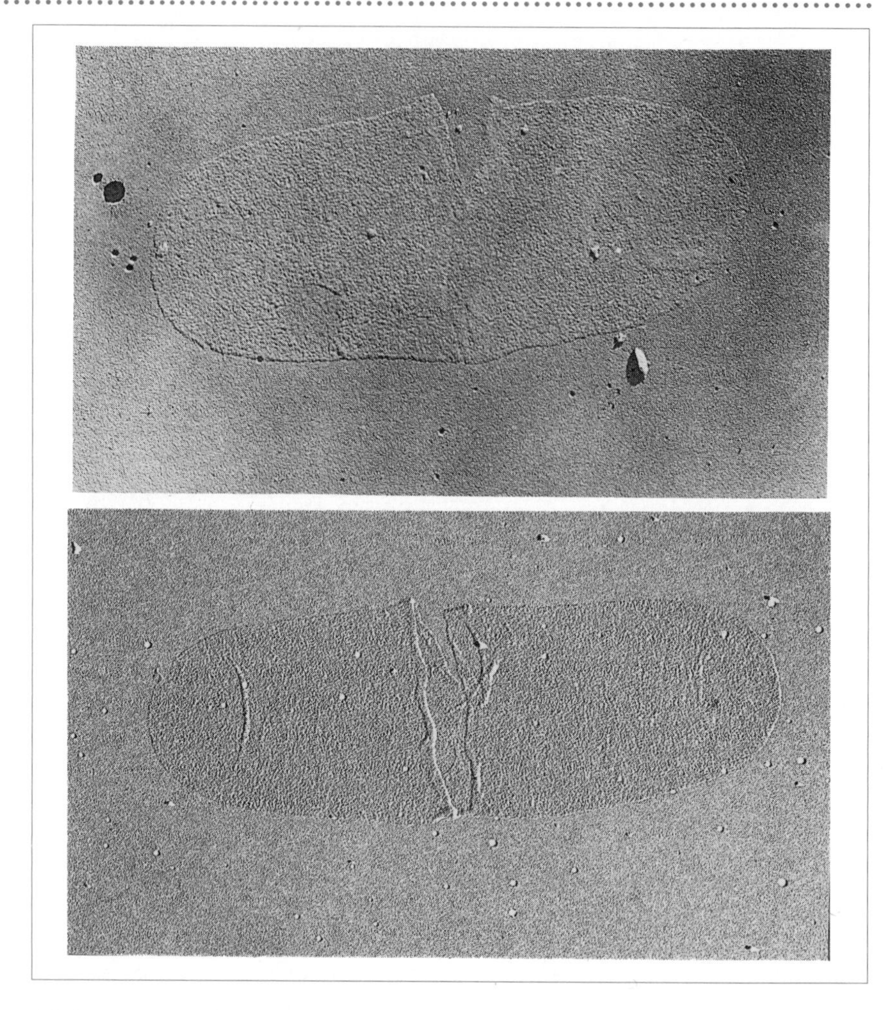

**(d) Streptomycin.** Streptomycin (Figure 11.9) is an example of a group of antibiotics called **aminoglycosides** which contain amino derivatives of sugars. Streptomycin is produced by *Streptomyces griseus*. It is a **broad-spectrum** antibiotic and is one of the relatively small number of antibiotics active against *Mycobacterium tuberculosis*, the bacterium responsible for tuberculosis. Although streptomycin is non-toxic in the short term, prolonged treatment can lead to deafness, so its use is limited. It is *bacteriocidal* and acts by inhibiting protein synthesis.

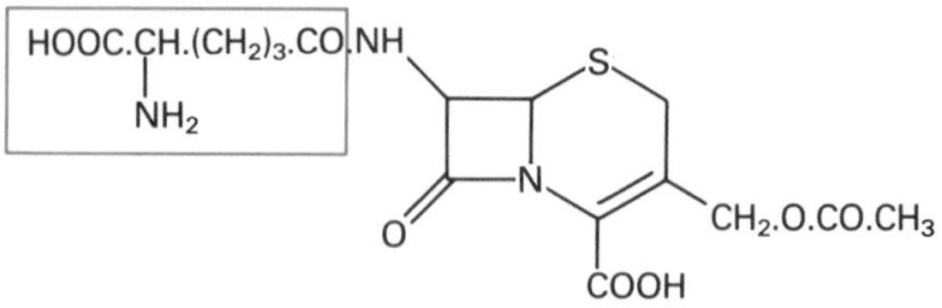

A **broad-spectrum** antibiotic is one which is active against a wide range of microorganisms.

**Figure 11.7** The structure of cephalosporin C.

$HOOC.CH(NH_2).(CH_2)_3.CO.NH$ — S — N — O — $CH_2.O.CO.CH_3$ — COOH

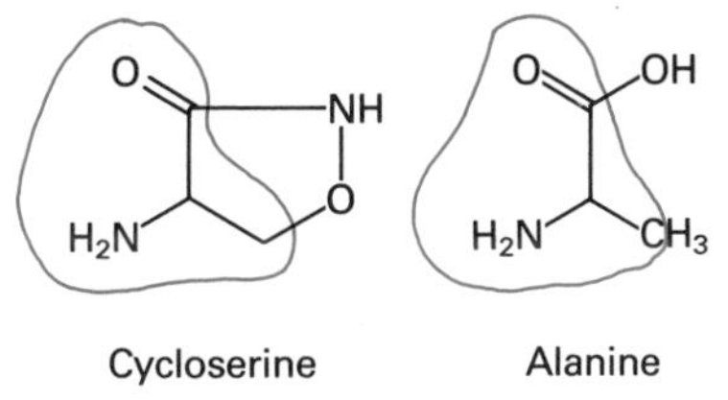

Cycloserine Alanine

**Figure 11.8** The structure of cycloserine (notice the similarity with alanine).

**(e) Tetracyclines.** The tetracycline group includes tetracycline, chlortetracycline and oxytetracycline (Figure 11.10), closely related antibiotics produced by species of *Streptomyces.* They all have similar clinical properties. They are active against a very wide range of bacteria. Toxicity against humans is low. They inhibit protein synthesis and are *bacteriostatic* rather than bacteriocidal.

**(f) Chloramphenicol.** Chloramphenicol (Figure 11.11), produced by *Streptomyces venezualae,* is *bacteriostatic* against a wide range of organisms. It inhibits protein synthesis. It is useful against typhoid and some types of meningitis. In a small proportion of patients it causes **aplastic anaemia,** a disease of the bone marrow. Because of its simple chemical structure, chemical synthesis is the economical method of production.

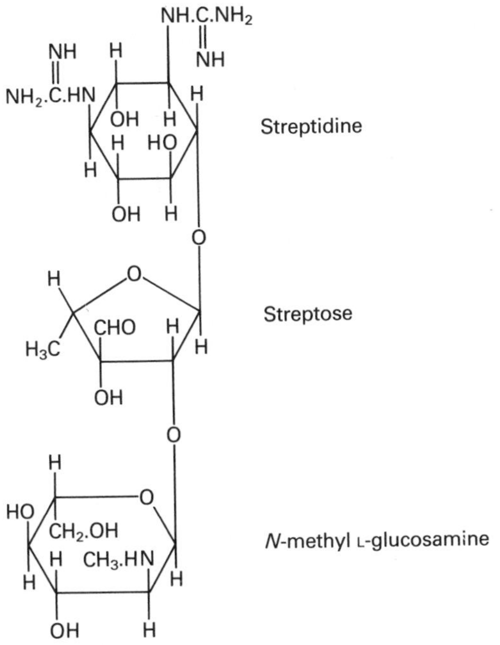

Streptidine

Streptose

*N*-methyl L-glucosamine

**Figure 11.9** The structure of streptomycin.

**Figure 11.10** The tetracyclines.

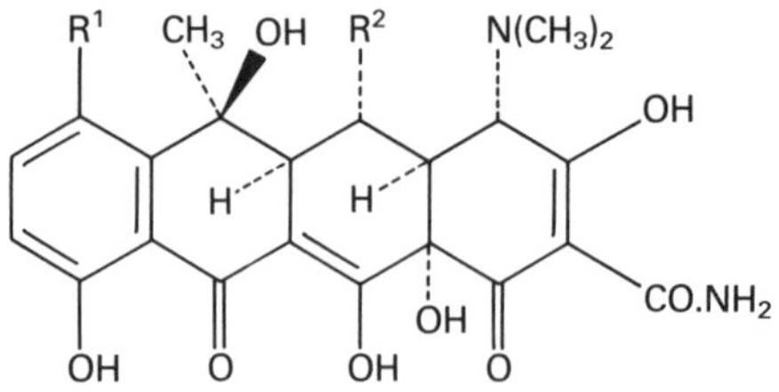

| $R^1$ | $R^2$ | |
|---|---|---|
| H | H | tetracycline |
| C1 | H | chlortetracycline (aureomycin) |
| H | OH | oxytetracycline (terramycin) |

**Figure 11.11** The structure of chloramphenicol.

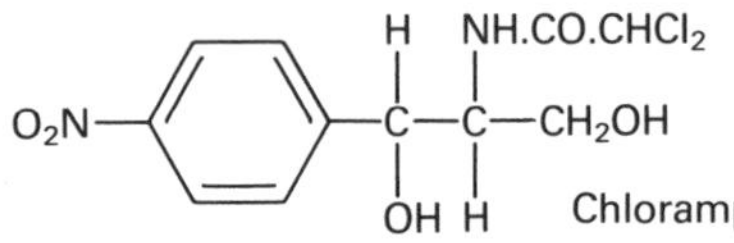

Chloramphenicol

**(g) Erythromycin.** Erythromycin (Figure 11.12), which is produced by *Streptomyces erythreus*, consists of two sugars linked to a large ring structure. It acts by inhibiting protein synthesis. It is useful against the same types of bacteria as the penicillins, but because its structure and mode of action are quite different it can be used against organisms which have become resistant to penicillin and for patients who are allergic to penicillin.

**(h) Antibiotics active against membranes.** *Bacillus* spp. produce a number of circular polypeptides with antibiotic activity. These are the **tyrocidins, gramicidins** (Figure 11.13) and **polymixins**. The tyrocidins

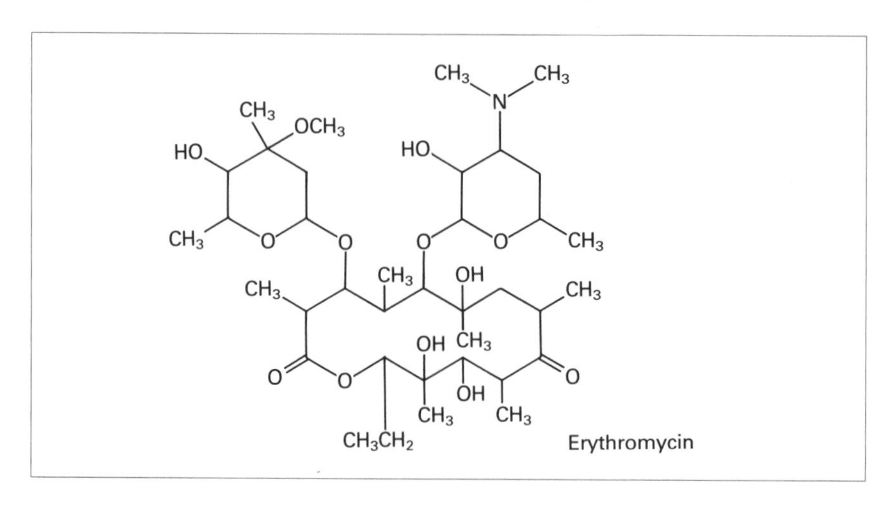


**Figure 11.12** The structure of erythromycin.

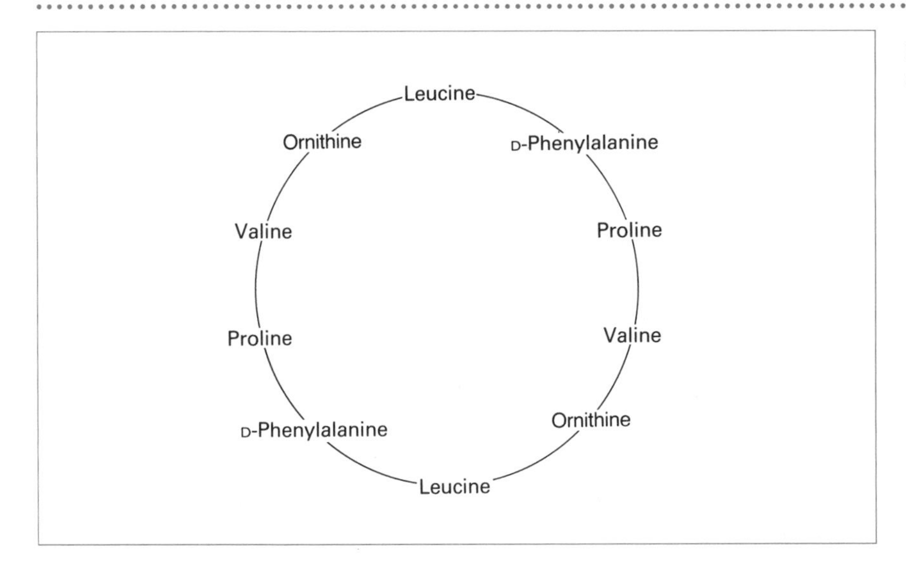


**Figure 11.13** The structure of gramicidin S.

and gramicidins are more active against Gram-negative bacteria while the polymixins are more active against Gram-positives. They all act by making the bacterial cell wall permeable so that the cell contents can leak out. They are *bacteriocidal.* The **polyene** antibiotics, which are produced by streptomycetes, consist of large rings with numerous double bonds. They act against fungi and cause leakage of cell contents by making the membranes permeable. Examples are **nystatin** and **amphotericin**.

## Agricultural uses of antibiotics

Many farm animals grow up to 50% faster if small quantities of broad-spectrum antibiotics such as penicillins and tetracyclines are included in their diets. Such treatments may suppress the growth of bacteria which cause minor infections without obvious symptoms. For example, pigs grow faster if oxytetracycline is included in their feed, possibly because it inhibits the growth of *Clostridium perfringens* in the intestine. The use of antibiotics for this purpose has now been restricted because of fears that it would lead to the widespread selection and dissemination of bacteria resistant to antibiotics.

## Commercial production of antibiotics

The production of penicillin will be described as an example. Much of what is said applies to other antibiotics. In Fleming's experiments, only a few micrograms of penicillin were produced per ml of culture fluid, and this could not be purified. Figure 11.14 shows how penicillin yields increased over a number of years, while Figure 11.15 shows how the price of penicillin changed over the same period. You might like to consider why

BOX 11.17

**Deep liquid culture**

Penicillin was initially produced commercially in **surface mat culture**. The mycelium grew as a thick surface film (mat) on trays containing nutrient medium. Production was inefficient because the mycelium grew in only a small fraction of the culture volume. Synthesis was relatively slow because the contact between the mycelium, the air and the liquid was limited. Large numbers of containers had to be used, so the process was very prone to contamination during inoculation.

In **deep liquid culture** the mycelium grows fully submerged in the medium and is distributed throughout the vessel. Sterile air is forced through the fluid to provide oxygen, and the fluid is vigorously mixed to keep nutrients and mycelium evenly distributed. This technique makes much more effective use of the culture volume. One 50 000-litre tank could provide a volume of broth equivalent to 100 000 of the containers used for surface culture. It is less labour-intensive because one large vessel replaces thousands of small ones, and the chances of contamination are reduced.

*Penicillium chrysogenum* is now used for all commercial penicillin production: this can be grown successfully in deep liquid culture because the mycelium grows in a dispersed form which permits easy access of nutrients. *Penicillium notatum* is unsuitable for deep liquid culture because the mycelium forms dense pellets, and oxygen and nutrients cannot penetrate readily.

BOX 11.18

**Development of penicillin production at Oxford and in the USA**

A group of research workers at Oxford University, led by Sir Howard Florey, followed up Fleming's early work on penicillin. They managed to concentrate penicillin by **freeze-drying** it (rapidly freezing a solution of it and evaporating the water under vacuum without allowing the ice to melt), and achieved a partial purification. They tested penicillin on laboratory animals infected with various pathogenic bacteria and found it highly effective. Doses of up to 30 mg injected into mice had no harmful effect. Using primitive purification methods, they processed large volumes of culture fluids and accumulated enough penicillin to use for a clinical trial.

The first human patient was an Oxford policeman who had contracted an infection with *Staphylococcus aureus*. Starting from a small scratch on his mouth, the infection had spread throughout his body. Treatment with sulphonamides had failed and he was on the point of death. Intravenous injections of penicillin, given every 3 hours, began on 12th February 1941. Within a few days the infection was on the retreat. The patient was sitting up in bed and eating. His temperature had dropped to normal. It was clear that the treatment was proving successful, but the supplies of penicillin were rapidly diminishing. The team worked hard to produce more penicillin, even extracting it from the patient's urine, but eventually their supplies ran out, and injections stopped before the infection was fully cured. The patient lived for a few days, but the infection soon reasserted itself and he died.

A few more clinical trials were more successful and convinced Florey that penicillin was an extremely effective drug if methods could be developed to produce it in useful quantities. Britain was then engaged in fighting a desperate war and there was no chance of getting the enormous resources which would be needed, so Florey decided to enlist help from the USA.

He visited the Northern Regional Research Laboratory (NRRL) of the US Department of Agriculture at Peoria, Illinois. Charles Thom, an expert on the penicillia, conducted a survey of all the strains in his collection, looking for higher yields. From all round the world, scientists and US government employees sent him cultures and soil samples. The best culture came from nearby. A mouldy melon, purchased in a local market, yielded a strain of *Penicillium chrysogenum* which produced more penicillin than any other strain tested. This was developed by a programme of selection and mutation to give the high-yielding strains used in industry today.

BOX 11.18 *continued*

One of the long-term tasks of the NRRL was to find uses for **corn steep liquor**. This is a waste product generated in large quantities when maize (known as corn in the USA) is soaked in water, before it is ground up to separate the oil and starch. Because it has a high BOD (biochemical oxygen demand, page 386), it cannot be discharged into rivers, and it is normally concentrated to keep it from going bad. Fermentation experts at NRRL tried adding corn steep liquor to penicillin fermentations as a source of nitrogen and carbon, and found that yields were doubled. Corn steep liquor concentrate is now a standard ingredient in the media used to produce antibiotics.

For several years universities and pharmaceutical companies collaborated and shared information freely. Strains of *P. chrysogenum* were passed from one laboratory to another. At each stage yields were improved by selection and mutation.

Improvement of *Penicillium* strains by mutation and selection. Reprinted with permission from Riviere (1975) *Industrial Applications of Microbiology*, p. 200. Copyright © M.O. Moss and J.E. Smith.

| *Date* | *Strain identification and origin* | *Yield (units/ml)* | *Comments* |
|---|---|---|---|
| 1929 | *P. notatum* (Fleming) | 2–20 | Wild-type isolate |
| 1941 | *P. notatum* (NRRL 832) | 40–80 | Wild-type isolate |
| 1943 | *P. chrysogenum* (NRRL 1951) | 80–100 | Wild-type isolate from a melon |
| | ↓*selection* | | |
| 1944 | NRRL 1951 B25 | 100–200 | (NRRL = Northern Regional Research Laboratories, Peoria) |
| | ↓*X-ray treatment* | | |
| 1944 | X 1612 | 300–500 | |
| | ↓*UV irradiation* | | |
| 1945 | Q 176 (Wisconsin) | 800–1000 | |
| | ↓*UV irradiation* | | |
| 1947 | BL 3D 10 (Wisconsin) | 800–1000 | Lacks the yellow pigment – chrysogenin |
| | ↓*nitrogen mustard* | | |
| 1949 | 49–133 (Wisconsin) | 1500–2000 | Poor growth |
| | ↓*nitrogen mustard* | | |
| 1951 | 51–20 (Wisconsin) | 2400 | Very poor growth |
| | ↓*selection* | | |
| 1953 | 53–399 (Wisconsin) | 2700 | Selected for improved growth |
| | ↓*selection* | | |
| 1960 | Commercial strains | *c*5000 | Continuous selection for improved behaviour under commercial conditions |
| | ? ↓ | | |
| 1970 | Commercial strains | *ca* 10 000 | |

Today, pharmaceutical companies compete strongly. They can make much more penicillin than is needed and the price is relatively low. The days of collaboration are over. Each company develops its own strains and production methods and the details are closely guarded secrets.

the pharmaceutical companies charged less for penicillin as their yields increased. Do you think it was altruism? Today, yields probably exceed 50 mg/ml of highly purified penicillin. The following factors were especially important in making this enormous change:

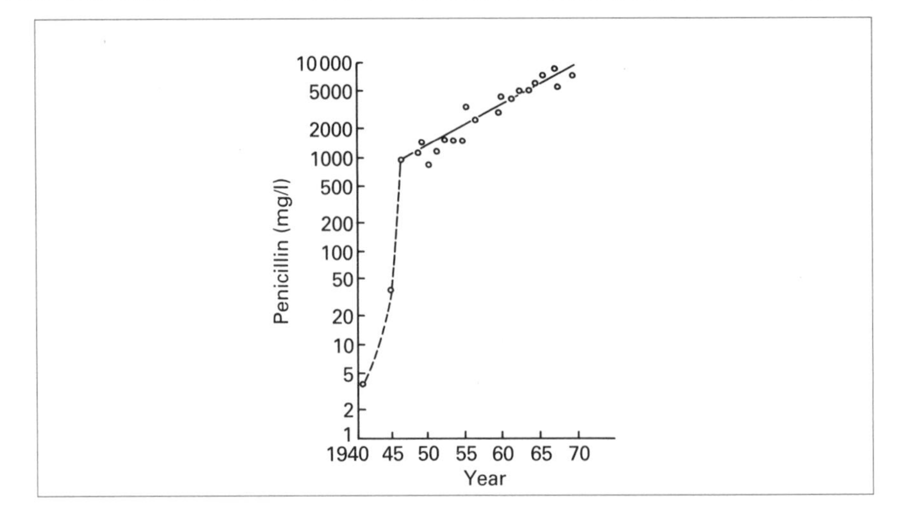


**e 11.14** How the yields of penicillin fermentation have increased with time. Reprinted with permission from Ghose *et al.* (eds), *Advances in Biochemical Engineering 1*, p.113. Copyright © 1971, Springer-Verlag.

1. Improvements in the organisms used for synthesis, including a change to a different species of *Penicillium.*
2. Improvements in medium composition.
3. Improvement in culture techniques, especially the development of **deep liquid culture** (page 368).
4. Much better techniques for extracting penicillin from the culture broth and purifying it.

Fleming's original strain of *Penicillium notatum* was abandoned in favour of a strain of *P. chrysogenum* which produced a higher yield of penicillin and could be grown in deep liquid culture. Over several years this strain was improved in several different university and industrial laboratories in the USA. Selection for desired properties (especially increased yield) was aided by a variety of mutagenic techniques (page 368). One undesirable feature, the production of a yellow pigment which was difficult to separate from the penicillin, was eliminated by mutation. The greatly improved strain was made widely available and is the ancestor of the strains now used for commercial production.

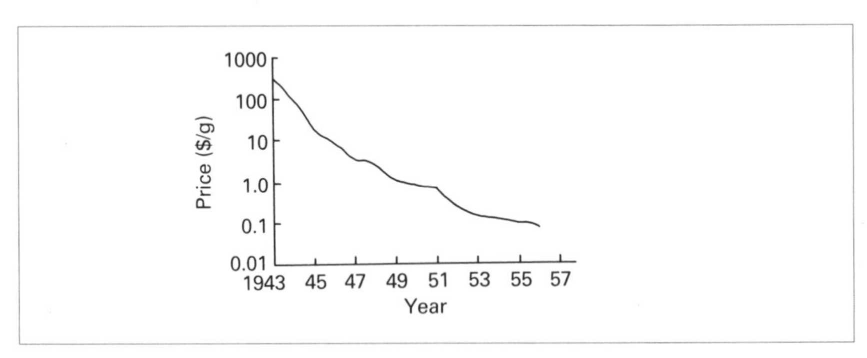


**Figure 11.15** How the cost of penicillin has decreased with time. Taken from King (1982) *Journal of Chemical Technology and Biotechnology*, **32**, 2–8.

The composition of the medium is critical. Glucose gives very rapid mycelial growth but penicillin production is very low because glucose causes **catabolite repression** (page 161) of the enzymes responsible for penicillin biosynthesis. Penicillin yields are greatly increased by using a mixture of glucose and lactose. The glucose is used for rapid mycelial growth during the first 30–40 hours. The lactose, which is metabolized more slowly and does not cause catabolite repression, is used later on to support penicillin production. Lactose is expensive so the preferred technique now is to use only glucose. Sufficient is added at the start to support the rapid phase of mycelial growth. When this is exhausted more glucose is added, either in small amounts or as a continuous feed, but always keeping the concentration low so that catabolite repression is avoided.

Nitrogen is supplied by cheap protein sources such as soya-bean flour, fish-meal and corn steep liquor. Various mineral salts are present including inorganic phosphate and calcium carbonate, which help to control pH. A variety of agents including oils, waxes, octadecanol, silicones and polypropylene glycols are used to prevent the medium forming foam as it is aerated and agitated. **Phenylacetic acid**, or a related compound, is added to act as a precursor of the penicillin side-chain. Supplying it in the medium increases penicillin production, and diverts synthesis in favour of penicillin G rather than less favoured types of penicillin.

The inoculum is prepared from a stock culture. Starting with a few thousand spores the culture is grown up in several stages, increasing the volume at each stage until it represents about 10% of the volume of the production vessel. About 3–5 tonnes of wet mycelial mass will be used to inoculate a 50 000-litre fermenter. There are advantages to this procedure. Checks can be made to ensure that the inoculum is uncontaminated and that antibiotic production and other properties have not changed. Also, the large inoculum makes the best use of the expensive production vessel by minimizing the delay before the culture is fully grown and production is at a maximum.

The production stage takes place in closed tanks, with air bubbled in to supply oxygen, and vigorous stirring to disperse the air and keep the medium, air bubbles and mycelium intimately mixed. A large fermenter will consume sugar at up to half a tonne per day. The heat generated by metabolism is enough to raise the temperature by up to 2°C per hour, so the fermenters are cooled by pumping water through internal coils or external jackets. Penicillin production is very sensitive to temperature: this is critical to within half a degree, so sophisticated control systems are necessary. pH is also carefully controlled between 6.5 and 7.0.

There are two clear phases in antibiotic fermentations. Rapid mycelial

**BOX 11.19**

**Extraction and purification of penicillin**

Penicillin is a weak acid. In alkaline solution its carboxyl group is ionized and it is water soluble. In acid solution, the carboxyl group is un-ionized and it is soluble in organic solvents. This behaviour is used to purify it. By acidifying the culture broth it can be extracted into organic solvents. It can then be extracted back into dilute alkali. The process is repeated using a different organic solvent. At each stage impurities are left behind. Finally, penicillin is concentrated by removal of water by vacuum evaporation at low temperature, and crystallized as the sodium or potassium salt.

growth takes place in the first 30–40 hours. After this, little further growth takes place. Penicillin production starts towards the end of the first phase. Glucose, penicillin side-chain precursor and other nutrients are added continuously to keep production going for up to 15 days. This method of adding nutrients to keep production going as long as possible is known as **fed batch fermentation**. It may be necessary to remove some of the culture fluid from time to time to make way for more nutrients. The mycelium is removed by filtration and the penicillin is extracted and purified.

**(a) Making different penicillins**. Penicillin G is made by putting **phenylacetic acid** (or a derivative) into the medium as the side-chain precursor. If **phenoxyacetic acid** is used, penicillin V will be produced. Other useful penicillins cannot be made this way because their side-chain precursors are unstable, are too toxic, or will not penetrate into the cells. These are made by first producing penicillin G or penicillin V, and removing the side-chain with penicillin acylase (Figure 11.5). The 6-aminopenicillanic acid formed in this way is converted to other penicillins by chemical addition of the side-chains.

## Production of human insulin by genetic manipulation

**Insulin** is used in the treatment of **diabetes mellitus** (diabetes), a disorder in the metabolism of carbohydrates, fats and proteins, which causes malfunctions of the vascular system. The disease is fatal if not treated, but daily injections of insulin, together with a controlled diet, can mean that most sufferers can lead almost normal lives.

A **disulphide bridge** is formed when sulphur atoms in two cysteine residues become covalently linked.

Insulin is a peptide hormone secreted by the **β cells** of the **Islets of Langerhans** in the pancreas. It consists of two amino acid chains, A (21 amino acids long) and B (30 amino acids long). The two chains are joined together by **disulphide bridges** between **cysteine** residues.

Natural human insulin is not available in useful quantities, so bovine (cow) and porcine (pig) insulin have been used routinely. The two animal insulins are very similar to human insulin: bovine insulin differs only by three amino acid residues, and porcine insulin only by one. Humans treated with bovine insulin produce antibodies which bind the hormone and make it less effective. This is less of a problem with porcine insulin.

The number of diabetics treated with insulin has increased steadily in the developed countries, and is expected to double by the end of the century. It seemed likely that a shortage of insulin might develop, and Eli Lilly & Co., a major pharmaceutical company, decided to produce human insulin from bacteria by genetic manipulation. Apart from being able to produce insulin in any required quantity, the bacterial product was not expected to cause antibody formation because it was chemically identical to the human product, and it would not be contaminated with animal peptides, which are difficult to remove from natural insulins.

**BOX 11.20 (A)** How to make DNA, starting from pancreatic messenger RNA.

**Isolation of the insulin gene**

Insulin is produced in the pancreas so pancreatic tissue contains messenger RNA to make insulin. Messenger RNA was extracted from human pancreatic tissue. **Reverse transcriptase** (page 309, 336), an enzyme which uses an RNA template to make a single DNA strand, was used to make a DNA strand **complementary** to the RNA. DNA polymerase was then used to convert this to double-stranded DNA. We shall refer to this as **pancreatic DNA**, although it was not actually isolated from the pancreas.

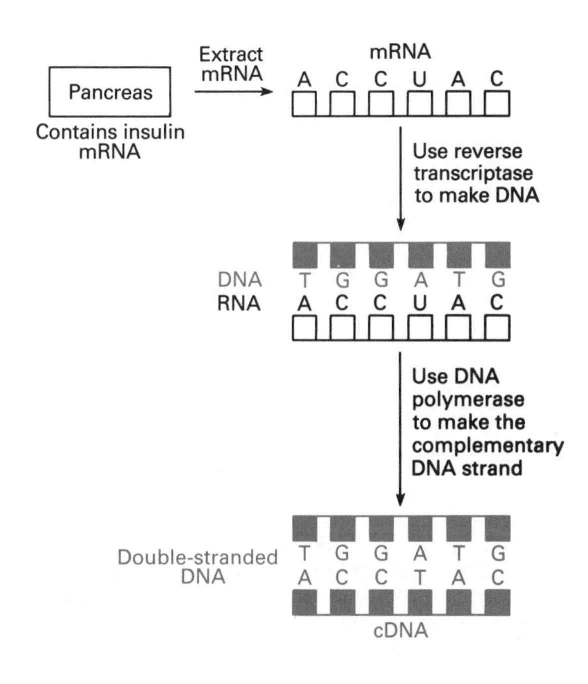

**BOX 11.20 (B)** How to get bacterial cells which contain pancreatic DNA. *Amp*$^r$ is the gene for ampicillin resistance. *Tet*$^r$ is the gene for tetracycline resistance.

The pancreatic DNA was cloned in the bacterium *E. coli*, strain K12. This is a deficient strain which is able to survive only under laboratory conditions, so there is no danger of it escaping from the laboratory or factory and infecting humans or animals.

The plasmid vector used (page 231) was **pBR322**, a small plasmid which replicates in *E. coli* and carries genes for resistance to tetracycline and ampicillin. The pancreatic DNA was cut into pieces with restriction enzymes, and inserted into pBR322 at a site in the tetracycline marker. *E. coli* K12 was transformed with this DNA, and the cells were transferred to a medium containing ampicillin. *E. coli* K12 is ampicillin-sensitive, so the colonies which grew *must have contained pBR322.* Replica plates were made onto a medium which contained tetracycline. This made it easy to detect which of the colonies carried *pancreatic DNA. E. coli* K12 is tetracycline-sensitive, so it can only grow on tetracycline medium if it carries pBR322 with an active gene for tetracycline resistance. Colonies containing pancreatic DNA *could not grow on tetracycline medium* because the DNA was inserted in the tetracycline resistance gene of pBR322, and made it inactive.

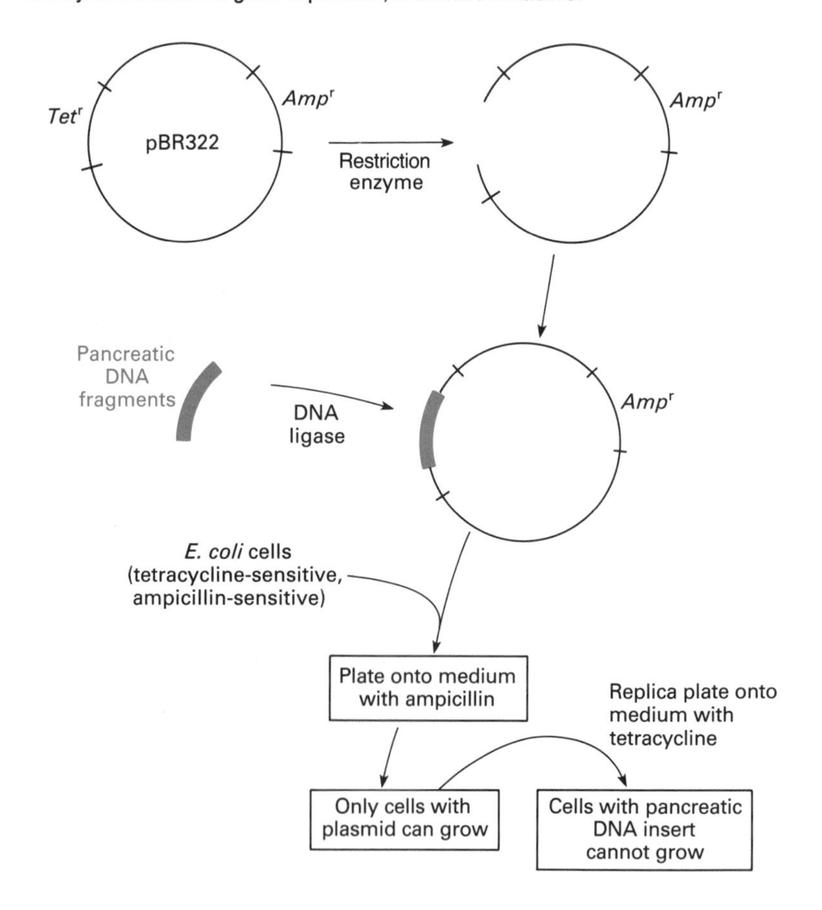


The pancreatic DNA which had been cloned contained many different genetic sequences. It was now necessary to isolate *E. coli* cells carrying a *complete copy of the insulin gene.* This was done by detecting colonies which could make insulin. Antibodies to insulin were bound to a membrane, which was placed on a plate in contact with growing colonies. Insulin produced by some of the colonies was bound by the antibodies. The membrane was removed and immersed in a solution of

**BOX 11.20 (C)** How to identify bacterial colonies which produce insulin.

radioactive insulin antibodies. Where insulin was present, radioactive antibodies were bound to the insulin so they became fixed in position on the membrane. The membrane was washed and dried, then stored in the dark in contact with photographic film. When the film was developed, dark spots showed which colonies had produced insulin.

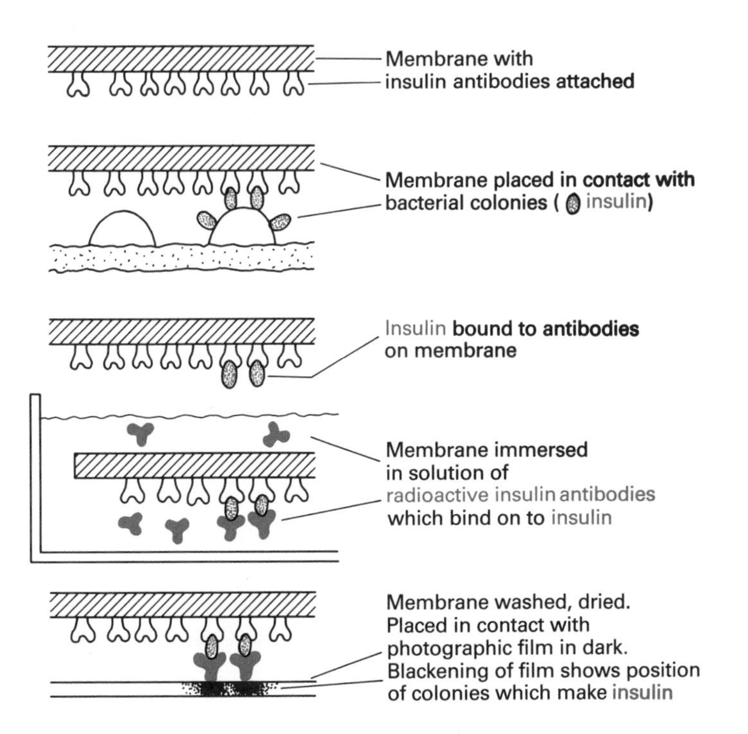


The DNA was isolated, and trimmed with restriction enzymes to remove unwanted sequences. To obtain rapid synthesis of insulin messenger RNA, the insulin was joined to the structural gene for **tryptophan synthetase**. The two genes were linked by a group of nucleotides coding for the amino acid methionine. When the genes were inserted into the plasmid, a hybrid protein was synthesized: **tryptophan synthetase–methionine–insulin**. The tryptophan synthetase gene has a very powerful **promoter** (page 119, 160), ensuring that numerous copies of the messenger RNA are made.

## Synthesis of insulin

To make human insulin, the insulin gene was isolated, joined to the gene for tryptophan synthetase, and cloned into *E. coli* strain K12.

When the insulin molecule is first synthesized in humans it contains four peptide chains. **Chains A** and **B**, the peptides present in mature insulin, are separated by **chain C**. A **leader sequence** is attached to chain A. The leader sequence and chain C are removed by enzymes to make active insulin, leaving peptides A and B held together by disulphide bridges.

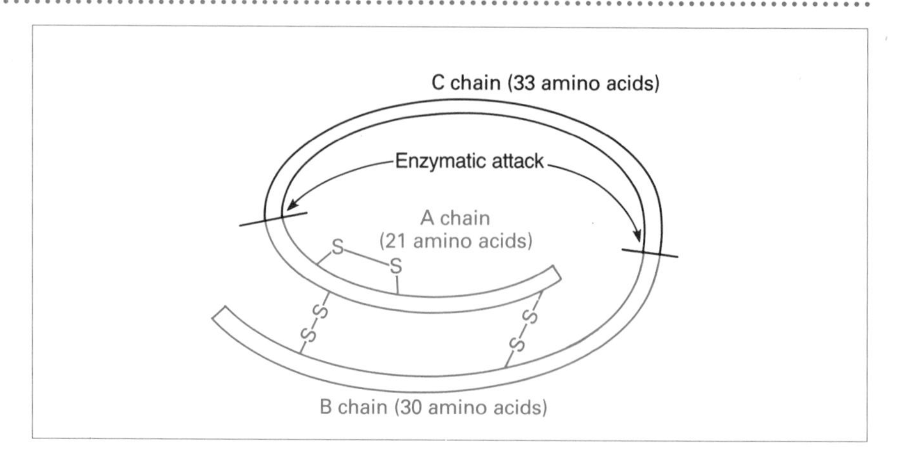


**Figure 11.16** The structure of pro-insulin. Insulin is made by removing chain C with the enzymes trypsin and carboxypeptidase. S—S represent disulphide bridges between cysteine residues. The A and B chains are joined by two disulphide bridges. There is a third bridge between two cysteines within chain A.

In the manufacture of insulin by gene manipulation, the part of the gene coding for the leader sequence has been removed, so the product synthesized has only chains A, B and C; this is known as **pro-insulin** (Figure 11.16). The pro-insulin is attached to tryptophan synthetase by the amino acid methionine. To free it from the tryptophan synthetase, the hybrid protein is treated with the reagent **cyanogen bromide**, which breaks proteins at methionine residues, so the pro-insulin is released (Figure 11.17). There are no methionine residues in pro-insulin, so this peptide is not attacked.

Because the pro-insulin is synthesized as a hybrid protein, the disulphide bridges holding chains A and B together form between the wrong cysteine residues, and the insulin has the wrong shape. This is corrected after the tryptophan synthetase has been removed, by chemically breaking the disulphide bridges and allowing them to reform in the correct manner. After purification of the pro-insulin, the unwanted C chain is removed by treatment with trypsin and carboxypeptidase (Figure 11.17).

Final purification of the insulin involves gel filtration, ion-exchange chromatography and crystallization. Only very small amounts of *E. coli* peptides survive the purification process and these appear to have no harmful effects.

Production of insulin by this procedure is efficient. *E. coli* grows rapidly to form a dense culture. Each cell contains several dozen copies of the plasmid, and each plasmid carries the insulin gene. The **tryptophan synthetase promoter** is very powerful, so plenty of insulin messenger RNA is made, and large amounts of insulin are synthesized.

Human insulin produced in this way is safe and effective, although some patients find it more difficult to control their blood sugar levels than with animal insulin. Human insulin appears to have some advantages for patients who have high levels of antibodies against animal insulin. The real advantage is that supplies are unlimited.

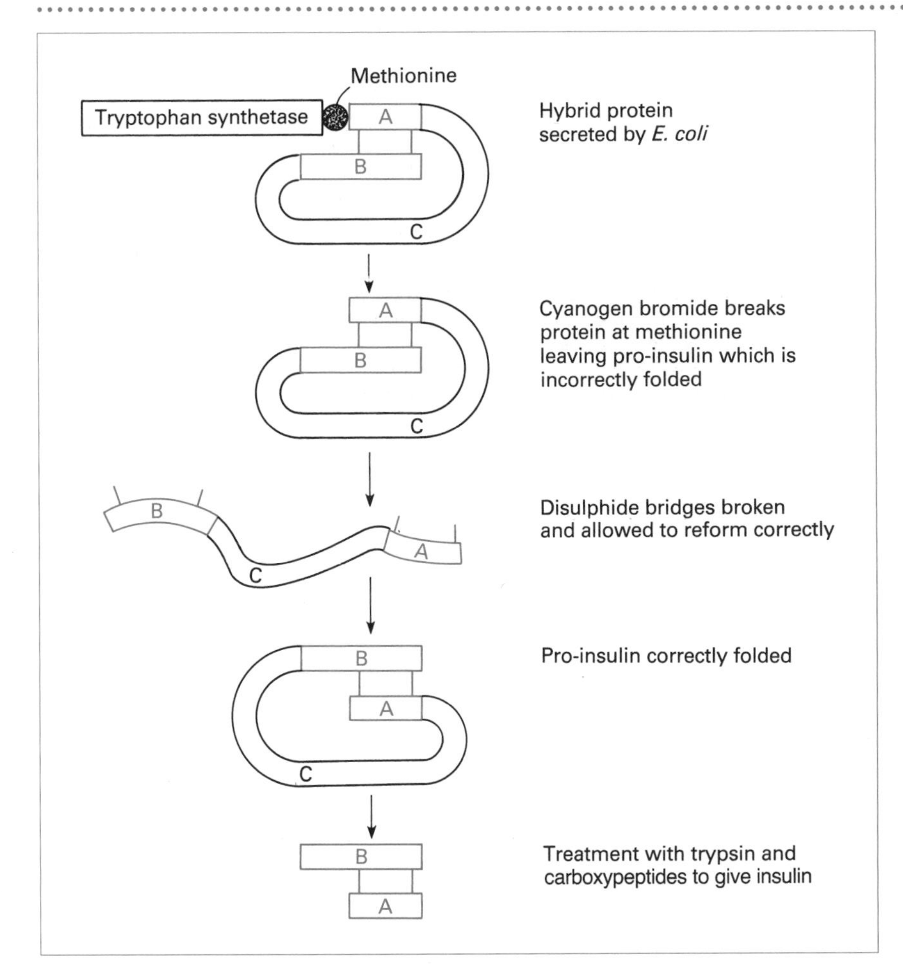


**Figure 11.17** How to get insulin from the hybrid protein with tryptophan synthetase.

## Production of hepatitis B vaccine by gene manipulation

Hepatitis B virus is responsible for the most important infectious disease of the liver. The patient may recover after a period of illness with complete elimination of the virus. Alternatively the patient may become a **chronic carrier** of the virus, sometimes with only mild symptoms, sometimes with progressive liver damage and hepatic cancer. It is estimated that there are 250 million chronic carriers, which is about 5% of the world's population.

There is no cure for the disease, but large amounts of the **surface antigen protein** of the virus (known as **HBsAG**) circulate in the blood of carriers. This antigen causes a powerful immune response, so an effective vaccine can be prepared from it. Considerable purification and testing are necessary to ensure its safety. Because it comes from the blood of carriers, patients are often reluctant to be injected with it.

**BOX 11.21** Cloning the gene for hepatitis B surface antigen protein (HBsAG) in yeast.

**Isolation of the HBsAG gene and cloning it in yeast**

DNA was extracted from the hepatitis B virus and cloned in *E. coli.* The cloned DNA was cut into fragments with restriction enzymes and the fragment containing the HBsAG gene was isolated by electrophoresis (page ???).

The HBsAG gene was joined to an active yeast **promoter** gene on one side (to make sure that large amounts of messenger RNA were synthesized), and a yeast **terminator** gene on the other (to make sure that transcription did not continue beyond the HBsAG gene). The three genes were then inserted into plasmid pBR325 and cloned in *E. coli.* Plasmid DNA carrying the extra genes was isolated.

Yeast plasmid pC1/1, which maintains itself at a level of 100 copies per cell, was used as the vector for synthesis of the protein. pC1/1 was cut open with restriction enzyme and the pBR325 DNA, carrying the promoter, HBsAG, and terminator sequences, was inserted using the enzyme DNA ligase.

DNA extracted from hepatitis B virus and cloned.

Cloned in *E. coli*
Purified. Cut into fragments with restriction enzymes

Fragment containing HBsAG gene isolated by electrophoresis and cloned in *E. coli*

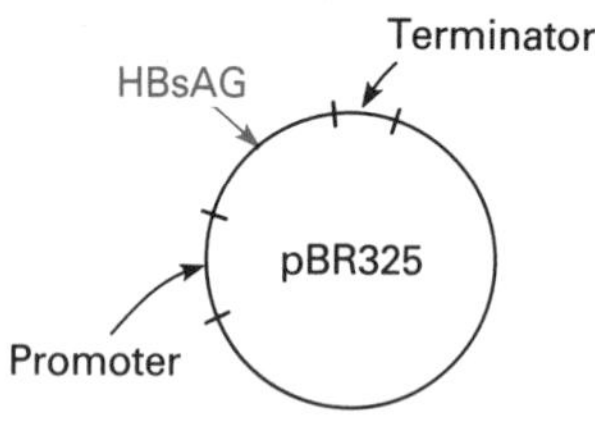


HBsAG gene joined to yeast promoter and yeast terminator, inserted into plasmid pBR325, cloned in *E. coli*

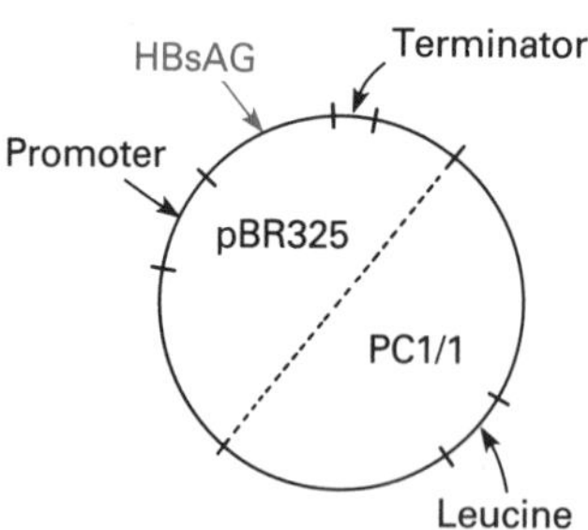


pBR325 DNA with HBsAG, promoter and terminator, joined to plasmid PC1/1. Cloned in yeast

Yeast grown to synthesize HBsAG protein

The DNA was then used to transform a yeast strain which was unable to synthesize leucine. pC1/1 carries a gene for leucine synthesis, so yeast cells which had been transformed were detectable because they could grow on medium without leucine.

The virus particle is a sphere 42 nm in diameter containing a circular DNA molecule 3200 nucleotides long, part of which is single-stranded. The surface layer of the virus consists of HBsAG embedded in lipid.

### Manufacture of HBsAG

Yeast is used to make the hepatitis surface antigen because it is considered to be a very safe organism, having been used in the production of food and drink for thousands of years. Its genetic stability can easily be monitored, it produces high yields of the protein, and production can easily be scaled up to any required level.

Stock cultures of yeast carrying the HBsAG gene are kept at −20°C. They are grown in broth culture for 2 days, and the cells are harvested by centrifugation. The cells are broken open by high pressure to release the HBsAG protein. The protein is purified by column chromatography, and filtered through a membrane filter to remove any microbial cells. It is then treated with formaldehyde to inactivate any virus which might be present as a contaminant. Finally, it is adsorbed onto aluminium hydroxide. This acts as an **adjuvant** (makes it much more effective at generating antibodies), and **thiomersal** is added as a preservative.

## Enzyme production

Purified or concentrated enzymes are used for many industrial, scientific and medical purposes. (Some examples of industrial uses are shown in Table 11.2.)

**Table 11.2** Some industrial uses of microbial enzymes

| *Enzyme* | *Organisms* | *Uses* |
|---|---|---|
| α-Amylase | *Bacillus licheniformis*<br>*B. amyloliquifasciens* | Hydrolysis of starch to dextrins<br>Textile manufacture<br>Paper manufacture<br>Baking<br>Ethanol production from starch |
| Glucoamylase | *Aspergillus niger* | Hydrolysis of dextrins to glucose |
| Glucose isomerase | *Actinomyces missouriensis*<br>*B. coagulans*<br>*Streptomyces* spp. | Glucose converted to glucose–fructose mixture |
| Alkaline protease | *B. licheniformis*<br>*B. subtilis* | Household detergents |
| Neutral proteases | *B. amyloliquifasciens* | Breakdown of protein in brewing materials to prevent beer hazes |
| Acid proteases | *Aspergillus niger* | Flavour improvement in cheese |
| Pectinase | *Aspergillus niger*<br>*B. subtilis* | Fruit juice processing |
| β-Galactosidase | *Aspergillus oryzae*<br>*Saccharomyces lactis* | Hydrolysis of lactose in milk products to increase sweetness and improve digestibility |
| Penicillin acylase | *Escherichia coli* | Removal of side-chain from penicillin G or V for manufacture of other penicillins |

**BOX 11.22**

**Enzyme immobilization**

Enzymes prepared for industrial, scientific and medical uses are easier to handle if they are **immobilized**. Immobilized enzymes are usually more resistant to denaturation and to attack by proteases. These advantages commonly justify the extra costs of making them.

Enzymes are often immobilized on a solid support such as cellulose, glass, silica gel, nylon or polystyrene. A common procedure is to immobilize the enzyme on beads of a support material which are then packed into a column. Substrates in solution are passed through the column, and the products of the reaction emerge at the other end. This has the advantage that the products are enzyme-free, and one batch of enzyme can be used to treat a large quantity of substrate.

Today, most of these enzymes are obtained from microorganisms. This has several advantages. Any required amount can be made by growing more cells. By choosing the right microbial strain, and perhaps improving it by mutation and selection, the enzyme can be obtained in a high starting concentration. Many enzymes are secreted from the cell so in these cases purification is easier. Usually, enzymes with a range of properties such as thermal stability and pH optimum are available in different organisms. The properties of an enzyme can be changed by random mutagenesis and selection, or by **site-directed mutagenesis**.

### High-fructose syrups

An important use of enzymes is in the production of **high-fructose syrups** (Figure 11.18) which can be used as a cheap substitute for sucrose in a wide range of foods and drinks. More than 5 million tonnes are produced annually. Large quantities of starch are produced as a by-product when oil is extracted from maize (known as corn in the USA). Glucose, made by hydrolysis of starch, is cheap but is less sweet than sucrose. Glucose can be converted by the enzyme **glucose isomerase** to a mixture of fructose and glucose (Figure 11.18) which has less sweetness than sucrose. Some of the glucose can be removed by chromatography to give a mixture which is as sweet as sucrose but costs less, and this is used to sweeten soft drinks.

## Steroid transformations

In 1949 it was reported that treatment with the steroid hormone **cortisone** could relieve the pain of patients suffering with rheumatoid arthritis. Cortisone had to be isolated from the adrenal glands of cattle, and an alternative source of supply was urgently needed. A route for chemical synthesis was known, but this required 31 separate steps and was uneconomic.

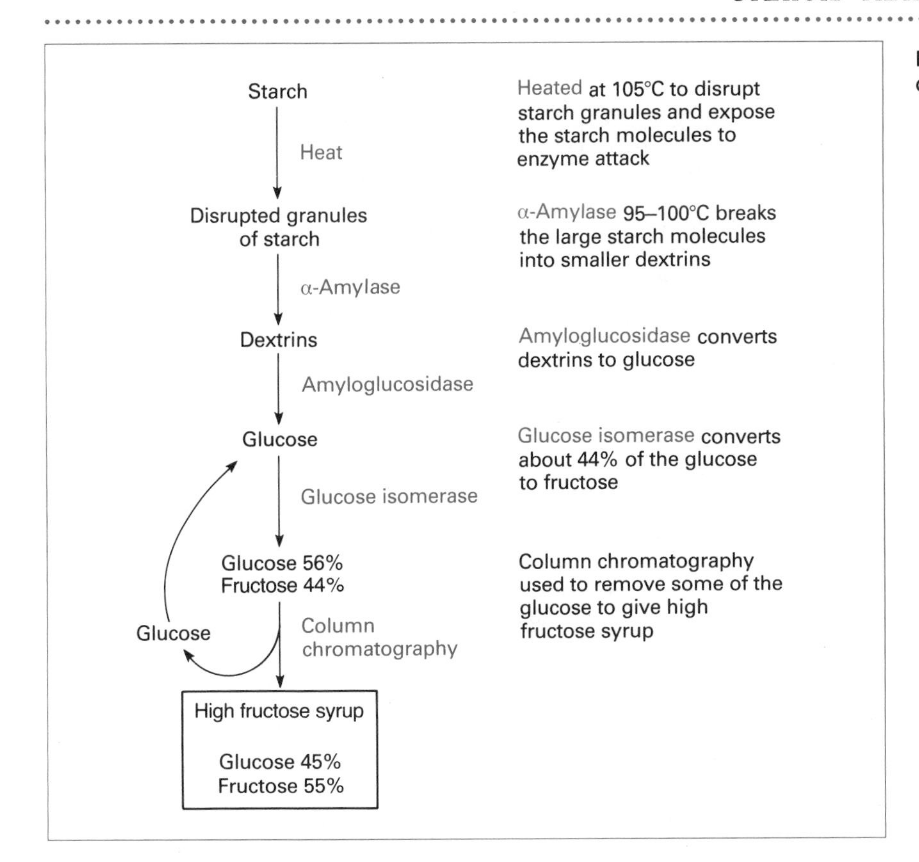


**Figure 11.18** How starch is converted to high-fructose syrup.

In 1952 it was reported that the fungus *Rhizopus arrhizus* could attach a hydroxyl group to the steroid **progesterone** at a particular position on the molecule. Two relatively cheap and abundant starting materials, **diosgenin** from the Mexican yam and **stigmasterol** from the soya bean, can readily be transformed into progesterone. *Rhizopus arrhizus* converts the progesterone to **11α-hydroxyprogesterone** with a yield of over 85%. This discovery provided the key to economical synthesis of cortisone and related compounds (Figure 11.19).

Four more chemical steps make **hydroxycortisone** and one further step makes cortisone. The closely related compounds **prednisolone** and **prednisone**, which are therapeutically even more effective, can be made by using the fungus *Septomyxa affinis* to remove hydrogen atoms from positions 1 and 2 (Figure 11.19).

Many organisms are known which can carry out conversions of steroid molecules. Each organism carries out one particular type of change at one position in the molecule. Microorganisms are useful in steroid conversions because: (1) the changes they bring about are very specific; (2) they are often particularly difficult to achieve by chemical means; and (3) the yields of the transformations are very high.

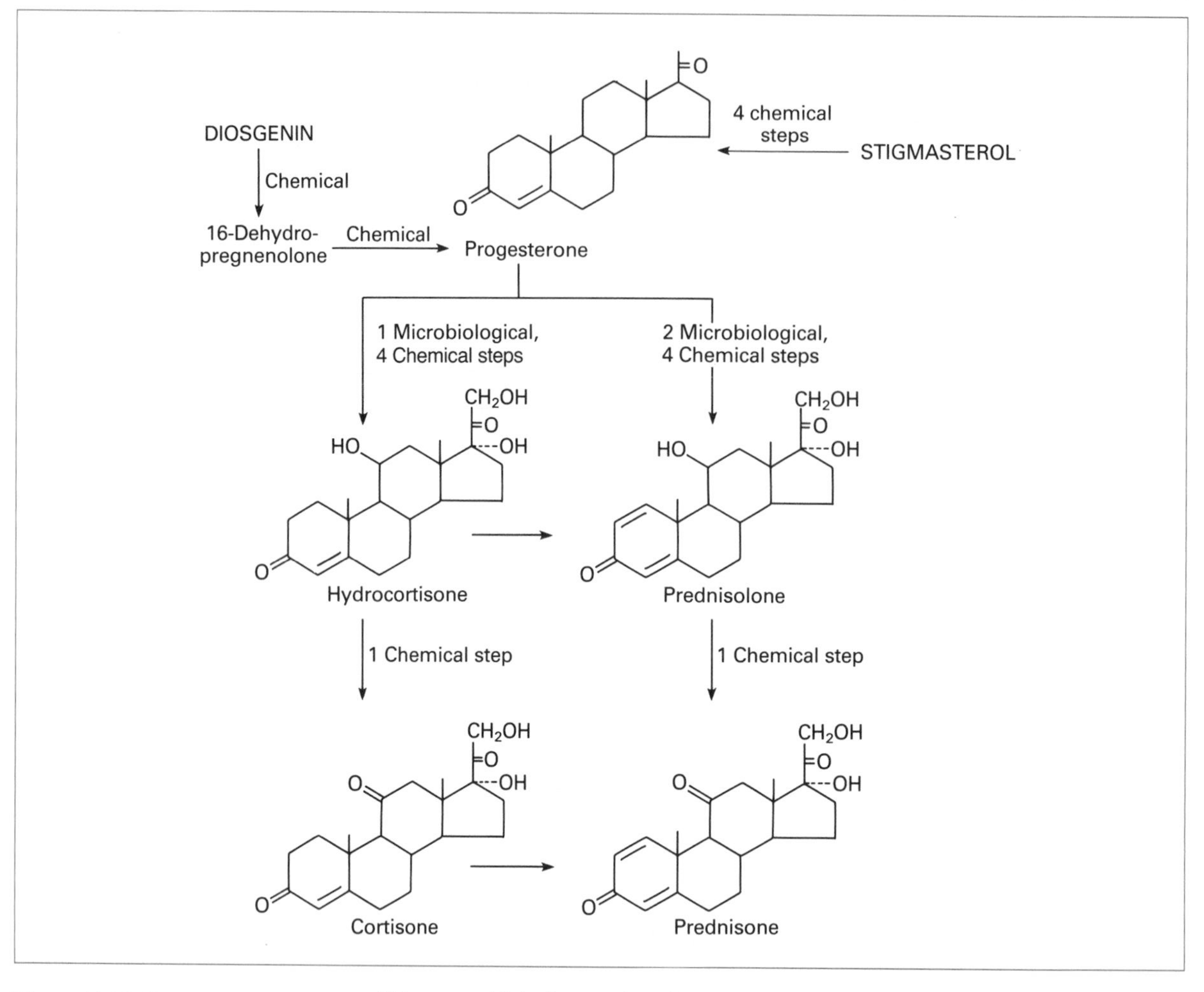


**Figure 11.19** How cortisones are made with the aid of microorganisms. Redrawn from Bu'Lock, J. and Kristiansen, B. (1987) *Basic Biotechnology*, Academic Press, London.

The steroid industry is of great medical importance and also of some economic value. Its products, manufactured by combined chemical and biological means, sell for more than 1 billion US dollars each year.

## Organic acids

Several organic acids are made on a large scale using microorganisms (Table 11.3). We shall look at **citric acid** as an example. Citric acid is used to flavour foods, especially jam, sweets and soft drinks. It is a good metal-binding agent and this makes it useful for a variety of industrial purposes such as keeping electroplating metals in solution. It is used to prevent blood from clotting, and in household detergents instead of polyphosphates (which encourage algal growth when the detergent finds its way into rivers).

**Table 11.3** Some organic acids produced by microorganisms

| *Acid* | *Organism* | *Uses* |
|---|---|---|
| Citric | *Aspergillus niger* | Flavouring foods and drinks. Industrial Medical |
| Gluconic | *A. niger* | Detergents |
| Itaconic | *A. terreus* | Polymers for fibres and paints |
| 2-Ketogluconic | *Serratia marcescens* | Starting material for chemical synthesis |
| L (+)-Tartaric | *Saccharomyces cerevisiae* | Flavouring foods (obtained as by-product of wine manufacture, crystallizes in vats) |
| Lactic | *Lactobacillus* spp. | Food, medical, industrial |

It can be added to foods to stop undesirable oxidizing reactions. Roughly 400 000 tonnes of citric acid are produced every year. About 75% of this is used in food production, 10% in the pharmaceutical industry, and the rest for other industrial purposes.

Citric acid was originally obtained from lemons but, in 1929, the American drug company Pfizer started production in the USA using the fungus *Aspergillus niger*. The industry is very competitive because manufacturers can produce far more citric acid than is needed. Several methods of chemical synthesis are known but they are more expensive than the microbiological method, which accounts for 99% of production.

The growth medium for *Aspergillus niger* is usually based on purified molasses with ammonium nitrate and other salts added. The pH starts between 5 and 7 but quickly falls to 3 or lower. If it remains above pH 3, too much oxalate will form. The medium composition is critical. High concentrations of citrate will form only if the levels of metal ions (copper, iron, magnesium, manganese, molybdenum and zinc) are very low so that enzymes which normally remove citrate can only function slowly. The unwanted metals are removed from the medium by ion-exchange treatment or by chemical precipitation. We can see how important this is. In one experiment, in which manganese was added back to purified medium at the rate of 1 μg/litre, the yield of citrate was reduced by 10%. Good yields are obtained only if mycelial growth and sporulation are restricted. This is done by limiting the concentration in the medium of either phosphate or nitrogen compounds. Citric acid is corrosive, so the equipment used in its manufacture must be made from very high-grade stainless steel or must be lined with rubber or glass, as otherwise metal ions will leach into the medium.

Most citric acid is produced by deep liquid culture in aerated tanks. There is a rapid growth phase but most of the citrate is produced during

**BOX 11.23** The biochemical pathway for production of citrate from sugars. (a) phosphofructokinase; (b) condensing enzyme; (c) pyruvate carboxylase.

**Biochemistry of citric acid production**

Citrate is a key metabolic intermediate and its concentration in cells is tightly regulated. To produce large amounts of citric acid, the reactions which form it must be stimulated, while those which consume it must be reduced to the lowest possible level. These aims are achieved by careful control of the medium and the growth conditions, and by using mutant strains of *Aspergillus niger*.

The diagram shows the biochemical pathway by which citrate is formed.

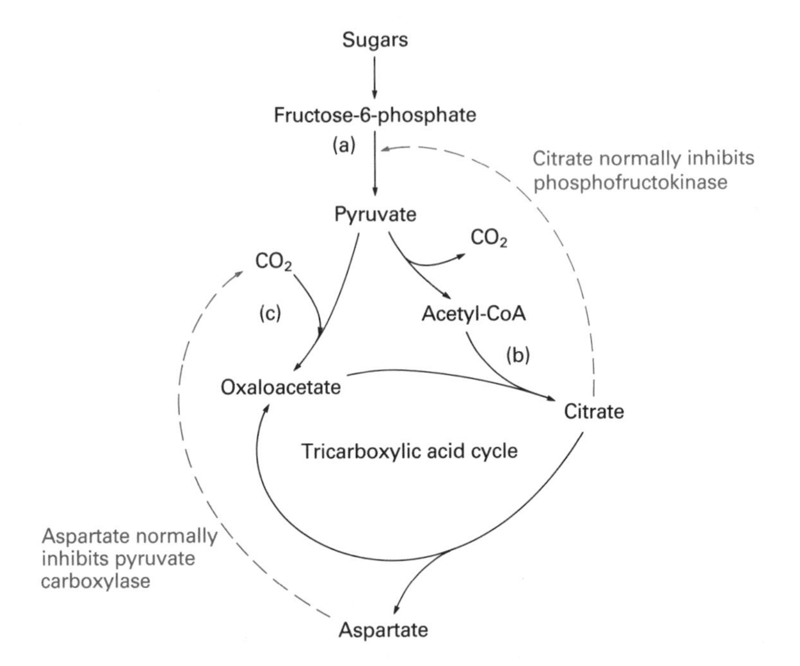


Sugars are the substrates. Citrate is formed from them by the reactions of glycolysis together with the enzyme **pyruvate carboxylase**. Citrate is synthesized from acetyl-CoA and oxaloacetate by **condensing enzyme**. The oxaloacetate is produced from pyruvate by pyruvate carboxylase. Citrate is normally oxidized by the remaining enzymes of the tricarboxylic acid cycle to form oxaloacetate.

During citrate manufacture, the medium is deficient in metal ions and the reactions of the tricarboxylic acid cycle are very slow so only a little citrate is oxidized. **Aspartate** normally inhibits pyruvate carboxylase, but very little of this is formed when the cycle activity is low. Normally, citrate inhibits the activity of **phosphofructokinase**, a key control point in glycolysis, so if citrate accumulates, the reactions which form it will slow down. If manganese levels are very low, ammonia ions accumulate in the cell and citrate does not inhibit phosphofructokinase.

**Oxalate** and **itaconate** are formed from citrate as unwanted by-products. The amounts formed are kept to a minimum by strain-selection and control of the culture conditions.

the slow growth phase which follows. The temperature is kept at about 30°C: if it rises above 33°C too much of the citrate is converted into oxalate. Surface mat culture, in which *Aspergillus* is grown in trays in ventilated chambers, is used to a much smaller extent. A process in which citric acid was produced by yeast from hydrocarbons is no longer economic and has been abandoned.

## Amino acids

Several types of microorganisms are able to grow on inorganic sources of nitrogen and excrete amino acids into the growth medium. In some cases this is an economic way to produce the amino acids. One advantage is that the biologically active forms (the L-isomers) are produced rather than the mixture of L- and D-isomers produced by chemical synthesis.

Most amino acids can exist as both L- and D-isomers which are mirror images of each other. The body can use only the L-isomers, which are those used in proteins.

**(a) Glutamic acid.** L-Glutamic acid, in the form of its salt monosodium glutamate, is widely used as a flavouring agent for foods. Several kinds of bacteria can be used to make it, including *Corynebacterium glutamicum*, and species of *Arthrobacter, Brevibacterium* and *Microbacterium*.

It is essential that the glutamic acid should leak out of the bacterial cells and into the growth medium: a high cellular concentration causes feedback inhibition. Originally this leakage was ensured by using a medium based on glucose and inorganic salts, which contained low levels of biotin. (Biotin-deficient cells became leaky, so glutamic acid escaped to the medium.) Today, a cheaper medium is used, based on molasses, and containing a higher level of biotin. Detergent substances are added to the medium to make the cells leaky, so a biotin-deficient medium is not necessary.

**(b) Lysine.** L-Lysine is one of the **essential amino acids**, which must be present in the human diet and those of many animals for correct biological functions to be maintained. Cereal proteins are usually deficient in lysine, so lysine is used as a supplement in bread and other foods, and in animal feeds. One method of production uses mutants derived from the strains of *Corynebacterium glutamicum* used to produce glutamic acid. These mutants are deficient in homoserine synthesis and overproduce lysine. The biotin level in the medium is critical: if there is too little, glutamic acid is produced; if too much, lactate and succinate are produced.

## Purification of organic wastes

In densely populated communities it is important that adequate arrangements are made for disposal of sewage. Untreated sewage contains millions of bacteria per ml. These will include the normal human gut flora such as coliforms, streptococci, anaerobic spore-forming rods and *Proteus*. Pathogenic bacteria, such as those responsible for typhoid, shigellosis or cholera, and viruses that cause infectious hepatitis and poliomyelitis, and coxsackie viruses, may also be present from the faeces or urine of diseased individuals. Rivers polluted with untreated sewage are unsafe for bathing, and may endanger supplies of drinking water. Freshwater or marine shellfish living in polluted water may concentrate pathogens and cause disease when they are eaten.

If sewage is discharged into rivers, the breakdown of organic compounds by aerobic microorganisms removes oxygen dissolved in the water, and kills aquatic organisms such as fish. Further breakdown by anaerobes produces toxic end-products such as ammonia and hydrogen sulphide, and offensive odours. Other types of organic wastes, such as those from slaughterhouses,

**BOX 11.24**

**BOD, COD and TOC**

BOD (biochemical oxygen demand) is a measure of the strength of an organic waste; that is, the concentration of carbon compounds which can readily be oxidized by bacteria. It estimates how much oxygen dissolved in river water will be needed by bacteria to oxidize the waste over a 5-day period. We need such a measure for three reasons:

1. So that we can decide what quantity of a particular organic waste can be added to a river without making it anaerobic and harming living organisms in the river.
2. To predict what treatment will be needed for a waste to make it safe for discharge into a river, and to check that treatment has been successful.
3. To decide how much should be charged for treating the organic waste from, say, a slaughterhouse or a brewery.

BOD is estimated by making a series of dilutions of the waste in stoppered bottles with aerated river water. If necessary, suitable bacteria are added. After 5 days, the oxygen concentration in each bottle is measured. In bottles containing undiluted waste there may not have been enough oxygen for complete oxidation, so we cannot tell how much would have been needed. However, some of the dilute samples will be fully oxidized, and will still contain some oxygen. From these, the oxygen which would be used by the undiluted waste can be calculated as follows:

$$\begin{aligned} \text{BOD} &= \text{oxygen needed for undiluted waste (mg/l)} \\ &= \text{oxygen used by diluted waste (mg/l)} \times \text{dilution factor} \end{aligned}$$

BOD is the most useful measure of the strength of a waste but it takes 5 days to estimate. Rapid estimates are often required. **COD** (chemical oxygen demand) and **TOC** (total organic carbon) are used for this purpose. COD is estimated by measuring the amount of an oxidizing agent (dichromate for example) needed to oxidize the sample. TOC is estimated by catalytic oxidation of the sample and measuring the carbon dioxide formed. Although these methods provide much quicker estimates of oxygen demand, they do not estimate how much of the organic material can easily be oxidized by microbes. They usually give very different values from the BOD estimate, so some experience is needed to interpret them.

breweries, food-processing factories and pig farms, cause similar problems and can be treated by the methods used for domestic sewage.

Microbiological treatment of waste aims to speed up the natural processes of breakdown and to carry them out in a treatment works. Dense microbial populations are used. These are not pure cultures but mixtures of bacteria, fungi, protozoans and sometimes algae and invertebrates. They are obtained by allowing effective combinations to establish themselves from organisms present in the wastes. The end-products are gases such as carbon dioxide, liquids and inoffensive solids. The liquids have greatly reduced BOD values. They can be discharged into rivers with little undesirable effect. Luckily, the treatment to reduce BOD also substantially reduces pathogen counts. Chlorination can be used to make the liquids still safer.

Treatment on a large scale usually involves a combination of several methods (Figure 11.20). Some of the most common are described here. Grit and gravel are first removed by sedimentation. The waste is then allowed to sediment, so the solids rise to the surface or sink to the bottom and can be separated from the liquids. Liquids are then treated by aerobic processes and solids by anaerobic digestion.

## Types of organic waste treatment

**(a) The trickling filter process.** A typical trickling filter (Figure 11.21) consists of a cylindrical tank many metres in diameter, open at the top and

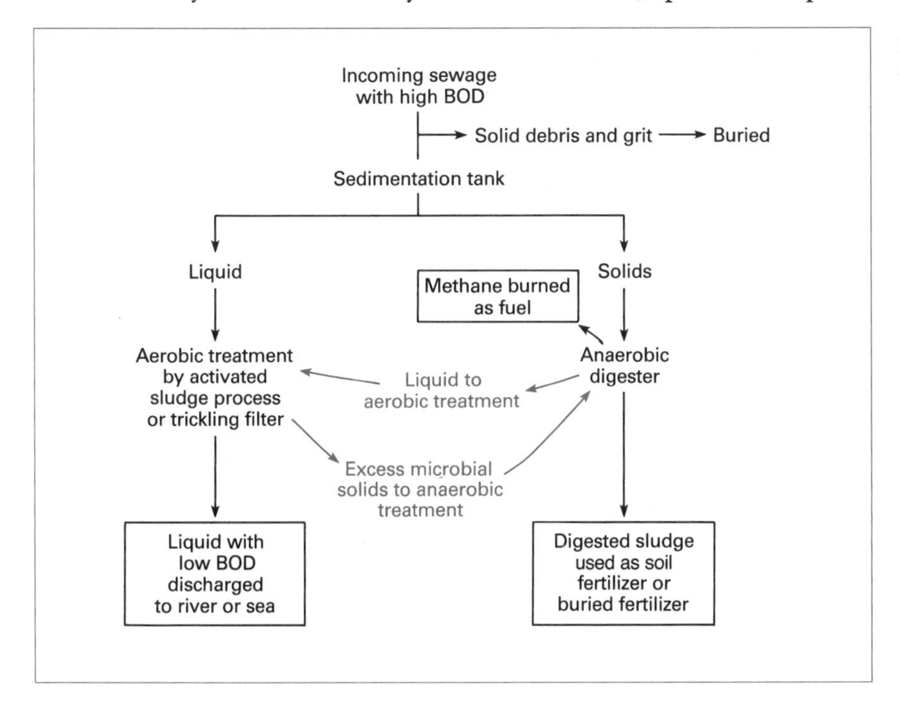


**Figure 11.20** Flow diagram showing how sewage is treated.

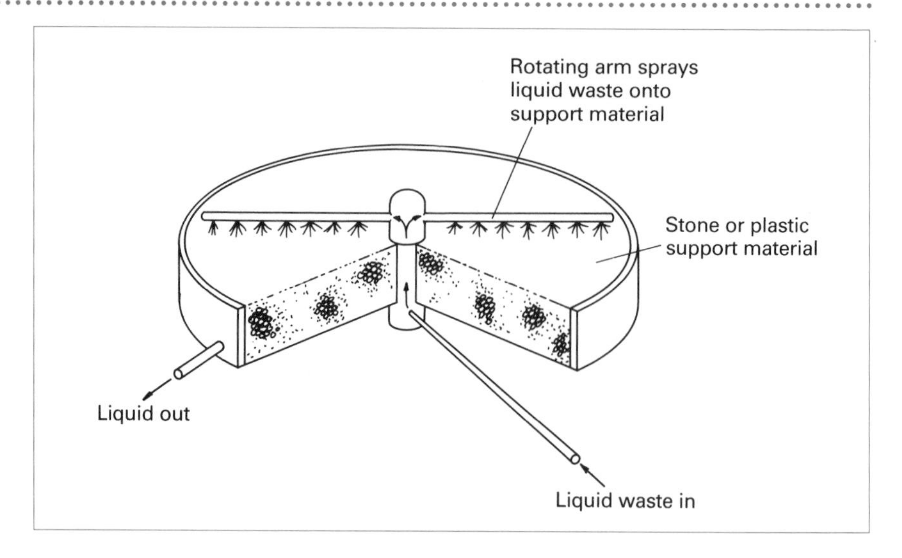


**Figure 11.21** The trickling filter system for treatment of liquid organic waste.

filled with support material. The support material may be crushed stone, gravel, clinker or plastic. Air can penetrate freely through the spaces between the support material. Liquid sewage is sprayed evenly onto the top of the support material by a rotating arm. It becomes saturated with oxygen and trickles down over the surface of the support material. A film of microorganisms, the **zoogleal film**, develops on the surface of the support material. It consists of bacteria, fungi, protozoans and algae. Invertebrates browse on the film and help to prevent it blocking the spaces. As the sewage flows slowly over the surface, organic compounds are removed and oxidized by the zoogleal film. Ammonia is oxidized first to nitrite and then to nitrate.

No microbial inoculum is added: the zoogleal film develops from organisms already present in the waste. It takes several weeks for the film to develop fully so that the trickling filter operates at full efficiency. Usually, two or more trickling filters will be used in series, to reduce the BOD to a low enough value.

The trickling filter is an example of **continuous culture**. The microbial population is not in suspension in the fluid but is fixed on a surface.

**(b) The activated sludge process.** In this process liquid waste is mixed with **activated sludge**, a complex microbial culture, and is aerated vigorously by injecting compressed air. By adding a heavy inoculum of activated sludge and aerating it, the waste is oxidized rapidly. Aeration takes place in a rectangular tank about 4 metres deep and at least three times as long as its breadth. Waste is added at one end of the tank, flows slowly along, and leaves the other end 4 to 8 hours later with a much lower BOD value. This is another continuous process.

**BOX 11.25**

**Activated sludge**

When organic wastes such as sewage are aerated, many organisms clump together with colloidal material from the waste to form **flocs** 0.2–5.0 mm in diameter. These flocs contain many kinds of different microorganisms – bacteria, fungi and protozoans – and are known as **activated sludge**. Waste leaving the aeration tank in the activated sludge process is passed through a sedimentation tank. The liquid flows on for further treatment but most of the flocs settle to the bottom, and the activated sludge is pumped back to inoculate the incoming waste. Excess sludge is passed to the anaerobic digesters for treatment. A strong selection pressure is applied in favour of organisms which form flocs because these are continually reinoculated into the aeration tank, while those that do not are allowed to escape.

**(c) Anaerobic digestion.** Solid materials are digested in cylindrical metal tanks with lids to exclude air. The temperature is maintained at 30 to 35°C and digestion takes about 3 weeks. A portion of digested sludge is mixed in to inoculate the waste with microorganisms. A mixed population of anaerobes and facultative anaerobes develops. About 50% of the organic material is broken down to **methane, carbon dioxide** and small amounts of **hydrogen**. Digestion occurs in three main stages:

1. Proteins, fats and carbohydrates are broken down to amino acids, fatty acids, glycerol and sugars. These are fermented to short-chain organic acids, esters, alcohols, carbon dioxide and hydrogen.
2. Products of the first stage are converted into acetic acid, carbon dioxide and hydrogen.
3. Anaerobic bacteria known as methanogens produce methane and carbon dioxide from acetic acid and hydrogen.

The solid waste (**sludge**) produced by anaerobic digesters is less smelly than raw sewage and most of the pathogens have been removed. Sludge is buried, burned or dumped at sea. The last option is now regarded as unacceptable on environmental grounds. The methane can be burned to provide heat or used to generate electricity. In less-industrialized countries, small anaerobic digesters are used by households to dispose of human and animal waste. The methane generated provides an alternative source of fuel.

## Chapter summary

In this chapter we have looked at ways in which man can use microorganisms for his own purposes.

We looked first at the production of foods and drinks. Alcoholic beverages are made using the yeast *Saccharomyces cerevisiae* to ferment sugars to ethanol and to produce the trace components which give the beverages their flavour and aroma. Alcohol produced by fermentation can be oxidized by bacteria to acetic acid in order to make vinegar. Yeast is also used in baking bread to improve the texture of the dough. In the production of cheese, yoghurt and other fermented foods, bacteria and fungi are used to improve the keeping qualities and flavours of perishable starting materials.

Medical products were then considered. We described the properties of some of the more important antibiotics and used the penicillins as an example of how antibiotics are manufactured. The use of gene manipulation was illustrated by looking at the production of human insulin and of a vaccine against hepatitis B virus.

We saw how a variety of products such as enzymes, steroids, organic acids and amino acids can be produced by selecting the right organisms (and sometimes mutating them) and providing them with the correct growth media and conditions.

In the last section we looked at ways of using microbes to oxidize or reduce sewage and other organic wastes to less harmful products. The aim is to reduce the biochemical oxygen demand of the waste. At the same time the numbers of pathogenic microorganisms are reduced. Here we are getting rid of unwanted materials rather than making something that we want.

## Further reading

Various authors (1981) Industrial Microbiology. *Scientific American*, **245**(3) – A special issue of the magazine which provides an excellent account of eight important topics.

I.J. Higgins, D.J. Best and J. Jones (1985) *Biotechnology: Principles and Applications*. Blackwell Scientific Publications, Oxford.

J. Bu'Lock and B. Kristiansen (1987) *Basic Biotechnology*. Academic Press, London – this book and the previous one give a more detailed and advanced treatment than that given here.

G. Turnock (editor) (1991) *Biotechnological Innovations in Health Care*. Butterworth-Heinemann, Oxford – very detailed accounts of how several biotechnological products were developed. Special emphasis on making sure that products of gene manipulation are really safe. Very careful, step-by-step explanations. Lots of self-test questions – with answers.

V. Moses and R.E. Cape (1991) *Biotechnology: The Science and the Business.* Harwood Academic Publishers, London – a very wide-ranging account of biotechnology with a lot of emphasis on future possibilities. It also shows how the non-scientific considerations of costs, benefits, investment, patents and government regulations affect the biotechnologist.

## Questions

**1.** Fill in the gaps:

(i) The organism used to make most alcoholic beverages is a yeast, _______ _______. The main metabolic activity is the conversion of hexose sugars to two main end-products, _______ and _______ _______ by the _______ pathway. _______ molecule(s) of ATP is/are produced for each molecule of hexose converted. The main starting material for making beer is _______ which contains _______ as its main carbohydrate. Yeast cannot metabolize this carbohydrate so a preliminary treatment stage is necessary in which a group of enzymes known as _______ break it down to fermentable sugars. Vinegar is a dilute solution of _______ acid. It can be produced by oxidizing an alcoholic beverage with the bacterium _______.

(ii) Cheese is made from _______ which contains _______ as its main sugar and _______ as its main protein. A starter culture of bacteria is used to convert the sugar to _______ acid. The enzyme _______ is added to hydrolyse a single bond in the protein.

(iii) Fleming discovered penicillin which was produced by the fungus _______ _______ in surface mat culture. Today, the fungus _______ _______ is used for commercial production by _______ (one or two words) culture. Many different types of penicillin have been produced. They all have the same penicillin nucleus, _______ (more than one word), but each has a different side-chain which, chemically, is an _______ (one or two words). In the manufacture of penicillin, it is necessary to keep the concentration of glucose in the medium low. This is because the enzymes involved in penicillin synthesis are subject to _______ _______.

(iv) Citric acid was originally obtained by extracting it from _______ but most of it is now produced using the fungus _______. To obtain good yields of citrate it is essential to keep the concentrations of ________ (one or two words) in the medium to very low levels. It is also necessary to restrict mycelial growth and sporulation. This is done by limiting the concentration in the medium of either _______ or _______ compounds.

**2.** Which processes require oxygen?

(i) Penicillin synthesis; (ii) ethanol formation; (iii) citric acid production; (iv) trickling filter process; (v) activated sludge process; (vi) yoghurt production.

**3.** Short-answer question:

(i) Why is it advantageous to use gene manipulation to make a vaccine against hepatitis B?

# Answers to selected questions

## Chapter 1

1. Short-answer questions:
   (i) First, the study of the organisms concerned requires a common technology, such as the use of microscopes to see details of structure. Second, some of the organisms are not easily classified as either animals or plants and do not fall within the remit of either botanists or zoologists. Finally, the small sizes of microbes bely their economic importance as agents of disease, in the food industry as antibiotic producers and so on.
   (ii) They are of microscopic dimensions.
   (iii) For example: beer, wine, vinegar, yoghurt, cheese, butter, pulque.
   (iv) **Advantages**: rapid uptake of nutrients, rapid growth and multiplication producing enormous populations with more variants (compared with larger organisms) which can adapt to different environmental conditions.
   **Disadvantages**: individuals are primitive, simple cells and very much at the mercy of the environment.
   (v) *Volvox*
   (vi) *Nitrosomonas* and *Nitrobacter*
   (vii) The use of gaseous (or molecular) nitrogen ($N_2$) as a nitrogen source.

2. Multiple-choice questions:
   (i) c, (ii) a, (iii) b, (iv) b, (v) b, (vi) d, (vii) b

3. Fill in the gaps:
   Words required as they occur: experimental, rapidly, results, crosses, large, mutant, unicellular, homogeneous, heterogeneous, dissection.

## Chapter 2

1. **Multiple-choice questions:**
   (i) b, (ii) b, (iii) c, (iv) d, (v) e

2. **Short-answer questions:**
   (i) One endospore is produced inside a bacterial cell and they are resistant to a variety of adverse environmental conditions including heat. Endospores are therefore structures for survival. Conidia are produced in astronomical numbers and are reproductive structures.
   (ii) The progenitors of chloroplasts and mitochondria were once free-living prokaryotes. Over the course of evolution they formed a symbiotic relationship with the progenitor of, what is now, the eukaryotic cytoplasm. The prokaryotes lost their independent existence and became organelles within the (now) eukaryotic cytoplasm.
   (iii) Prokaryotic flagella have a simple structure and are composed of the protein flagellin – they cannot be seen under the light-microscope unless special stains are used to thicken them. Eukaryotic flagella have a more complex arrangement of microtubules – the typical 9 + 2 organization – and can be seen under the light-microscope without staining.
   (iv) A bacterial chromosome which, unlike eukaryotic chromosomes, is not surrounded by a nuclear envelope.
   (v) Conjugation tube and for adhesion to a surface.

3. **Fill in the gaps:**
   Words required as they occur: chemotaxis, required, towards, toxic, away, flagella, rotate, runs, twiddles, anticlockwise, receptors, thicker, light, 9, 2, whiplash, energy.

## Chapter 3

1. **Multiple-choice questions:**
   (i) a, (ii) d, (iii) c, (iv) d, (v) b, (vi) b

2. **Short-answer questions:**
   (i) Impurities in water.

(ii) (a) An antibiotic (such as penicillin, streptomycin or tetracycline) is an antimicrobial agent produced by another microbe. An antiseptic is a mild disinfectant which can safely be used in contact with living tissue.
(b) Pasteurization is a heat treatment which kills pathogenic microorganisms only. Sterilization kills or removes all living organisms present at particular locations.
(c) Microbiostatic agents *inhibit* the growth of a microbial population. Microbiocidal agents *kill* microorganisms.
(d) Ionizing radiations (such as X-rays) are so powerful that they can break chemical bonds in DNA. Non-ionizing radiations (such as UV light) have less energy and cause new chemical bonds to be formed between adjacent thymine residues in DNA.
(iii) The cells in the inoculum may be depleted of nutrients and/or lack the enzymes required for growth. There will be no increase in cell number until nutrients have been taken up from the medium and/or enzymes synthesized.
(iv) They either become starved of an essential nutrient and/or they are poisoned by an accumulation of toxic waste products in the medium.
(v) Suspend a small sample of soil in sterile buffer, water or saline and streak over the surface of a solid culture medium. Following incubation, use an isolated colony to re-streak a new plate of medium. Repeat this procedure twice more to produce a pure culture.
(vi) Enrichment culture uses a liquid medium with a chemical composition favourable to the growth of the desired microorganism. Conditions of incubation, such as aeration, pH, temperature and illumination, are also chosen to favour the required organism.

**3.** Fill in the gaps:
Words required as they occur:
(i) selective, microbial, patient, chemotherapeutic, sulphonamide, antagonists, structures, natural, PAB, folic, cofactor, competitively, humans.
(ii) increases, increases, biochemical, enzyme, increases, catalysts, proteins, dimensional, weak, linkages, broken, structure, activity, enzyme, inhibited, activity, optimum.

## Chapter 4

**1.** Multiple-choice questions:
(i) c, (ii) b, (iii) d, (iv) a, (v) d, (vi) e

**2.** Short-answer questions:
(i) (a) $CO_2$ and $NO_2^-$ (nitrite)
(b) $CO_2$ and sunlight
(c) Organic compounds in both cases
(d) Usually organic molecules and sunlight but in the dark can obtain energy from the oxidation of organic molecules
(e) $CO_2$ and sunlight
(ii) Oxygenic (oxygen producing): two photosystems means water is used as a reductant. Following splitting of the water molecule by Photosystem II, oxygen is evolved.
Anoxygenic (oxygen not produced): only one photosystem (I). Because there is no Photosystem II, water cannot be used as a reductant. There is no photolysis of water and oxygen is not evolved.
(iii) (a) Loop mechanism and Q-cycle
(b) Proton motive force or proticity can be used to do work in a parallel fashion to electron motive force or electricity.
(iv) Synthesis of reduced coenzymes for oxidative phosphorylation and provision of precursors for biosynthesis.
(v) (a) Usually oxidative phosphorylation but occasionally substrate level phosphorylation
(b) reverse electron flow
(c) Calvin cycle
(vi) (a) chloroplasts, (b) cytoplasmic membrane, (c) invaginations of the cytoplasmic membrane, (d) chlorobium vesicles (chlorosomes)

**3.** Fill in the gaps:
Words required as they occur:
(i) free, energy, biosynthetic, wasted, cannot, organisms
(ii) energy, reduction, electron, electrons, donor, acceptor, pair, couple, electrons, sulphur, hydrogen, $NAD^+$, $NADP^+$
(iii) reduction, redox, donate, electrons, negative, $NADH_2$, $H_2O$, positive, electrons, negative, positive

## Chapter 5

1. Multiple-choice questions:
(i) c, (ii) a, (iii) c, (iv) d, (v) d, (vi) b

2. Short-answer questions:
(i) Replenishing reactions required to replace intermediates (of the TCA cycle for example) used for biosynthesis.
(ii) Teichoic acids.
(iii) Messenger, transfer and ribosomal.
(iv) Introns are the intervening sequences of eukaryotic genes which are not reflected in the active mRNA molecule. Exons are the coding sequences of eukaryotic DNA, separated from each other by the introns.
(v) To seek 'nicks' in a DNA strand, ligation occurs only where adjacent nucleotides are present which have not been joined by a phosphodiester linkage.

3. Fill in the gaps:
Words required as they occur: strand, helix, enzymes, polymerases, pol III, deoxyribonucleotides, primer, 3′-hydroxyl, 5′-deoxyribonucleotides, 3′-hydroxyl, deoxyribonucleotide, 3′-hydroxyl, strand, 3′-hydroxyl, 5′-phosphate, 5′→3′

## Chapter 6

1. Multiple-choice questions:
(i) b, (ii) a,b,d, (iii) d

2. Fill in the gaps:
Words required as they occur: repression control, induction, co-repressor, inducer, (apo)repressor, negative, attenuation, leader, tryptophan, codons, tryptophan, tryptophanyl-tRNA, stem, loop, transcription, structural, stem, loop.

3. True or false?
(i) false, (ii) false, (iii) true, (iv) false, (v) true, (vi) false, (vii) true, (viii) false, (ix) true, (x) false, (xi) false

**4.** Short-answer questions:

(i) Prevents wastage of energy synthesizing enzymes when they are not required.

(ii) Attenuation and repression.

(iii) A number of structural genes which are controlled by the same regulatory gene but which are not next to each other on the chromosome – the structural genes are not clustered as they are in an operon.

(iv) The end-product of a biosynthetic pathway (the allosteric effector or feedback inhibitor) binds to an earlier enzyme at a site (its allosteric site) distinct from the active site. Binding of effector distorts the enzyme so that the affinity of the substrate for the active site is reduced and the enzyme is inhibited.

## Chapter 7

**1.** Fill in the gaps:

Words required in the order they occur: recombination, mutation, sexual, exchange, spontaneously, mutagens, ultraviolet light, analogues, nitrous acid, base, DNA, substitution mutation, amino acid, polypeptide, gene, missense, nonsense, codon, chain, triplet, shortened, deleted, frame-shift, base sequence, amino acids, phenotype, enzyme, catalytic activity, frame-shift.

**2.** Multiple-choice questions:

(i) b, (ii) a, (iii) a, (iv) b, (v) a, (vi) d

**3.** True or false?

(i) true, (ii) true, (iii) false, (iv) true, (v) false, (vi) false, (vii) true, (viii) true

## Chapter 8

**1.** Fill in the gaps:

Words required in the order they occur: F factor, double, closed loop, non-essential, accessory, antibiotic, antibiotic, factor, cytoplasm, $F^-$, sex pili, cell to cell, donor, $F^-$, conjugation tube, donor, Hfr, conjugation tube, integrated, linear, linear, *oriT*, origin of transfer, fragile, break.

**2.** Multiple-choice questions:
(i) d,e, (ii) e, (iii) c, (iv) a,d

**3.** Short-answer questions:
(i) Incorporation of foreign (for example human) genes into a suitable host (for example *Escherichia coli*) with a view to producing valuable proteins (for example vaccines and hormones) by the bacterial cells.
(ii) Plasmids and viruses.
(iii) To provide an origin of replication so that a desired gene may be replicated (cloned) in a host cell.
(iv) They cleave DNA molecules at specific sites. Many produce 'sticky ends' which facilitate joining the vector DNA and the DNA of a desired gene before introduction into a host cell.
(v) Insertional inactivation of an antibiotic-resistance gene.
(vi) Transformation.
(vii) Valuable (human) proteins can be synthesized by bacterial cells. By introducing foreign genes into crop plants, they may be engineered to possess useful characteristics. For example introduction of genes for nitrogen fixation would permit the growth of cereals without need to add nitrogenous fertilizers.

## Chapter 9

**1.** Obligate pathogen: a pathogen that is obliged to live on a living host.
Phototrophy: the ability to use light, carbon dioxide and water to form carbon compounds by photosynthesis.
Osmoregulation: the control of uptake and excretion of water.
Holozoic nutrition: the incorporation of large particles of food in phagocytotic vesicles.
Binary fission: mitotic division of the nucleus followed by division of the cell into two.
Autogamy: gametes of both mating types produced from one parent cell.
Dikaryon: a mycelium containing two types of nuclei.
Oogamy: a process in which a motile gamete fuses with a non-motile gamete.
Shizogony: multiple mitotic divisions of the nucleus to produce many asexual infective cells.

2. You will find Table 9.1 on page 245 helpful in compiling your answer to this question.

3. In an essay on this topic you should give details of the similarity and differences between the structure and physiology of *Euglena* and the non-photosynthetic species from the protozoa that closely resemble it. You should then provide a reasoned argument as to whether or not the possession of chloroplasts makes the photosynthetic flagellate so different to the non-photosynthetic that they should be placed in the separate groups.

4. In this essay you should list the various types of reproduction seen in the protozoa, binary fission, budding and multiple fission. You should provide details of all processes and quote named examples. Remember to cover both sexual and asexual reproduction.

5. (i) refer to page 242
(ii) refer to Figures 9.5 and 9.6
(iii) refer to page 259
(iv) refer to page 266
(v) refer to Figure 9.22
(vi) refer to Figure 9.24

## Chapter 10

1. Refer to *Distinctive properties of viruses*. The answer could include: small size, filterability, the fact that some virus particles can be crystallized, non-cellular, simple construction (genome, multiple copies of coat protein, possible envelope), distinctive morphology, their genome can be RNA or DNA (according to the virus), they are all obligate intracellular parasites.

2. Refer to *Structure of virus particles*. Capsomeres are the structural subunits that collectively make up the protein shell or capsid. (Capsomeres can contain one polypeptide of perhaps three or four different polypeptide chains.) The viral RNA or DNA and capsid together form the nucleocapsid.

3. Refer to *The viral genome*. Viroids contain only RNA and have no proteins associated with them (either coat protein or virus-specific enzymes). Their genomes are much smaller in size than those of viruses. Viroid RNA is in the form of a covalently closed circular molecule, base-paired together (unlike any plant viral RNAs).
4. Refer to Box 10.11. The list could include: filterable infectious agent, causes damage in CNS tissue, long incubation period, unique agent unlike any virus or viroid, apparently a proteinaceous agent because infectivity is destroyed by proteases, agent is relatively resistant to inactivation by chemicals, heat and radiation.
5. TMV, infective; influenza virus, non-infective; MS2, infective; poliovirus, infective; LNYV, non-infective.
6. Refer to *Virus multiplication (Baltimore classification, classes IV and V)*. Plus-strand viruses have RNA in their particles that functions directly in the cell as mRNA. During replication this has to be copied from a template strand of complementary minus-strand RNA. Minus-strand viruses have RNA in their particles which does not act directly as mRNA in the cell. It is a complementary copy of the viral mRNA. During infection mRNA has to be copied from such a minus-strand template.
7. Efficiency of infection = 1.0 (Table 10.2). So in this case one virus particle should give rise to one plaque.
8. Infectivity $= 139 \times 10^9 \times 10 = 1.39 \times 10^{12}$ pfu/ml
9. Anywhere where large numbers of yeast cells live. For example, in fermentation vessels in breweries, or in industrial plants growing vast amounts of baker's yeast.
10. (i) true; (ii) false!; (iii) false [there are not many DNA plant viruses, one example is cauliflower mosaic virus]; (iv) false; (v) true.

## Chapter 11

1. Fill in the gaps:
   Words required as they occur: (i) *Saccharomyces cerevisiae*, ethanol, carbon dioxide, glycolytic, two, barley, starch, amylases, acetic, *Acetobacter*.
   (ii) Milk, lactose, casein, lactic acid, chymosin (rennin is an alternative name).
   (iii) *Penicillium notatum*, *Penicillium chrysogenum*, submerged or deep liquid, 6-aminopenicillanic acid, acid or acyl group, catabolite repression.

(iv) Lemons, *Aspergillus niger*, metals or metal ions, nitrogen, phosphorus.

**2.** Which processes require oxygen?
(i), (iii), (iv), (v).

**3.** Short-answer question:
Because there is no need to use blood from carriers of the disease, unlimited amounts can be produced. Also, there is no risk of infection for patients, and those preparing and handling the vaccine.

# Index

Note: entries in *italics* refer to tables and entries in **bold** to figures.


Abutilon mosaic virus 299, **300**
Acceptor
  electron 67
    external 80
    internal 76
Acetic acid bacteria, in alcoholic fermentations 350
*Acetobacter*, in vinegar production 353–5
Acidity 52
Acidophile 52
Acid protease *379*
Acridine orange 194
*Actinomucor elegans 359*
*Actinomyces missouriensis 379*
Actinomycetin 361
Adaptation, environmental 150
Adenosine triphosphate 71–2
Adenovirus 304, 334
Adjuvant in production of hepatitis B surface antigen protein 379
Adsorption, of virus 330
Agar 37
Aerobe 53
  hydrogen peroxide and 53
  obligate 53
  oxygen and 53
  superoxide and 53
AIDS, *see* Acquired immune deficiency syndrome
Algae
  cell walls *251*
  chlorophyll *251*
  chloroplasts 251, **253**
  colonial 251
  dinoflagellates 250
  euglenids 250
  filamentous 250
  golden brown 250
  green 250
  movement in 254, 255
  nutrition of 252, 253
  photosynthesis in 252, 253
  reproduction of 275–80
  structure of **252**
  taxonomy of *251*
Alkaline protease *379*
Alkalinity 52
Aldolase **74**
Alkalophile 52
Allele
  dominant 177
  recessive 177
Allolactose 156
Allostery 151
  allosteric site and 151
  effector and 151
  inhibition and **152**
Ames test **199**
Amino acids 113, 385
  amination and 113
  biosynthesis of 113
  families of **114**
  glutamic dehydrogenase and 114
  transamination and 115
p-Aminobenzoic acid 55
6-Aminopenicillanic acid 363, 372
  *see also* Penicillin nucleus
2-Aminopurine 192, **193**
Ammonification 8
*Amoeba* 1, 2
cAMP 161
Amphotericin 367
Ampicillin 362, **362**
α-Amylase, industrial use of *379*, 381
Amyloglucoside **381**
Anaerobe
  facultative 53, 89
  obligate 53, 89
Analogue
  base 192
  metabolic 55
Anaplerosis **111**
  glycolysis and **111**
  glyoxylate shunt and 112
  PEP carboxylase and 112
  pyruvate carboxylase and 112
  tricarboxylic acid cycle and **111**
Antibiotics 56, 359–72
  agricultural uses of 367
  bacteriocidal 361
  bacteriostatic 361
  broad spectrum 58, 364
  chloramphenicol 126
  commercial production 367–72
  erythromycin 126
  mode of action *58*
  penicillin 56
  puromycin 126
  semisynthetic 56
  streptomycin 126
  tetracycline 126
Anticodon 122
Antigenic
  drift 316
  shift 317

Antimicrobial activity
  agar diffusion technique and **57**
  minimum inhibitory concentration and 57
  tube dilution technique and **57**
Antiseptics 55, *56*
Aporepressor, *see* Repressor
Arbuscules 247
Archaebacteria, cell wall of *146*
*Arthrobacter* 385
Ascomycetes
  asexual reproduction of 269
  sexual reproduction of 269–71
  spore discharge by **266**, **270**
  structure of 245
*Aspergillus*
  *niger 379*, 383, 384
  *oryzae 379*
  species 366
  *terreus* 383
Assay, microbiological **47**
Assembly of virus 332
ATP synthesis 83
ATP synthetase 87, 88
  $F_0$ 87, **88**
  $F_1$ 87, **88**
  of lactic acid bacteria 87
  subunits of 87
Attenuation
  antiterminator loop and **168**
  leader polypeptide and 164, 166, *167*, 168
  pause loop and 168
  physiology of 167
  terminator loop and **168**
  *trp* codons and 167
  *trp* operon and 164, 165
  tryptophanyl-tRNA and 167
Aureomycin, *see* Chlortetracycline
Autoclave 27, 54
Autotroph 2
  ammonium 106
  chemosynthetic 3, 95, 105
  hydrogen 106
  iron 106
  nitrite 106
  photosynthetic 3, 95
  sulphur 106
Auxotroph 143, 196, 214
  multiple 214, **220**

*Bacillus*
  *amyloliquefaciens 379*
  *coagulans 379*
  *licheniformis 379*
  species 366
  *subtilis 359*, *379*
Bacteria
  lactic acid 87
  photosynthetic 95
    Chlorobiaceae 95
    Chloroflexaceae 95
    Chromatiaceae 95
    Halobacteriaceae 104
    Rhodospirillaceae 95
Bacteriochlorophyll 92
  bacteriophaeophytin and **97**
  special pair 96
  voyeur 96
Bacteriophaeophytin **97**
Bacteriophage lambda (λ)
  attachment sites of **226**
  in cheese production 357
  integration of **226**
  *bio* genes and 226, **227**
  *gal* genes and 226, **227**
  restricted transduction and 225, **226**, **227**
Bacteriorhodopsin **105**
Bacteriochlorophyll 92
Bactoprenol **141**
  disaccharide pentapeptide and 140, **141**
  in peptidoglycan synthesis 140
Baltimore classification 333
Barley 346
Barley yellow dwarf virus 298
Base analogues
  2-aminopurine 92, *193*
  5-bromouracil 192, *193*
Basidiomycetes
  fruiting bodies of **272**
  sexual reproduction of 272–3
  spore discharge by 272–3
  structure of 245
*Bdellovibrio* 292
Beer 343, 346–9
  Bavarian beer laws 344
  stages in production **346**
Benzer, Seymour
  hot spots and 191
  phage T4 and 191
Benzylpenicilloic acid **363**
Beverages
  alcohol concentrations of 344
  alcoholic 343–52
  flavours in **349**
Bilayer, lipid 14
Binary fission 43, 249, 261, 262
  daughter cells and 43
Biochemical oxygen demand (BOD) 369, 386, 387, 388
Bioluminescence 279
Biosynthetic load 111
Boiling 346, 348
Bovine spongiform encephalopathy 320
*Brevibacterium* 385
BSE, *see* Bovine spongiform encephalopathy
5-Bromouracil 192, **193**
Broth, nutrient 37
Budding 249, 262, **271**
Buffers, pH and 52
Burst size 329

Calorie 62
Calvin cycle 17, 101, **102**, **103**
Candicidin 16
*Candida 359*
Canine parvovirus 287, 291, **311**, 334
Cap, of mRNA 127, 314
Capsid 291, 338
Capsomere 303

Carbon dioxide
  fixation 101
    radioisotopes and 101
    Calvin cycle and 101, **102**, **103**
    *Chlorella* and 101
    phosphoribulokinase and **102**, **103**
    reductive tricarboxylic acid cycle and 103, **104**
    ribulose bisphosphate carboxylase and **102**
Carcinogenicity, Ames test and **197**
Carriers
  electron 70
    $NAD^+$ 70
    $NADP^+$ 70
Casein 356, 357, 358
Catabolic infallibility 33
Catabolite repression 161
  in penicillin biosynthesis 371
  in yeast 345
Cell division **261**, 262, 275–9, 281
Cell surface
  bacterial 26
  division and 176
  fimbriae and 26
  glycocalyx and 26
  phagocytosis and 26
  sex pilus and 26
Cellulose 21, **21**
Cell wall
  algal 21, 251
  archaebacterial 135, *146*
  cellulose 21, **21**
  eubacterial structure of 134, **136**
  fungal 245, **249**
  Gram staining reaction of 134
  growth of
    *Escherichia coli* **145**
    *Streptococcus faecalis* **145**
  hemicellulose 21
  protozoal 255
  pseudopeptidoglycan 146
Cephalosporins 363
*Cephalosporium acremonium* 363
Chain, Sir Ernst 360
Champagne 350
Cheese 356
Chemical control 55
Chemical oxygen demand, *see* COD
Chemiosmosis
  asymetrical arrangement and 84
  electrical potential and 85
  loop mechanism and **85**
  membranes and 85
  mitrochondria and 84
  oxidative phosphorylation and 83
  pH gradient and 85
  proticity and 84
  protonmotive force and 84
  proton translocation and 84
  vectorial arrangement and 85
Chemolithotroph 95
Chemolithotrophy 105
Chemotaxis 23, **24**
  chemoreceptors and 23
  in *Escherichia coli* 23
  flagellar bundle and 23
  runs and twiddles and 23
  in *Salmonella typhimurium* 23
Chemotherapeutic agents 55
  sulphonamides 55
Chitin 245
Chlamydiae 292
Chloramphenicol 126, **366**
*Chlorella* 101
Chlorobiaceae 95
Chloroflexaceae 95
Chlorophyll 251
  absorbtion spectrum of 93
Chloroplast 16, **17**, 251, 253
  grana of 17
  thylakoids and 17
Chlortetracycline 365, 366
Chromatiaceae 95
Chromatophore **18**, 96
Chromosome 260, **261**, 262
  bacterial 19
  of *Escherichia coli* **221**
  eukaryotic 19, 20
  size 128
Chymosin 356, 357, 358
Chytrids, *see* Water moulds
Cider 343
Cilia **25**, 254, 255
Ciliates 255
Cistron 118
Citric acid 382–5, *383*, **384**
Citric acid cycle, *see* Tricarboxylic acid cycle
Classification 241
Cloning vectors
  cDNA and 233, **234**
  cleavage site 231
  complementation and 234
  copy number 231, 232
  eukaryotic genes and 233
  ideal characteristics 231
  identification of desired genes and 234, 235, **236**
  immunoprecipitation test and 234
  insertional inactivation and **232**
  pBR322 231, **232**, **374**
  pBR325 in cloning of hepatitis B surface antigen gene 378
  pC1/1 and hepatitis B DNA 378
  phage and 232
  plasmids as 231
  probes and 235–6
  reverse transcriptase and 233
  selectable phenotype 231, **232**
*Clostridium*
  *botulinum* 359
  *perfringens* 367
COD 386
Code
  degeneracy of 186, **188**
  genetic *187*
Codon 122
  nonsense 124
Colony forming unit 40
Commensalism 247
Condensing enzyme **384**
Conditioning 349, 350

Conjugation 26, 277
- bacterial 214, **217**, **219**
- conjugation tube 216
- Davis U-tube and 215, **216**
- donor and 214
- F factor 216, **217**
- Hfr and 217, **218**
- interrupted mating experiment 218, **220**, **221**
- Lederberg and Tatum and 214, **215**
- *oriT* and **217**
- pseudosexual 214
- recipient and 214
- sex pilus and 216
- tube 216
- unidirectional transfer and **217**, 218

Continuous culture
- chemostat **48**
- turbidostat 48

Contractile vacuole 258
Control
- of branched pathways 152, 154
- of enzyme activity 151
- of enzyme synthesis 155
- feedback inhibition and 151
- isozymes 155
- negative **160**
- positive 161
- ribosomal protein and *170*, **171**
- translational 170
- *see also* Inducible enzyme

Copy number 231
Corn steep liquor 369
Cortisones 380–2, **382**
*Corynebacterium glutamicum* 385
Coulter counter 41, **42**
Counts
- absorbance and 42
- bacterial 41
- direct microscopic method 41
- electrical impedence and 43
- membrane filtration method **40**, 41
- total 41, **41**
- turbidity and 42
- viable 38

Covalent modification 151
Creutzfeldt-Jacob disease 320
Cristae 18
Culture
- batch 43
- continuous 46, **48**, 353, 383
- enrichment 38, *40*
- laboratory 37
- pure 37
- surface mat 368

Culture medium 37
- complex 37
- defined (synthetic) 37, **38**
- McConkey's 38

Curing, *see Euglena*
Cyanogen bromide in synthesis of insulin 376, **377**
Cycle
- Calvin 17, 101, **102**, **103**
- citric acid, *see* Tricarboxylic cycle
- Krebb's, *see* Tricarboxylic acid cycle
- nitrogen 7
- oxidative pentose phosphate **79**
- tricarboxylic acid 80, **81**

Cytochromes 82, **83**
Cycloserine 363, **365**
Cytochrome oxidase 86
Cytopathic effects 325
Cytoskeleton 16
- of *Amoeba* 16
- microfilaments and 16
- microtubules and 16
- of spirochaetes 16
- of *Trypanasoma* 16

Deep liquid culture 354, 368, 370
Dehydrogenase, $NADH_2$ 82
Denitrification 8
Deuteromycetes *245*
Diabetes 372
Diabetes mellitis, *see* Diabetes
Diabetics 373
Diaminopimelic acid **137**
Diatoms 277, **278**
Digestion, anaerobic 389
Dilution rate
- maximum growth rate and 47
- steady state and 47
- washout and 47

Dimer, thymine 191, **192**
Dinophytes 244
Dipicolinic acid 27
Direct virus particle count 321, 339
Disease 9
- gonorrhoea 9
- microorganisms and 9

Disinfectants 55, *56*
Distillation 344, **352**
DNA
- antiparallel strands of 130
- complementary 233, **234**
- discontinuous synthesis and 131
- genetic engineering and 233
- as genetic material 177
- Griffith and 179
- gyrase 129
- helicase 129
- Hershey and Chase and **180**
- lagging strand and **133**
- leading strand and 132
- ligase **133**
  - in cloning hepatitis B DNA 378
  - in cloning insulin gene 374
- in meiosis 262
- in mitosis 261
- Meselson and Stahl and **131**
- methylation of 230
- of phage T2 179, **180**
- Okasaki fragments of 129
- polymerases 130, 133
- polyribonucleotide primer 129
- replication fork 129
- semi-conservative replication and 129, **130**, 131
- synthesis 128
  - displacement loop and 129
  - replication bubble and 129
- transformation and 177
- viruses 310

DNA repair systems
  dark 203, **205**
  daughter strand gap repair and 205
  endonuclease complex and 204
  error-free 203
  error-prone 206
  excision **205**
  inducible 206
  long patch 207
  photolyase and 204
  photoreactivation and **203**, 204
  post dimer synthesis and 204
  recombination and 203, **206**
  sister strand exchange and 204
  SOS 192, 203, 205
  transdimer synthesis and 207
Donor
  electron 70
    primary 70
Dung fungi **266**

Eclipse period 329
Efficiency of infection *327*
Electron
  acceptor 67
    external 80
    internal 76
  carriers 70
  tower 68
Electron transport chain 82, **86**
  ATP synthesis 83
  cytochromes 82, **83**
  flavoproteins 82
  iron–sulphur proteins 82–3, **84**
  $NADH_2$ dehydrogenase 82
  photosynthetic 90
  quinones 83
  respiratory 83
  in yeast 345
Encapsulation, of yeast 350
Endonuclease, restriction 230, *230*
Endospore 27, **27**
  of *Bacillus* 27
  and calcium ions 27
  of *Clostridium*
  and dipicolinic acid 27
  test strip 27
Endosymbiotic theory 12, 14
End-point titrations 326
Energy
  activation 64
  calorie and 62
  enthalpy and 63
  entropy and 63
  free 63
  joule and 62
  standard free 63, *64*
  *see also* Thermodynamics
Enhancers 183
Enrichment, penicillin **143**
Entner–Doudoroff 78
  *see also* Pulque; *Zymomonas mobilis*
Enthalpy 63
Entropy 63
Enveloped virus 305, *307*
Enzymes
  activation energy of 64, **65**
  adaptive 156
  commercial production of 379–80
  constitutive 156
  control 150
  enzyme–substrate complex 65, **65**
  immobilization of 380
  inducible 156
  repressible 156, 163, **165**
  substrate 65
Equilibrium constant 63
Erythromycin 126, 366, **366**
*Escherichia coli*
  in cloning hepatitis B DNA 378
  in cloning insulin gene **374**, 375, **377**
Ethanol, *see* Beverages, alcoholic
Ethidium bromide **194**
*Euglena* 3
  curing 242
Eukaryote 12
  comparison with prokaryote *13*
Eukaryotic microorganisms 241–85
Exon **127**, 319
Extrachromosomal genetic elements, *see* Plasmids

FAD 31
Fairy rings 249
Feedback inhibition 151
  allosteric effector and 151
  allosteric enzymes and 151
  allosteric site and 151
  concerted 152
  covalent modification and 151
  isozymes and 154, **155**
  multivalent 152, **155**
  sequential 154, **155**
  sigmoid kinetics and 151
Fermentation 72
  alcoholic 72
  anaerobic conditions and 349
  and beer 348
  Entner–Doudoroff **77**, 78
  fedbatch 372
  formic **73**
  glycolysis and 74
  heterolactic **76**, 77
  homolactic 75
  lactic 75
  penicillin 367–72
  phosphoketolase **76**
  secondary 350
  vessels **347**, **348**
  vinegar 353
  wine 350–1
  whisky 351–2
Ferredoxin 99
Fertility factor 216, **217**
Field process 353
Filipin 15
Filter, trickling 387–8, **388**
Fimbriae 26
Flagellates 254, 255
Flagellin 22
Flagellum 25, *251*, 254, 255
Flavoprotein 82
Fleming, Sir Alexander **360**, 367
Florey, Sir Howard 360
Fluid mosaic 15
Fluorescence 93
Folic acid 55
Foods, fermented 356–9
Foot and mouth disease virus 297

N-formylmethionyl-tRNA 123
Frameshift mutation 187, **189**
Freeze-drying 368
Fructose-1,6-bisphosphate **74**
Fungi
  cell walls *245*
  growth 249–50
  hyphae 245
  reproduction of 263–75

β-Galactosidase 156, *379*
Gametes
  of algae 275–80
  of protozoa 280–4
Genes
  eukaryotic
    exons of **127**, 183
    introns of **127**, **183**
  naming 158
  regulatory 159, **164**, 166
  split **127**, 183
  structural 159, **164**, 166, 170
  *see also* Genotype; Phenotype
Genetic engineering 229
  cloning and 229, **232**
  in eukaryotes 236
  host for 232, 236
  insert and 232
  recombinant DNA technology and 229
  stages of **232**
  vector for 229, 231
  *see also* Restriction endonuclease
Genome 181
  *Escherichia coli* **181** 182
  eukaryotic 182, 183
  gene map of 183
  genotype and 184
  junk DNA of 182
  nucleoid and 181, 182
  nucleus and 181
  operon and 182
  prokaryotic 181
  promoter and 183
  scaffold protein and 181
  supercoiled DNA and 181
Genotype 184
Geochemical cycle 7
  *see also* Nitrogen cycle
Glucoamylase *379*
Gluconic acid 383
Glucose
  effect, *see* Catabolite repression
  isomerase *379*, 380, **381**
Glutamic
  acid 385
  dehydrogenase 114
Glycocalyx 26
Glycolysis 73, **74**
  aldolase **74**
  ATP synthesis **74**
  energetics 74
  fructose-1,6-bisphosphate **74**
  $NAD^+$ 74
  substrate level phosphorylation 74
  in yeast 73–4, 345
Glyoxylate shunt 112, **113**
  isocitrate lyase 112
  malate synthase 112
Gramicidins 366, 367, **367**
Gram-negative bacteria, susceptibility to
  cephalosporins 363
  gramicidins 367
  penicillins 362
  tyrocidins 367
Gram-positive bacteria, susceptibility to
  cephalosporins 363
  penicillins 362
  polymixins 367
Gram stain 134–5
Grana 17
Grapes 350–1
Gratuitous inducer 159
Griffith, Fred 177, **178**
Growth
  and acidity 52
  of algae 252, 254, 261, 275–80
  and alkalinity 52
  control 53
  exponential 43
  of filamentous organisms 46
  of fungi 249–50
  logarithmic 43
  phases of 44, **45**
  of protozoa 256, 260, 280–4
  rate 6, 44
  synchronous 49, **50**
  and temperature 51, **51**
  and water 52
Growth factors 34, *35*

Haemagglutination assay **322**, 339
Haemoglobin 118
Halobacteriaceae 104
*Halobacterium salinarium* **105**
Halophile 53
Hartig net 247
HBsAG, *see* Hepatitis B surface antigen protein
Headful hypothesis 225
Helical symmetry 305, 339
Hepatitis B
  surface antigen protein 377–9, **378**
  vaccine 377–9
  virus 377, 378
Herpes viruses 306
Hershey and Chase 179, **180**
Heterotroph 3
  chemosynthetic 3, 95
  photosynthetic 3, 95
Histone, and DNA 177
HIV-1, *see* Human immunodeficiency virus
HIV-2, *see* Human immunodeficiency virus
HMC, *see* 5-hydroxymethylcytosine
Holliday model **201**
Hops 346
Human hepatitis delta virus 319
Human immunodeficiency virus (HIV) 305, 336, 337, 338
Human T cell lymphotropic virus 336
Human wart virus 295, **296**, 333
*Humulus lupulus*, *see* Hops
Hydrogen peroxide 53
Hydroxylamine **195**

5-Hydroxymethylcytosine **313**, 331
Hyphae 245

Icosahedral symmetry 303, 339
Immunoprecipitation test 234
Inducer **157**, 158
Inducible enzyme
  basal level 156, 158
  β-galactosidase 156
  β-galactoside permease 156
  β-galactoside transacetylase 156
  inducer **157**, 158
  lactose operon 159
Infection
  asymptomatic 295
  generalized 295
  lysogenic 299
  lytic 299
  silent 295
  systemic 295
Infectious dose ($ID_{50}$) **326**
Infectivity assay 319, 323, 339
Influenza virus 295, 305, 308, 316, 322, 335
Inhibition
  competitive 56
  feedback 151–5
Insert 232
Insulin 372–7
  structure 372, **376**
Intercalating agent 92, **94**
  acridine orange **194**
  ethidium bromide **194**
  proflavine **194**
Intracellular virus accumulation period 330
Intron **127**, 319
IPTG 158
Isozyme 154, **155**
Itaconic acid **383**

Jacob and Monod, operon model 158–9

Ketogluconic acid **383**
Krebb's cycle, *see* Tricarboxylic acid cycle

Kuru 320
β-Lactam ring 362, 363
β-Lactamase, *see* Penicillinase
Lactic acid 34, **383**
  bacteria 34
    in alcoholic fermentations 350
    in cheese production 356
    in production of
    foods fermented with salt 358
    in yoghurt production 358
    *see also Lactobacillus*; *Leuconostoc*; *Pediococcus*; *Streptococcus*
*Lactobacillus*
  in cheese production 357
  species **359**, 383
  in yoghurt production 358
Lactose
  catabolism 156, **157**
  operon 159
  structure **157**
Latent period 329
Leader polypeptide 164, 166, *167*, **168**
Lettuce necrotic yellows virus 294, 335
*Leuconostoc* 76, *359*
Life cycle
  of *Chlamydomonas* **275**
  of diatoms **278**
  of dinoflagellates **279**
  of *Giardia intestinalis* **282**
  of *Plasmodium falciparum* 284
  of slime moulds **281**
  of *Spirogyra* **276**
  of *Ulothrix* **276**
  of water moulds **267**
  of yeast **271**
Local lesion **268**, 297
  assay 325
Locomotion
  of algae 255
  of protozoa 252
Logarithms 43
Loop
  antiterminator **168**
  mechanism 85
  pause 168
  terminator **168**
Lyophilization, *see* Freeze-drying
Lysine 385
  and *Brevibacterium flavum* **153**
  production **153**
Lysogeny 222, **223**, **226**
Lysozyme 136, **138**, 333
  and protoplasts **138**
  and sphaeroplasts **138**
Lytic cycle **179**

Macromolecules, and precursors 110, **110**
Major chemical elements 31, *32*
Malt 346, 351, 353
Malting, *see* Malt
Mashing 346–8, 351
Maturation
  of whisky 352
  of wine 351
Mean generation time 44
Meiosis 262–3
Melanin 245
Membrane
  cytoplasmic 14
  fluid mosaic 15
  lipid bilayer 14
  photosynthetic
    bacterial 198
    and *Chlorobium* vesicles **18**
    chromatophores as **18**
  and sterols 16
Meselson and Stahl **131**
Mesosome **182**
Methicillin **362**, 363
*n*-Methyl-L-glucosamine **365**
Microaerophile 53
*Microbacterium* 385
Microfilament 16
Microorganisms, advantages for genetic studies 176
Microtubule 16, **25**, 254
Milling 346
Minor chemical elements 31
Mitchell, Peter, *see* Chemiosmotic theory

Mitochondrion 17, **18**
  cristae 18
  matrix 19
Mitosis **261**
Modification, covalent 151, **154**
*Monascus purpureas 359*
Mosaic symptoms 288
Motility 22–6
  amoeboid **25**
  cilia **25**
  flagellum **25**
  and metachronal rhythm 25
*Mucor*, in cheese production 357
Multiple fission 262
Murein sacculus 136
Mutagen
  2-aminopurine 192, **193**
  base analogues 192
  5-bromouracil 192, **193**
  hydroxylamine **195**
  intercalating agent 192, **193**
  nitrous acid **194**
  physical 191
  and repair mechanisms 191
  and thymine dimers 191, **192**
  ultraviolet light 191, **192**
Mutation 175
Mutagenicity testing, Ames test **197**
Mutant
  auxotrophic 143, **196**
  and increased lysine synthesis **153**
  and minimal medium 196
  and penicillin enrichment 196
  and prototrophy 196
  regulatory 153
  selection of 195
Mycelium 245
*Mycobacterium tuberculosis* 364
Mycoplasmas 16
Mycorrhizae 247

$NAD^+$ 31, 70
*Nitrobacter* 8
Negative staining 302
Neutral protease *379*
Nitrogen cycle 7, **8**
  ammonification 8
  denitrification 8
  nitrifying bacteria 8
  *Nitrobacter* 8
  *Nitrosomonas* 8
  Nitrogen fixation 7
  *Rhizobium* 7
*Nitrosomonas* 8
Nucleocapsid 291
Nucleoid 19, 20, 181–2
  and bacterial chromosome 19
Nucleus 177, 181
  protozoan 255
Nutrition
  in algae 253
  in fungi 247
  in protozoa 256, 259, 260
Nystatin 16, 367

Okasaki fragment 129
Oncogenes 336
One-step growth curve 328, 339
ONPG 158
Oogamy 277
Oomycetes, *see* Water moulds
Operator 159, 162, **163**
Operon 156, **157**, *170*
  and catabolite repression 161
  lactose 159
  and operator 159, 162, **163**
  tryptophan **164**, 167
Organic acids 382–5
Organic wastes, purification of 385–9, **387**
Orleans process 353
Osmoregulation 254, 258
Osmotic pressure 52
  and protoplast 52
Oxidation–reduction reactions 66
  coupled 67, 68, **68**
  and electron tower 68
  and half reactions 67
Oxytetracycline 365, **366**, 367

Parasite 3
Parvoviruses 287, 334
Pasteur, Louis 349
Pasteurization 54
  in beer production 349
  flash 54
  of milk 54
  and Pasteur 54
Pathogenic conversion 301
Pectinase *379*
*Pediococcus 359*
Pellicle 251
Penetration by virus 331
Penicillin 56, 142
  acylase **363**, 372, *379*
  agricultural uses of 367
  allergy to 366
  binding protein *142*
  commercial production of 367–72
  costs of **370**
  discovery of **360**
  effect of enzymes on 363, **363**
  enrichment technique **143**
  extraction and purification of 372
  mode of action 363, **364**
  nucleus 362
  production of different types of 372
  structure 362, **362**
  yields of fermentations **370**
  *see also* 6-aminopenicillanic acid
Penicillinase 363, **363**
*Penicillium*
  in antibiotic production **360**
  in cheese production 357
  *chrysogenum* 368, 369, 370
  effects on bacteria **364**
  *notatum*
    strain improvement and *369*
Pentose phosphate cycle
  in *Acetobacter* 80
  functions of 79
  in *Gluconobacter* 80
  and $NADP^+$ 79
  oxidative **79**
  reductive 80
  in yeast 345
PEP carboxylase 112
Peploma, *see* Spike

Peptidoglycan 135, 363, **364**
  amino acids in **137**
  biosynthesis 138, **139**
  cross-bridges in 138
  and lysozyme 136, **138**
  and murein sacculus 136
  and transpeptidase 142
Phaeophytin 99
Phage 294
  β 301
  ε 300
  and Hershey and Chase 179, **180**
  lambda 300, **301**, 307, 312, **313**
  and lysogeny 222, **223**, **226**
  and lytic cycle 179, 222, **226**
  MS2 312, 335
  P22 223
  prophage 222, **226**
  Qβ 312, 335
  and superinfection immunity 222
  T2 179, **290**, **292**, 299, 307, 334
  T4 299, 307, 310, 328, **332**, 334
  temperate 222, 300
  transducing 223
  virulent 222
  ϕX174 312, 334
Phagocytosis 26, 254
Phase
  of decline 44, **45**
  exponential 44, **45**
  growth 44, **45**
  logarithmic 44, **45**
  of retardation 44, **45**
  stationary 44, **45**
Phenotype 184
Phenoxy acetic acid 372
Phenylacetic acid **363**, 371, 372
Phocine distemper virus 297
Phosphofructokinase **384**
Phosphoketolase **76**
Phosphoribulokinase **102**, **103**
Phosphorylation
  oxidative 87
  photosynthetic 90
  substrate level 74
Photolithotroph 95
Photolysis 91
Photophosphorylation
  cyclic **97**, 98, 101
  non-cyclic 100
Photoreactivation **203**, 204
Photosynthesis 252
  accessory pigments and 92
  anoxygenic 91, 95–6
  archaebacterial 104
  and bacteriorhodopsin **105**
  and carotenoids 92
  and chloroplasts 94
  electron flow of 96, **97**
  in *Halobacterium salinarium* **105**
  and hydrogenase 98
  light rections of 91
  oxygenic 91, 98
  and photolysis 91
  phycobilisomes and 93
  photosystem I of 99
  photosystem II of 99
  pigments of 92
  proton pump and 98
  $NADP^+$ reduction and 98
  and retinal 105
  and reverse electron flow 98
  thylakoid and **94**
  Z-scheme of 99, **100**
Photosystem 99
Phycobiliprotein 93
Phycobilisome 93
*Phylloxera vastatrix* 351
Phylogenetic tree **244**
Pigments *251*
  photosynthetic 92–5
Pilus, sex 26, 216
Plaque assay 323, 324
Plasmid
  bacterial 211
  classification of 212
  and conjugation 212, **216**
  colicinogenic *212*
  cryptic 213
  degradative 213
  ssDNA and 212
  fertility 212
  functions 212
  incompatibility groups of 213
  properties *212*
  virulence 213
Ploidy 260, 263
Poliovirus **289**, **290**, 335
Poly(A) tail of mRNA 315
Polyenes 367
Polymixins 366
Polynucleotide
  polydeoxyribonucleotide 128
  polyribonucleotide 126
  synthesis 126
Polypeptide antibiotics 366
Population doubling time 44
Potato spindle tuber viroid **318**
Potential, reduction 67, *69*
Poxviruses 287, 291, 307, 312, 334
Pribnow box 119, **120**
Primer, DNA synthesis and 129, **132**
Prions 320, 339
Probe, nucleic acid 235, 236
Proflavine **194**
Prokaryote 12
  compared with eukaryotes 13
Pro-insulin **376**, 376, **377**
Promoter **120**, 160, 183
  and open promoter complex 119, **120**
  Pribnow box and 119, **120**
Proof reading, DNA polymerase and 132
*Propionibacterium*, in cheese production 357
Proteases *379*
Protein
  amino acid sequence and 118
  antibiotics and 126
  iron–sulphur 82–3, **84**
  polypeptide and 118
  and post-translational modification 125
  secretory 124
  signal hypothesis and 124–5
  synthesis 118
  transcription and 118, **119**
  translation 122
Proticity 84
Protista 241

Proton
  motive force 84
  pump 98
  translocation 84
Protoplast 52, **138**
  osmotic pressure and 52
Prototroph 196
Protozoa
  locomotion of 255
  nuclei of 255
  nutrition of 259
  reproduction of 280–4
  structure of 256–7, **256**
  taxonomy of 255
Provirus 336
Pseudopeptidoglycan 146
Pseudopodium 26, 258, **259**
  and amoeboid movement 26
  and cytoplasmic streaming 26
Pulque 78
Purification, of mixed bacterial cultures **39**
Purines, biosynthesis of 115, **115**
Puromycin 126
Pyocyanase 361
Pyrimidines
  biosynthesis of 117, **117**
  phosphoribosylpyrophosphate and 117
Pyruvate 74
  carboxylase 112, **384**

Quinone 83
  cycle 86, **87**

Radiations
  cobalt-60 and 54
  ultraviolet light and 54
  X-rays and 54
Radiorespirometry
  Entner–Doudoroff and 78
  glycolysis and 78

Reactions
  atoms and 64
  endergonic 63
  enzymes and 64
  exergonic 63
  molecules and 64
  oxidation–reduction 66–7
  rate of 64
  redox 66–7
Recombination 175
  chi form and **202**
  general **201**
  heteroduplex and **201**
  Holliday model of **201**
  site specific 201
Red tides 279
Release of virus 333
Rennin, *see* Chymosin
Reoviruses 315, 335
Replica plating, auxotrophic mutants and 196
Replication fork 129, **132–4**
Repressor
  equilibrium dialysis and 163
  isolation of 161
  lactose 161
  phosphocellulose binding of 163
  of *Saccharomyces cerevisiae* 171
  translational 171
Repression
  cAMP and 161
  catabolite 161
  catabolite receptor protein and 161
  corepressor and 165
  *trp* operon and 164, **165**, **166**
Reproduction
  of algae 275–80
  of fungi 262–75
  of protozoa 280–4
Respiration 72
  aerobic 80
  anaerobic 89
  citric acid and **81**
  electron transport chain and 82
  energetics of 88, *89*
  nitrate 89, **90**
  oxaloacetate and **81**
  and oxidative decarboxylation 80
  and tricarboxylic acid cycle 80, **81**
Restriction endonuclease *230*
  blunt ends and 231
  for cloning insulin gene **374**
  for cloning hepatitis G surface antigen gene **374**
  cohesive termini and **231**
  flush ends and **231**
  gene cloning and 231
  sticky ends and 231
Restriction enzymes, *see* Restriction endonuclease
Restriction nuclease, *see* Restriction endonuclease
Retinal 105
*Rhizobium* 7
Retroviruses 335, **336**, 339
Reverse transcriptase 309, 336
Rhabdoviruses 294, 335
*Rhizopus*
  *arrhizus* 380–1
  species *359*
Rhodospirillaceae 95
Rhythm, metachronal 25
Ribosome **123**, **169**
  initiation complex of **123**
  preinitiation complex of **123**
  protein synthesis and 123
  A site **123**
  P site **123**
  subunits of 124
  Svedberg units and 124
  translation and 123
  translocation and 124, **125**
Ribozyme 128
Ribulose bisphosphate carboxylase **102**
Ringspots 288
Ringworm 249
Rise period 329

RNA
- messenger
  - cap of 127
  - polycistronic 160, **166**
  - primary transcript and 127
  - tail of 127
- processing 127
- polymerase 119, 120, **121**, 127
- sequence analysis 242, **244**
- synthesis 118, **119**
- viruses 310, 314

Rotaviruses 316

Rous sarcoma virus 336

*Saccharomyces*, budding of 46
- *carlsbergensis* 343
- *cerevisiae* 75
  - in production of alcoholic beverages 343, 351
  - in production of tartaric acid **383**
- *lactis* *379*
- species *359*
- taxonomy 343
- *uvarum* 343
- *see also* Yeast

Saprophyte 2

Schizogony **283**

Scrapie 320

Secretory protein 124

Segmented genome 315, 339

Selective toxicity 55, 361
- chemotherapeutic agents and 55
- metabolic analogues and 55
- sulphonamides and 55

Septa *245*

*Septomyxa affinis* 381

*Serratia marsescens* **383**

Sewage purification, *see* Organic wastes

Sexduction
- F′ and 227, **227**
- F′-*lac* 228

Shadow casting **302**

Shine–Dalgarno sequence 123, **124**

Signal hypothesis 124

Single cell protein, continuous culture and 49

Size
- of *Amoeba* 1, **2**
- evolutionary success and 5
- of foot and mouth disease virus 1, **2**
- of microorganisms 1, *4*
- and surface area : volume ratio *3*
- and units of measurement 1

Sludge, activated 388–9

Smallpox 293
- virus 296, **306**

Sources of energy *66*

Specific growth rate 44

Spike
- haemagglutinin (HA) 305, 317
- neuraminidase 305, 317

Sphaeroplast **138**, 363

Spirits 344

Sporogony 283

Sporozoans 255

*Staphylococcus aureus* 360, 368

Sterilization
- endospore and 54
- filtration and 55
- heat and 54

Steroid transformations 380–2, **382**

Sterols 16
- candicidin 16
- filipin 16
- and mycoplasmas 16
- in yeast 345
- nystatin 16

Streptidine 365

*Streptococcus* 357, 358, *359*
- *pneumoniae*
  - transformation in 177

*Streptomyces*
- *erythreus* 366
- *griseus* 364
- species 365, *379*
- *see also* Streptomycetes

Streptomycetes 359, 363
- *see also Streptomyces*

Streptomycin 126, 364, **365**

Streptose **365**

Stunts 288

Sub-viral agent 319

Sulphonamides
- *p*-aminobenzoic acid and 55, **116**
- competitive inhibition and 56
- folic acid and 55, **116**
- metabolic analogue and 55
- mode of action **116**

Superoxide 53

Suppression 197
- and back mutation 197
- of frameshift 199
- intergenic 198–9
- intragenic 198, **199**
- of missense **198**
- of nonsense **200**

Surface area
- and growth rate *5*
- to volume ratio *3*

Svedberg unit 124

Symbiosis 247

Syrups, high-fructose 380, **381**

Tartaric acid 383

Tautomerism 189, *189*, **190**

Taxonomy
- of algae *251*
- of fungi *245*
- of protozoa 255

Teichoic acid 143, **144**
- function 144
- membrane 144
- polyglycerol phosphate and **144**
- polyribitol phosphate and **144**
- wall 144

Termination sequence 120, **121**
- hairpin configuration of 120
- intrastrand base pairing of 120

Terramycin, *see* Oxytetracycline

Tetracycline 126, 365, **366**, 367

Thermodynamics
- first law 62
- second law 62

Thom, Charles 368

Thylakoid 17

TMV, *see* Tobacco mosaic virus

Tobacco mosaic virus
assay 321, 326
discovery 288, 289
RNA 314
structure **289**, 302, 305, 306
survival 291
Tobacco rattle virus 315
TOC 386
*Torulopsis 359*
Total organic carbon, *see* TOC
Trace elements 32, *33*
Transamination 115
Transcription 118–19
Transduction
abortive 225
complete **224**
Davis U-tube and 223
headful hypothesis and 225
lambda and 222, 225, **226**
lysogeny and 222, **223**, **226**
phage P22 and 223
restricted 223, 225, **226**
superinfection immunity and 222
vector for 223
Transfer RNA
aminoacyl-tRNA synthetase and 122
anticodon and 122
codon for 122
protein synthesis and 122
Transformation
Avery, MacLeod, McCarty and 178
bacterial 177, **178**, 228
in genetic engineering 178, 233
Griffith and 177, **178**
*Streptococcus pneumoniae* and 177, **178**
transforming principle and 177, **178**
Transition mutation 184, **186**
Translation 122
aminoacyl-tRNA synthetase and 122
codon and 122
ribosome and 123, **123**
of mRNA 122
tRNA and 122
Translocation 124, **125**
Transversion mutation 184, **186**
Tricarboxylic acid cycle 80, **81**
reductive 103, **104**
Trickling filter process 387–8, **388**
Tryptophan
codon 167
operon 164–5
synthetase
hybrid protein with insulin **377**
promoter in cloning insulin gene **375**
Tryptophanyl-tRNA 167
Tulip break 293
Turnip yellow mosaic virus 304
Tyrocidins 366

Ultraviolet light
as sterilant 54
thymine dimers and 191, **192**

Vesicle, *Chlorobium* **18**
Vinegar 352–5, 354
Vines, *see* Grapes
Viral genome 291, *311*, 312
Virion 287
Viroids 316, 339
Virus
foot and mouth 1, **2**
particle 294
Virus-specific enzyme 309
Vitamin 34, **35**
*Vitis*, *see* Grapes
Voges–Proskauer 73, **73**
butanediol fermentation and 73
formic fermentation and 73
mixed acid fermentation and 73
*Volvox* 2

Washout, chemostat and 47
Water 52
activity 53
halophiles and 53
moulds **267**
Watson and Crick 180
Whisky 351–2
Wild-type 196
Wine 343, 350–1
fortified 344
Wort 346, 348

X-ray crystallography 302
X-rays, as mutagen 54

Yeast
baker's 355–6
brewer's 75, 346
in cloning hepatitis B surface antigen **378**, 379
in manufacture of hepatitis B surface antigen protein 378
metabolism 344–5
oxygen in metabolism 345
promoter in cloning hepatitis B DNA 378
terminator in cloning hepatitis B DNA 378
in vinegar production 352
in whisky fermentation 345, 351
wild 350
in wine making 351
*see also Saccharomyces*
Yoghurt 358

Zoogleal film 388
Zygomycetes 244, *245*
*Zymomonas mobilis* **77**, 78